AF616547

FLUID MECHANICS

Meares Glacier, Unikwak Bay, Prince William Sound, Alaska. The flow of the glacier ice is an example of a non-Newtonian fluid in motion. (Photograph by Austin Post, University of Washington.)

FLUID MECHANICS

Theodore Allen Jr., M.S.
College of Engineering Sciences
Arizona State University

Richard L. Ditsworth, Ph.D.
College of Engineering Sciences
Arizona State University

McGRAW-HILL BOOK COMPANY *New York San Francisco St. Louis Düsseldorf Johannesburg Kuala Lumpur London Mexico Montreal New Delhi Panama Rio de Janeiro Singapore Sydney Toronto*

This book was set in Monophoto Times Roman by Holmes Typography, Inc., and printed and bound by The Maple Press Company. The designer was Janet Bollow; the drawings were done by David A. Strassman. The editors were B. J. Clark and Marge Woodhurst. Charles A. Goehring supervised production.

FLUID MECHANICS

Printed in the United States of America.

Library of Congress catalog card number: 75–167552

1234567890 MAMM 798765432

07–001095–1

CONTENTS

PREFACE

This book is intended as a text for a first course in fluid mechanics offered to engineering students. We feel that an economy of time and effort has been achieved by the arrangement and selection of subject matter as presented. Increased needs and interests in presenting the various engineering sciences have encouraged us to consider the following objectives:

1. To introduce vector field operations and theorems early in the text, and use these where they confer advantages with respect to brevity and generality of formulations.
2. To preserve a consistent method of deriving control-volume equations from corresponding system equations.
3. To emphasize the need of constitutive relations as well as the basic equations of fluid mechanics in the solution of physical problems.

The book begins with an introduction to basic definitions, field concepts, and pertinent field theorems. Chapters 2 through 6 present basic laws such as conservation of matter, momentum, and energy as well as the concept of state of stress and its relation to forces in a fluid field. Chapter 7 discusses the motion of an ideal fluid, while Chapter 8 introduces a simplified approach to the solution of problems through an introduction to dimensional analysis and similarity. The latter chapter also presents the idealization of one-dimensional flows and their utility in solving simple fluid flow problems of engineering relevance.

The remainder of the text is concerned with some beginning concepts of turbulent and boundary-layer flows. Fluid machinery, controls, and instrumentation have not been included except as examples and assigned problems. The coverage of fluid statics, dimensional analysis, and simplified frictional flows has been reduced relative to that found in existing texts.

The text includes more material than would be covered normally in a course of three semester hours. Chapters 1 through 6 would be a necessary part of any elementary course in fluid mechanics presented from a vector point of view. Since there is little interdependence of Chapters 7, 8, and 9, these may be used to satisfy a given course objective.

The text is written for students who have had courses in vectorial mechanics and thermodynamics. A number of examples are included and these should be considered as an *essential part* of the presentation. There are adequate problems and self-study questions at the end of each chapter to provide different assignments for several semesters.

The authors wish to acknowledge with gratitude the assistance given to them by the faculty and administration of their college. Special appreciation is due to Dr. Daniel F. Jankowski, a colleague, for valuable assistance and suggestions, and Dean Lee P. Thompson for continued encouragement and support. Any shortcomings of this work are due entirely to the authors.

Theodore Allen Jr.
Richard L. Ditsworth

LIST OF MOST USED SYMBOLS

a	linear acceleration	**curl**	curl operator
A	cross-sectional area	D	diameter
A	a field variable	div	divergence operator
A^*	reference area at $N_M = 1$	div **grad**	laplacian operator
B (**b**)	an extensive continuous vector function (specific quantity)	D_H	hydraulic diameter
		D/Dt	material derivative operator
		e	surface-roughness height
B (b)	a continuous scalar function (specific quantity)	e_{ii}	component of longitudinal rates of strain
B_i (b_i)	ith component of extensive vector variable (specific quantity)	e, exp	exponential
		E (e)	energy (specific quantity)
		E	modulus of elasticity
c	sonic celerity	f	friction factor
c_v, c_p	specific heat at constant volume, at constant pressure	$\mathbf{F}_B$ ($\mathbf{f}_B$)	body force (specific quantity)
		$\mathbf{F}_S$	surface force
		$\mathbf{F}_w$	wall force
C_f	surface-resistance coefficient	**g**	gravitational force per mass
C_D	drag coefficient	**grad**	gradient operator

h_L	head loss
h	specific enthalpy
$\mathbf{H}$	moment of momentum
$\mathbf{i}, \mathbf{j}, \mathbf{k}$	unit vectors in x, y, z directions, respectively
K_L	loss coefficient
$d\mathbf{l}$	differential line vector
l	mixing length
l, m, n	direction cosines
L	length
L_{crit}	choking length for Fanno flow
$L^{\dagger}$	transition length
m	mass
$\dot{m}$	mass rate of flow
$\mathbf{M}_B$	moment of body forces
$\mathbf{M}_S$	moment of surface forces
$\mathbf{n}$	outward normal unit vector to an area
n	roughness parameter
n	flow-behavior index
N_C	Cauchy number
N_E	Euler number
N_F	Froude number
N_M	Mach number
N_R	Reynolds number
N_S	Strouhal number
p	pressure
p_0	isentropic pressure stagnation
Δp_L	pressure loss due to friction
p^*	reference pressure at $N_M = 1$
P	perimeter
$\mathbf{P}$ ($\mathbf{p}$)	linear momentum vector (specific quantity)
$\dot{\mathbf{q}}$	rate-of-heat-transfer vector per area
Q	quantity of heat transfer
r, θ, z	cylindrical coordinates
$\mathbf{r}$	position vector
R_H	hydraulic radius
R	radius
R	specific gas constant
R	function of r only
$d\mathbf{S}$	differential surface-area vector
S	surface-area magnitude
S_0	slope
$\$$ ($\not{s}$)	entropy (specific quantity)
t	time
$\mathbf{t}$	unit tangent vector
T	absolute temperature
T_0	stagnation temperature
U (u)	internal energy (specific quantity)
V_x, V_y, V_z	components of velocity in x,y,z
V_r, V_θ, V_z	components of velocity in r,θ,z
$\dot{\not{V}}$	volume flow rate
$\not{V}$	volume
V_*	shearing stress velocity
$\mathbf{V}$	velocity
$\bar{V}$	time-mean average speed
$\tilde{V}$	time-fluctuating speed
V_{SA}	space-average velocity
W	work
$\frac{\not{d}W_s}{dt}$	rate of shaft work
x, y, z	spatial coordinates
y_{lS}	laminar sublayer thickness
$\bar{z}$	z coordinate of centroid of an area
γ	specific-heat ratio
γ_{ij}	component of rate of shear strain
Γ	circulation
δ	boundary-layer thickness
δ^*	boundary-layer-displacement thickness
Δ	finite quantity or change of a quantity
ε_r, ε_θ, ε_z	unit vectors in cylindrical coordinates

η defined nondimensional variable
η_p plastic viscosity
Θ function of θ only
κ thermal conductivity
μ absolute viscosity
ν kinematic viscosity
ρ mass density
ρ_0 isentropic stagnation density
σ normal stress
$\bar{\sigma}$ mean bulk stress
τ shearing stress
τ_0 yield shear stress; also, wall shear stress
ϕ velocity potential
Φ_B specific body-force potential
ψ stream function
$\boldsymbol{\omega}$ angular velocity
$\dot{\boldsymbol{\omega}}$ angular acceleration
$\boldsymbol{\Omega}$ vorticity

Special notation

$\simeq$ approximately equal
$\bar{(\ )}$ time-mean average (used with p, $\mathbf{V}$, $\mathbf{f}_B$, and ρ)
$(\ ')$ time-fluctuating part of a variable in turbulent flow
$|\ |$ absolute value
$\oint_S$ integration over a closed surface S
$\oint_C$ integration around a closed path C in direction shown
0( ) order of magnitude
$\sum$ summation

1

BASIC DEFINITIONS AND INTRODUCTION TO FIELD CONCEPTS

1.1 Introduction

A basic knowledge of fluid mechanics is essential to engineers and applied scientists because they will likely become involved directly or indirectly in problems involving the flow of fluids. A diversity of application is evident in the following list of prediction and design activities associated with fluids in motion:

aerodynamic surfaces for desired lift
structural surfaces to withstand temperatures and forces of a fluid
propulsion systems
energy conversion systems
transport of fluids
bioengineering
fluid control systems
fluid computers

climatology
oceanography

Although these examples are by no means exhaustive, they do serve to establish the varied applications of a single fascinating discipline.

The study begins with some basic definitions and concepts used to represent the observed behavior of fluids; emphasis is placed on the physical meaning of each representation made in mathematical language.

1.2 Fluids and Continuum Concepts

A *fluid* is defined as any substance deforming continuously when subjected to a shear stress regardless of how small the shear stress may be. This means that fluids will "flow" when subjected to a shear stress; and, conversely, flowing fluids will generally exhibit the presence of shear stresses. (A detailed discussion of stresses is deferred until Chapter 3.) It is of interest to note the difference between a fluid and a solid deformed in the elastic range. A solid deformed by a shear stress is capable of resisting deformation within limits by generating an internal stress proportional to the deformation. If the externally applied shear stress is not too great (so that it does not exceed the limit of elastic behavior for the material), then the internal stress developed can become large enough to equalize the external stress. In such a case deformation ceases; that is, the solid does not deform continuously. The above definition could be used as a criterion to distinguish fluids from "nonfluids" by experiment.

Since this study will not consider fluid behavior from the point of view of kinetic theory, a conceptual model is needed that can be used in a physical as well as a mathematical sense. This model is referred to as the *continuum*. Instead of defining a continuum at the outset, a hypothetical experiment is proposed. Consider a rather large and finite region in a fluid whose mass is not uniformly distributed in a given space, and let this region be completely bounded by a spatial volume $\mathcal{V}_1$. Next imagine a point to be fixed in position inside this volume. The density ρ_1 of the fluid inside this space may be thought of as simply the mass m_1 of this fluid divided by $\mathcal{V}_1$. Such a density would obviously be a gross or average description. At one instant of time one may imagine taking successively smaller regions (that is, $\mathcal{V}_1 > \mathcal{V}_2 > \mathcal{V}_3$, etc.) with each region still surrounding the point but lying wholly inside the previous region, as shown in Fig. 1.1. As sample size is decreased, one would witness an approach to a limiting value because the sample would tend to become more uniform in mass distribution. After reaching a certain volume $\mathcal{V}_s$, any reduction in the sample size beyond this value would yield fluctuations in the calculated value of the gross density. If $\mathcal{V}$ were permitted to approach zero, i.e., reduced to the point it surrounds, the density might approach a very

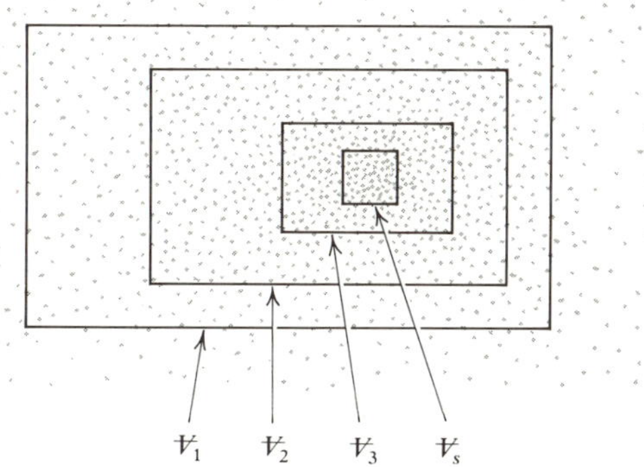

Fig. 1.1 *Variation in mass with sample size at a given instant.*

large value or become zero depending on whether the point was or was not coincident with a molecule of fluid on arriving at the limit. Thus V_s is the minimum volume that lends a unique value to the calculated density. The *continuum concept* imagines the extrapolation of the value of the property ρ from its statistically significant value at V_s to $V = 0$ with ρ remaining constant. Hence the definition of density ρ at a point is

$$\rho = \rho_s = \frac{m_s}{V_s} \qquad \text{extrapolated at constant value } \rho_s \text{ to zero sample size} \tag{1.1}$$

This experiment could be carried out at various points in the fluid region and a value of ρ could be determined for each point; hence one could have a functional variation of ρ in space. The region would then be referred to as a *fluid continuum* or, simply, a continuum.

The concept of a continuum appears to involve a physical region of size V_s surrounding a mathematical point, and this is so since the extrapolation has been made from V_s to $V = 0$. However, it should be kept in mind that any theory using any given model or idealized concept must eventually yield results that are subject to measurement or observation. Such measurement can involve the insertion of a probe or sensing element into the medium; this element should be small enough so that the experiment is not "appreciably changed" by insertion. It was noted earlier that extremely small samples ($V < V_s$) could give fluctuations that cannot be interpreted with respect to the continuum attitude. This leads to an inquiry concerning the molecular population in the fluid. Some insight can be gained by considering a perfect gas at atmospheric conditions in a cubical sample of 0.0001 in. on an edge. The molecular population of this sample exceeds 400 million. It seems reasonable that with such a large number of molecules in the given space, one could expect that a time-averaged measurement could be made that would have some physical significance and that the physical region $\mathcal{V}_s$ can be considered a mathematical point with regard to measurement.

Before closing the discussion of this concept, one should note that the molecules in $\mathcal{V}_s$ are in motion and hence migrate from one small region to another. It is this transfer of molecules that gives rise to mass diffusion. There is also a transfer of momentum and energy because of this motion, which leads to the concepts of viscosity and thermal conductivity discussed later in the text. In addition to these transfers due to well-ordered molecular motion, there may exist exchanges resulting from macroscopic random fluctuations, a concept that is introduced and discussed in later chapters where turbulence effects are considered. These effects are generally treated from the continuum point of view, even though the identity of mass, momentum, and energy transfers can be partially obscured when the identity of molecules in the physical region $\mathcal{V}_s$ is changing very rapidly in a random manner. These last few statements will become more apparent as one's study of fluid mechanics progresses.

It has been shown that a characteristic defined as density may be attributed to a continuum. This may be considered as the density at the point associated with the physical region $\mathcal{V}_s$. Since the mass per unit volume has been given significance, it would follow that any property that is proportional to the amount of mass would also have meaning. Such properties are referred to as *extensive properties*; examples of such are enthalpy, internal energy, and entropy, to list a few. Volume is also extensive in that the volume of a mass small enough to be homogeneous may be obtained by multiplying its specific volume by the mass. The specific volume is simply the reciprocal of the mass density, or volume per unit mass. Association of intensive properties, such as normal stress and temperature, with the continuum is also possible; however, in the case of normal stress, an area about a point is reduced to a continuum limit. When the instantaneous arithmetic average of the molecular speeds in the small region $\mathcal{V}_s$ is not zero, this speed, which is continuous in space and time, is imagined to be associated with the point the small region surrounds. *Hereafter, the term fluid particle will be used to describe a fluid in a continuum sense at a point, and the state of the fluid particle is described by values of its properties at that point.*

1.3 Fields

A *field* is defined as a continuous distribution of a quantity in space and time. The quantity may be scalar, vector, or tensor. The scalar density field is thought of as the continuous variation of the quantity given by

$$\rho = \rho(x,y,z,t) \tag{1.2}$$

which indicates that density is considered as the dependent variable and is a function of three space coordinates and time. However, no preconceived notion should be fostered that ρ has to be a dependent variable in every case. As an example, the equation of state for a perfect gas, $p = \rho RT$, is functionally expressed to indicate that density is an independent variable. The velocity of *fluid particles* referred to at the end of the previous section is a vector field, which may be formulated as

$$\mathbf{V} = \mathbf{V}(x,y,z,t) \tag{1.3}$$

This states that velocity is a vector function of space coordinates and time. Equation (1.3) could be written as three scalar equations expressing each of the three velocity components in terms of space coordinates and time. The moment of inertia, a tensor field of rank two, was introduced in engineering mechanics. For a general configuration, the moment of inertia has nine components, and in rectangular cartesian coordinates these are represented by I_{xx}, I_{yy}, I_{zz}, I_{xy}, I_{xz}, I_{yx}, I_{yz}, I_{zx}, and I_{zy}. Each of these moments of inertia is a function of space coordinates and time, though one may not have thought of these from the field point of view when the moment of inertia was introduced.

Another tensor of rank two will be introduced in Chapter 3 where state of stress is discussed. Since this text does not present fluid mechanics from a tensor point of view, no further comment will be made concerning the properties of tensors, except for points where a tensor quantity is introduced.

1.4 Material and Space Coordinates

The analysis of fluid mechanics problems from the continuum point of view has evolved two methods of description. First, it is possible to ask, "Where in space does a given fluid particle reside at a given instant of time?" This question implies a functional relationship of the form

$$\mathbf{r} = \mathbf{r}\,(\text{fluid particle}, t) \tag{1.4}$$

which states that *the position vector is a function of the particle being observed and time*. A more convenient form of Eq. (1.4) is obtained if the fluid particle can be identified in such a manner as to refer it to a certain position, say $\mathbf{r}_0$, at the time $t = 0$. The equation for the position vector now becomes

$$\mathbf{r} = \mathbf{r}(\mathbf{r}_0, t) \tag{1.5}$$

and it is understood that $\mathbf{r}_0$ is one of the independent variables. Figure 1.2 shows several fluid particles at $t = 0$ with the material coordinates indicated.

In this manner each of the fluid particles in the field may be assigned an identification number $\mathbf{r}_0$; and this number does not change as the particle changes position in space since it is still the same particle that was coincident with the point $\mathbf{r}_0$ at time $t = 0$. The assignment of the identification number at zero time is entirely arbitrary, for any time may be used with the corresponding particle position specified at this time. Zero time will normally be used throughout this text. This leads to the definition that *material coordinates* are the coordinates of the fluid particle at zero time. The use of material coordinates is one method of description. The assumption of a continuum (Section 1.2) allows expressions of the form in Eq. (1.5) since this idealization endows very small regions of space in the fluid with characteristics that are not allowed on a microscopic or molecular basis.

Second, one may ask, "What fluid particle resides at a given point in space at a given instant of time?" If it is assumed that Eq. (1.5) is continuous and single-valued and that its inverse exists, then it is permissible to write

$$\mathbf{r}_0 = \mathbf{r}_0(\mathbf{r}, t) \tag{1.6}$$

which states that the *particle being observed is a function of the position being viewed and time.* The physical implications of continuity and single-valuedness of these functions are the requirements that a group (or string) of fluid particles are continuous in space and that only one fluid particle occupies a given

Fig. 1.2 Three different fluid particles at $t = 0$ identified by material coordinates $\mathbf{r}_{0_1}$, $\mathbf{r}_{0_2}$, and $\mathbf{r}_{0_3}$.

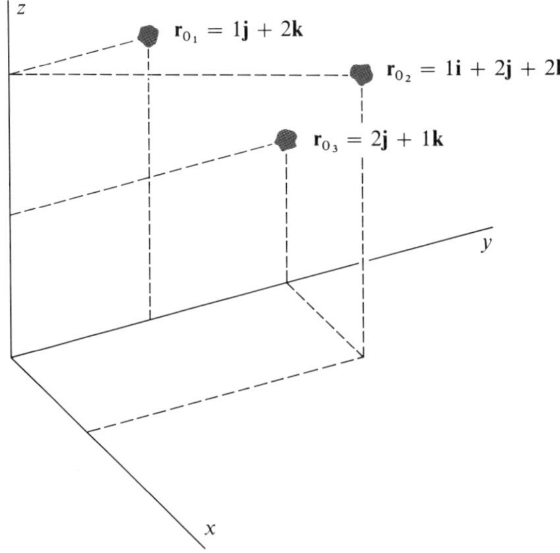

position in space at a given instant of time. The position in space is given by *space coordinates*, which are the same as the coordinates referred to in Section 1.3 or the ordered coordinate triples of both scalar and vector algebra. In the description given by Eq. (1.6), the space coordinates and time are used to locate a fluid particle.

The use of material and space coordinates in connection with the velocity field discussed in the next section will assist in gaining familiarity with these methods of description. The velocity at a point in space is the velocity of the particle that instantaneously occupies this position. Conversely, the velocity of the particle at a point in space is the field velocity for that point.

1.5 Streamlines, Pathlines, and Streaklines

Streamlines represent the loci that are tangent to the velocity vectors in the flow field at a given instant of time. Figure 1.3 shows several of these lines with the tangent velocity vectors. There is obviously no limit to the number of streamlines one may draw in a given flow field, since they are lines and have no thickness. If the streamlines are to be tangent to the velocity vectors, then the differential equation determining these lines is

$$\mathbf{V} \times d\mathbf{l} = 0 \tag{1.7}$$

where $d\mathbf{l}$ is a small displacement vector *along a streamline*. Equation (1.7) gives a valid expression for any coordinate system in which cross multiplication of vectors is defined. If the differential equations in an x, y, z coordinate

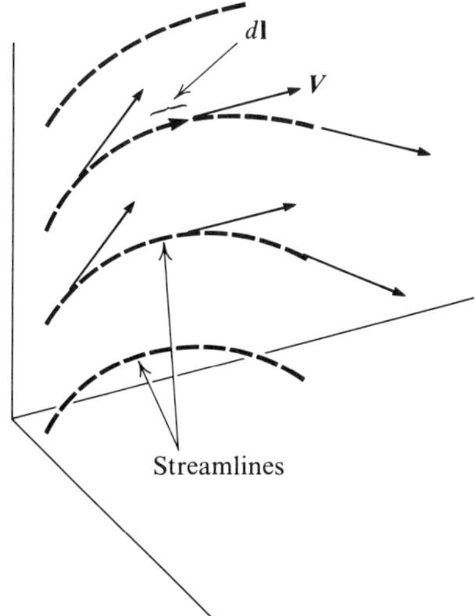

Fig. 1.3 Streamlines at a given instant.

system are desired, then one simply inserts the velocity and displacement in Eq. (1.7) to obtain

$$(V_x\mathbf{i} + V_y\mathbf{j} + V_z\mathbf{k}) \times (dx\,\mathbf{i} + dy\,\mathbf{j} + dz\,\mathbf{k}) = 0$$

carries out the cross multiplication, and simplifies to

$$\frac{dx}{V_x} = \frac{dy}{V_y} = \frac{dz}{V_z} \tag{1.8}$$

Solution of the system of equations given by Eq. (1.8) requires knowledge of the velocity components of the fluid particles as a function of the space coordinates and time. The integration of the system of differential equations yields the equation for the streamline family. It is apparent that if two streamlines corresponding to the same velocity field intersect each other at the same instant of time, the velocity vector would have two different directions at this point. Since physical significance is not ascribed to a velocity with two different directions, streamlines do not intersect in a physical problem. The only exception to this is at a location where the velocity is zero, since the zero or null vector has no inherent direction. Points in a fluid flow field where the velocity is zero are referred to as *stagnation points* or *critical points*. Points A and B in Fig. 7.1a are examples of stagnation points.

A *pathline* is defined as the locus or trajectory followed by any given fluid particle for a given time interval. The path of the particle initially at $\mathbf{r}_0$ is given by

$$\mathbf{r} = \mathbf{r}(\mathbf{r}_0, t) \tag{1.5}$$

This form uses the first method of description in Section 1.4 and may be readily obtained from a spatial description of the velocity field $\mathbf{V} = \mathbf{V}(\mathbf{r}, t)$, by noting that the velocity of the fluid particle is given by the time derivative of the position vector, that is,

$$\mathbf{V} = \frac{d\mathbf{r}}{dt} = \frac{\partial[\mathbf{r}(\mathbf{r}_0, t)]}{\partial t} \tag{1.9}$$

where the partial derivative of the preceding equation indicates that the differentiation is to be carried out for a given particle, that is, holding $\mathbf{r}_0$ constant. The equations for pathlines are then the solutions obtained by integrating $d\mathbf{r}/dt = \mathbf{V}(\mathbf{r}, t)$ and setting $\mathbf{r} = \mathbf{r}_0$ at $t = 0$.

A *streakline* is identified by the "string" of fluid particles that has passed through a given point in space at a given instant of time. This concept is important in flow visualization techniques employed in the laboratory. If a liquid dye is introduced at a point in a fluid stream, then the dye trace in the fluid can closely represent a streakline provided the dye does not disperse or mix with the fluid to any great extent. In order to determine a streakline at a given instant t, the equation of the pathlines may be used to specify which particles have passed through the given point as a function of time. The string of particles that pass through point P at a, b, c in the time interval t is shown in Fig. 1.4. The pathline of each fluid particle is given by $\mathbf{r} = \mathbf{r}(\mathbf{r}_0, t)$ and, conversely, the identity of each particle in terms of spatial coordinates at any time is given by $\mathbf{r}_0 = \mathbf{r}_0(\mathbf{r}, t)$. Since each particle has different material coordinates (location at $t = 0$), it is convenient to change the identification from material coordinates to the time τ (when the particle resides at point P), hence $\mathbf{r}_0 = \mathbf{r}_0[(a,b,c),\tau]$. Upon substitution of this into the equation of a pathline for a particle, one obtains

$$\mathbf{r} = \mathbf{r}\{\mathbf{r}_0[(a,b,c),\tau],t\} \qquad \text{for } 0 \leq \tau \leq t$$

A streakline can then be plotted for a given instant of time by determining the spatial location of each particle (identified by its τ value). The following examples will illustrate procedures for finding streamline, pathline, and streakline loci from a spatial description of the velocity fields. In the case of a velocity field that is not a function of time, the equations are the same for the streamline, pathline, and streakline.

Example 1.1

The components of a velocity field are given by

$$V_x = K_1 x e^{-K_3 t} \qquad V_y = K_2 y \qquad \text{and} \qquad V_z = 0 \qquad \text{for } t \geq 0$$

Find the equation that represents the family of streamlines when $K_1 = K_2 = K_3 = 1$. Each of these constants has a dimension of reciprocal time.

Fig. 1.4 *Location of fluid particles that pass through a given point P at several instants of time.*

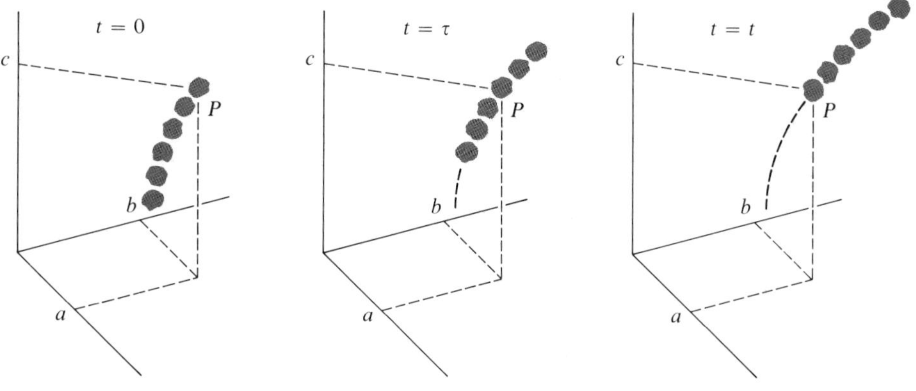

Solution. Using $dx/V_x = dy/V_y$ of Eq. (1.8) and substituting for the velocity components, one obtains

$$\frac{dx}{xe^{(-t)}} = \frac{dy}{y}$$

The preceding expression may be integrated with time considered constant since streamlines describe the tangent loci to the velocity vectors at a given instant of time. Carrying out this integration gives

$$y = Cx^{e^t}$$ ◀

in which C is an integration constant. The last equation represents the streamline family corresponding to the given velocity field. It should be noted that all of the streamlines pass through the origin at the same instant of time. From what has been stated about the intersection of streamlines at a given instant of time, one might suspect that the origin is a stagnation or critical point. Examination of the given velocity components indicates that the velocity is zero at the origin. A specific set of streamlines may be found by obtaining the integration constant required by the coordinates of some point. For example, the streamlines passing through the point a, b at any time t are given by

$$y = b\left(\frac{x}{a}\right)^{e^t}$$

A plot of this last equation for several instants of time is shown in Fig. 1.5.

Example 1.2

For the velocity field given in Example 1.1, what is the pathline for the fluid particle coincident with the point x_0, y_0 at time $t = 0$?

Fig. 1.5 *A plot showing a few streamlines in the first quadrant. Each streamline is shown for different instants of time and all pass through the point a, b.*

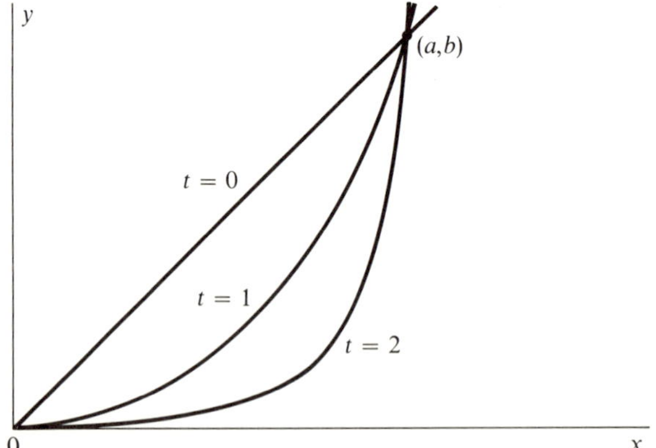

Solution. To find a pathline equation of the form of Eq. (1.5), a description of the velocity field in the form of Eq. (1.9) is considered in which the coordinates of the fluid particle are expressed as functions of time. This leads to the following differential equations for $x(t)$ and $y(t)$:

$$\frac{dx}{dt} = V_x = xe^{-t} \qquad \frac{dy}{dt} = V_y = y$$

in which it is understood that x and y are coordinates of a given particle and as such are functions of time. Integrating these equations yields

$$\int_{x_0}^{x} \frac{dx}{x} = \int_0^t e^{-t}\,dt \qquad \text{or} \qquad x = x_0 e^{1-e^{-t}}$$

and

$$\int_{y_0}^{y} \frac{dy}{y} = \int_0^t dt \qquad \text{or} \qquad y = y_0 e^t$$

The desired pathline may be found from the parametric equations for $x(t)$ and $y(t)$ by elimination of the parameter time. This results in

$$y = \frac{y_0}{1 - \ln(x/x_0)}$$ ◀

Figure 1.6 shows the pathline of the fluid particle that was coincident with the point x_0, y_0 at time $t = 0$.

Fig. 1.6 *Pathline for* $0 \leq t \leq \infty$. *The line* $x = ex_0$ *is an asymptote.*

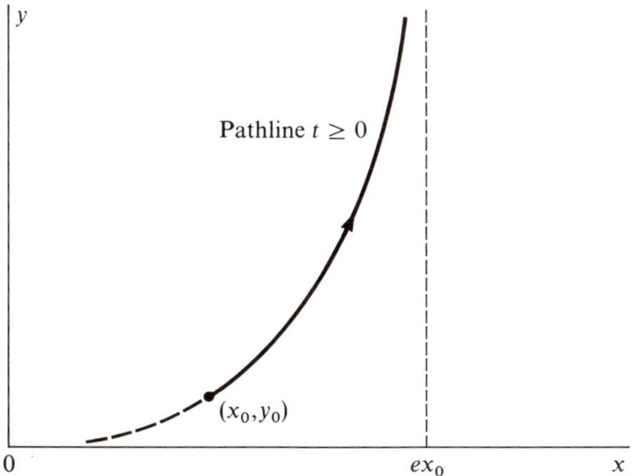

Example 1.3

Obtain the streamline passing through the point x_0, y_0, and the pathline of the fluid particle coincident with the point x_0, y_0 at time $t = 0$, if the velocity field is given by

$$\mathbf{V} = (K_1 x e^{-K})\mathbf{i} + (K_2 y)\mathbf{j} + (0)\mathbf{k}$$

where K is a dimensionless constant, and constants K_1 and K_2 each have a value of 1 and a dimension of reciprocal time.

Solution. For the streamline,

$$\frac{dx}{V_x} = \frac{dy}{V_y} \qquad \text{and} \qquad \frac{dx}{x e^{-K}} = \frac{dy}{y}$$

and for the given point x_0, y_0,

$$y = y_0 \left(\frac{x}{x_0} \right)^{eK}$$ ◀

For the pathline,

$$\int_{x_0}^{x} \frac{dx}{x} = \int_0^t e^{-K}\, dt \qquad \text{or} \qquad x = x_0\, e^{(e^{-K} t)}$$

and

$$\int_{y_0}^{y} \frac{dy}{y} = \int_0^t dt \qquad \text{or} \qquad y = y_0 e^t$$

Elimination of t between the two parametric equations results in

$$y = y_0 \left(\frac{x}{x_0} \right)^{eK}$$ ◀

which is the same as the equation for the streamline. The important observation here is that *streamlines and pathlines are the same locus for a velocity field that is not a function of time*. This is generally true for all velocity fields which are not functions of time, and the inquisitive reader is invited to verify this statement using the basic definitions of these loci given in Section 1.5.

Example 1.4

For the velocity field given in Example 1.1, find the equation of the streakline passing through the point given by $\mathbf{r} = a\mathbf{i} + b\mathbf{j}$.

Solution. The parametric equations for the pathlines in Example 1.2 are

$$x = x_0 e^{1 - e^{-t}} \qquad \text{and} \qquad y = y_0 e^t$$

Recalling that x_0 and y_0 are material coordinates of the fluid particle at $t = 0$, then as they are assigned different pairs of values at $t = 0$, the pairs identify different fluid particles. Solving for the material coordinates from the pathline equations gives

$$x_0 = \frac{x}{e^{1-e^{-t}}} \qquad \text{and} \qquad y_0 = \frac{y}{e^t}$$

The fluid particle at the position $\mathbf{r} = a\mathbf{i} + b\mathbf{j}$ at time τ is given by

$$x_0 = \frac{a}{e^{1-e^{-\tau}}} \qquad \text{and} \qquad y_0 = \frac{b}{e^\tau} \qquad \text{for } 0 \leq \tau \leq t$$

and these are the parametric forms of the vector equation $\mathbf{r}_0 = \mathbf{r}_0[(a,b),\tau]$. To find the equation of the streakline, it is necessary to substitute the values of x_0 and y_0 into the pathline equations. The desired result is

$$x = \frac{ae^{1-e^{-t}}}{e^{1-e^{-\tau}}} \qquad \text{and} \qquad y = \frac{be^t}{e^\tau}$$

This solution is of the form: $\mathbf{r} = \mathbf{r}\{\mathbf{r}_0[(a,b),\tau],t\}$. Plotting the locus of a streakline for a given time t requires the calculation of x and y, which are the space coordinates for values of τ ranging from zero to the value of t. Upon simultaneous solution of the above equations for x and y, the following equation for the streakline is obtained:

$$\ln\frac{x}{a} = \left(1 - \frac{b}{y}\right)e^{-\tau}$$ ◀

Example 1.5

Obtain a streakline through the point a,b for the velocity field given in Example 1.3.

Solution. The equation locating each fluid particle $\mathbf{r}_0$ for any time t was found to be

$$x = x_0 e^{te^{-K}} \qquad \text{and} \qquad y = y_0 e^t$$

For each particle passing through point a, b at time $t = \tau$,

$$a = x_0 e^{\tau e^{-K}} \qquad \text{and} \qquad b = y_0 e^\tau$$

The identification of each particle can now be made in terms of τ by substitution for x_0 and y_0, so that

$$x = ae^{e^{-K}(t-\tau)} \qquad \text{and} \qquad y = be^{t-\tau} \qquad \text{for } \tau \leq t$$

Solving these equations simultaneously to eliminate $t - \tau$ yields

$$y = b\left(\frac{x}{a}\right)^{e^K}$$ ◀

This result demonstrates that the streaklines are coincident with the pathlines and streamlines for a given velocity field that is not a function of time.

Additional definitions in field analysis abbreviate the description of a field by specifying whether the dependent variables are functions of time or not and how many space coordinates are independent variables. These descriptions are steady, unsteady, uniform, one-, two-, and three-dimensional fields. Consider a field expressed by $\mathbf{A} = \mathbf{A}(x,y,z,t)$. If this field is *steady*, then the variable is not a function of time, so $\mathbf{A} = \mathbf{A}(x,y,z)$. Mathematically this means that $\partial\mathbf{A}/\partial t = 0$; that is, if one "stands" at a given point in the field (at a given set of coordinates), then no time rate of change in $\mathbf{A}$ is observed at this point. If the field is *unsteady*, then $\partial\mathbf{A}/\partial t \neq 0$. If a field is *uniform*, then the variable is not a function of space coordinates, so $\mathbf{A} = \mathbf{A}(t)$. For example, this means that $\partial\mathbf{A}/\partial x = \partial\mathbf{A}/\partial y = \partial\mathbf{A}/\partial z = 0$. A field is said to be *one dimensional* if it is a function of only one space coordinate and time. A one-dimensional and steady velocity field could be noted by $\mathbf{V} = \mathbf{V}(x)$; if the field were one dimensional and unsteady, this could be noted by $\mathbf{V} = \mathbf{V}(x,t)$. A *two-dimensional* field can be expressed by two spatial coordinates and time, such as $\mathbf{V} = \mathbf{V}(x,y,t)$. A *three-dimensional* field is a function of three space coordinates and time. Of course, two- and three-dimensional fields may be steady or unsteady. The term *flow field* will mean velocity field, and steady flow field will mean $\partial\mathbf{V}/\partial t = 0$ at every point in the field. A representation of multidimensional effects in the velocity field is given by a *velocity profile*, in which the speeds of fluid particles at a cross section in the flow are represented by the lengths of the arrows. One- and two-dimensional flow through parallel flat plates are shown in Fig. 1.7.

Example 1.6

Classify the following velocity fields as one, two, or three dimensional: (*a*) The flow depicted is between two large plates parallel to each other. The velocity profiles shown in Fig. 1.8 are identical for any value of x. (*b*) The flow shown in Fig. 1.9 is the passage

Fig. 1.7 *Representation of one- and two-dimensional flow by velocity profiles (no variation in z direction).*

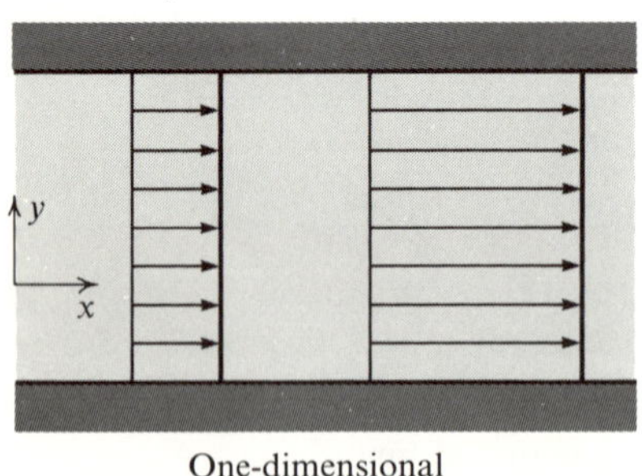

One-dimensional
(variation in x only)

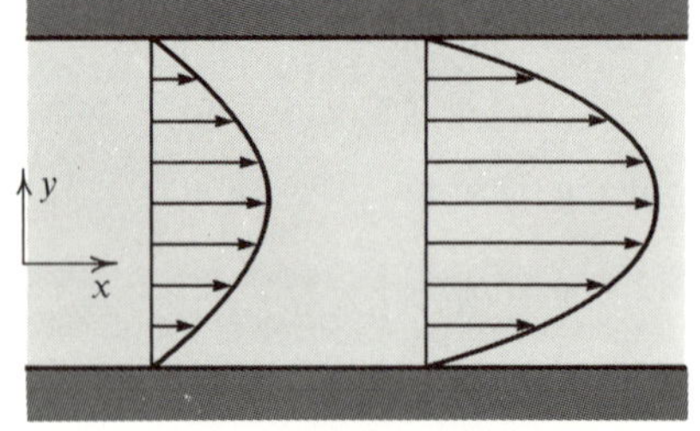

Two-dimensional
(variation in x and y)

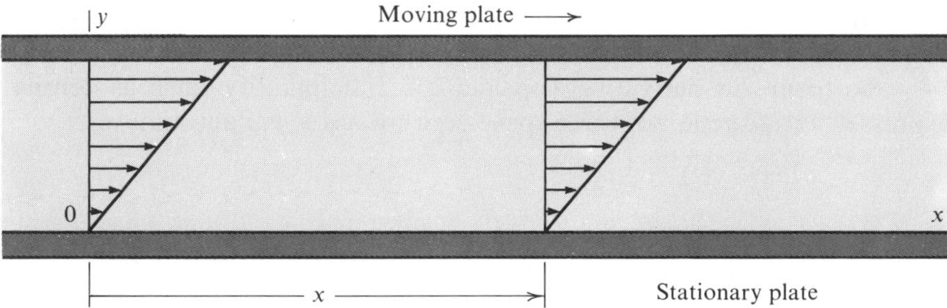

Fig. 1.8

of a lubricant through a slipper bearing. There is no velocity component in the z direction (a very small y component of velocity will not be considered).

Discussion. *a.* This field is one dimensional since the velocity varies only with the y coordinate. This could be noted by $V_x = V_x(y)$, or $\mathbf{V} = \mathbf{V}(y)$.

b. This flow is two dimensional since there is an obvious dependence of V_x on the y coordinate as well as a dependence on x. Note the profiles shown at $x = 0$ and $x = L$. This may be indicated by $V_x = V_x(x,y)$.

1.6 Material Derivative

As previously mentioned, the continuum idea permits expressing the variables of interest in fluid mechanics as continuous and single-valued functions of space and time; hence these functions may be differentiated and integrated. A frequent need for differentiation arises where the change in properties of the fluid particle is desired. While following the particle, such operations are

Fig. 1.9

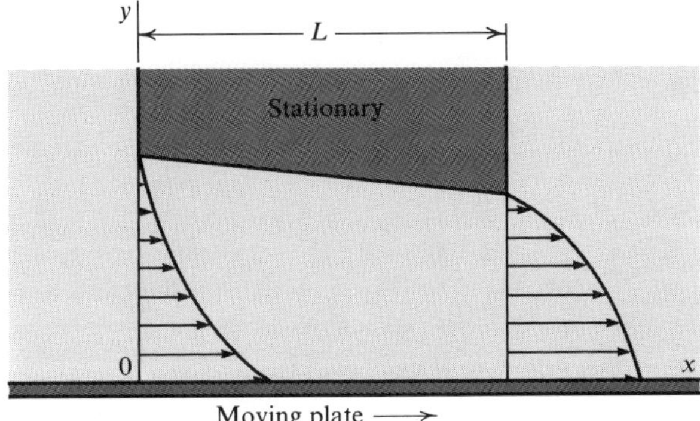

referred to as *differentiation following the motion of the particle*, and the derivatives obtained are referred to as the *material derivatives*.

To form this derivative, consider a scalar quantity, such as density, expressed as a function of three space coordinates x, y, z and time,

$$\rho = \rho[x(t), y(t), z(t), t]$$

Each of these space coordinates is a function of time since the particle is in motion. During the time interval Δt, the particle moves from the position x, y, z to $x + \Delta x, y + \Delta y, z + \Delta z$, and the density changes by $\Delta \rho$ which is approximately given by

$$\Delta \rho \simeq \frac{\partial \rho}{\partial x} \Delta x + \frac{\partial \rho}{\partial y} \Delta y + \frac{\partial \rho}{\partial z} \Delta z + \frac{\partial \rho}{\partial t} \Delta t$$

The change in each of the space coordinates may be expressed in terms of the respective velocity components V_x, V_y, and V_z as shown in Fig. 1.10; and these expressions may be substituted into the expression for $\Delta \rho$ to give

$$\Delta \rho \simeq \frac{\partial \rho}{\partial x} V_x \Delta t + \frac{\partial \rho}{\partial y} V_y \Delta t + \frac{\partial \rho}{\partial z} V_z \Delta t + \frac{\partial \rho}{\partial t} \Delta t$$

Dividing the preceding expression by Δt and limiting the result as Δt approaches zero yields

$$\lim_{\Delta t \to 0} \left(\frac{\Delta \rho}{\Delta t} \right) = \frac{\partial \rho}{\partial x} V_x + \frac{\partial \rho}{\partial y} V_y + \frac{\partial \rho}{\partial z} V_z + \frac{\partial \rho}{\partial t}$$

Fig. 1.10 *A typical fluid particle with two different positions (for the time interval Δt).*

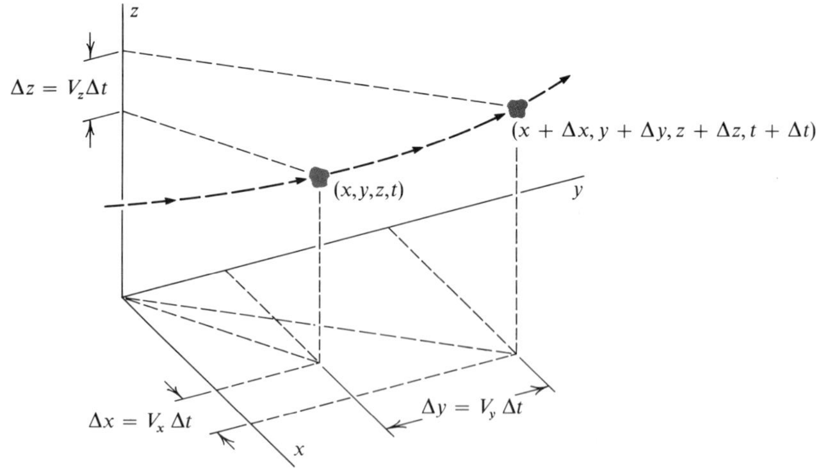

The limit on the left-hand side of the last equation is noted by $D\rho/Dt$; it is called the material derivative of the quantity ρ since the particle (material) identity was held constant during the differentiation process. The *material derivative* may now be written as

$$\frac{D\rho}{Dt} = \underbrace{\frac{\partial \rho}{\partial x} V_x + \frac{\partial \rho}{\partial y} V_y + \frac{\partial \rho}{\partial z} V_z}_{\text{convective}} + \underbrace{\frac{\partial \rho}{\partial t}}_{\text{local}} \tag{1.10}$$

This expression is written as an equality, instead of an approximation, after the limit has been taken; and all of the quantities in Eq. (1.10) are evaluated at the position x, y, z at the time t. The terms labeled convective are so named since each of these three represent a change associated with the particle being "conveyed" or changing position. The last term is referred to as the "local" change since it represents the change in ρ due to any "unsteadiness" at the point x, y, z. In this respect Eq. (1.10) is remarkable in that it relates a change of a quantity associated with a given particle (material identity held constant) to a change of the same quantity with locality (position in space) held constant —the connection between the two being the convective change. This statement may be emphasized by differentiation of the given expression for ρ while the fluid particle is held constant. The operation may be carried out by recognizing that

$$\rho = \rho(\mathbf{r},t) = \rho[\mathbf{r}(\mathbf{r}_0,t), t] \tag{1.11}$$

which simply means that the value of the density at a given position in space at a given time, $\rho(\mathbf{r},t)$, is the value of the density of the fluid particle at this position at the given time, $\rho[\mathbf{r}(\mathbf{r}_0,t), t]$, a statement that applies to any quantity (as well as density) that can be attributed to the fluid through the continuum idealization.

The differentiation may be indicated alternatively as

$$\frac{D\rho}{Dt} = \left.\frac{d\rho}{dt}\right|_{\substack{\text{following}\\ \text{fluid particle}}} = \left.\frac{\partial \rho}{\partial t}\right|_{\mathbf{r}_0,\ \substack{\text{fluid particle}\\ \text{identity held constant}}}$$

Using the last functional expression in Eq. (1.11)

$$\rho[\mathbf{r}(\mathbf{r}_0,t), t] = \rho[x(\mathbf{r}_0,t), y(\mathbf{r}_0,t), z(\mathbf{r}_0,t), t]$$

Recognizing that the independent variables are $\mathbf{r}_0$ and t, and employing the rules of partial differentiation, the following is obtained:

$$\left.\frac{\partial \rho}{\partial t}\right|_{\mathbf{r}_0} = \frac{\partial \rho}{\partial x}\left(\frac{\partial x}{\partial t}\right)_{\mathbf{r}_0} + \frac{\partial \rho}{\partial y}\left(\frac{\partial y}{\partial t}\right)_{\mathbf{r}_0} + \frac{\partial \rho}{\partial z}\left(\frac{\partial z}{\partial t}\right)_{\mathbf{r}_0} + \frac{\partial \rho}{\partial t}$$

Now the derivatives $(\partial x/\partial t)_{\mathbf{r}_0}$, $(\partial y/\partial t)_{\mathbf{r}_0}$, and $(\partial z/\partial t)_{\mathbf{r}_0}$ represent the respective velocity components of the particle $\mathbf{r}_0$, so that

$$\frac{D\rho}{Dt} = \frac{\partial \rho}{\partial x} V_x + \frac{\partial \rho}{\partial y} V_y + \frac{\partial \rho}{\partial z} V_z + \frac{\partial \rho}{\partial t} \tag{1.12}$$

which is the same as Eq. (1.10). This second method of obtaining $D\rho/Dt$ certainly gives meaning to the name material derivative.

In the analysis of fluid motion, the forces acting on a fluid particle are related to the acceleration through the use of Newton's law of motion; and this requires an expression of acceleration (or differentiation of the velocity vector) following the motion of the fluid particle. It is then obvious that the desired derivative would be $D\mathbf{V}/Dt$. The acceleration, following the fluid particle, is

$$\mathbf{a} = \frac{D\mathbf{V}}{Dt} = \frac{\partial \mathbf{V}}{\partial x} V_x + \frac{\partial \mathbf{V}}{\partial y} V_y + \frac{\partial \mathbf{V}}{\partial z} V_z + \frac{\partial \mathbf{V}}{\partial t} \tag{1.13}$$

In r, θ, z coordinates,

$$\mathbf{a} = \frac{D\mathbf{V}}{Dt} = \frac{\partial \mathbf{V}}{\partial r} V_r + \frac{1}{r}\frac{\partial \mathbf{V}}{\partial \theta} V_\theta + \frac{\partial \mathbf{V}}{\partial z} V_z + \frac{\partial \mathbf{V}}{\partial t} \tag{1.14}$$

Example 1.7

Following the fluid particle, calculate the y component of acceleration for a particle whose velocity vector is given by $\mathbf{V} = (3z - x^2)\mathbf{i} + yt^2\mathbf{j} + xz^2\mathbf{k}$ in ft/sec at the point $x = 1$ ft, $y = 1$ ft, $z = 9$ ft, and $t = 2$ sec.

Solution.

$$a_y = \frac{DV_y}{Dt} = V_x\frac{\partial V_y}{\partial x} + V_y\frac{\partial V_y}{\partial y} + V_z\frac{\partial V_y}{\partial z} + \frac{\partial V_y}{\partial t}$$

$$= (3z - x^2)\frac{\partial (yt^2)}{\partial x} + yt^2\frac{\partial (yt^2)}{\partial y} + xz^2\frac{\partial (yt^2)}{\partial z} + \frac{\partial (yt^2)}{\partial t}$$

$$= (3z - x^2)(0) + yt^2(t^2) + xz^2(0) + 2yt = yt^4 + 2yt$$

$$a_y(1,1,9,2) = (1)(2)^4 + 2(1)(2) = 20 \text{ ft/sec}^2$$ ◀

Example 1.8

Generate an expression for the acceleration of a particle in plane polar coordinates following the fluid particle.

Solution. The velocity vector is given by $\mathbf{V} = V_r\boldsymbol{\varepsilon}_r + V_\theta\boldsymbol{\varepsilon}_\theta$, where $\boldsymbol{\varepsilon}_r$ and $\boldsymbol{\varepsilon}_\theta$ are unit vectors in the r and θ directions, respectively, as shown in Fig. 1.11.

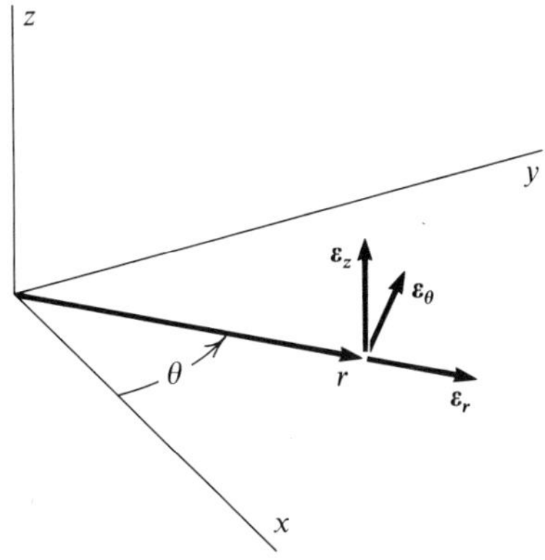

Fig. 1.11

$$\mathbf{a} = \frac{D\mathbf{V}}{Dt} = V_r \frac{\partial \mathbf{V}}{\partial r} + \frac{1}{r}\frac{\partial \mathbf{V}}{\partial \theta} V_\theta + \frac{\partial \mathbf{V}}{\partial t}$$

$$= V_r \frac{\partial}{\partial r}(V_r \boldsymbol{\varepsilon}_r + V_\theta \boldsymbol{\varepsilon}_\theta) + \frac{V_\theta}{r}\frac{\partial}{\partial \theta}(V_r \boldsymbol{\varepsilon}_r + V_\theta \boldsymbol{\varepsilon}_\theta) + \frac{\partial}{\partial t}(V_r \boldsymbol{\varepsilon}_r + V_\theta \boldsymbol{\varepsilon}_\theta)$$

Since[1] $\partial \boldsymbol{\varepsilon}_r/\partial r = 0$, $\partial \boldsymbol{\varepsilon}_\theta/\partial r = 0$, $\partial \boldsymbol{\varepsilon}_r/\partial t = 0$, $\partial \boldsymbol{\varepsilon}_\theta/\partial t = 0$, $\partial \boldsymbol{\varepsilon}_r/\partial \theta = \boldsymbol{\varepsilon}_\theta$, $\partial \boldsymbol{\varepsilon}_\theta/\partial \theta = -\boldsymbol{\varepsilon}_r$,

$$\mathbf{a} = \left(V_r \frac{\partial V_r}{\partial r} + \frac{V_\theta}{r}\frac{\partial V_r}{\partial \theta} - \frac{V_\theta^2}{r} + \frac{\partial V_r}{\partial t}\right)\boldsymbol{\varepsilon}_r + \left(V_r \frac{\partial V_\theta}{\partial r} + \frac{V_\theta}{r}\frac{\partial V_\theta}{\partial \theta} + \frac{V_\theta V_r}{r} + \frac{\partial V_\theta}{\partial t}\right)\boldsymbol{\varepsilon}_\theta \quad \blacktriangleleft$$

Example 1.9

A two-dimensional velocity field is given as

$$\mathbf{V} = (2x^2 - y)\mathbf{i} + (3xy + x^2)\mathbf{j}$$

a. Is this field steady?
b. Obtain an expression for the material derivative of **V**.

Solution. *a.* Since the velocity vector is not a function of time, the field is steady. There is no need to form $\partial \mathbf{V}/\partial t$; it is obviously zero.

b.
$$\frac{D\mathbf{V}}{Dt} = V_x \frac{\partial \mathbf{V}}{\partial x} + V_y \frac{\partial \mathbf{V}}{\partial y} + \frac{\partial \mathbf{V}}{\partial t}$$

$$= (2x^2 - y)[4x\mathbf{i} + (3y + 2x)\mathbf{j}] + (3xy + x^2)[-\mathbf{i} + 3x\mathbf{j}] + 0$$

$$= (8x^3 - 7xy - x^2)\mathbf{i} + (7x^3 - 2xy + 15x^2y - 3y^2)\mathbf{j} \quad \blacktriangleleft$$

[1] Irving H. Shames, "Engineering Mechanics-Dynamics," 2d ed., pp. 314–316, Prentice-Hall, Inc., Englewood Cliffs, N.J., 1966.

1.7 Some Vector Concepts

The gradient, divergence, and curl are presented in this section because of their utility in representing the quantities that appear in the basic equations of fluid mechanics. The definitions offered are given in forms that are independent of a given coordinate system and therefore are rather flexible in their application to a particular problem.

Gradient of a scalar field. The *gradient* of a scalar field is defined as

$$\mathbf{grad}\,A = \lim_{\Delta V \to 0} \frac{1}{\Delta V} \oint_S A\,d\mathbf{S} \tag{1.15}$$

where A represents the scalar field whose gradient is desired. Other symbols are interpreted as follows: ΔV is some infinitesimal volume in coordinate space; S is the surface enclosing ΔV; $d\mathbf{S}$ is a vector that is normal to the surface S at a point on the surface with magnitude $|d\mathbf{S}|$ (see Fig. 1.12). The normal to the surface is drawn outward with respect to the volume ΔV. Adherence to this convention will be observed throughout this text. The meaning of the integral is that the scalar function A is known at the surface area $d\mathbf{S}$, and the product $A\,d\mathbf{S}$ is summed by integration of the components over the closed surface S. Thus the gradient A is this result divided by the volume and limited to a point.

Since Eq. (1.15) does not infer any particular coordinate system, the analyst may make his formulation in the coordinate system best suited to his problem. In order to formulate the gradient of a scalar into a derivative operator, it is necessary to relate the scalar function A at any surface point x,y,z to the scalar value A_1 at a specific point x_1, y_1, z_1. This location is usually chosen at the corner of an infinitesimal volume element closest to the origin

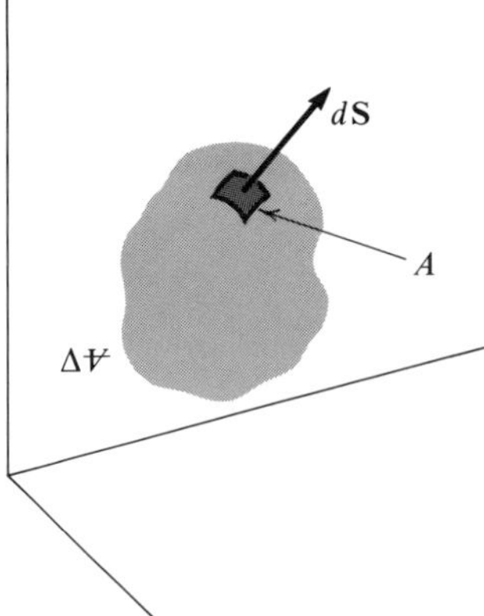

Fig. 1.12

of the axes so that any other point on the surface can be reached by increasing coordinate values; for example, Δx represents a positive value. A Taylor series for several variables provides the relation needed:

$$A = A_1 + \frac{\partial A}{\partial x}(x - x_1) + \frac{\partial A}{\partial y}(y - y_1) + \frac{\partial A}{\partial z}(z - z_1) + \frac{1}{2!}\left[\frac{\partial^2 A}{\partial x^2}(x - x_1)^2 + \cdots\right.$$

where all derivatives are evaluated at x_1, y_1, z_1. An approximation of A is made for small values of $x - x_1, y - y_1, z - z_1$, using

$$A \simeq A_1 + \frac{\partial A}{\partial x}(x - x_1) + \frac{\partial A}{\partial y}(y - y_1) + \frac{\partial A}{\partial z}(z - z_1) \tag{1.16}$$

The value of the integral $\int A\, dS\,\mathbf{j}$ for the two infinitesimal faces in the y direction (before limiting the size of the infinitesimal to zero) shall be approximated by summing the products of the values of A given by Eq. (1.16) and the areas as shown in Fig. 1.13. A reduction in the number of terms occurs on summation, and only $(\partial A/\partial y)\Delta y\,\Delta x\,\Delta z\,\mathbf{j}$ survives. This result permits the further simplification of noting the terms on each face by $-A_1\Delta x\,\Delta z\,\mathbf{j}$ and $[A_1 + (\partial A/\partial y)\Delta y]\,\Delta x\,\Delta z\,\mathbf{j}$. Of course, as the infinitesimal volume is decreased in size, the approximation becomes more correct. In later derivations only the simplified terms above will be used in formulating an expression at

Fig. 1.13 *Approximate values for $\int A\, dS\,\mathbf{j}$ for infinitesimal faces in y direction.*

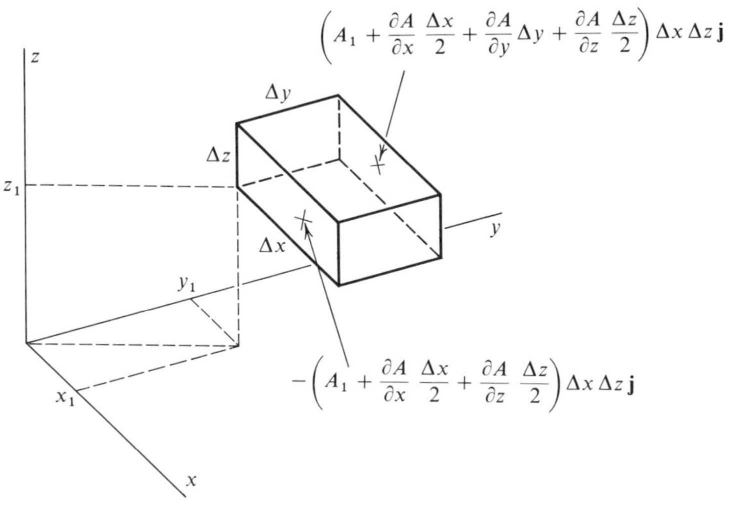

a "point" in a fluid; in addition, the subscript that indicates where the pertinent variable is evaluated will be omitted.

To express the gradient A using the definition in x, y, z coordinates, the approximate values of $A\,d\mathbf{S}$ are noted on each of the faces in Fig. 1.14. Summing the values on all six faces yields an approximate value for $\oint A\,d\mathbf{S}$, and dividing by $\Delta V\!\!\!/$ gives

$$\frac{1}{\Delta V\!\!\!/}\oint_S A\,d\mathbf{S} \simeq \frac{1}{\Delta x\,\Delta y\,\Delta z}\left[\left(A + \frac{\partial A}{\partial x}\Delta x\right)\Delta y\,\Delta z\,\mathbf{i} - A\,\Delta y\,\Delta z\,\mathbf{i} + \left(A + \frac{\partial A}{\partial y}\Delta y\right)\Delta x\,\Delta z\,\mathbf{j} - A\,\Delta x\,\Delta z\,\mathbf{j} + \left(A + \frac{\partial A}{\partial z}\Delta z\right)\Delta x\,\Delta y\,\mathbf{k} - A\,\Delta x\,\Delta y\,\mathbf{k}\right]$$

When the limit is taken, the approximate value becomes exact, so

$$\mathbf{grad}\,A = \lim_{\Delta V\!\!\!/ \to 0}\left(\frac{\partial A}{\partial x}\mathbf{i} + \frac{\partial A}{\partial y}\mathbf{j} + \frac{\partial A}{\partial z}\mathbf{k}\right)$$

If it appears curious that there is nothing inside the parentheses to limit as $\Delta V\!\!\!/ \to 0$, one can note what would have been left to limit by retaining a larger number of terms in the Taylor series. In view of this, one simply writes

$$\mathbf{grad}\,A = \frac{\partial A}{\partial x}\mathbf{i} + \frac{\partial A}{\partial y}\mathbf{j} + \frac{\partial A}{\partial z}\mathbf{k} \tag{1.17}$$

Fig. 1.14 *Approximate values of $\int A\,d\mathbf{S}$ for formulation of* **grad** *A.*

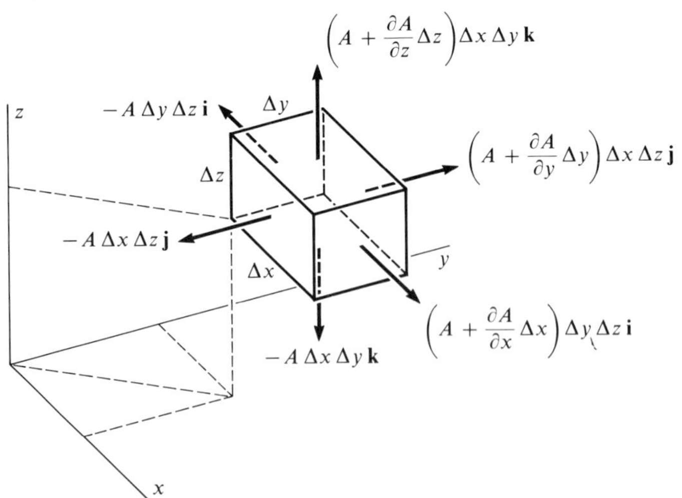

A comparable form of the gradient in cylindrical coordinates is

$$\mathbf{grad}\, A = \frac{\partial A}{\partial r}\boldsymbol{\varepsilon}_r + \frac{1}{r}\frac{\partial A}{\partial \theta}\boldsymbol{\varepsilon}_\theta + \frac{\partial A}{\partial z}\boldsymbol{\varepsilon}_z \tag{1.18}$$

The gradient of a scalar function has a magnitude equal to the maximum value of the directional derivative of the function and a direction normal to a surface over which the scalar field is constant. A paramount advantage of the gradient lies in the fact that once the maximum change of a function is known, it is a simple matter to find a component of this change in any arbitrary direction.

Example 1.10

A function is given as $A = x^2 + y^2 t - 2z$ in lb_f/ft^2 with x, y, and z in units of feet and t in seconds. (*a*) What is the gradient of A? (*b*) What is the magnitude of the gradient of A at the point $x = 4$ ft, $y = 1$ ft, $z = 10$ ft, and $t = 2$ sec? (*c*) What is the component of the gradient of A in the direction of $\mathbf{r} = 6x\mathbf{i} - \mathbf{j} + y\mathbf{k}$ at the given point and time?

Solution.

a.
$$\mathbf{grad}\, A = \frac{\partial(x^2 + y^2 t - 2z)}{\partial x}\mathbf{i} + \frac{\partial(x^2 + y^2 t - 2z)}{\partial y}\mathbf{j} + \frac{\partial(x^2 + y^2 t - 2z)}{\partial z}\mathbf{k}$$
$$= 2x\mathbf{i} + 2yt\mathbf{j} - 2\mathbf{k}\ \text{lb}_f/\text{ft}^3 \quad ◀$$

b.
$$\mathbf{grad}\, A = (2)(4)\mathbf{i} + (2)(1)(2)\mathbf{j} - 2\mathbf{k}$$
$$= 8\mathbf{i} + 4\mathbf{j} - 2\mathbf{k}\ \text{lb}_f/\text{ft}^3 \quad ◀$$

$$\text{Magnitude of } \mathbf{grad}\, A = |\mathbf{grad}\, A| = [8^2 + 4^2 + (-2)^2]^{1/2}$$
$$= (64 + 16 + 4)^{1/2} = (84)^{1/2}$$
$$= 9.17\ \text{lb}_f/\text{ft}^3 \quad ◀$$

c. To find the component of the gradient in the direction $\mathbf{r} = 6x\mathbf{i} - \mathbf{j} + y\mathbf{k}$, one forms the scalar product of **grad** A and a unit vector in the direction of $\mathbf{r}$, that is,

$$\mathbf{grad}\, A \cdot \frac{\mathbf{r}}{|\mathbf{r}|} = (2x\mathbf{i} + 2yt\mathbf{j} - 2\mathbf{k}) \cdot \frac{6x\mathbf{i} - \mathbf{j} + y\mathbf{k}}{[(6x)^2 + (-1)^2 + y^2]^{1/2}}$$

$$\mathbf{grad}\, A \cdot \frac{\mathbf{r}}{|\mathbf{r}|} = \frac{12x^2 - 2yt - 2y}{(36x^2 + 1 + y^2)^{1/2}}$$

$$\mathbf{grad}\, A \cdot \frac{\mathbf{r}}{|\mathbf{r}|} = \frac{(12)(4)^2 - (2)(1)(2) - (2)(1)}{[(36)(16) + 1 + (1)^2]^{1/2}} = 7.74\ \text{lb}_f/\text{ft}^3 \quad ◀$$

It is useful to formulate the material derivative of scalar variables in terms of the gradient. Consider an expression for a scalar variable B in terms of x, y, z coordinates:

$$\frac{DB}{Dt} = \frac{\partial B}{\partial x} V_x + \frac{\partial B}{\partial y} V_y + \frac{\partial B}{\partial z} V_z + \frac{\partial B}{\partial t}$$

One notes that the gradient takes on a form involving the sum of partial derivatives of a scalar with respect to spatial coordinates that are components of the vector called **grad** B. The above expression, in part, appears as a dot product of velocity and **grad** B. So with

$$\mathbf{V} = V_x\mathbf{i} + V_y\mathbf{j} + V_z\mathbf{k}$$

and

$$\mathbf{grad}\, B = \frac{\partial B}{\partial x}\mathbf{i} + \frac{\partial B}{\partial y}\mathbf{j} + \frac{\partial B}{\partial z}\mathbf{k}$$

one sees that

$$\frac{DB}{Dt} = \mathbf{V} \cdot \mathbf{grad}\, B + \frac{\partial B}{\partial t} \tag{1.19}$$

Divergence of a vector field. The *divergence* of a vector field is given by

$$\operatorname{div} \mathbf{A} = \lim_{\Delta V \to 0} \frac{1}{\Delta V} \oint_S \mathbf{A} \cdot d\mathbf{S} \tag{1.20}$$

The symbols in this definition are shown in Fig. 1.15 and are interpreted as follows: ΔV is some small volume in coordinate space; S is the surface enclosing ΔV; and $d\mathbf{S}$ is a vector normal to the surface S at a point on this

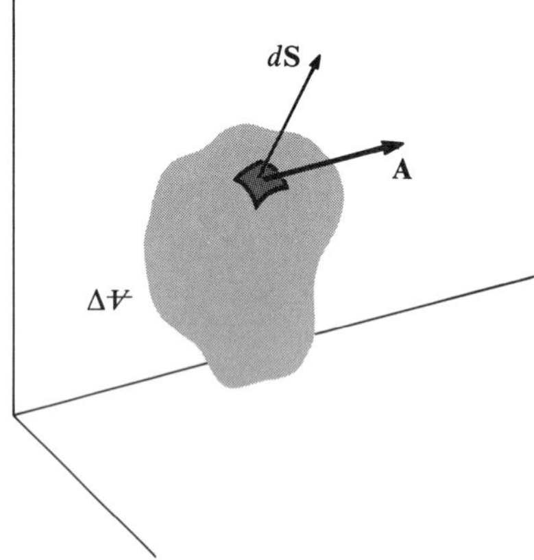

Fig. 1.15

surface. **A** is a vector evaluated at $d\mathbf{S}$. The integral is a sum of the dot product terms of **A** and $d\mathbf{S}$ over the closed surface. The divergence of **A** is this result divided by the volume and limited to a point.

When approximate values of **A** are formulated into a scalar product with corresponding surfaces $d\mathbf{S}$ on a small element of volume $\Delta x \Delta y \Delta z$, Eq. (1.20) for the divergence of **A** simplifies to

$$\operatorname{div} \mathbf{A} = \lim_{\Delta V \to 0} \frac{1}{\Delta V}\left(\frac{\partial A_x}{\partial x} + \frac{\partial A_y}{\partial y} + \frac{\partial A_z}{\partial z}\right)\Delta x \Delta y \Delta z$$

and, on limiting, reduces to

$$\operatorname{div} \mathbf{A} = \frac{\partial A_x}{\partial x} + \frac{\partial A_y}{\partial y} + \frac{\partial A_z}{\partial z} \tag{1.21}$$

The result of formulating the divergence in cylindrical coordinates is

$$\operatorname{div} \mathbf{A} = \frac{A_r}{r} + \frac{\partial A_r}{\partial r} + \frac{1}{r}\frac{\partial A_\theta}{\partial \theta} + \frac{\partial A_z}{\partial z} \tag{1.22}$$

Example 1.11

A velocity field is given as $\mathbf{V} = (x^2 + y^2)\boldsymbol{i} + yzt\mathbf{j} + z^2t\mathbf{k}$ in units of ft/sec with x, y, z in units of feet and t in seconds. (*a*) Obtain the divergence of this vector field. (*b*) What is the value of div **V** at the point $x = 5$ ft, $y = 17$ ft, $z = 1$ ft, and $t = 2$ sec?

Solution.

$$a.\ \operatorname{div} \mathbf{V} = \frac{\partial V_x}{\partial x} + \frac{\partial V_y}{\partial y} + \frac{\partial V_z}{\partial z} = \frac{\partial}{\partial x}(x^2 + y^2) + \frac{\partial}{\partial y}(yzt) + \frac{\partial}{\partial z}(z^2t)$$

$$= 2x + zt + 2zt = 2x + 3zt \quad ◀$$

$$b.\ \operatorname{div} \mathbf{V} = (2)(5) + (3)(1)(2) = 16\ \text{sec}^{-1} \quad ◀$$

A very useful physical interpretation is made of the divergence definition in fluid flow by considering the velocity field. The scalar product $\mathbf{V} \cdot d\mathbf{S}$ or $V_n\, dS$, where V_n is the velocity component normal to the area dS, represents the volume flow rate through the stationary area $d\mathbf{S}$. Hence $\oint \mathbf{V} \cdot d\mathbf{S}$ is the net volume flow rate through a closed stationary surface S and the divergence of **V** represents the net volume flow rate through a fixed surface per unit volume at a point as the volume is shrunk to that point. The requirement of the area being fixed in space (or stationary) may be relaxed by considering the velocity **V** to be relative to the point on the surface where the vector $d\mathbf{S}$ is localized.

It is helpful to have available the following identity involving the divergence of the product of a scalar and vector:

$$\operatorname{div} B\mathbf{V} = B \operatorname{div} \mathbf{V} + \mathbf{V} \cdot \mathbf{grad}\, B \tag{1.23}$$

The correctness of this identity may be readily checked.

Curl of a vector field. The *curl* of the vector field is defined by

$$\mathbf{curl\,A} = -\lim_{\Delta\mathcal{V} \to 0} \frac{1}{\Delta V} \oint_S \mathbf{A} \times d\mathbf{S} \tag{1.24}$$

The symbols in the definition and shown in Fig. 1.16 have the following meaning: **A** is any vector field evaluated at $d\mathbf{S}$; $d\mathbf{S}$ is the usual area vector; and S is the surface enclosing the small element of volume $\Delta\mathcal{V}$. The cross product of **A** and $d\mathbf{S}$ is then taken at each position on the surface S; and these vector products are summed and then divided by $\Delta\mathcal{V}$. The limit is carried out in the usual manner, that is, reducing $\Delta\mathcal{V}$ to a point. Obviously the curl of a vector is another vector.

Application of the definition of **curl A** to a volume element $\Delta x \Delta y \Delta z$ yields

$$\mathbf{curl\,A} = \left(\frac{\partial A_z}{\partial y} - \frac{\partial A_y}{\partial z}\right)\mathbf{i} + \left(\frac{\partial A_x}{\partial z} - \frac{\partial A_z}{\partial x}\right)\mathbf{j} + \left(\frac{\partial A_y}{\partial x} - \frac{\partial A_x}{\partial y}\right)\mathbf{k} \tag{1.25}$$

The formulation of the **curl A** in cylindrical coordinates is

$$\mathbf{curl\,A} = \left(\frac{1}{r}\frac{\partial A_z}{\partial \theta} - \frac{\partial A_\theta}{\partial z}\right)\boldsymbol{\varepsilon}_r + \left(\frac{\partial A_r}{\partial z} - \frac{\partial A_z}{\partial r}\right)\boldsymbol{\varepsilon}_\theta + \frac{1}{r}\left(\frac{\partial (rA_\theta)}{\partial r} - \frac{\partial A_r}{\partial \theta}\right)\boldsymbol{\varepsilon}_z \tag{1.26}$$

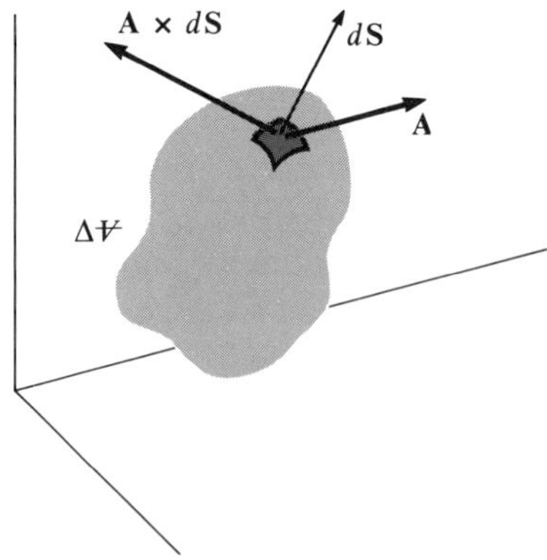

Fig. 1.16

A vector field whose curl vanishes identically throughout the field is said to be an *irrotational field*. With respect to fluid mechanics, the terms *rotational* and *irrotational* will gain some significance on examination of the velocity field of the fluid. The rotation indicated in Fig. 1.17 is shown about the z axis only for purposes of clarity. The rotational rate about this axis is defined as the average angular speed of two lines in the element such that these lines are each perpendicular to the axis of rotation and to each other. The angular velocity of the line Δx is given by

$$\boldsymbol{\omega}_{\Delta x} = \frac{[V_y + (\partial V_y/\partial x)\,\Delta x] - V_y}{\Delta x}\mathbf{k} = \frac{\partial V_y}{\partial x}\mathbf{k}$$

and that of the Δy is

$$\boldsymbol{\omega}_{\Delta y} = \frac{[V_x + (\partial V_x/\partial y)\,\Delta y] - V_x}{\Delta y}(-\mathbf{k}) = -\frac{\partial V_x}{\partial y}\mathbf{k}$$

The $\mathbf{k}$ and $-\mathbf{k}$ are employed to be consistent with the arbitrarily chosen positive directions of the coordinate axes indicated in Fig. 1.17. It should be recalled from engineering mechanics that infinitesimal angular displacements, and hence angular velocities, have vectorial characteristics so that arbitrary resolution into components, as undertaken here, is permissible.

The average of $\boldsymbol{\omega}_{\Delta x}$ and $\boldsymbol{\omega}_{\Delta y}$ is the rotational rate about the z axis and is given by

$$\omega_z = \frac{1}{2}(\omega_{\Delta x} + \omega_{\Delta y}) = \frac{1}{2}\left(\frac{\partial V_y}{\partial x} - \frac{\partial V_x}{\partial y}\right)$$

By comparison with Eq. (1.25) it is seen that ω_z is one-half the z component

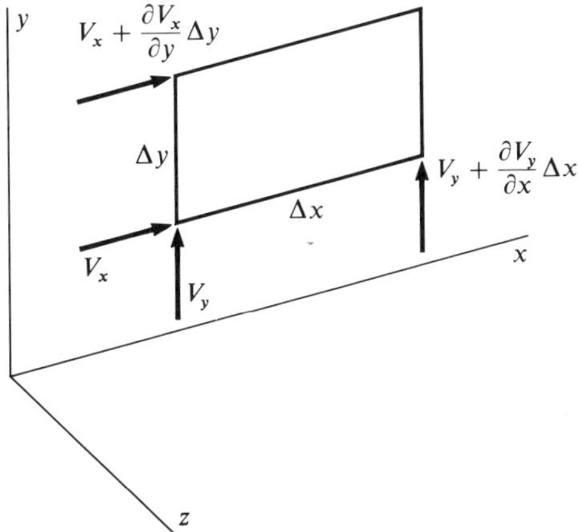

Fig. 1.17 Fluid element at time t.

of the curl of the velocity vector. Similarly, a relationship could be obtained for ω_x and ω_y. Summarizing these results,

$$\boldsymbol{\omega} = \frac{1}{2}\,\mathbf{curl\ V} \tag{1.27}$$

The vorticity vector $\boldsymbol{\Omega}$ is defined as

$$\boldsymbol{\Omega} = \mathbf{curl\ V} \tag{1.28}$$

and this quantity is important in the study of lifting forces on airfoils.

Example 1.12

Is the field given by $\mathbf{A} = (x^2 + y^2)\mathbf{i} - 2xy\mathbf{j} + 4z^2\mathbf{k}$ irrotational?

Solution.

$$\mathbf{curl\ A} = \left[\frac{\partial(4z^2)}{\partial y} - \frac{\partial(-2xy)}{\partial z}\right]\mathbf{i} + \left[\frac{\partial(x^2 + y^2)}{\partial z} - \frac{\partial(4z^2)}{\partial x}\right]\mathbf{j} + \left[\frac{\partial(-2xy)}{\partial x} - \frac{\partial(x^2 + y^2)}{\partial y}\right]\mathbf{k}$$

$$= (0 - 0)\mathbf{i} + (0 - 0)\mathbf{j} + (-2y - 2y)\mathbf{k}$$

$$= -4y\mathbf{k} \neq 0$$ ◀

Since the **curl A** does not vanish at all points in the field, the field is not irrotational. In preference to using the double negative, the field is said to be rotational.

Example 1.13

Find the vorticity of the velocity field given by $\mathbf{V} = 3x\mathbf{i} - 6xy^2t\mathbf{j}$ in feet per second with x, y, and z in feet and t in seconds at the point $x = 2$ ft, $y = 1$ ft, $z = 3$ ft, for $t = 2$ sec.

Solution.

$$\boldsymbol{\Omega} = \mathbf{curl\ V} = \left[\frac{\partial(0)}{\partial y} - \frac{\partial(-6xy^2t)}{\partial z}\right]\mathbf{i} + \left[\frac{\partial(3x)}{\partial z} - \frac{\partial(0)}{\partial x}\right]\mathbf{j} + \left[\frac{\partial(-6xy^2t)}{\partial x} - \frac{\partial(3x)}{\partial y}\right]\mathbf{k}$$

$$= (0 - 0)\mathbf{i} + (0 - 0)\mathbf{j} + (-6y^2t - 0)\mathbf{k} = -6y^2t\mathbf{k}$$

At the point 2, 1, 3 for $t = 2$,

$$\boldsymbol{\Omega} = -6(1)^2(2) = -12\ \text{sec}^{-1}$$ ◀

1.8 Some Vector-Field Theorems

Before undertaking the study of fluid mechanics, it is expedient to formalize a few integral relations pertaining to vector fields. The first of these is called the *divergence theorem* of Gauss. It may be stated as: "The integral of the divergence of a vector field taken over a volume is equal to the integral of the normal component of this field taken over the surface enclosing the volume," and mathematically it is

$$\int_{\mathcal{V}} \operatorname{div} \mathbf{A}\, d\mathcal{V} = \oint_S \mathbf{A} \cdot d\mathbf{S} \tag{1.29}$$

The derivation of this theorem is readily available,[1] and follows directly from the definition of the divergence given by Eq. (1.20). The physical importance of this theorem is that it relates the net efflux of a quantity through a closed surface to a function evaluated throughout the region enclosed by this surface. This is of considerable significance in fluid mechanics and other physical sciences where the concept of flux or flow of quantity must be used in analysis. The mass flux of a fluid is considered in Chapter 2, and it should be recalled from elementary physics that the concept of electric flux was employed in expressing Gauss' law for the electrostatic field.

Example 1.14

Given a velocity field $\mathbf{V} = 2x\mathbf{i} - 3y\mathbf{j}$, check the divergence theorem for the unit cube shown in Fig. 1.18.

[1] Parry Moon and Domina Eberle Spencer, "Vectors," pp. 222–223, D. Van Nostrand Company, Inc., Princeton, N.J., 1965.

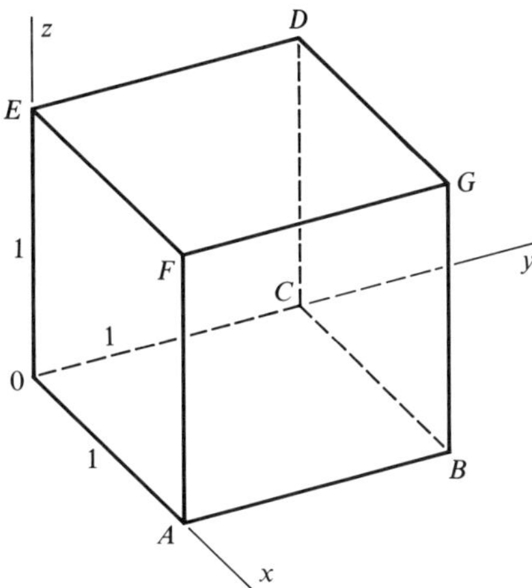

Fig. 1.18

Solution. Evaluating the volume integral of the div **V**,

$$\int_{\mathcal{V}} \text{div}\, \mathbf{V}\, d\mathcal{V} = \int_{\mathcal{V}} (2 - 3)\, d\mathcal{V} = -1 \text{ (vol of cube)} = -1 \qquad \blacktriangleleft$$

The surface integral of the normal component of **V** may be evaluated as the sum of six integrals as follows:

$$\oint_S \mathbf{V}\cdot d\mathbf{S} = \int_{S_{OAFE}} \mathbf{V}\cdot d\mathbf{S} + \int_{S_{OCDE}} \mathbf{V}\cdot d\mathbf{S} + \int_{S_{OABC}} \mathbf{V}\cdot d\mathbf{S} + \int_{S_{ABGF}} \mathbf{V}\cdot d\mathbf{S} + \int_{S_{GBCD}} \mathbf{V}\cdot d\mathbf{S} + \int_{S_{FGDE}} \mathbf{V}\cdot d\mathbf{S}$$

$$= \int_0^1 (2x\mathbf{i})\cdot(1)(-dx\mathbf{j}) + \int_0^1 (-3y\mathbf{j})\cdot(1)(-dy\mathbf{i})$$

$$+ \int_{S_{OABC}} (2x\mathbf{i} - 3y\mathbf{j})\cdot(-dx\, dy\, \mathbf{k}) + \int_{S_{ABGF}} (2x\mathbf{i} - 3y\mathbf{j})\cdot(dy\, dz\, \mathbf{i})$$

$$+ \int_{S_{GBCD}} (2x\mathbf{i} - 3y\mathbf{j})\cdot(dx\, dz\, \mathbf{j}) + \int_{S_{FGDE}} (2x\mathbf{i} - 3y\mathbf{j})\cdot(dx\, dy\, \mathbf{k})$$

$$= 0 + 0 + 0 + \int_{S_{ABGF}} 2(1)\, dy\, dz - \int_{S_{GBCD}} 3(1)\, dx\, dz + 0$$

$$= 2 \text{ (area of } ABGF) - 3 \text{ (area of } GBCD)$$

$$= 2 - 3 = -1 \qquad \blacktriangleleft$$

which checks with the evaluation made by using the volume integral of the divergence of **V**.

Example 1.15

A velocity is uniform and constant in the z direction, with a magnitude of V_∞. Check the divergence theorem using this field and a sphere of unit radius.

Solution. Evaluation of the volume integral yields

$$\int_{\mathcal{V}} \text{div}\, \mathbf{V}\, d\mathcal{V} = \int_{\mathcal{V}} \text{div}\, (V_\infty \mathbf{k})\, d\mathcal{V} = 0 \qquad \blacktriangleleft$$

Using Fig. 1.19 and relating unit vectors $\boldsymbol{\varepsilon}_r$ and $\boldsymbol{\varepsilon}_\phi$, one finds

$$\mathbf{V} = V_\infty \mathbf{k} = V_\infty \cos\phi\, \boldsymbol{\varepsilon}_r - V_\infty \sin\phi\, \boldsymbol{\varepsilon}_\phi$$

As shown in Fig. 1.20, the surface element $d\mathbf{S}$ is $r^2 \sin\phi\, d\theta\, d\phi\, \boldsymbol{\varepsilon}_r$, and for $r = 1$,

$$d\mathbf{S} = \sin\phi\, d\theta\, d\phi\, \boldsymbol{\varepsilon}_r$$

Evaluation of the surface integral yields

$$\oint_S \mathbf{V}\cdot d\mathbf{S} = \oint_S (V_\infty \cos\phi\, \boldsymbol{\varepsilon}_r - V_\infty \sin\phi\, \boldsymbol{\varepsilon}_\phi)\cdot(\sin\phi\, d\phi\, d\theta\, \boldsymbol{\varepsilon}_r)$$

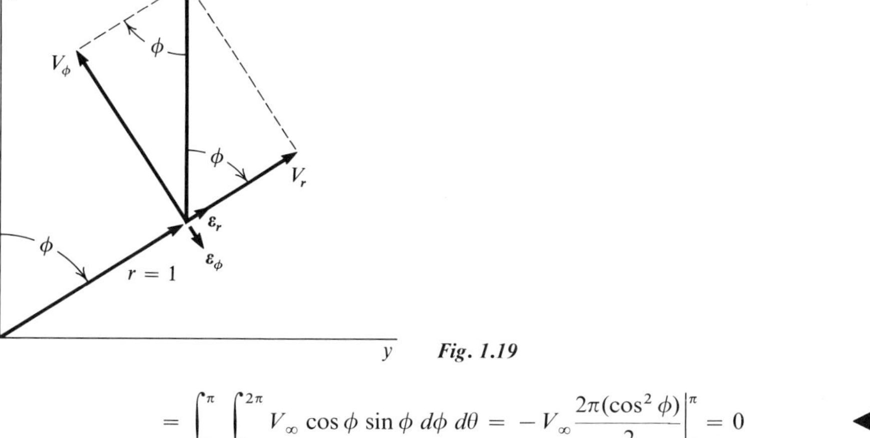

Fig. 1.19

$$= \int_0^{\pi} \int_0^{2\pi} V_\infty \cos\phi \sin\phi \, d\phi \, d\theta = -V_\infty \frac{2\pi(\cos^2\phi)}{2}\bigg|_0^{\pi} = 0$$ ◀

The two integrals are equal and illustrate the validity of the divergence theorem.

A second field theorem of great utility in fluid mechanics is *Stokes' theorem*. This may be stated as: "The line integral of the tangential component of a vector field taken around any closed curve is equal to the surface integral of the normal component of the curl of the vector field taken over any surface bounded by the curve." Formalized mathematically, this theorem may be written as

$$\oint_C \mathbf{A} \cdot d\mathbf{l} = \int_S \mathbf{curl\,A} \cdot d\mathbf{S} \tag{1.30}$$

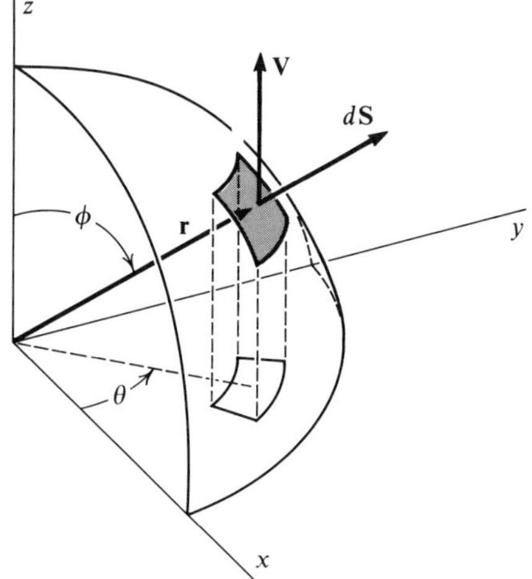

Fig. 1.20

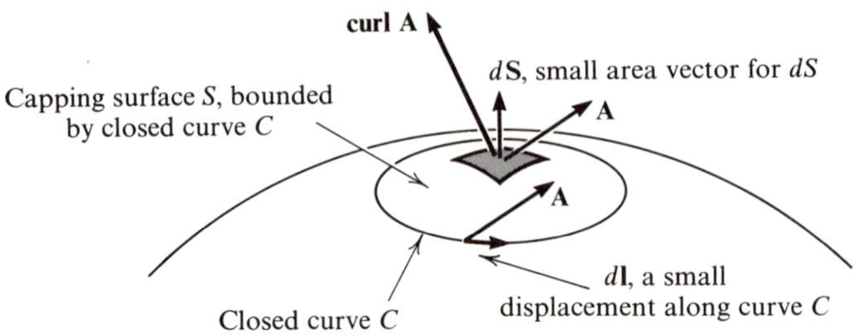

Fig. 1.21 Capping surface S with bounding curve C.

in which S is a capping surface of the bounding curve C. The nomenclature involved in this formalization is shown in Fig. 1.21.

It should be understood that each integrand in Stokes' theorem must exist; that is, **A** is defined at all points on C as well as **curl A** being defined at all points on S, for the theorem to have meaning. The direction of integration along the curve C is taken with the area to the left, as indicated on the figure. A derivation of this theorem is readily available.[1] Its bearing on fluid mechanics is important in the study of irrotational flows. If the curl of the velocity vector is zero at all points in the field, then the theorem requires that $\oint \mathbf{V} \cdot d\mathbf{l} = 0$. This integral is called the *circulation of the velocity* or simply the *circulation*, and is noted by

$$\Gamma = \oint \mathbf{V} \cdot d\mathbf{l} \tag{1.31}$$

In flow fields where the circulation is zero around all closed paths, it can be shown that the velocity vector is given by the gradient of a scalar field, which is called the *velocity potential* field. Irrotational flows are considered later in Chapter 7.

Example 1.16

Evaluate the $\oint \mathbf{B} \cdot d\mathbf{l}$ around the curve $(x - 9)^2 + (y - 5)^2 = 9$ for $\mathbf{B} = 3y\mathbf{i} + 4x\mathbf{j}$. The direction of integration is indicated on Fig. 1.22. The path of integration is a circle of radius 3 and the center is located at the point 9, 5.

Solution. It would appear that the line integration around the circle might be more difficult than integrating the normal component of the curl over the area of the circle, since

$$\mathbf{curl\ B} = \left[\frac{\partial B_y}{\partial x} - \frac{\partial B_x}{\partial y}\right]\mathbf{k} = 1\mathbf{k}$$

[1] *Ibid.*, pp. 232–233.

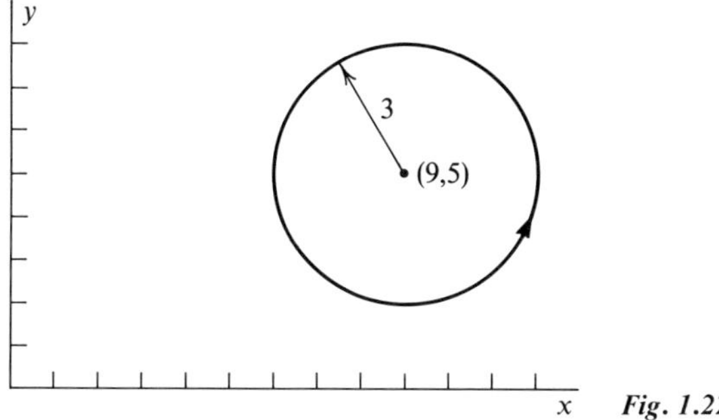

Fig. 1.22

If one does this, the capping surface can be considered as the area enclosed by the path of integration. This is always possible in the case of a plane curve since an enclosed plane area can always be identified with a closed plane curve. Proceeding with the integration by using Stokes' theorem,

$$\oint_C \mathbf{B} \cdot d\mathbf{l} = \int_S \mathbf{curl}\,\mathbf{B} \cdot d\mathbf{S} = \int_S \left[\frac{\partial B_y}{\partial x} - \frac{\partial B_x}{\partial y} \right] \mathbf{k} \cdot d\mathbf{S} = \int_S \mathbf{k} \cdot dS\,\mathbf{k}$$

$$= \int_S dS = S = \text{circle area}$$

$$= 9\pi$$

◀

The application of Stokes' theorem above presupposes that **B** and **curl B** are defined at all points on C and S. This is the case for the given problem.

Example 1.17

A vector field is specified in plane polar coordinates by $\mathbf{V} = 5/r\ \boldsymbol{\varepsilon}_\theta$, where $\boldsymbol{\varepsilon}_\theta$ is a unit vector in the θ direction. Evaluate the circulation of this vector around the circle that has its center at the origin and a radius of 4 as shown in Fig. 1.23.

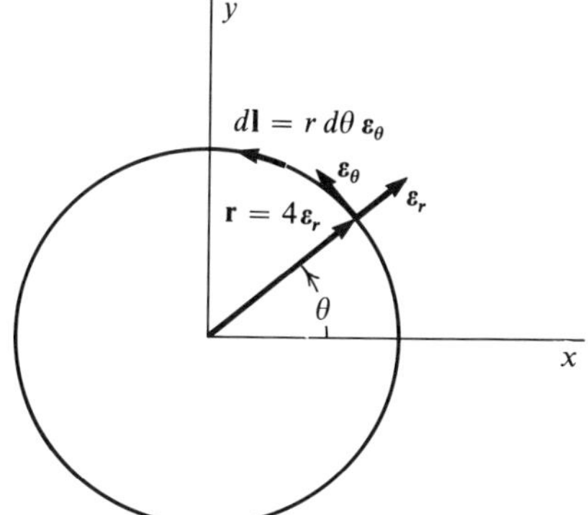

Fig. 1.23

Solution.

$$\oint \mathbf{V} \cdot d\mathbf{l} = \int_0^{2\pi} \frac{5}{r} \boldsymbol{\varepsilon}_\theta \cdot r\, d\theta\, \boldsymbol{\varepsilon}_\theta = 5 \int_0^{2\pi} d\theta$$

$$= 10\pi$$ ◀

If one were tempted to check this answer by using Stokes' theorem, the finding would be

$$\textbf{curl V} = \left[\frac{1}{r}\frac{\partial(rV_\theta)}{\partial r} - \frac{1}{r}\frac{\partial V_r}{\partial \theta}\right]\boldsymbol{\varepsilon}_z = 0$$

and it might be concluded $\int_S$ **curl V** $\cdot\, d\mathbf{S} = 0$; hence the circulation might be considered zero. This would be erroneous since **curl V** and **V** are not defined at the origin that is a point on the surface bounded by the curve C. The theorem does *not* apply.

1.9 Control Volume and System

A *control volume* is a region fixed in position, size, shape, and orientation relative to an observer.[1] The surface enclosing the control volume is referred to as a *control surface*. Figure 1.24 represents the flow of a fluid through a nozzle with entrance and exit sections designated by 1 and 2, respectively.

One could consider that the nozzle is fixed in the x', y', z' coordinate frame of reference, which, in turn, is "fixed in space." The volume between section 1, section 2, and the nozzle walls could be taken as a control volume for purposes of analyzing the flow.[2] The choice of a control volume is

[1]Some writers define a control volume that changes in size, shape, etc. However, the definition above, with an appropriate "observer," is sufficient to allow analysis of most elementary fluid mechanics problems.

[2]For example, one may desire to find the force exerted by the flowing fluid on the nozzle walls or the exiting velocity and temperature.

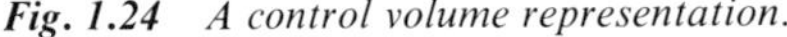

Fig. 1.24 *A control volume representation.*

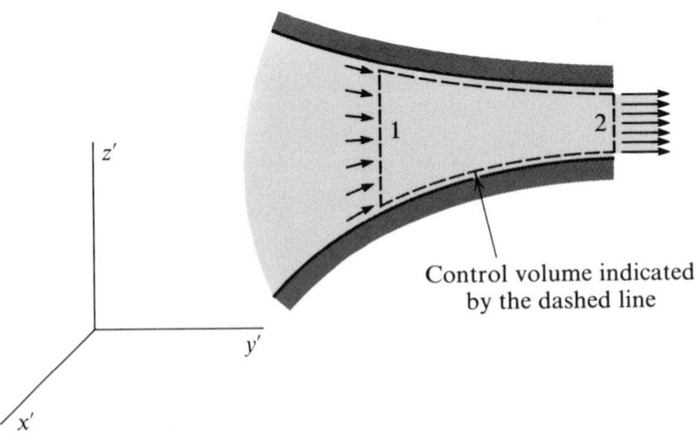

generally dictated by the particular problem, which in principle is similar to the selection of a system in the analysis of a thermodynamic problem or the choice of a free-body diagram in mechanics.

Figure 1.25 shows the same nozzle on a vehicle that has some general motion with respect to the set of fixed axes x, y, z. The analysis of this flow is more easily effected by considering a control volume similar to that in Fig. 1.24 and if such a control volume is used, then the observer[1] is fixed in the x', y', z' frame of reference which in general could be experiencing an angular acceleration and motion as well as a linear acceleration. The characteristics of the fluid in the control volume referred to the primed axes could be related to unprimed axes by accounting for the motion of the moving axes relative to the fixed ones. This procedure is used in solving some of the more involved problems in dynamics of rigid bodies. Further, the time changes of vector quantities referred to a rotating set of axes reflect the motion of the moving coordinate system when these time changes are in turn referred to a fixed coordinate system. Coriolis acceleration is an example of the last statement since it includes the angular velocity of the rotating coordinate system measured relative to a fixed coordinate system as well as the velocity of a given point relative to the rotating coordinate system.

A *system* is an identifiable collection of matter. Most of the basic laws of physical science are expressed in forms applicable to a system; for example: the first law of thermodynamics relates the effect of heat and work transfer to the change in energy of a system; Newton's law of motion expresses the

[1] The "observer" here means (per definition of control volume) one who views the control volume such that this region is fixed in position, shape, size, and orientation.

Fig. 1.25 *Nozzle in motion relative to x, y, z frame of reference. The axes x', y', z' are fixed in the moving vehicle.*

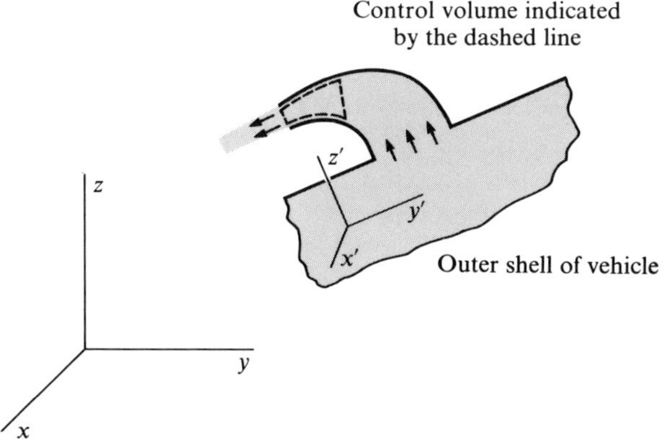

relationship between external forces acting on a given mass (system) to the time rate of change of momentum of the mass. This presents a need for expressing the basic laws of physics in forms that apply to fluid particles flowing through a control volume. Section 1.10 develops a means whereby rates of change of extensive properties of a system, which instantaneously occupies a control volume while passing through it, are related to rates associated with the control volume and surface. This relation and variations of it are called *transport theorems*.

1.10 Transport Theorem

The task is to relate the two ways of describing fluid flow by formally connecting changes in a characteristic of a system with corresponding rates of the same characteristic of the material in and passing through a control volume. The characteristic will have to be "something" that can be associated with the mass, and for this reason the selection is some extensive property in general. This extensive property is designated by the symbol **B** when such refers to any given finite amount of matter, and is symbolized by the symbol **b** when referring to a unit of mass specifically. The relation between **B** and **b** for a given mass is

$$\mathbf{B} = \int_m \mathbf{b}\, dm = \int_{\forall} \mathbf{b}\, \rho\, d\forall \tag{1.32}$$

in which dm is a suitably small mass such that it is uniform or homogeneous with respect to the variable **b**. The integration is carried out so as to sum over the entire mass m for a given instant of time. The second expression for **B** simply consists of the substitution of $\rho d\forall$ for dm; and the integration is carried out by summing over the entire volume $\forall$ occupied by mass m. The general extensive property selected here is attributed a vectorial sense since there is an interest in such quantities as linear and angular momenta in fluid mechanics; yet, at times, it is more convenient to deal with components of vectors.

When relating the time rate of change of **B** for a system to the time rate of change of **B** associated with material instantaneously occupying and passing through a control volume, the control volume will be fixed in an x, y, z frame of reference as shown in Fig. 1.26. For any fluid particle that makes up the system, the magnitude of a component in the ith direction of the extensive property **B** for the mass Δm occupying an infinitesimal volume $\Delta\forall$ is $b_i \rho\, \Delta\forall$. The time rate of change of this property for the fluid particle is expressed by

$$\frac{D(b_i\,\Delta m)}{Dt} = \frac{D}{Dt}(b_i\rho\,\Delta\mathcal{V}) = b_i\rho\frac{D(\Delta\mathcal{V})}{Dt} + \Delta\mathcal{V}\frac{D(b_i\rho)}{Dt}$$

$$= \left[b_i\rho\frac{1}{\Delta\mathcal{V}}\frac{D(\Delta\mathcal{V})}{Dt} + \frac{D(b_i\rho)}{Dt}\right]\Delta\mathcal{V} \tag{1.33}$$

In searching for the identity of $\dfrac{1}{\Delta\mathcal{V}}\dfrac{D(\Delta\mathcal{V})}{Dt}$, the physical interpretation is helpful. At any instant, the time rate of change of volume of a system is equal to the net rate at which its boundaries are "sweeping out" volume, that is,

$$\frac{D(\Delta\mathcal{V})}{Dt} = \oint_S \mathbf{V}_s \cdot d\mathbf{S}$$

where S represents the boundaries of the system at a given instant of time, and $\mathbf{V}_s$ is the velocity of a general point on the boundary referred to a frame of reference of arbitrary motion. If $\mathbf{V}_s$ is measured relative to a fixed surface S, then $\mathbf{V}_s$ becomes the absolute velocity of the fluid, $\mathbf{V}$, at points on S relative to the fixed surface. The previous expression, divided by $\Delta\mathcal{V}$, may then be written

$$\frac{1}{\Delta\mathcal{V}}\frac{D(\Delta\mathcal{V})}{Dt} = \frac{1}{\Delta\mathcal{V}}\oint_S \mathbf{V} \cdot d\mathbf{S}$$

Fig. 1.26 *Control volume fixed in frame of reference shown.*

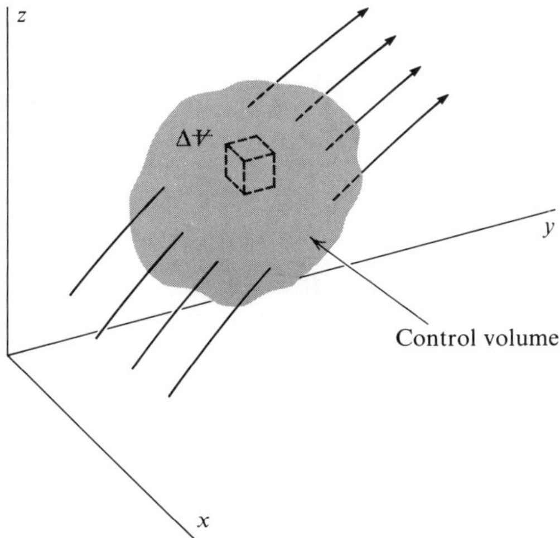

which is, by definition, the divergence of **V** at a point interior to a suitably small neighborhood, $\Delta \mathcal{V}$. Hence, from the interpretation of div **V** in Section 1.7,

$$\frac{1}{\Delta \mathcal{V}} \frac{D(\Delta \mathcal{V})}{Dt} = \operatorname{div} \mathbf{V} \tag{1.34}$$

Using this identity in Eq. (1.33) and expanding the material derivative according to Eq. (1.19), one obtains

$$\frac{D(b_i \, \Delta m)}{Dt} = \left[b_i \rho \operatorname{div} \mathbf{V} + \mathbf{V} \cdot \mathbf{grad}\, (b_i \rho) + \frac{\partial}{\partial t}(b_i \rho) \right] \Delta \mathcal{V}$$

which, on using the vector identity given by Eq. (1.23), becomes

$$\frac{D(b_i \, \Delta m)}{Dt} = \left[\operatorname{div}(b_i \rho \mathbf{V}) + \frac{\partial}{\partial t}(b_i \rho) \right] \Delta \mathcal{V} \tag{1.35}$$

This expression applies to all points in the flow field, and the summing procedure at a given instant of time for all fluid particles of the system may be achieved by integration (where $\Delta \mathcal{V}$ is suitably small, as previously assumed), so

$$\sum \frac{D(b_i \, \Delta m)}{Dt} = \int_{\mathcal{V}} \operatorname{div}(b_i \rho \mathbf{V}) \, d\mathcal{V} + \int_{\mathcal{V}} \frac{\partial (b_i \rho)}{\partial t} \, d\mathcal{V} \tag{1.36}$$

The left-hand term sums over the mass of the system that is not a function of time, and so the term is equal to the material derivative of the integral of $b_i \, dm$. With this information and using the divergence theorem of Gauss, Eq. (1.29), the following *transport equation*[1] is obtained:

$$\underset{(1)}{\frac{DB_i}{Dt}} = \frac{D}{Dt} \int_m b_i \, dm = \underset{(2)}{\int_{\mathcal{V}} \frac{\partial (b_i \rho)}{\partial t} \, d\mathcal{V}} + \underset{(3)}{\oint_S b_i \rho \mathbf{V} \cdot d\mathbf{S}} \tag{1.37}$$

A physical interpretation for each of the terms in the transport equation is now given:

(1) represents the instantaneous rate of change of the extensive property B_i associated with the system that occupies $\mathcal{V}$ at a given instant of time

[1] The transport theorem is developed by using a component that is a scalar quantity by definition. If a vector quantity had been used, the divergence theorem of Gauss in a form applicable to dyadics or tensors would be needed in the transformation corresponding to that used between Eqs. (1.36) and (1.37). This is beyond the scope of this text. However, Eq. (1.37) applied to each of three cartesian components of a vector will yield the correct formal result for $D\mathbf{B}/Dt$ in component form. Furthermore, the ith direction does not specifically refer to the x direction, but is used to note any of the three components of **B** or **b**.

(2) represents the rate of change of the value B_i within the volume $\mathcal{V}$ at a given instant of time
(3) represents the total (net) rate of efflux of the property B_i passing through the closed surface S of the volume $\mathcal{V}$ at a given instant of time

Example 1.18

Internal energy U is known to be an extensive property in thermodynamics. (*a*) Express the time rate of change of internal energy of a system using the transport Eq. (1.37). (*b*) Give a word explanation of each of the integrals. (*c*) Express the time rate of change of internal energy of a system using the transport equation (1.37) for steady u and ρ fields.

Solution. *a*. Since U is a scalar quantity, $B_i = U$ and, on a specific mass basis, $b_i = u$. That is, $U = \int_{\mathcal{V}} u\rho \, d\mathcal{V}$. The transport equation can be expressed as

$$\frac{DU}{Dt} = \int_{\mathcal{V}} \frac{\partial(u\rho)}{\partial t} d\mathcal{V} + \oint_S \rho u \, \mathbf{V} \cdot d\mathbf{S} \quad ◀$$

b. The first integral represents the rate of change of internal energy of the matter within the control volume at a given instant of time. The second integral describes the net efflux of internal energy through the control surface at a given instant of time. The internal energy "flows" because it is associated with the mass flowing through the control surface.

c. Since internal energy and density fields are steady, i.e., not a function of time, the integrand of the volume integral vanishes, and so

$$\frac{DU}{Dt} = \oint_S \rho u \, \mathbf{V} \cdot d\mathbf{S} \quad ◀$$

1.11 Dimensions and Units

It is desirable to recognize that equations representing physical phenomena must be homogeneous in dimensions and units. Often the incorrectness of derived expressions can be found quickly by such a check. When dealing with specific problems, equations are sometimes given with numerical values inserted for dimensional constants, and one must recognize that these numbers carry dimensions and units. It is important to be familiar with[1] and capable of solving problems in various systems of units. The material presented here will primarily use the lb_f-lb_m-ft-sec and the lb_f-slug-ft-sec system of units, and proportionality factors relating units will not be in-

[1]George N. Hatsopoulos and Joseph H. Keenan, "Principles of General Thermodynamics," pp. 99–106, John Wiley & Sons, Inc., New York, 1965.

cluded in statements of basic laws and derived results. Unit conversion factors and proportionality constants are listed in Appendix C.

1.12 Closing Remarks

With a few basic concepts of vector analysis and kinematics of the fluid flow field understood, the next step is the formulation of physical laws into expressions suitable for analysis. The following chapter begins this task by considering one of the most rudimentary principles of physical analysis—the law of conservation of matter.

Self-Study Questions

1.1 State a definition of a fluid.

1.2 Does the definition of a fluid include only substances in a liquid phase? Briefly explain.

1.3 Briefly discuss the continuum concept and why it is used in fluid mechanics.

1.4 Briefly discuss a measurable difference in a characteristic of nonfluids and fluids, i.e., solids and liquids.

1.5 State a definition of density at a "point."

1.6 What is the meaning of a "fluid particle" in this text?

1.7 Is a continuum concept needed in order to sum extensive properties of finite-sized systems?

1.8 *a.* Briefly discuss the differences of the two methods of description of a fluid flow field.
b. What is meant by a material coordinate?
c. What is meant by a space coordinate?

1.9 Which of the following use material coordinates?
a. $x = At + x_0$ and $y = y_0$
b. $\mathbf{r} = (At + x_0)\mathbf{i} + y_0\mathbf{j}$
c. $x_0 = x - Ct$ and $y_0 = y + Ct$
d. $\mathbf{r}_0 = (x - Ct)\mathbf{i} + (y + Ct)\mathbf{j}$
e. $\mathbf{r}_0 = (xe^{-Ct})\mathbf{i} + (ye^{Ct})\mathbf{j}$

1.10 Given a flow field where the position vector is $\mathbf{r} = (x_0 + Ct)\mathbf{i} + (y_0 - Ct)\mathbf{j}$ and C is a dimensional constant, sketch the position of a specified particle with material coordinates of 0, 1
a. At $t = 1/C$.
b. At $t = 2/C$.
c. At $t = 3/C$.

1.11 Given a flow field where $x_0 = xe^{-Ct}$, $y_0 = ye^{Ct}$, and C is a dimensional constant, sketch the position of a specified particle with material coordinates of 1, 6 for $t > 0$.

1.12 *a.* State a word and a mathematical definition of a streamline.
b. Can the equation of a streamline be a function of time; i.e., can a plot of a family of streamlines at a given instant of time change to a different plot at a different instant of time?

1.13 *a.* Briefly discuss the meaning of a pathline.
b. Briefly discuss the meaning of a streakline.
c. Under what condition(s) will the pathline be the same as the streakline for a velocity field?
d. Under what condition(s) will the pathline and streamline be the same for a velocity field?

1.14 What is meant by steady flow field in this text?

1.15 Write a functional description of
a. A steady two-dimensional velocity field.
b. An unsteady two-dimensional density field.
c. A two-dimensional field of steady flow and an unsteady density field.

1.16 Verify that streamlines and pathlines have the same loci for steady flow by using the basic definitions of Section 1.5.

1.17 Make a sketch of a family of streamlines for a steady flow field of a fluid moving
a. Around a ball or sphere that is translating only relative to the fluid.
b. Around a 90° corner.
c. Over the top of a dam.
d. Through a pipe with an abrupt enlargement.

1.18 Which of the following represent steady and two-dimensional fields?
a. $\rho = Ax^2 + Byt + \rho_0$, where A and B are dimensional constants

b. $\mathbf{V} = 3x\mathbf{i} + (2z + 3x^2)\mathbf{j} - (3z + 2t)\mathbf{k}$, in ft/sec

c. $\mathbf{V} = V_0(\cos\theta\, \boldsymbol{\varepsilon}_r - \sin\theta\, \boldsymbol{\varepsilon}_\theta)$, where V_0 is a constant

d. $\phi = \dfrac{x}{r}\mathbf{i}$, where r is the value of the position vector

e. $\mathbf{A} = 2y\mathbf{i} + 3x\mathbf{j} + 6x\mathbf{k}$

f. $V_x = \dfrac{Axy}{(x^2 + y^2)} e^{-Bz}$

1.19 The expression $\mathbf{V} \mathbf{x}\, d\mathbf{l} \equiv 0$ is stated to be useful in providing a differential equation for streamlines.
a. Explain the terms $\mathbf{V}$ and $d\mathbf{l}$ which make up the vector product.
b. Explain what is meant by $\equiv 0$, that is, identically equal to zero.
c. What is the slope of a streamline equal to in terms of velocity components?
d. For a streamline, if $V_x \neq 0$, $V_y \neq 0$, and $V_z = 0$, why is $dz = 0$ in $d\mathbf{l}$?

1.20 Why is $D(\ \ )/Dt$ called a material derivative operator?

1.21 Consider the following derivative operators: $\partial(\ \ )/\partial r$, $\partial(\ \ )/r\partial\theta$, $\partial(\ \ )/\partial\theta$, and $(1/r \sin\theta)\partial(\ \ )/\partial\phi$
a. What are the basic dimensions of each?
b. Which of these represent the differentiation with respect to a spatial displacement?

1.22 What is an expression for a fluid-particle acceleration if spatial coordinates x, y, z are interpreted as the location of the fluid particle?

1.23 *a.* What is the physical meaning of $\partial\mathbf{V}/\partial x$ and what are the implied independent variables in Eq. (1.13)?

b. What is the physical meaning of $\frac{1}{r}\frac{\partial \mathbf{V}}{\partial \theta} V_\theta$ in Eq. (1.14)?

1.24 Which of the following are stationary (local) changes?

a. $\frac{\partial \mathbf{V}}{\partial t}$ *b*. $\frac{\partial \rho}{\partial x}$ *c*. $\frac{\partial V_x}{\partial y}$ *d*. $\frac{\partial (V^2/2)}{\partial t}$

1.25 Briefly explain why certain terms are called convective terms.

1.26 Give a word statement of the meaning of the following:
a. Gradient of a scalar field ϕ.
b. Divergence of vector velocity **V**.
c. Curl of vector velocity **V**.

1.27 Show that $\mathbf{grad}\, A \cdot d\mathbf{l}/dl = dA/dl$ by expansion in x, y, z coordinates.

1.28 Check the correctness of Eq. (1.19) by using the identities for a material derivative and the gradient operator in x, y, z coordinates.

1.29 Two lines of constant density are shown for a density field in Fig. 1.27. Which path ($d\mathbf{l}$) is applicable for the following cases:
a. $\mathbf{grad}\, \rho \cdot d\mathbf{l} = 0$
b. $0 < \mathbf{grad}\, \rho \cdot d\mathbf{l} <$ maximum value
c. $\mathbf{grad}\, \rho \cdot d\mathbf{l} =$ maximum value

1.30 Obtain an expression for the acceleration in the x direction by using Eq. (1.19).

1.31 State in words a physical meaning of
a. $\mathbf{A} \cdot d\mathbf{S}$
b. $\oint \mathbf{A} \cdot d\mathbf{S}$

1.32 Check the correctness of Eq. (1.23) by using the identities for **grad** and div operators in x, y, z coordinates.

1.33 *a*. Are the forms of the derivative operators for the gradient, divergence, and curl the same in cartesian and cylindrical coordinates?
b. What advantage is gained by defining the gradient, divergence, and curl operators as integral operators?

1.34 State in words the meaning of
a. $\mathbf{A} \cdot d\mathbf{l}$
b. $\oint \mathbf{A} \cdot d\mathbf{l}$

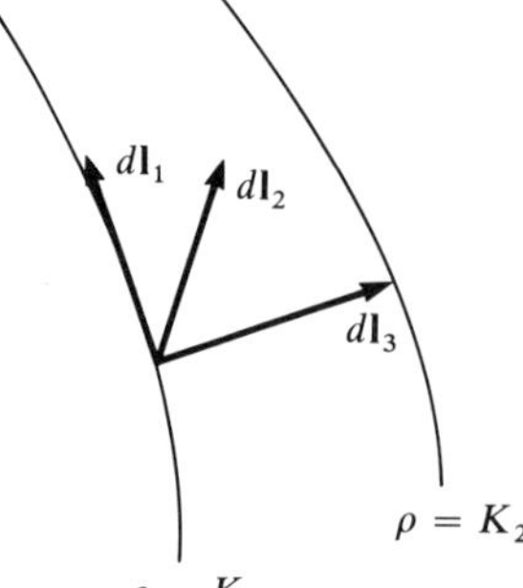

Fig. 1.27

1.35 In what way is circulation related to the **curl V**?

1.36 Consider that velocity **V** is used for **A** in Eq. (1.29).
a. Briefly discuss a physical interpretation of both of the integrals in this equation.
b. Does this coincide with the interpretation of divergence of **V** given in Section 1.7?

1.37 Consider variable **A** and its derivatives to exist as continuous functions in the field except in the solid region shown in Fig. 1.28. The variable **A** is such that **curl A** is zero everywhere except in the darker region. Determine that only two of the cases have the circulation of zero for the path shown as C.

1.38 Briefly discuss the difference(s) of the system and the control volume forms of analysis.

1.39 Briefly explain which of the following are extensive properties and thus can be substituted for B_i in Eq. (1.37):

a. Kinetic energy.
b. Momentum.
c. Mass.
d. Pressure.
e. Shear stress.
f. Temperature.
g. Heat.
h. Entropy.
i. Volume.
j. Any vector.
k. Any scalar.

1.40 Describe the physical meaning of the following:

a. $\dfrac{DB_i}{Dt}$ *b.* $b_i\,dm$ *c.* $\displaystyle\int_m b_i\,dm$ *d.* $\displaystyle\frac{D}{Dt}\int_m b_i\,dm$

1.41 Describe the physical meaning of the following:

a. $\dfrac{\partial}{\partial t}(b_i\rho)\,d\mathcal{V}$ *b.* $\displaystyle\frac{\partial}{\partial t}\int_{\mathcal{V}}(b_i\rho)\,d\mathcal{V}$

1.42 Describe the physical meaning of the following:
a. $\mathbf{V}\cdot d\mathbf{S}$
b. $\rho\mathbf{V}\cdot d\mathbf{S}$
c. $b_i\rho\,\mathbf{V}\cdot d\mathbf{S}$
d. $\displaystyle\oint_S b_i\rho\,\mathbf{V}\cdot d\mathbf{S}$

1.43 State mathematically and in words the transport equation.

1.44 The entropy of a given system of fluid particles is not changing with time. Express this in terms of the transport equation.

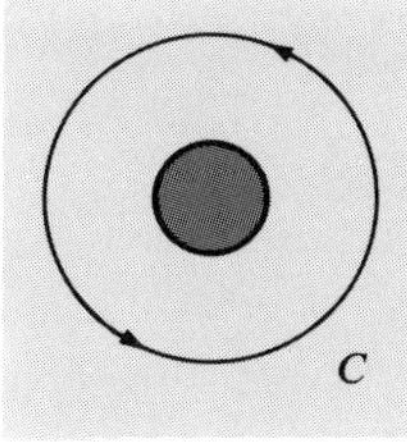

a

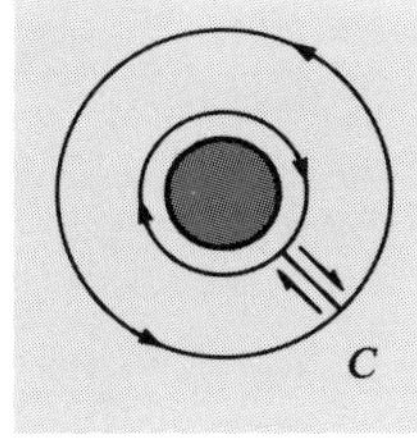

b

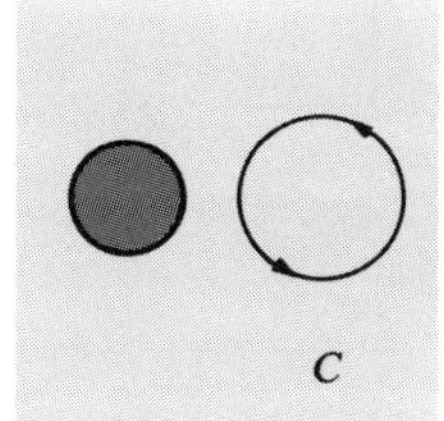

c

Fig. 1.28

1.45 There are many fluid flow situations in which cyclic changes in property values occur at various locations within the control volume with time while values of the properties of the fluid moving through the control surface are essentially steady. In such a situation, show that a time average of DB_i/Dt over the interval Δt for such a cycle $(1/\Delta t)\int_t^{t+\Delta t}(DB_i/Dt)\,dt$ yields $\oint_S b_i\rho\,\mathbf{V}\cdot d\mathbf{S}\big|_{\text{at any }t}$.

Problems

1.1 The following scalar density field is given: $\rho = Axy + C$ where A and C are dimensional constants.

a. What is the density variation in the z direction? With time?
b. What are the basic dimensions of A?
c. Sketch three different constant-density lines on an xy plane.

1.2 The scalar density field is given by $\rho = Ax^2 + Byt + C$ with A, B, and C as dimensional constants.

a. What is the density variation in the z direction?
b. Sketch a solid line of constant density equal to 9 on the xy plane for $t = 1$. Also, sketch a dashed line of the same density for $t = 2$. Let $A = 1$, $B = 1$, and $C = 5$.
c. What is the value of $\partial\rho/\partial t$ at 2, 0? At 0, 2?
d. What are the basic dimensions of A? Of B? Of C?

1.3 The following scalar density field is given by $\rho = At/x$ where A is a dimensional constant and $x > 0$.

a. Evaluate $\partial\rho/\partial x$ for $x > 0$.
b. Give a word description of the meaning of the derivative evaluated in (*a*).
c. Evaluate $\partial\rho/\partial t$ at $x = 10$.
d. Give a word description of the meaning of the derivative evaluated in (*c*).
e. Sketch in ρ, x, t space the following:
  i. A variation of density when $x = 10$.
  ii. A variation of density when $t = 10$.

1.4 Consider a density that varies as given in Prob. 1.1.

a. What is a differential expression for the mass contained in a fluid element of volume (1) $dx\,dy$, that is, considering unit depth in the z direction?
b. What is the mass of fluid in the volume of unit depth shown in Fig. 1.29?

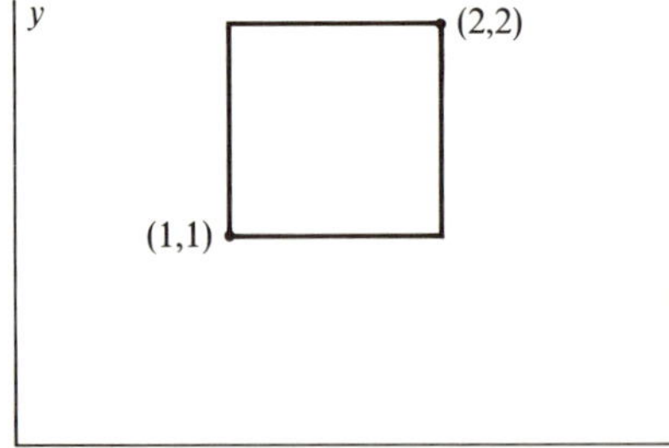

Fig. 1.29

1.5 Consider a density variation in a fluid flow field as given in Prob. 1.2. Obtain an algebraic expression for the mass of a fluid in the volume shown in Fig. 1.30 at time t in terms of dimensional constants A, B, and C.

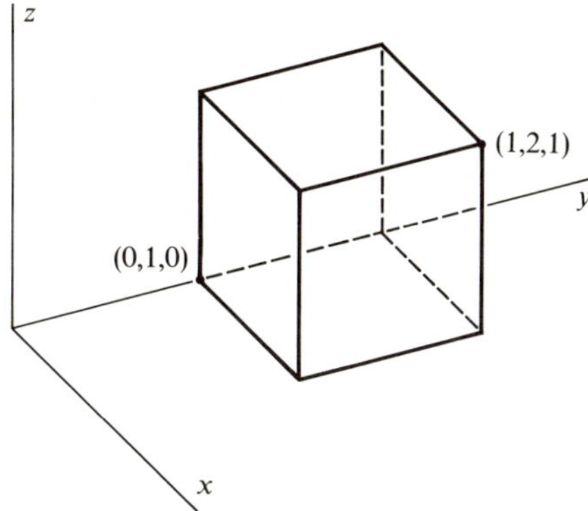

Fig. 1.30

1.6 A scalar field is given as $\rho = Axy + C$ where A and C are dimensional constants.
a. Obtain a differential expression for the mass of fluid contained in a unit depth (in the z direction) of area $r\,d\theta\,dr$.
b. Obtain an algebraic expression for the mass of fluid contained in a unit depth of the circular area shown ($R = 2$) in Fig. 1.31.

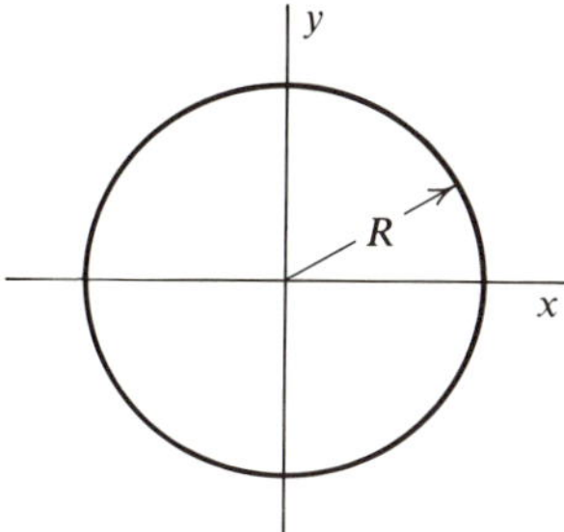

Fig. 1.31

1.7 Calculate the mass of fluid located in the rectangular box shown in Fig. 1.32. The density varies as $\rho = Axy + Bz^2t$ in mass/length^3

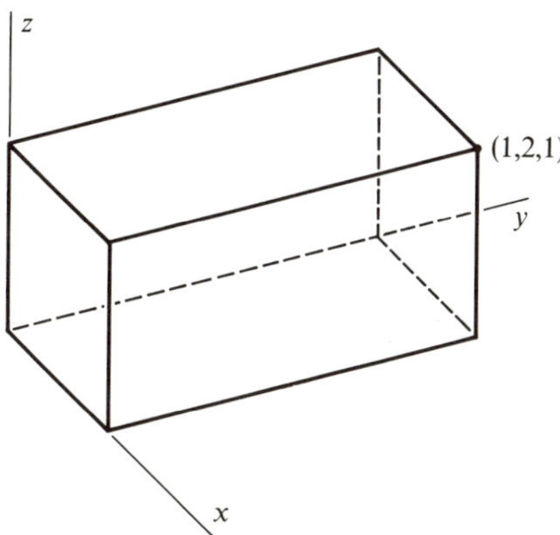

Fig. 1.32

1.8 The given velocity field is represented by

$$\mathbf{V} = (3x + 2y)\mathbf{i} + (2z + 3x^2)\mathbf{j} + (-3z + 2t)\mathbf{k}$$

where the numerical coefficients have appropriate dimensions.
a. What is V_x, V_y, and V_z at any point in the flow field?
b. What is the speed at $(0,0,0,t)$?
c. What is the speed at $(1,1,1,0)$?
d. What are the direction cosine values (l, m, n, with respect to the x, y, z axes, respectively) for the velocity at $(1,1,1,0)$?

1.9 A fluid is conceived to be flowing radially outward in the xy plane from a point, as if there is a fluid source at the point. The magnitude of velocity is inversely proportional to the radius from the point of origin. There is no variation in the flow field in the z direction. The origin is excluded as part of the flow field.
a. Obtain an expression for speed in cylindrical coordinates.
b. Sketch the flow field (by using the tips of arrows to indicate the direction of flow and the length of the arrow to indicate the magnitude of velocity) at various points of several fixed radii.
c. Obtain an expression for velocity using unit vectors $\boldsymbol{\varepsilon}_r$ and $\boldsymbol{\varepsilon}_\theta$ as needed.
d. Express the velocity in terms of x, y, z coordinates.

1.10 The components of velocity in a fluid flow field at one point, expressed in cylindrical coordinates, are $V_r = 5$, $V_\theta = 10$, and $V_z = 3$.
a. What is the speed of the fluid at this point?
b. Evaluate the direction cosines for the velocity at this point, that is, l, m, n (with respect to the x, y, z axes, respectively).

1.11 Consider a uniform flow field with a velocity $\mathbf{V} = V\mathbf{i}$. Show that this velocity in cylindrical coordinates is

$$\mathbf{V} = V(\cos\theta\, \boldsymbol{\varepsilon}_r - \sin\theta\, \boldsymbol{\varepsilon}_\theta)$$

1.12 Obtain the direction cosines l, m, n of the tangent to the streamline for the following velocity fields.
a. $\mathbf{V} = 3x\mathbf{i} + (2z + 3x^2)\mathbf{j} - (3z + 2t)\mathbf{k}$ at $(1,1,1,0)$.
The coefficients are dimensional numbers.
b. $\mathbf{V} = V(\cos\theta\, \boldsymbol{\varepsilon}_r - \sin\theta\, \boldsymbol{\varepsilon}_\theta)$ at $(1.414, \pi/4)$.

1.13 Find an equation for the streamlines in the following velocity fields (the K's are dimensional constants).
a. $V_x = K_1 \quad V_y = K_2 \quad V_z = 0$
b. $V_x = K_1 \quad V_y = K_2 \quad V_z = K_3$
c. $V_x = K_1 \quad V_y = 0 \quad V_z = 0$

1.14 Obtain an equation for the streamlines for the following velocity field where K_1 and K_2 are dimensional constants: $\mathbf{V} = K_1 x\mathbf{i} - K_2 y^2\mathbf{j}$.

1.15 The identification of a given particle is given by $\mathbf{r}_0 = x(1 + Kt)^{1/2}\mathbf{i} + y(1 + Kt)\mathbf{j}$ where K is a dimensional constant.
a. What is the position vector $\mathbf{r}$ in terms of material coordinates and time?
b. Obtain an equation for the family of pathlines and show a sketch of this family in the first quadrant and the direction of motion of a particle on a pathline.
c. Obtain an equation for the family of streamlines and compare it to your answer to (*b*).

1.16 The position of a given fluid particle, in the two-dimensional flow within the region of $0 \leq y \leq 1$, is given by

$$\mathbf{r} = K_2[e^{K_1 t}]i$$

where the constant K_1 has dimensions of reciprocal time and K_2 of length.

a. Obtain expressions for the velocity components.
b. Obtain an equation for a particle on its pathline in terms of material coordinates.
c. What is the description for the loci of the pathlines?

1.17 The identification of a given particle is given by

$$\mathbf{r}_0 = x(1 - Kt)\mathbf{i} + \frac{y}{1 - Kt}\mathbf{j}$$

where K is a dimensional constant.

a. What is the position vector **r** in terms of material coordinates and time?
b. Obtain expressions for the velocity components.
c. Obtain an expression for a pathline.
d. Make a sketch of the pathline family in the positive xy quadrant.

1.18 Obtain the equation for the locus of a streamline where

$$V_r = \frac{A}{r} \qquad V_\theta = \frac{A}{r} \qquad \text{and} \qquad V_z = 0 \qquad \text{for } r > 0$$

A is a dimensional constant. Show a sketch of one streamline.

1.19 Obtain the equation of the streamlines in the velocity field

$$V_r = \left(A - \frac{B}{r^2}\right)\cos\theta \qquad V_\theta = -\left(A + \frac{B}{r^2}\right)\sin\theta \qquad V_z = 0 \qquad \text{for } r > 0$$

The constants A and B have dimensions.

1.20 Find the equation for the family of streamlines for the field $x > 0$, $y > 0$ where

$$V_x = \frac{-1x}{(x^2 + y^2)^{3/2}} \qquad \text{and} \qquad V_y = \frac{1y}{(x^2 + y^2)^{3/2}}$$

The unit coefficient has dimensions.

1.21 Obtain an equation of the streakline through the point a, b for the flow field in Prob. 1.14.

1.22 Obtain an equation of the streamline passing through a, b for the flow field given in Prob. 1.15.

1.23 Obtain an equation for the pathline passing through r_0, θ_0 for the flow field in Prob. 1.18.

1.24 Obtain an equation for a streakline passing through a,b for the flow field described in Prob. 1.15.

1.25 A flow field is described by (for $0 \leq \theta \leq 3/2\,\pi$)

$$V_r = -A\cos\left(\frac{2\theta}{3}\right)/r^{1/3} \qquad V_\theta = A\sin\left(\frac{2\theta}{3}\right)/r^{1/3}$$

where A is a positive dimensional constant. Obtain an equation for the streamlines and show a sketch of several streamlines and the direction of flow.

1.26 A velocity field is given by

$$V_r = K_1 K_2 e^{K_1 t} \qquad V_\theta = r K_1 K_2 e^{K_1 t} \qquad \text{and} \qquad V_z = 0$$

Obtain an equation for a streamline, a pathline, and a streakline passing through the point r_1, θ_1.

1.27 In a given flow field, $V_x = 10y$ and $V_y = -10x$, and the numerical coefficients have dimensions.
a. Obtain an equation of the streamlines.
b. Show several of these streamlines on an xy plot.
c. What is the acceleration at any point?
d. What is the speed at any point?

1.28 A fluid field has a velocity $\mathbf{V}$ of $(2x + y)\mathbf{i} + (3x^2 + 2zt)\mathbf{j} + 3z\mathbf{k}$ where the coefficients are dimensional numbers.
a. What is the acceleration of a fluid particle? Where does the acceleration have a zero value?
b. Express $V_y(\partial V_x/\partial y)$ as $f(x,z,t)$.
c. Obtain an expression for the local acceleration.
d. What is the acceleration at $(0,1,0,1)$?

1.29 Obtain the acceleration for a flow field where $\mathbf{V} = r\varepsilon_r + r^2 \cos\theta\, \varepsilon_\theta$ in length/time
a. In cylindrical coordinates.
b. In cartesian coordinates.

1.30 *a.* Express the velocity $\mathbf{V} = [At(x^2 + y^2)^{1/2}/x]\boldsymbol{i}$ in terms of plane polar coordinates. The constant A has dimensions.
b. Obtain the acceleration for the velocity field in (*a*) in terms of plane polar coordinates.

1.31 A scalar density field of $\rho = Axy + C$ is given where A and C are dimensional constants.
a. Obtain an expression for the time rate of change of density for a fluid particle (in terms of an unspecified velocity field).
b. Obtain an expression for the time rate of change of density at a fixed point.

1.32 A density field is given as $\rho = A/x$ for $x > 0$. The position of a given particle (in terms of material coordinates) is given by $x = x_0 e^{Bt}$, $y = y_0$, and $z = z_0$. The symbols A and B represent dimensional constants.
a. Obtain an expression for the velocity components.
b. Obtain an expression for $D\rho/Dt$.

1.33 A density field is given as $\rho = A/x$ for $x > 0$. The position of a given particle (in terms of material coordinates) is given by $x = x_0\,[1 + \ln\,(Bt + 1)]$, $y = y_0$, and $z = z_0$. The constants A and B have dimensions. Obtain an expression for the time rate of change of density for a fluid particle.

1.34 A density field is given as $\rho = Ax^2 + Byt + C$ where A, B, and C are dimensional constants.
a. What is the gradient of ρ at any point in the flow field?
b. What is the value of the gradient of ρ at $(1,1,1,0)$?
c. Obtain a unit vector in the direction of the gradient of ρ.

1.35 A density field is described as $\rho = Axy + C$ where A and C are dimensional constants.
a. What is the gradient of ρ at any point in the flow field?
b. What is the gradient of ρ at 1, 1? What is the magnitude of the gradient of ρ? What is the value of $\partial\rho/\partial y$ at 1,1? What is the value of the change of ρ along the line in the field where xy is a constant?

c. Where in the field is the gradient of ρ equal to $Ay\mathbf{i}$?
d. What is the value of the gradient of ρ along the x axis? What is the value of $\partial\rho/\partial y$ along the x axis?

1.36 Evaluate **grad** $(-\rho gz)$ when ρ and g are taken as dimensional constants. What are the basic dimensions of **grad** $(-\rho gz)$?

1.37 A scalar function is given as $\phi = \ln r/\theta$ for $r > 1$ where r and θ are cylindrical coordinates.
a. Obtain an expression for the gradient of ϕ in cylindrical coordinates.
b. Sketch a constant ϕ line and the **grad** ϕ passing through the point $2, \pi/2$.
c. What is the value of **grad** ϕ at point $2, \pi/2$?

1.38 Show that the gradient operator as a differential operator in cylindrical coordinates takes the form

$$\mathbf{grad} = \frac{\partial}{\partial r}\boldsymbol{\varepsilon}_r + \frac{1}{r}\frac{\partial}{\partial \theta}\boldsymbol{\varepsilon}_\theta + \frac{\partial}{\partial z}\boldsymbol{\varepsilon}_z$$

by changing variables from the corresponding expression in cartesian coordinates.

1.39 A velocity field is given as $\mathbf{V} = Kx\mathbf{i} + Ky\mathbf{j} + Kz\mathbf{k}$ where K is a dimensional constant.
a. Obtain the divergence of $\mathbf{V}$ by using the integral operator for the volume shown in Fig. 1.33 and limiting to a point at x, y, z.
b. Check your answer for the divergence of $\mathbf{V}$ by using Eq. (1.21).

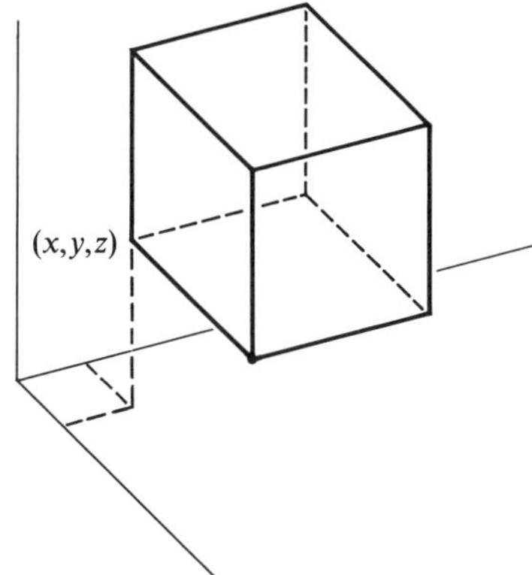

Fig. 1.33

1.40 A vector field is given as $\mathbf{A} = (xt/r)\mathbf{i}$ where $\mathbf{A} = \mathbf{A}(x,y,t)$.
a. Obtain div $\mathbf{A}$ by using Eq. (1.21).
b. Obtain div $\mathbf{A}$ by using Eq. (1.22).

1.41 A vector field is given as $\mathbf{A} = (x^2\mathbf{i} + y^2\mathbf{j})/y$.
a. Obtain div $\mathbf{A}$ by using Eq. (1.21).
b. Obtain div $\mathbf{A}$ by using Eq. (1.22).

1.42 A fluid moves with uniform speed V in a 45° direction to the x axis. The flow moves through the hypothetical surface of unit depth shown as a dashed line in Fig. 1.34.
a. Give an integral scalar expression for calculating the volume flow rate through the hypothetical surface.
b. Obtain an algebraic expression for volume flow rate through the hypothetical surface.
c. Calculate the net volume flow rate through the closed surface 0-*a*-*b*-*c*-0.

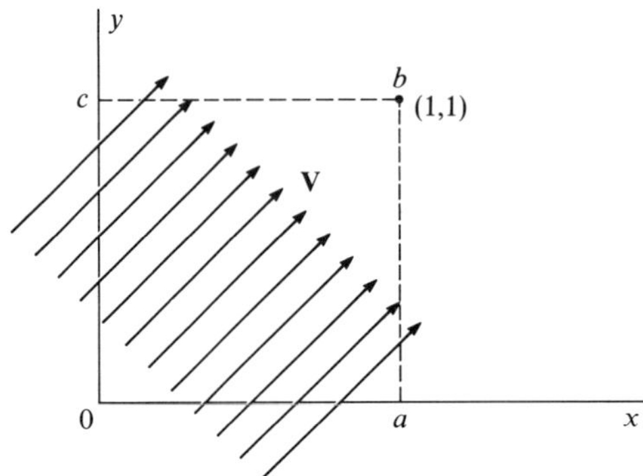

Fig. 1.34

1.43 A fluid moves at uniform velocity in the xy plane with $\mathbf{V} = V_\infty \cos\theta\, \boldsymbol{\varepsilon}_r - V_\infty \sin\theta\, \boldsymbol{\varepsilon}_\theta$. The hypothetical surface of unit depth, shown as a dashed line in Fig. 1.35, has a circular shape of radius R.

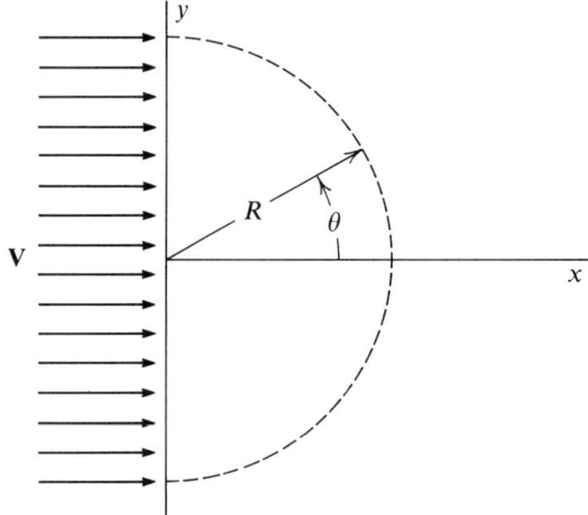

Fig. 1.35

a. Obtain an expression for a unit normal vector to this surface.
b. Obtain an expression for $d\mathbf{S}$ for the hypothetical surface.
c. Give an integral expression for the volume flow rate through the semicircular surface.
d. Give an integral expression in terms of scalar quantities for the volume flow rate through the semicircular surface.
e. Obtain an algebraic expression for volume flow rate through the semicircular surface in terms of V_∞ and R.

1.44 Consider a flow field between two walls. The dashed line in Fig. 1.36 represents a hypothetical surface. At point P, the unit normal vector $\mathbf{n}$ of the area and the velocity $\mathbf{V}$ are in the same plane as shown.
a. What does the product of $\mathbf{V} \cdot \mathbf{n}$ represent?
b. Consider dS to be the differential area of the surface at point P. Express in words the meaning of $\mathbf{V} \cdot dS\, \mathbf{n}$ in a physical sense.
c. If the density at point P is ρ, then what is the physical meaning of $\rho\mathbf{V} \cdot dS\, \mathbf{n}$?

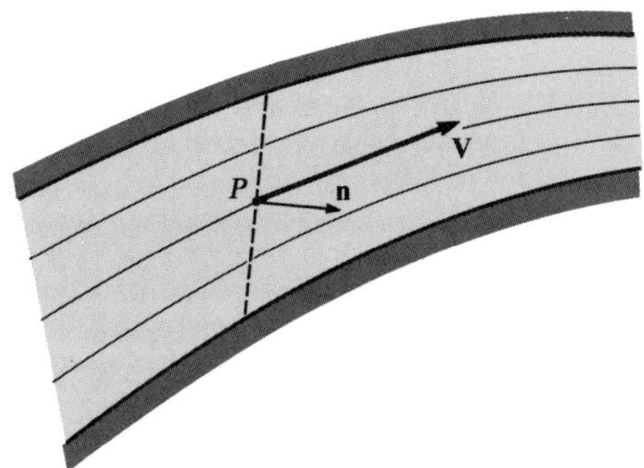

Fig. 1.36

1.45 Show that the divergence in cylindrical coordinates of vector **A** is

$$\text{div}\,\mathbf{A} = \frac{1}{r}\frac{\partial}{\partial r}(rA_r) + \frac{1}{r}\frac{\partial A_\theta}{\partial \theta} + \frac{\partial A_z}{\partial z}$$

by starting with Eq. (1.21) and changing variables from cartesian coordinates.

1.46 Evaluate $\oint \mathbf{A} \cdot d\mathbf{S}$ for a unit cube surface in Fig. 1.37 for **A** as given in Prob. 1.41.

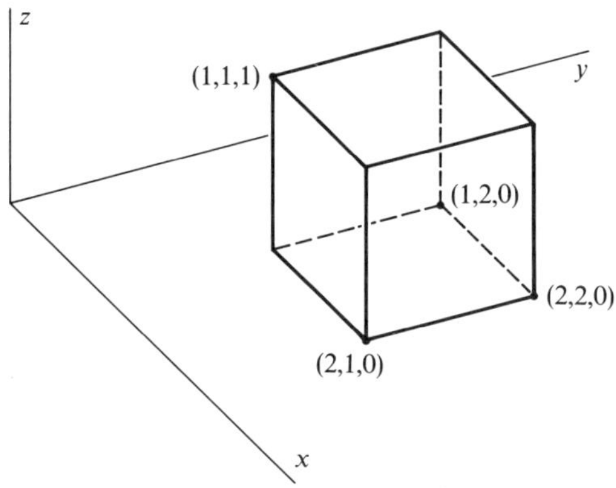

Fig. 1.37

1.47 Obtain the **curl A** where $\mathbf{A} = K_1 y\mathbf{i} + K_2 x^2 t\mathbf{j} + K_3 z\mathbf{k}$, and K_1, K_2, and K_3 are dimensional constants.

1.48 Obtain the curl differential operator for cylindrical coordinates starting with Eq. (1.25) and using transformation formulas.

1.49 Obtain the **curl V** for the flow field given in Prob. 1.14.

1.50 Show whether the flow field with $V_x = 10y$, $V_y = -10x$, and $V_z = 0$ is rotational. The numerical coefficients have dimensions.

1.51 Show whether the flow field in Prob. 1.18 is rotational.

1.52 Obtain the average rotational rate in a fluid field where $\mathbf{V} = K(x^2/y)\mathbf{i} + Ky\mathbf{j}$, and K is a dimensional constant.

1.53 Evaluate the vorticity for the flow field in Prob. 1.19.

1.54 Evaluate the vorticity for the flow field in Prob. 1.18.

1.55 Check the divergence theorem (Gauss' theorem) for the velocity field given in Prob. 1.8 for a unit cube with one corner at the origin of the coordinate axes.

1.56 Given $\mathbf{B} = 6x\mathbf{i} + 7y\mathbf{j}$, evaluate

a. $\oint \mathbf{B} \cdot d\mathbf{l}$ by integration for a unit radius circular path.

b. $\oint \mathbf{B} \cdot d\mathbf{l}$ for the same path as (*a*), using Stokes' theorem.

1.57 Evaluate $\oint \mathbf{B} \cdot d\mathbf{l}$ for the path shown in Fig. 1.38 (where $z = 0$) using Stokes' theorem where $\mathbf{B} = 2x\mathbf{i} + y^2z\mathbf{j} + x^3yt\mathbf{k}$ and the coefficients are dimensional constants.

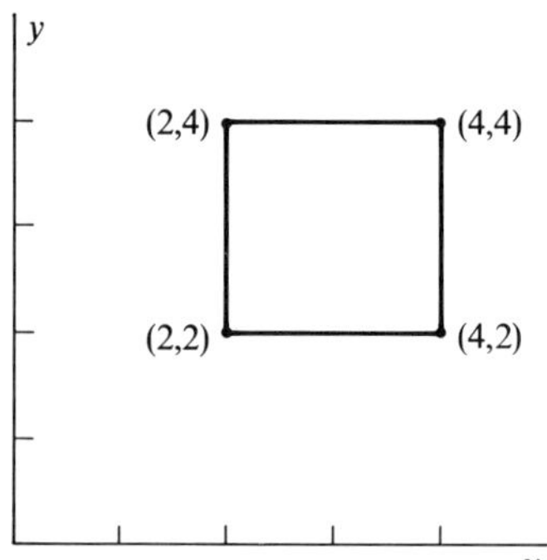

Fig. 1.38

1.58 Evaluate $\oint \mathbf{B} \cdot d\mathbf{l}$ for the path shown in Prob. 1.57 using Stokes' theorem with $\mathbf{B} = 2xy\mathbf{i} - 4x^2y\mathbf{j}$. The numerical coefficients have dimensions.

1.59 Evaluate the circulation of the velocity field given in Prob. 1.14 for a square path in the xy plane. Is the answer the same for any closed path in this plane?

1.60 Consider a unit cube located in the first octant with one corner at the origin of the coordinate axes. Evaluate the circulation of the velocity field given in Prob. 1.28 for each of the following:

a. A closed path on the cube surface where $y = 0$.

b. A closed path on the cube surface where $x = 0$.

c. A closed path on the cube surface where $x = 1$.

1.61 Kinetic energy, $(m\mathbf{V} \cdot \mathbf{V})/2$, is an extensive property.

a. Express the transport equation for kinetic energy.

b. Is the result in (*a*) a scalar or vector equation?

1.62 Express the transport equation for

a. Total energy, $E = me$.

b. Enthalpy, $H = mh$.

c. Momentum, $m\mathbf{V}$.

d. Moment of momentum, $m\mathbf{r} \times \mathbf{V}$.

1.63 A flow system is isolated from heat and work interactions at its surface. If the fluid is liquid water with a density of 1.94 slugs/ft^3, and the net energy efflux through the control surface is 100 Btu/sec, what is the instantaneous rate of energy change within a control volume of 10 ft^3?

1.64 Use cylindrical coordinates to obtain an expression for the energy U that a fluid has at any given time in the annulus indicated in Fig. 1.39. The specific energy of the field varies according to $u = C_1 r + C_2$; and the density ρ is expressed by

$$\rho = \frac{C_3 z + C_4}{C_1 r + C_2}$$

where C_1, C_2, C_3, and C_4 are dimensional constants.

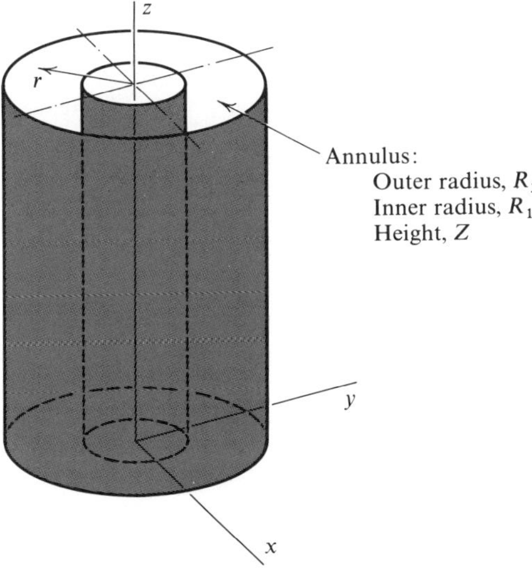

Fig. 1.39

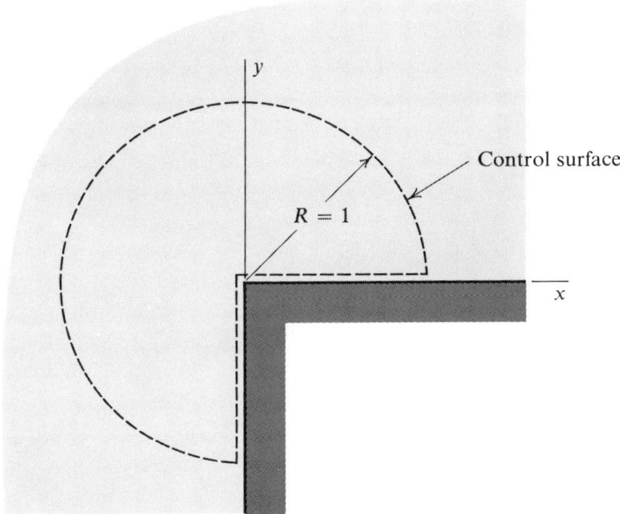

Fig. 1.40

1.65 Obtain an algebraic expression for the x component of the momentum flux (that is, $\oint V_x \, \rho\mathbf{V} \cdot d\mathbf{S}$) through a control surface of unit depth shown in Fig. 1.40, for the velocity field given in Prob. 1.25. The density of the fluid is constant.

1.66 Check the following identities by carrying out the operations in cartesian coordinates.

a. $\mathbf{V} \cdot \mathbf{grad}\, b_i = V_x \dfrac{\partial b_i}{\partial x} + V_y \dfrac{\partial b_i}{\partial y} + V_z \dfrac{\partial b_i}{\partial z}$

b. $\rho\mathbf{V} \cdot \mathbf{grad}\, b_i = \text{div}\,(\rho b_i \mathbf{V}) - b_i \,\text{div}\, \rho\mathbf{V}$

1.67 Consider a fluid with a uniform and steady velocity field of $\mathbf{V} = V\mathbf{i}$ and a density variation of $\rho = Ay + C$. The constants A and C have dimensions. The control volume is a unit cube located as shown in Fig. 1.41. Obtain an expression for

$$\left[\frac{D(m\mathbf{V})}{Dt}\right]_{\text{syst}}$$

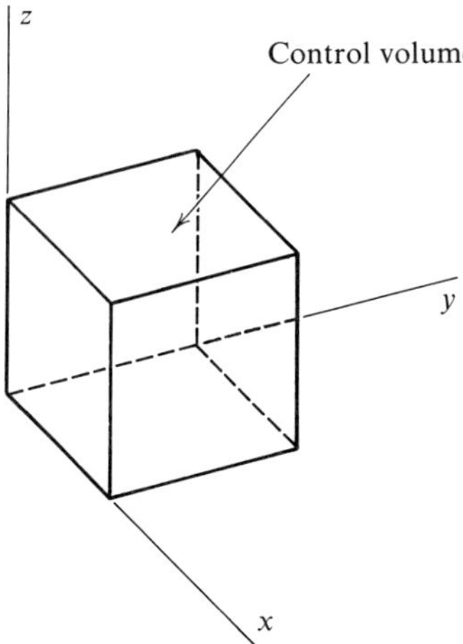

Fig. 1.41

2 CONSERVATION OF MATTER

2.1 Introduction

The law of conservation of matter may be stated as *matter cannot be created or destroyed.* This concept was certainly given experimental significance by A. L. Lavoisier, the French chemist, toward the end of the eighteenth century. Perhaps it would be well to point out that this is a law in a classical sense, since modern physics emphasizes the interplay between mass and energy and the conversion of one into the other. However, the engineer finds extensive use for the classical laws of physics, particularly in the analysis of problems in elementary fluid mechanics. Indeed, this classical principle not only demands satisfaction, but its application is necessary to obtain the solutions to many problems the engineer encounters.

It is the purpose of this chapter to translate the physical law into several forms that are readily applicable to problems of engineering interest by use of mathematics as a tool. The forms developed are somewhat general and apply to points or regions in the fluid where there are no conceptual mass sources or sinks, i.e., points where matter is not created or destroyed.

2.2 *Conservation of Matter in Control Volume Form*

Consider a region fixed in size, shape, orientation, and position relative to an observer in the x, y, z frame of reference. In addition, the mass of the system instantaneously occupying this region (or control volume) and passing through the control surface S is taken as the extensive property B_i as used in Eq. (1.37). This control volume is shown in Fig. 2.1. Then b_i would be the mass per unit mass which is unity. The statement of the *conservation of matter* insures that $DB_i/Dt = 0$; that is, the mass of the system does not change with respect to time as one follows the system. Substituting these observations into Eq. (1.37) results in

$$\int_{\mathcal{V}} \frac{\partial \rho}{\partial t} d\mathcal{V} + \oint_S \rho \mathbf{V} \cdot d\mathbf{S} = 0 \tag{2.1}$$

The first term represents the instantaneous time rate of change of mass within the control volume, and the second term represents the instantaneous and net rate of mass flow through the control surface.

Example 2.1

Figure 2.2 shows a tank into which two streams flow. One of these entering streams is uniform as indicated, while the other has a velocity profile that is a paraboloid of revolution. This profile is given by $V_x = 10[1 - 4(r/D)^2]$; V_x is in units of ft/sec, r is the radial distance in feet from the centerline of the 1-ft-diameter pipe to a point in the fluid. Apparently the number 10 has associated with it the units of ft/sec. The uniform velocity in the 6-in. pipe is 4 ft/sec, and the density of the fluid is uniform and steady at the value ρ. Calculate the uniform velocity V_3 at the exit section 3.

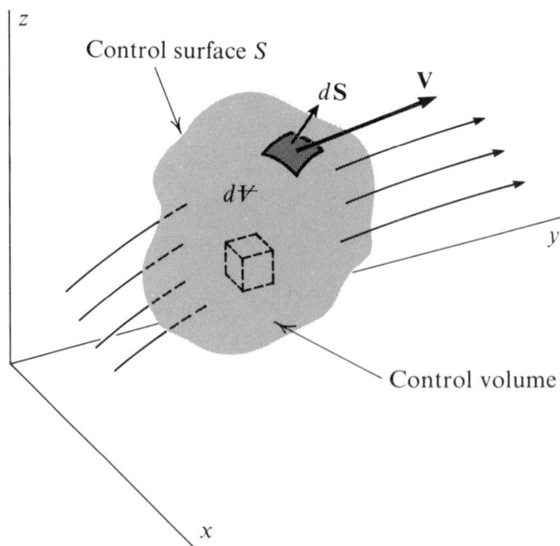

Fig. 2.1 A control volume with fluid passing through it.

Solution. Consider a control volume defined by the circular cross-sectional areas A_1, A_2, A_3, and the walls of the pipe and the tank. This choice of control surface is convenient for evaluating the surface integral because the expression for $d\mathbf{S}$ is readily known and because the direction of the velocity $\mathbf{V}$ is parallel to the direction of area $d\mathbf{S}$. Equation (2.1) requires that

$$\int_{\mathcal{V}} \frac{\partial \rho}{\partial t} d\mathcal{V} + \oint_S \rho \mathbf{V} \cdot d\mathbf{S} = 0$$

Since the density of the fluid is steady, $\partial \rho / \partial t$ is equal to zero for every point in the control volume, and the first integral is equal to zero. With density not varying over the control surface, ρ may be removed from the integrand of the second integral. One can now write, after division by ρ,

$$\oint_S \mathbf{V} \cdot d\mathbf{S} = 0 = \int_{A_1} \mathbf{V} \cdot d\mathbf{S} + \int_{A_2} \mathbf{V} \cdot d\mathbf{S} + \int_{A_3} \mathbf{V} \cdot d\mathbf{S}$$

Since the velocity is uniform at sections 1 and 3, the first and third integrals may be written as $-V_1 A_1$ and $V_3 A_3$, respectively. The minus sign in front of $V_1 A_1$ appears since the outward drawn normal and the velocity vector have an angle of 180° between them at section 1. The conservation of matter statement becomes

$$-V_1 A_1 + \int_{A_2} \mathbf{V} \cdot d\mathbf{S} + V_3 A_3 = 0$$

Substituting for $\mathbf{V}$, using an area element similar to that in Fig. 4.15, and performing the operation $\mathbf{V} \cdot d\mathbf{S}$ gives

$$-V_1 A_1 - \int_0^{1/2} 10(1 - 4r^2)\, 2\pi r \, dr + V_3 A_3 = 0$$

Integration yields

$$-V_1 A_1 - 20\pi \left(\frac{r^2}{2} - r^4 \right) \Bigg|_{r=0}^{r=1/2} + V_3 A_3 = 0$$

$$-V_1 A_1 - \frac{5\pi}{4} + V_3 A_3 = 0$$

Fig. 2.2

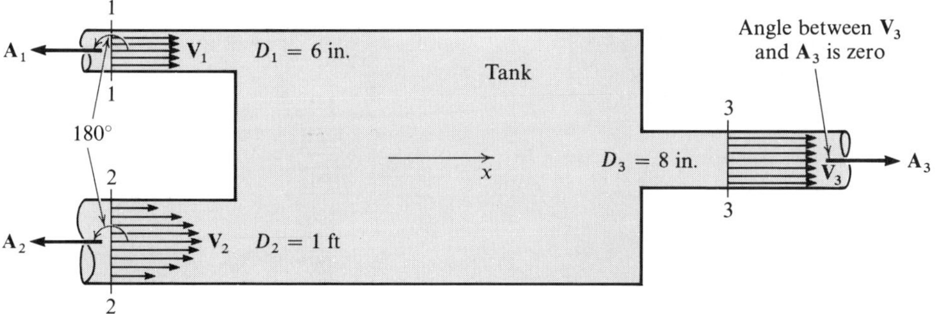

Solving for V_3,

$$V_3 = \frac{5\pi/4 + V_1 A_1}{A_3} = \frac{5\pi/4(\text{ft}^3/\text{sec}) + 4(\text{ft/sec})\,\pi/4\,(1/2)^2\,(\text{ft}^2)}{\pi/4\,(8/12)^2\,(\text{ft}^2)}$$

$$= 13.5 \text{ ft/sec}$$

◀

Example 2.2

In Fig. 2.3 a jet of steady and uniform density is shown leaving a nozzle. This jet continues to contract after leaving the nozzle. Just before the convergent section of the nozzle, the diameter D_1 is 1 in., and the velocity $\mathbf{V}_1 = 2.5\mathbf{k}$ ft/sec. At location 2, velocity $\mathbf{V}_2 = 10\mathbf{k}$ ft/sec. Calculate the diameter of the circular section at location 2 where the streamlines are essentially parallel.

Solution. The given information indicates that the velocities are uniform over the cross-sectional area at 1 and 2. A control volume is selected that is bounded by the two cross-sectional areas A_1 and A_2, the surface of the convergent wall, and the surface formed by the streamlines separating the jet from the atmosphere. This control volume is shown bounded by dashed lines representing the control surface in Fig. 2.4. Applying the conservation of matter, as stated in Eq. (2.1),

$$\int_{\forall} \frac{\partial \rho}{\partial t} d\forall + \oint_S \rho \mathbf{V} \cdot d\mathbf{S} = 0$$

The first integral is zero because of a steady density. The second integral can be written as the sum of four other surface integrals

$$\rho \int_{A_1} \mathbf{V} \cdot d\mathbf{S} + \rho \int_{S_{\text{wall}}} \mathbf{V} \cdot d\mathbf{S} + \rho \int_{S_{\text{atm}}} \mathbf{V} \cdot d\mathbf{S} + \rho \int_{A_2} \mathbf{V} \cdot d\mathbf{S} = 0$$

The uniform density has been removed from the surface integrals, and it should be noted that the second and third terms are zero since the second integral is for the wall area and the third is for the area comprised of streamlines adjacent to the atmosphere. One will recall that there is no flow across the streamlines since the velocity vector is

Fig. 2.3

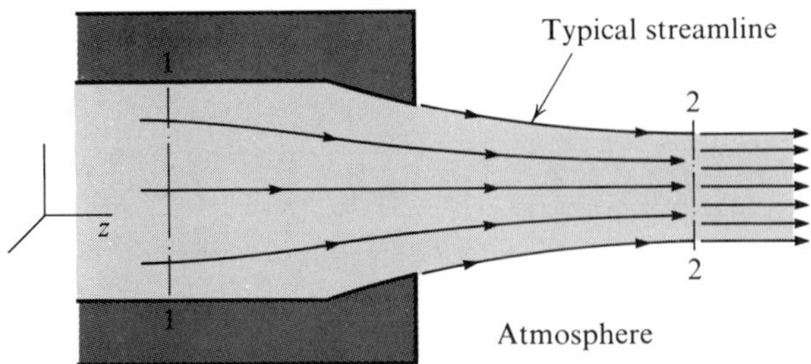

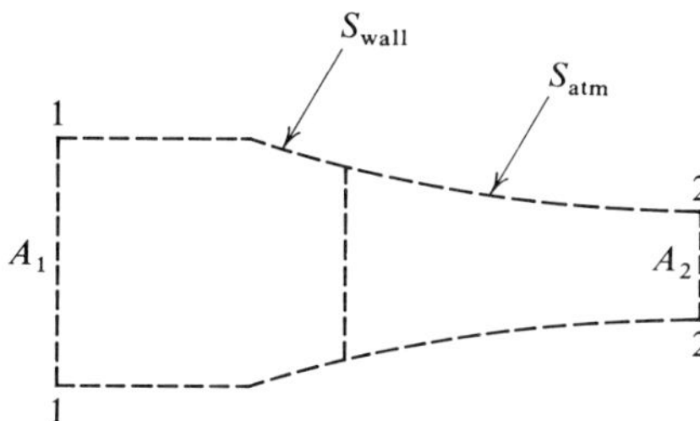

Fig. 2.4

everywhere tangent to these lines. Dividing the preceding equation by ρ, and noting that the velocity is uniform over the area A_1 and A_2,

$$-V_1A_1 + V_2A_2 = 0$$

Solving for D_2,

$$\frac{A_2}{A_1} = \frac{{D_2}^2}{{D_1}^2} = \frac{V_1}{V_2}$$

and

$$D_2 = \left[\frac{V_1}{V_2}\right]^{1/2} D_1 = \left[\frac{2.5 \text{ ft/sec}}{10 \text{ ft/sec}}\right]^{1/2} (1 \text{ in.}) = 0.5 \text{ in.}$$ ◀

Example 2.3

A perfect gas flows steadily through a nozzle with a 3-in. diameter at entrance and 1.5-in. diameter at exit. This is illustrated in Fig. 2.5. If the nozzle is designed properly, the gas will change state at "almost uniform and steady" entropy as it expands through the nozzle. The entrance pressure, velocity, and density are 100 $\text{lb}_f/\text{in.}^2$, 100 ft/sec, and 0.05 lb_m/ft^3, respectively, and these are uniform over the entrance section. The pressure at nozzle exit is 60 lb_f/in^2. Calculate the exit velocity assuming it is uniform over the exit area. The specific heat ratio for the gas may be considered constant at a value of $\gamma = c_p/c_v = 1.4$.

Fig. 2.5

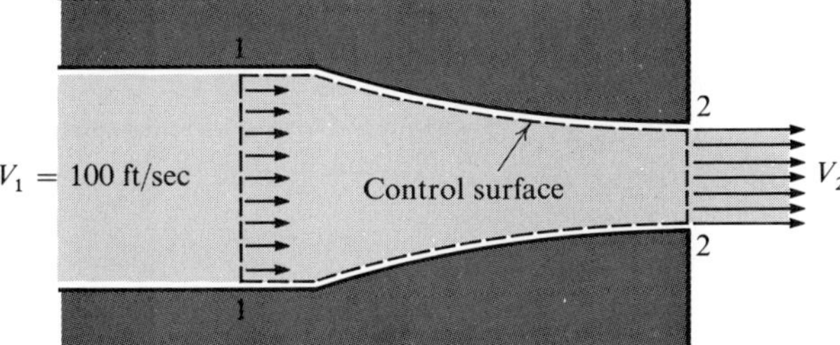

Solution. First a control volume is selected; this is indicated by dashed lines in Fig. 2.5. From Eq. (2.1),

$$\int_{\mathcal{V}} \frac{\partial \rho}{\partial t} d\mathcal{V} + \oint_S \rho \mathbf{V} \cdot d\mathbf{S} = 0$$

The first integral will be taken as zero by assuming that the density does not vary with time at a given location in the control volume. The second integral is replaced by the sum of two integrals, one over the entrance section and another over the exit section. This may be expressed by

$$\int_{A_1} \rho \mathbf{V} \cdot d\mathbf{S} + \int_{A_2} \rho \mathbf{V} \cdot d\mathbf{S} = 0$$

Since the velocities and densities are taken to be uniform over A_1 and A_2, the integrations may be carried out, and

$$-\rho_1 V_1 A_1 + \rho_2 V_2 A_2 = 0$$

Again, note the minus sign in front of the first term. This is a result of taking the product $\mathbf{V}_1 \cdot d\mathbf{S}_1$ where there is an angle of 180° between these two vectors at entrance. To emphasize this, one may consider that the *mass influx is negative and the mass efflux is positive*. Solving for the desired velocity yields

$$V_2 = V_1 \left[\frac{A_1}{A_2}\right] \left[\frac{\rho_1}{\rho_2}\right]$$

If ρ_2 is known, V_2 may be found from the given data. Since it is given that the flow is isentropic, recollection from thermodynamics gives $\rho_1/\rho_2 = (p_1/p_2)^{1/\gamma}$ for a perfect gas with a constant specific heat ratio. Substitution of this information into the expression for V_2 gives

$$V_2 = (100 \text{ ft/sec}) \left(\frac{3 \text{ in.}}{1.5 \text{ in.}}\right)^2 \left(\frac{100 \text{ lb}_f/\text{in.}^2}{60 \text{ lb}_f/\text{in.}^2}\right)^{1/1.4}$$

$$= 576 \text{ ft/sec}$$ ◀

The assumption of isentropic flow was based on a well-designed nozzle. Actually there is some "fluid friction" that precludes reversibility, and the velocity would be lower than that calculated in this problem. However, with good design it is possible to achieve a velocity in excess of 95 percent of the calculated figure.

2.3 Conservation of Matter in Partial Derivative Form

Equation (2.1) can be altered by exchanging the surface integral for a volume integral through application of the divergence theorem of Gauss. Doing this, one obtains

$$\int_{\mathcal{V}} \left(\frac{\partial \rho}{\partial t} + \text{div}\, \rho \mathbf{V}\right) d\mathcal{V} = 0 \tag{2.2}$$

Since Eq. (2.2) must vanish for any and all volumes, a volume can be selected suitably small such that the integrand ($\partial\rho/\partial t + \operatorname{div}\rho\mathbf{V}$) cannot change sign.[1] If the integral is to vanish when taken over this small volume, then the integrand will have to vanish at all points within this volume; hence

$$\frac{\partial\rho}{\partial t} + \operatorname{div}\rho\mathbf{V} = 0 \tag{2.3}$$

The preceding equation is also referred to as the *continuity equation*. Two forms of expression for the conservation of matter are now available:

1. Equation (2.1), which applies to a *region* in space.
2. Equation (2.3), which applies to a *point* in space.

It should be noted that there is no coordinate system expressed in either of these equations. Again it is emphasized that Eq. (2.1) does not apply to a region in which there are points where matter is created or destroyed. Hence Eq. (2.3) does not apply at these points either.

2.4 Incompressible Fluid

A fluid is defined as *incompressible* if there is no change of density of the fluid particle as its motion is followed. This definition is equivalent to requiring the material derivative of the density to vanish identically, that is,

$$\frac{D\rho}{Dt} = 0 \tag{2.4}$$

for all positions in space and at all times. Using Eq. (1.19),

$$\frac{D\rho}{Dt} = \frac{\partial\rho}{\partial t} + \mathbf{V}\cdot\mathbf{grad}\,\rho = 0 \tag{2.5}$$

Equation (2.5) clearly indicates that the condition of incompressibility does not imply that the density field is steady, that is, $D\rho/Dt = 0$ does not require that $\partial\rho/\partial t = 0$.

If Eq. (2.5) is used to substitute for $\partial\rho/\partial t$ in Eq. (2.3), the continuity equation becomes

$$\operatorname{div}\rho\mathbf{V} - \mathbf{V}\cdot\mathbf{grad}\,\rho = 0$$

The difference between the two terms in the preceding equation is also equal to $\rho\operatorname{div}\mathbf{V}$ according to Eq. (1.23), and the *continuity equation for an incompressible fluid* may be written as $\rho\operatorname{div}\mathbf{V} = 0$ or

$$\operatorname{div}\mathbf{V} = 0 \tag{2.6}$$

[1] It is always possible to select a volume so small that the integrand cannot change sign provided that it is continuous.

Recalling the interpretation of the divergence of the velocity vector given in Section 1.7, Eq. (2.6) states physically that the rate of change of volume of the fluid element per unit volume must be zero if matter is to be conserved for an incompressible fluid—which is so plausible that it appears obvious. If it is desired to convey that the density has the same value at all points in the field at all times, this can be insured by stating that the density is *steady and uniform*. It should be noted from Eq. (2.5) that steady and uniform density insures incompressibility, but the term incompressible does not require that the density be steady or uniform.

Example 2.4

Does the velocity field $\mathbf{V} = 3x^2y\mathbf{i} + 2yz\mathbf{j} - 3x^2z^3\mathbf{k}$ (in length/time dimensions) satisfy the conservation of matter requirement for an incompressible flow field?

Solution. For the incompressible fluid, conservation of matter requires that the div $\mathbf{V} = 0$. For the given field,

$$\text{div } \mathbf{V} = \text{div }(3x^2y\mathbf{i} + 2yz\mathbf{j} - 3x^2z^3\mathbf{k}) = 6xy + 2z - 9x^2z^2$$

The divergence of the velocity vector does not vanish at all points in the field, so the given vector could not represent the velocity of an incompressible flow field. It is true that the divergence of the given vector is equal to zero at the origin ($x = 0$, $y = 0$, $z = 0$) or anywhere on the locus $6xy + 2z - 9x^2z^2 = 0$; but the question asked is, "Does div $\mathbf{V}$ vanish identically, that is, for all values of x, y, z?" The answer is, "It does not!"

Example 2.5

Find a z component of velocity compatible with the x and y components of velocity given in Example 2.4 for an incompressible flow field.

Solution. Conservation of matter requires that div $\mathbf{V} = 0$, and so

$$\frac{\partial V_z}{\partial z} = -\frac{\partial V_x}{\partial x} - \frac{\partial V_y}{\partial y} = -6xy - 2z$$

Integrating this last expression with respect to z yields

$$V_z = -6xyz - z^2 + f(x,y)$$ ◀

Since $f(x,y)$ is an arbitrary function of x and y, the answer is not unique. Any differentiable function of x and y will satisfy conservation of matter requirements for V_z when added to the first two terms of the answer.

Example 2.6

A two-dimensional steady flow field has an x component of velocity given by

$$V_x = \frac{V_b}{2}\left(1 + \frac{y}{b}\right)$$

in which V_b and b are dimensional constants. Assume the fluid to be incompressible and demonstrate that the y component of velocity is a function of the x coordinate only, assuming V_y exists.

Solution. Conservation of matter for the two-dimensional constant-density flow requires that

$$\frac{\partial V_x}{\partial x} + \frac{\partial V_y}{\partial y} = 0$$

thus

$$\frac{\partial V_y}{\partial y} = -\frac{\partial V_x}{\partial x} = 0$$

since $V_x = V_x(y)$. Integrating with respect to y yields

$$V_y = \int \frac{\partial V_y}{\partial y}\bigg|_{x=\text{const}} dy + f(x) = \int (0)dy + f(x)$$

$$= f(x) = \text{arbitrary function of } x \text{ only} \quad ◀$$

The given velocity field could describe the velocity profile for the flow between two parallel plates as shown in Fig. 2.6. In obtaining the solution of the velocity field, it is assumed that $V_y = 0$. For this situation it can be shown that V_x is a function of y only by similar reasoning. Knowledge of the fact that $V_x = V_x(y)$ is of great help in obtaining a solution for V_x from the equations of motion for the fluid.

2.5 Closing Remarks

Conservation of mass is certainly one of the basic tools of analysis. If the fluid is incompressible, then the continuity equation involves only the derivatives of kinematic quantities with respect to space coordinates. How-

Fig. 2.6

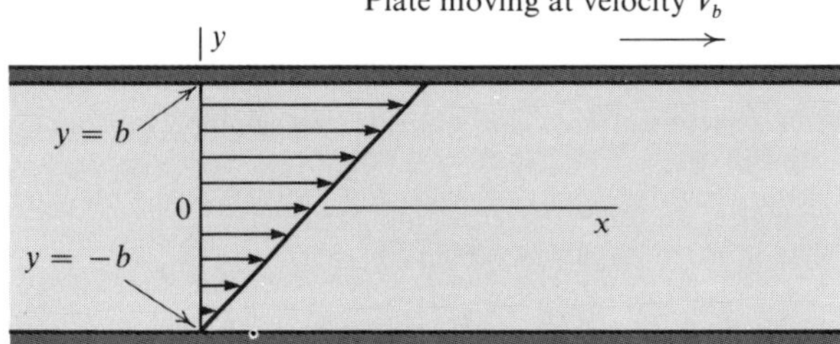

ever, for the case of the compressible fluid, such as the flow of a gas, the density variable must be considered. In either case, Eqs. (2.6) and (2.3) are inadequate, as single expressions, to determine the velocity components and density field. Recourse to other physical laws is necessary to provide further relations involving these variables. With this in mind, it is natural to look toward mechanics for additional physical information; and indeed Newton's laws involve forces and time derivatives of velocity. In order to better understand how forces are transmitted through a fluid, the next chapter will introduce the concepts of surface and body forces and then formulate these into the equations of motion.

Self-Study Questions

2.1 Sketch the streamlines a fluid might have in the region about a point that is a sink and for one that is a source.

2.2 State a physical meaning for

a. $\rho \, d\mathcal{V}$

b. $\int_{\mathcal{V}} \rho \, d\mathcal{V}$

c. $\int_{\mathcal{V}} \frac{\partial \rho}{\partial t} \, d\mathcal{V}$

2.3 Express a physical meaning for

a. div $\rho\mathbf{V}$

b. div $\rho\mathbf{V} \, d\mathcal{V}$

c. $\int_{\mathcal{V}} \text{div } \rho\mathbf{V} \, d\mathcal{V}$

2.4 What equation or expression may be used to establish that a mass source or sink "exists" in a given hypothetical fluid flow field?

2.5 Describe a density and velocity field that will reduce the continuity equation to

a. div $\mathbf{V} = 0$

b. $\frac{\partial(\rho V_x)}{\partial x} + \frac{\partial(\rho V_y)}{\partial y} = 0$

c. $\frac{\partial \rho}{\partial t} + \frac{\partial(\rho V_x)}{\partial x} = 0$

2.6 Obtain an expression for $\partial V_r/\partial r$, in terms of V_r, V_θ, r, and θ, for an incompressible fluid flow field in plane spherical coordinates.

2.7 State in words the physical meaning of the following terms from Eq. (2.1):

a. $\mathbf{V} \cdot d\mathbf{S}$ where S is a specified surface through which fluid may pass.

b. $\rho\mathbf{V} \cdot d\mathbf{S}$

c. $\oint_S \rho\mathbf{V} \cdot d\mathbf{S}$

d. $\int_{\mathcal{V}} \frac{\partial \rho}{\partial t}\, d\mathcal{V} + \oint_S \rho \mathbf{V} \cdot d\mathbf{S}$

2.8 In a finite and fixed volume, is

$$\frac{\partial}{\partial t}\int_{\mathcal{V}} \rho\, d\mathcal{V} \equiv \int_{\mathcal{V}} \frac{\partial \rho}{\partial t}\, d\mathcal{V}$$

2.9 Will the integral $\int_{\mathcal{V}} \partial\rho/\partial t\, d\mathcal{V}$ be equal to zero when a control volume contains a cyclic device that "operates steadily"?

2.10 Given that a control volume is fixed in a given reference space, and that the change of mass is desired for a finite time interval Δt within the control volume, explain how one could obtain this. What information would be needed to make the evaluation?

2.11 Is the definition of an incompressible fluid equivalent to stating that in a fluid density field, ρ is a constant (that is, $d\rho = 0$)?

2.12 Show that if the density field is
a. Uniform and steady, it is then incompressible.
b. Incompressible and steady, it is then uniform.
c. Incompressible and uniform, it is then steady.

Problems

2.1 Sketch the velocity profiles at area *OABC* in Fig. 2.7 and calculate the mass rate of flow through this area. Consider $dS\,\mathbf{i}$ to be positive along the $+x$ axis
a. When $\rho = 0.08$ lb$_m$/ft^3 and $\mathbf{V} = (2z^2 + 3)\mathbf{i}$ ft/sec.
b. When $\rho = 2.0 - 0.02z$ slug/ft^3 and $\mathbf{V} = 10z\mathbf{i}$ ft/sec.

2.2 Consider a flow through the nozzle as shown in Fig. 2.3. An observer is moving past the nozzle in the downstream direction at a velocity of 50 ft/sec. He measures a uniform zero velocity at cross section 1. What must he measure as a velocity at cross section 2 if the inlet diameter is 2 in. and the exit diameter is 1 in.? The density of the fluid is 62.4 lb$_m$/ft^3.

Fig. 2.7

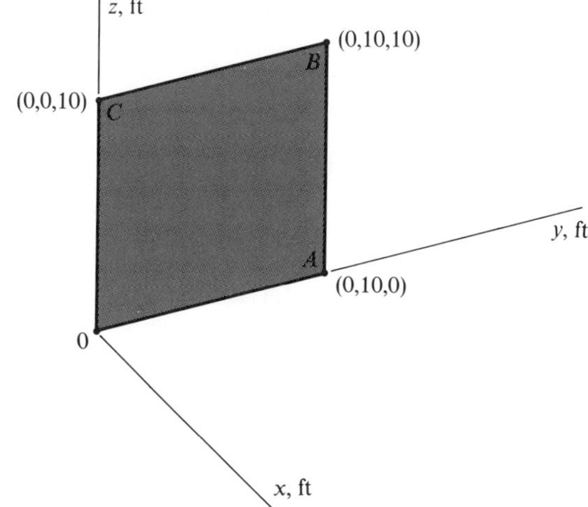

2.3 Refer to Example 2.1 and calculate the average speed of the fluid passing through cross section 2 so that the mass rate of flow may be evaluated from the product of density, cross-sectional area, and average velocity at that section.

2.4 A fluid in two-dimensional flow is indicated in Fig. 2.8 as passing through a diverging cross section with velocity $\mathbf{V} = (3/r)\boldsymbol{\varepsilon}_r$ ft/sec for $r > 0$. The density at each given surface is 0.15 lb_m/ft^3

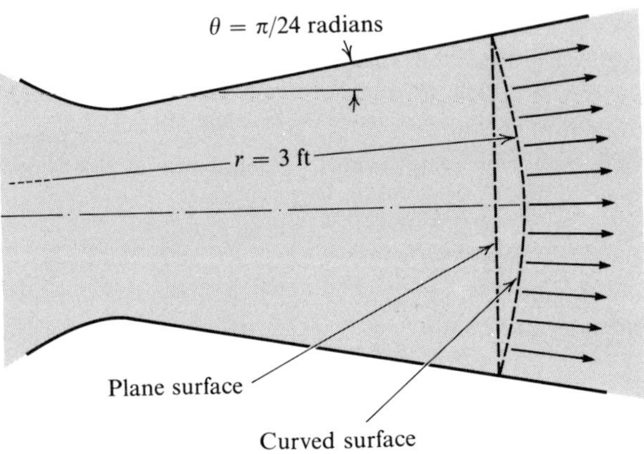

Fig. 2.8

a. Calculate the mass rate of flow per unit width through the curved surface.
b. Calculate the mass rate of flow per unit width through the plane surface.

2.5 A fluid moves uniformly and radially out through a spherical control surface and at a given instant of time at a speed $|\mathbf{V}| = 10r$ ft/sec, where r is the radius in feet. The radius of the control surface is 1 ft where the density is 0.1 lb_m/ft^3 Calculate the time rate of change of mass within the sphere if there are no "sources" within the sphere.

2.6 A fluid flows out of a tank at a steady rate at the one exit section for a specified time interval. The initial mass is 500 slugs. The mass rate of efflux from the tank is 100 lb_m/sec. The tank volume is 10 ft^3

a. What is the instantaneous time rate of change of density of the fluid in the tank if the density in the tank is uniform at any time?

b. What is the change of density in the tank after a 2 sec interval?

2.7 A vertical cylindrical tank of constant volume has a valve opening to the surrounding atmosphere. The density of the contained gas varies along the length of the tank (z, in feet) and with time (t, in seconds) when the valve is open. This variation is expressed by $\rho = C_1 z\, e^{-0.1t} + C_2$ where C_1 and C_2 are dimensional constants. The initial density at the bottom and top of the tank is 2.0 lb_m/ft^3 and 1.8 lb_m/ft^3, respectively. The cross-sectional area is 3 ft^2 The length is 10 ft.

a. What is the instantaneous mass rate of flow at the valve located in the bottom of the tank?

b. Would the exit velocity be the same if the valve were located either at the top or on the bottom of the tank?

2.8 Assume that the valve in Prob. 2.7 is located at the bottom of the tank and calculate the volume flow rate out of the valve when $t = 1$ sec.

2.9 A parallel flow of fluid with uniform and steady density exists between two surfaces (at $y = a$ and $y = -a$) as shown in Fig. 2.9. The velocity profile for this one-dimensional flow is

$$V_x = C_1\left(1 + \frac{y}{a}\right) - C_2\left(1 - \frac{y^2}{a^2}\right)$$

where $C_1 = 2C_2/3$, and C_1 and C_2 have dimensions of length/time.

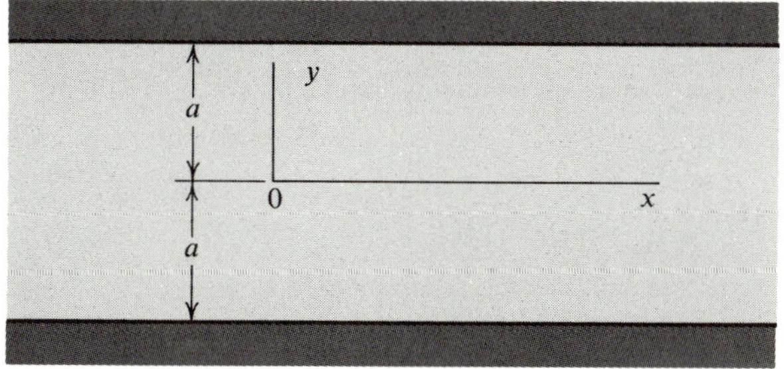

Fig. 2.9

a. Calculate the mass flow rate past a cross section of unit depth.
b. Sketch the velocity profile of V_x.

2.10 Show whether continuity requirements are satisfied for a proposed flow field whose density is a linearly decreasing function with y and whose velocity is
a. A constant in the x direction.
b. $\mathbf{V} = Cy^2\mathbf{i}$ where C is a constant.
c. $\mathbf{V} = C_1x\mathbf{i} - C_2y\mathbf{j}$ when $C_1 = 2C_2$.

2.11 Determine whether matter is conserved for the following selected functions of ρ and $\mathbf{V}$ which contain appropriate dimensional coefficients.
a. $\rho = 3t$ and $\mathbf{V} = -r\boldsymbol{\varepsilon}_r + r^2\theta\boldsymbol{\varepsilon}_\theta - rz\boldsymbol{\varepsilon}_z$

b. $\rho \neq f(r,\theta,z,t)$ and $\mathbf{V} = \dfrac{100}{r}\boldsymbol{\varepsilon}_r \neq f(\theta,z)$

2.12 A two-dimensional and steady flow of a fluid whose density is uniform and steady has a velocity field of $V_r = Ae^{Br}$ with B and A being dimensional constants. Find a function for V_θ consistent with conservation of matter.

2.13 Could the following velocity field (in length/time dimensions) represent the flow of a fluid continuum with a density field of $\rho = 4ty$ in mass/length^3 dimensions?

$$\mathbf{V} = -\frac{x}{t}\mathbf{i} + 3z^2\mathbf{j} - \left(\frac{z^3}{y} + y\right)\mathbf{k}$$

2.14 Select a velocity and density field and test to see if your choice satisfies continuity.

2.15 A flow field has a density variation of $\rho = Axy + Bz^2t + C$ where A, B, and C are dimensional constants and t is a time variable.

a. Obtain an expression for the local rate of change of density with time at $x = 1$, $y = 1$, $z = 1$.

b. Obtain an expression for the change of density in the y direction at $x = 1$, $y = 1$, $z = 1$, and $t = 0$.

c. Evaluate the time rate of change of density of a "fluid particle" at $x = 0$, $y = 0$, $z = 10$, and $t = 1$, and where $V_z(0,0,10,1) = 2$.

2.16 In a given fluid flow, the velocity field is $\mathbf{V} = 10\boldsymbol{i} - (20/x^2)\mathbf{j}$ for $x > 0$. Is it possible for

a. The density to be a constant (i.e., with respect to space and time)?

b. The density field to be steady?

c. The fluid to be incompressible?

2.17 Does the velocity field $\mathbf{V} = -x\mathbf{i} + y\mathbf{j}$ satisfy continuity requirements for a flow field where $\rho \neq \rho(x,y)$?

2.18 Does the following velocity function $\mathbf{V} = (1/x)\mathbf{i} + (y/x^2)\mathbf{j}$ for $x > 0$ satisfy continuity requirements for an incompressible fluid?

3

NATURE OF FORCES IN A FLUID FIELD AND THEIR EFFECTS

3.1 Introduction

Before one can analyze the motion of a fluid, it is necessary to understand what forces are involved in producing and maintaining the motion. This is a natural consequence of applying Newton's laws of motion since these laws contain force explicitly.

In this chapter the concept of stress will be introduced so that one may see how the effects of forces are transmitted through a fluid continuum. A discussion of body forces is presented next, since they are distributed throughout the mass of the fluid instead of over a surface within or bounding the fluid. Relations between the forces acting on a fluid particle and its acceleration are developed with the aim of analyzing some problems of engineering interest. It becomes apparent from these examples that analysis of more difficult problems involving shear stresses will require some information concerning the particular fluid involved. This leads to the consideration of constitutive relations in Chapter 4.

3.2 Surface and Body Forces

Consider the fluid instantaneously occupying the control volume shown in Fig. 3.1. A force $\mathbf{F}_S$ is exerted by the surroundings on ΔS, which is a typical small element of area on the control surface; $\mathbf{C}$ is a couple exerted by the surroundings on ΔS. The force $\mathbf{F}_S$ may be replaced by two components, one normal to ΔS and one tangential to ΔS. These are indicated by F_n and F_t, respectively. A *normal stress* is defined by the expression

$$\lim_{\Delta S \to 0} \frac{F_n}{\Delta S} = \sigma_{nn}$$

and a *shear stress* by

$$\lim_{\Delta S \to 0} \frac{F_t}{\Delta S} = \tau_{nt}$$

The couple $\mathbf{C}$ does not appear in the definition of stress since ΔS approaches zero in the limiting process. As this portion of the surface is made arbitrarily small, the forces acting on it approach a concurrent force distribution which may be replaced by a single force. Note that a double subscript notation is employed in the symbols defining stress. The first subscript (n in both defining equations above) denotes the orientation of the surface ΔS; this orientation is given by the direction of the outward-drawn normal to ΔS. The second subscript denotes the direction of the force: σ_{nn} in the case of F_n, and τ_{nt} in the case of F_t. One notes that stresses by definition transmit forces that act over surfaces. These forces are then called *surface forces*.

Next, consider the force due to gravity acting on the element of volume $\mathcal{V}$. This force would be

$$\mathbf{F}_B = \int_m \mathbf{f}_B \, dm = \int_{\mathcal{V}} \mathbf{g}\rho \, d\mathcal{V}$$

where ρ = density of fluid,

$\mathbf{g}$ = gravity field vector at some point inside the small element of volume $d\mathcal{V}$,

$\mathbf{f}_B$ = body force/unit mass at this point

Forces of this nature are termed *body forces*. Another example of a body force is that exerted on a charged particle or collection of charged particles moving in a magnetic field. As in the case of gravity, the force due to the magnetic field acts on each particle.

Body forces may be classified as conservative or nonconservative. A force distribution or field acting on a particle is said to be *conservative* if it depends on the position of the particle and can be expressed as the gradient of a scalar field. The force acting on a mass moving in a gravitational field

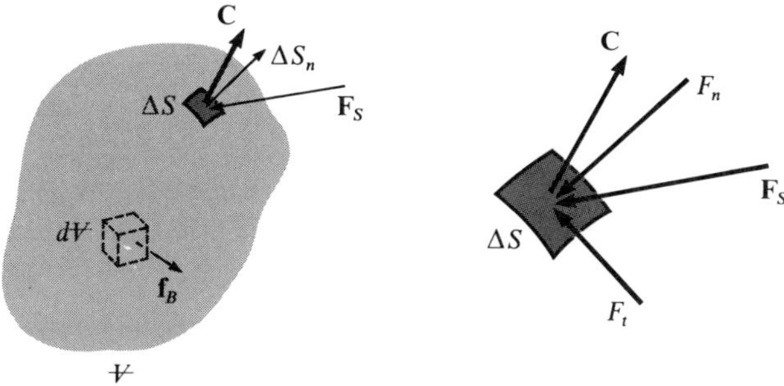

Fig. 3.1 *Control volume with surface and body forces acting on typical elements.*

can be obtained by taking the gradient of the gravitational potential, and thus is conservative. However, in the case of a charged particle moving in a magnetic field, the force on the particle depends on the velocity of the particle since this force is proportional to the vector product of the magnetic flux density vector and the velocity. Since the magnetic body force is not expressible as the gradient of a scalar potential, it is *nonconservative*.

Example 3.1

The column of gas illustrated in Fig. 3.2 has a density distribution given by $\rho = \rho_0 e^{-Cz}$ in which ρ_0 is the density of the gas at positions given by $z = 0$, and C is a constant with the dimension of reciprocal length. The cross-sectional area A of the column

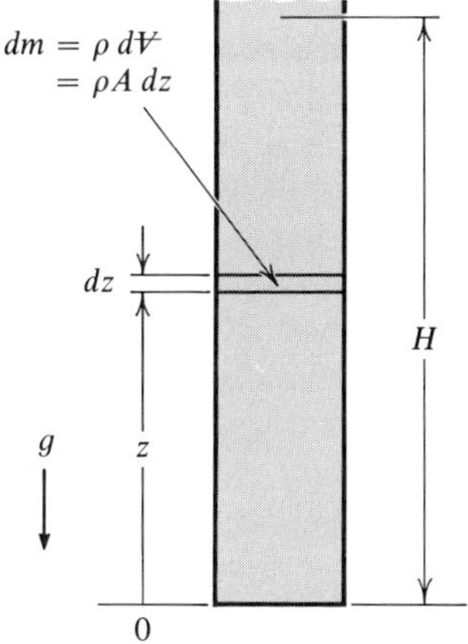

Fig. 3.2

(perpendicular to the z axis) is constant with respect to z. Neglecting variation of the acceleration due to gravity (g) with height, calculate the weight of a column of height H of this gas.

Solution. The weight of the column of gas is the body force due to gravity. To calculate this body force acting on the entire column, the force acting on a small typical element is integrated over the mass of gas in the column. The force acting on the element shown, which has a small height dz, is given by

$$d\mathbf{F}_B = -g\mathbf{k}\,dm = -g\mathbf{k}A\,(\rho_0 e^{-Cz})\,dz$$

in which A is the constant cross-sectional area. Integration over the mass of the gas may be effected by allowing z to range from zero to H. This yields

$$\mathbf{F}_B = \int_m d\mathbf{F}_B = -g\rho_0 A\mathbf{k}\int_0^H e^{-Cz}\,dz = -\frac{g\rho_0 A}{C}(1 - e^{-CH})\mathbf{k} \qquad \blacktriangleleft$$

The negative sign introduced into the problem as $-g\mathbf{k}$ indicates the force is acting in the negative z direction, that of gravity. Density variations frequently occur in the earth's atmosphere. These are due to partial absorption of energy transmitted through the atmosphere and the fact that the air closer to the earth is compressed by the weight of air more remote from the earth's surface.

3.3 Sign Convention for Stresses

In the previous section it was noted that the specification of a stress at a point required a double subscript notation. Consider a small element of fluid in a flow field with the positive coordinate directions indicated by axes as shown in Fig. 3.3. All of the stresses shown in the figure are positive. A stress is *positive* if it produces a force that acts in the positive coordinate direction on a positive face, or if it produces a force that acts in a negative coordinate direction on a negative face. A face is considered positive or negative according to the direction of its outward-drawn normal with respect to the coordinate direction. By way of illustration, consider the two stresses σ_{yy} and $\sigma_{yy} + (\partial\sigma_{yy}/\partial y)\,\Delta y$ in Fig. 3.3. The one indicated by an arrow in the positive y direction acts on a face that has an outward-drawn normal in the positive y direction. This is a positive stress. The other stress, designated by σ_{yy}, is indicated by an arrow in the negative y direction, and the direction of the outward-drawn normal to the face over which it acts is in the negative y direction. This stress is also positive by definition. If the direction of the force produced by a stress is positive (or negative) and the stress acts on a negative (or positive) face, then the stress is defined as a *negative* stress. According to these definitions, negative normal stresses are *compressive*, and positive normal stresses are *tensile*.

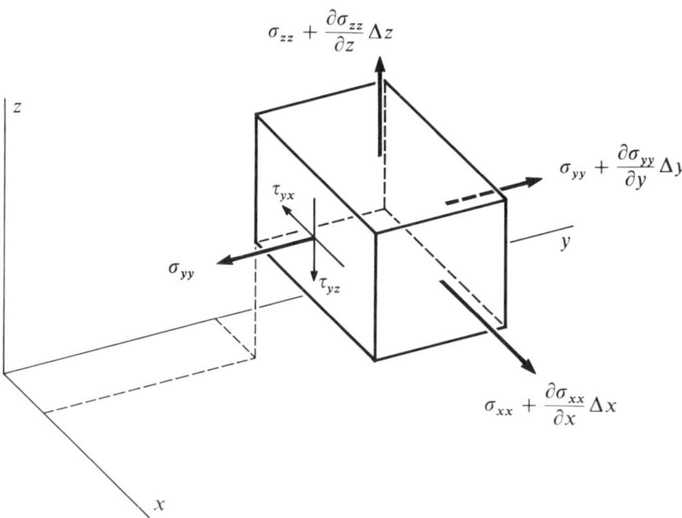

Fig. 3.3 Element in a fluid flow field with several stresses indicated.

3.4 State of Stress at a Point

The state of stress at a given point in a continuum may be specified by a normal stress and two shear stresses acting in given coordinate directions on each of three mutually perpendicular planes intersecting at a given point. This means that nine components[1] are required to describe the most general *state of stress*. These nine quantities are referred to as the *components of the stress tensor*. It is understandable that the state of stress would require something more than a simple vector description since stress depends on the orientation of a force vector as well as the orientation of an area vector.

The problem of finding the resultant shear and normal stresses acting on a plane of arbitrary orientation at a point in the continuum in terms of components of the stress tensor is considered next. The element shown in Fig. 3.4 is subject to surface forces due to the stresses shown. It is also subjected to a body force field (e.g., gravity) designated by $\mathbf{f}_B$. The direction cosines of the outward-drawn normal to the oblique face are l, m, and n with respect to the x, y, and z axes, and the area of the oblique face is A. The problem consists of expressing σ, τ, and the direction cosines of τ (designated l', m', and n', respectively) in terms of the given quantities

[1]The fact that these nine components are *not all distinct* for the fluid continuum is discussed in Chapter 4.

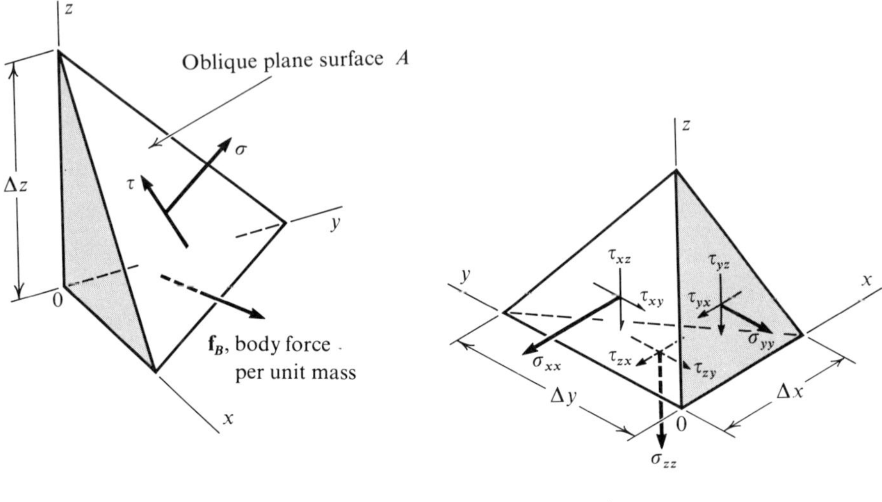

Fig. 3.4 *Element of fluid used for obtaining the resultant shear stress τ and normal stress σ on a plane of arbitrary orientation.*

l, m, n, and the components of the stress tensor. Summing forces in the x, y, and z directions, by using mean values of stresses with the appropriate areas and applying Newton's law of motion to the infinitesimal element, yields

$$(\mathbf{f}_B)_x\,\rho\left(\frac{\Delta x\,\Delta y\,\Delta z}{6}\right) + \sigma A l - \sigma_{xx}\left(\frac{\Delta y\,\Delta z}{2}\right) - \tau_{yx}\left(\frac{\Delta x\,\Delta z}{2}\right) - \tau_{zx}\left(\frac{\Delta x\,\Delta y}{2}\right) + \tau A l' = \rho\left(\frac{\Delta x\,\Delta y\,\Delta z}{6}\right)\frac{DV_x}{Dt} \tag{3.1}$$

$$(\mathbf{f}_B)_y\,\rho\left(\frac{\Delta x\,\Delta y\,\Delta z}{6}\right) + \sigma A m - \tau_{xy}\left(\frac{\Delta y\,\Delta z}{2}\right) - \sigma_{yy}\left(\frac{\Delta x\,\Delta z}{2}\right) - \tau_{zy}\left(\frac{\Delta x\,\Delta y}{2}\right) + \tau A m' = \rho\left(\frac{\Delta x\,\Delta y\,\Delta z}{6}\right)\frac{DV_y}{Dt} \tag{3.2}$$

$$(\mathbf{f}_B)_z\,\rho\left(\frac{\Delta x\,\Delta y\,\Delta z}{6}\right) + \sigma A n - \tau_{xz}\left(\frac{\Delta y\,\Delta z}{2}\right) - \tau_{yz}\left(\frac{\Delta x\,\Delta z}{2}\right) - \sigma_{zz}\left(\frac{\Delta x\,\Delta y}{2}\right) + \tau A n' = \rho\left(\frac{\Delta x\,\Delta y\,\Delta z}{6}\right)\frac{DV_z}{Dt} \tag{3.3}$$

in which DV_x/Dt, DV_y/Dt, and DV_z/Dt are the acceleration components of the center of mass of the element of fluid.

The area of the oblique face, A, may be expressed as a function of its projection on the coordinate planes as follows:

$$Al = \frac{1}{2}\Delta y\,\Delta z \qquad Am = \frac{1}{2}\Delta x\,\Delta z \qquad An = \frac{1}{2}\Delta x\,\Delta y$$

Substituting these relations into Eq. (3.1) gives

$$(\mathbf{f}_B)_x\,\rho\left(\frac{Al\,\Delta x}{3}\right) + \sigma Al - \sigma_{xx}Al - \tau_{yx}Am - \tau_{zx}An + \tau Al' = \rho\left(\frac{Al\,\Delta x}{3}\right)\frac{DV_x}{Dt}$$

If this equation is divided by A, and then limited by letting $\Delta x \to 0$, $\Delta y \to 0$, and $\Delta z \to 0$ in such a way that the orientation of the oblique face is preserved, then the mean values of stresses in an area take on values of stress at the point 0, and

$$\sigma l - \sigma_{xx}l - \tau_{yx}m - \tau_{zx}n + \tau l' = 0 \tag{3.4}$$

By similar reasoning Eqs. (3.2) and (3.3) become

$$\sigma m - \tau_{xy}l - \sigma_{yy}m - \tau_{zy}n + \tau m' = 0 \tag{3.5}$$

and

$$\sigma n - \tau_{xz}l - \tau_{yz}m - \sigma_{zz}n + \tau n' = 0 \tag{3.6}$$

The terms containing the body force and accelerations go to zero in the limit, since these contain infinitesimals of one order higher than the other terms. This results in three equations in the five unknowns σ, τ, l', m', and n'. The remaining two equations come from geometric considerations, and these are

$$(l')^2 + (m')^2 + (n')^2 = 1 \tag{3.7}$$

and

$$ll' + mm' + nn' = 0 \tag{3.8}$$

Equation (3.8) results from the orthogonality of the orientations described by the two sets of direction cosines, l, m, n and l', m', n'. Equations (3.4) through (3.8) will then yield the desired resultant normal and shear stresses and the direction of the resultant shear stress for a *given state of stress at a point* on a plane of arbitrary orientation.

An interesting and useful result accrues if a fluid is considered in which shear stresses are absent. With the shear stresses zero in Eqs. (3.4), (3.5), and (3.6),

$$\sigma - \sigma_{xx} = 0$$

$$\sigma - \sigma_{yy} = 0$$

$$\sigma - \sigma_{zz} = 0$$

or

$$\sigma = \sigma_{xx} = \sigma_{yy} = \sigma_{zz} \tag{3.9}$$

As a result of Eq. (3.9), it may be stated that the normal stress is isotropic (i.e., independent of direction at a point) in a fluid void of shear. One situation in which there are no shear stresses in a fluid is the case of a static fluid,[1] i.e., moving as a rigid body. This would follow from the definition of a fluid, for if there were a shear stress present, the fluid would begin to deform continuously and would no longer remain static. Throughout this study of fluid mechanics certain idealized fluid motions will be assumed free of shear (or almost free of shear). It should be observed that the nine given component stresses indicated in Fig. 3.4 are positive as indicated, and the equations that involve these given quantities, Eqs. (3.4) through (3.6), were formulated on this basis.

3.5 Equations of Motion

In relating the forces acting on a fluid element to the motion of a fluid, a small element is considered with surface and body forces acting on it. Application of Newton's law of motion relates these forces to acceleration of the fluid element. Figure 3.5 shows an element with stress components acting on the respective faces, and these stresses, as well as their respective partial derivatives, refer to the center of the fluid element. The specific body-force component, density, and the acceleration component are also referred to the center of the element. The stresses in Fig. 3.5 represent the approximate values to be used in conjunction with the areas of the corresponding faces, and the coordinate system shown is an inertial frame of reference.[2] Summing the approximate forces in the x direction and equating this to the product of the mass of the element and its corresponding acceleration yields

[1]For the static fluid or a fluid moving in the absence of shear, the pressure p is taken as the negative of the normal stress ($p = -\sigma$). The interpretation of pressure in a moving fluid with shear is deferred until Chapter 4.

[2]Irving H. Shames, "Engineering Mechanics—Statics," 2d ed., p. 13, Prentice-Hall, Inc., Englewood Cliffs, N.J., 1966.

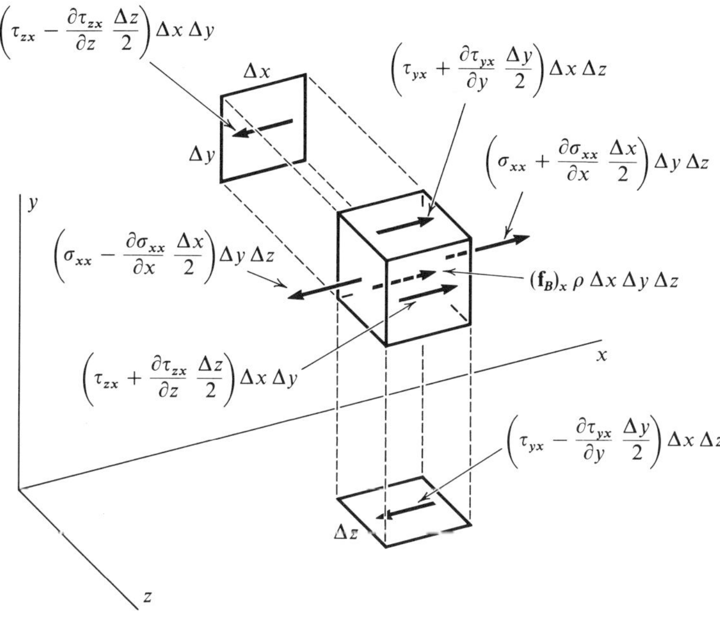

Fig. 3.5 *Fluid element selected for derivation of equations of motion with surface forces due to chosen positive stresses and body-force component. Only forces in x direction shown.*

$$
\begin{aligned}
&\left(\sigma_{xx} + \frac{\partial \sigma_{xx}}{\partial x}\frac{\Delta x}{2}\right)\Delta y\,\Delta z - \left(\sigma_{xx} - \frac{\partial \sigma_{xx}}{\partial x}\frac{\Delta x}{2}\right)\Delta y\,\Delta z \\
&\quad + \left(\tau_{yx} + \frac{\partial \tau_{yx}}{\partial y}\frac{\Delta y}{2}\right)\Delta x\,\Delta z - \left(\tau_{yx} - \frac{\partial \tau_{yx}}{\partial y}\frac{\Delta y}{2}\right)\Delta x\,\Delta z \\
&\quad + \left(\tau_{zx} + \frac{\partial \tau_{zx}}{\partial z}\frac{\Delta z}{2}\right)\Delta x\,\Delta y - \left(\tau_{zx} - \frac{\partial \tau_{zx}}{\partial z}\frac{\Delta z}{2}\right)\Delta x\,\Delta y \\
&\qquad\qquad + (\mathbf{f}_B)_x\,\rho\,\Delta x\,\Delta y\,\Delta z = \rho\,\Delta x\,\Delta y\,\Delta z\,\frac{DV_x}{Dt}
\end{aligned}
$$

The equation above may be simplified by carrying out the algebraic operations indicated and dividing through by the volume of the element, $\Delta x\ \Delta y\ \Delta z$. Upon limiting Δx, Δy, and Δz to zero, the equation becomes exact and reduces to

$$
\frac{\partial \sigma_{xx}}{\partial x} + \frac{\partial \tau_{yx}}{\partial y} + \frac{\partial \tau_{zx}}{\partial z} + \rho(\mathbf{f}_B)_x = \rho\frac{DV_x}{Dt} \qquad (3.10)
$$

Similarly,

$$\frac{\partial \tau_{xy}}{\partial x} + \frac{\partial \sigma_{yy}}{\partial y} + \frac{\partial \tau_{zy}}{\partial z} + \rho(\mathbf{f}_B)_y = \rho \frac{DV_y}{Dt} \tag{3.11}$$

$$\frac{\partial \tau_{xz}}{\partial x} + \frac{\partial \tau_{yz}}{\partial y} + \frac{\partial \sigma_{zz}}{\partial z} + \rho(\mathbf{f}_B)_z = \rho \frac{DV_z}{Dt} \tag{3.12}$$

which are the *equations of motion for the fluid particle.*

Example 3.2

What is the pressure distribution in the tank shown in Fig. 3.6 if it is filled with a liquid of uniform density ρ? The fluid is in static equilibrium. The only body force is gravity as shown, and the surface of the fluid is open to the atmosphere.

Solution. A selection of the coordinate axes is made as shown in the figure. Since the fluid is in static equilibrium, $V_x = V_y = V_z = 0$, for all times and at all points in the fluid. The shear stresses are known to be zero since there is no motion of the fluid. The absence of shear insures isotropy of the normal stress so that $\sigma_{xx} = \sigma_{yy} = \sigma_{zz} = -p$. The body force has only a z component in the negative direction, so $(\mathbf{f}_B)_z = -g$. Equations (3.10), (3.11), and (3.12) specialize to

$$\frac{\partial p}{\partial x} = 0 \qquad \frac{\partial p}{\partial y} = 0 \qquad \frac{\partial p}{\partial z} + \rho g = 0$$

Since p is not a function of x or y, it must be a function of z only, and

$$\frac{dp}{dz} + \rho g = 0$$

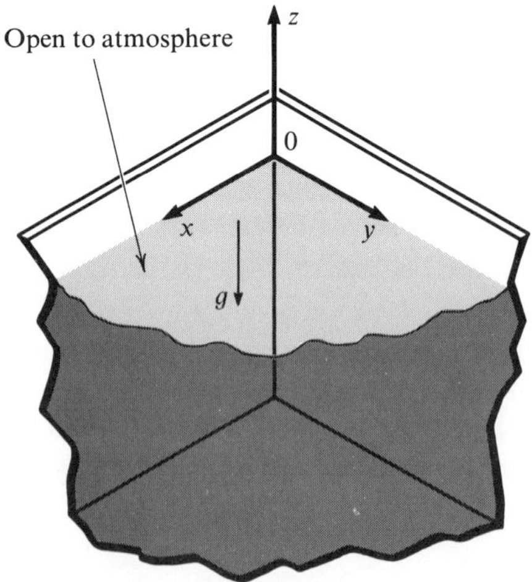

Fig. 3.6

Integrating this, assuming g independent of z, and recalling that ρ was given as uniform, one obtains

$$p = p_0 - \rho g z$$ ◀

in which p_0 is the value of the pressure at the free surface where $z = 0$. This is the required pressure distribution, which allows calculation of the pressure at any point in the fluid.

Example 3.3

What is the pressure distribution in a static isothermal atmosphere, if the pressure, temperature, and density at sea level ($z = 0$) are p_0, T_0, and ρ_0, respectively? Assume the local acceleration of gravity g to be independent of z; also, consider the atmosphere to be a perfect gas.

Solution. Equations (3.10), (3.11), and (3.12) again reduce to $dp/dz + \rho g = 0$ (refer to Example 3.2). Integration of this ordinary differential equation requires knowing the pressure as a function of ρ, since ρ was not given as a constant. For a perfect gas at constant temperature, it is known that $p/\rho = p_0/\rho_0 = C$, a constant, or $\rho = p/C$. Substituting for ρ in the differential equation, one obtains

$$\frac{dp}{p} = -\frac{g}{C}\,dz$$

Integrating between the limits $z = 0$ (where $p = p_0$) and z in general results in $\ln(p/p_0) = -gz/C$. From the equation of state for a perfect gas, $C = p_0/\rho_0 = RT_0$, where R is the specific gas constant. The pressure distribution may then be written as

$$p = p_0 e^{-gz/RT_0}$$ ◀

If the final result of this problem were applied to a situation where z is "very large," then g would have to be expressed as a function of z in the differential equation before the integration is made.

Example 3.4

Calculate the total resultant force of the water on the rectangular gate shown in Fig. 3.7. The dimensions of the gate are 5 ft by 10 ft. The density of the fluid may be taken as constant at 62.4 lb_m/ft^3, and the local acceleration due to gravity is constant at 32.2 ft/sec^2. The pressure at the free surface is 14.7 lb_f/in^2.

Solution. Equations (3.10), (3.11), and (3.12) reduce to $dp/dz - \rho g = 0$. For constant ρ and g, the integration yields $p = \rho g z + p_0$ where p_0 is the value of the pressure at $z = 0$, the free surface. Since the pressure varies over the surface of the gate (p is a function of z), a small element of area on the gate is considered such that the pressure "almost" remains constant over this area. This element is dA shown in Fig. 3.8, and dF is the elemental force acting on this typical strip of area. Hence

$$dF = p\,dA = (\rho g z + p_0)\,dA$$

To find the total force on the gate, all of the dF's need to be summed over the area of the gate, which requires integration of dF over A, the area of the gate. Each of these incremental forces is a vector, but each is perpendicular to the plane of the gate (a fluid at rest supports no shear); therefore the incremental forces are added algebraically since they constitute a parallel force system. This explains the absence of any vector designation of dF. Proceeding to integrate over the area A, and noting that $dA = 10\,dz/\sin\theta$,

$$F = \int_A dF = \int_A (\rho g z + p_0)\,dA = \int_6^{6+5\sin\theta} (\rho g z + p_0)\frac{10\,dz}{\sin\theta}$$

$$= \frac{10}{\sin\theta}\left[\frac{\rho g z^2}{2} + p_0 z\right]_6^{6+5\sin\theta}$$

Noting that $6 + 5\sin\theta = 6 + 2.5 = 8.5$,

$$F = \frac{10\text{ ft}}{0.5}\left[\frac{(62.4\text{ lb}_m/\text{ft}^3)(32.2\text{ ft/sec}^2)(8.5\text{ ft})^2}{2(32.2\text{ ft-lb}_m/\text{lb}_f\text{-sec}^2)} + \left(14.7\frac{\text{lb}_f}{\text{in}^2}\right)(8.5\text{ ft})\left(144\frac{\text{in}^2}{\text{ft}^2}\right)\right]$$

$$- \frac{10\text{ ft}}{0.5}\left[\frac{(62.4\text{ lb}_m/\text{ft}^3)(32.2\text{ ft/sec}^2)(6\text{ ft})^2}{2(32.2\text{ ft-lb}_m/\text{lb}_f\text{-sec}^2)} + \left(14.7\frac{\text{lb}_f}{\text{in}^2}\right)(6\text{ ft})\left(144\frac{\text{in}^2}{\text{ft}^2}\right)\right]$$

$$= 128{,}400\text{ lb}_f$$

The direction of this force is normal to the plate and has the same sense as dF shown in Fig. 3.8. The reaction of the gate on the water is equal and opposite to this, of course. The water is in equilibrium under the action of compressive stresses (pressure in this case) and body forces due to gravity.

The location of this resultant force may be found by summing moments of all incremental forces dF about some point and equating this sum to the moment of the resultant force about the same point, that is,

$$\mathbf{r}_R \times \mathbf{F} = \int_F \mathbf{r} \times d\mathbf{F} = \int_A \mathbf{r} \times p\,d\mathbf{A}$$

Fig. 3.7

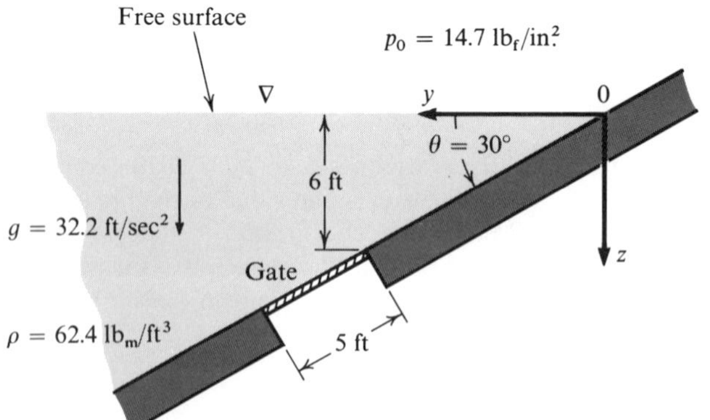

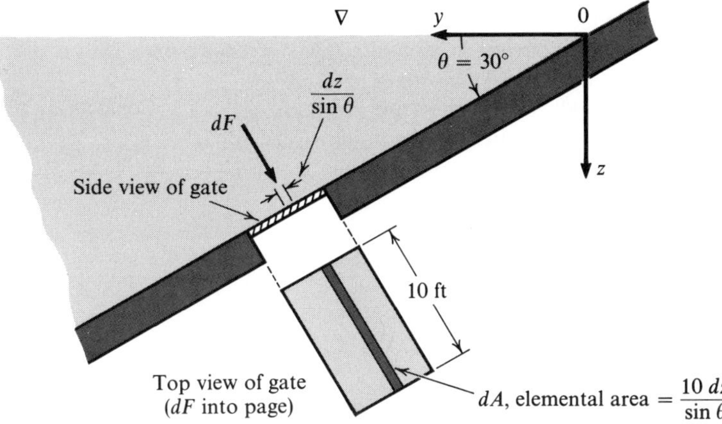

Fig. 3.8

where $\mathbf{r}_R$ is a position vector of the resultant force. Problems of this nature are generally classified as *fluid statics*. Since one has usually analyzed such problems in engineering mechanics in finding the resultant force due to a continuously distributed load, no special emphasis will be placed on this phase of fluid mechanics. The only way the fluid entered into this problem was in the observation that it could support no shear when at rest relative to its confining walls.

Example 3.5

Calculate the pressure difference $p_A - p_B$, when the total deflection of the indicating fluid is observed to be 4 in. as illustrated in Fig. 3.9. The acceleration due to gravity

Fig. 3.9

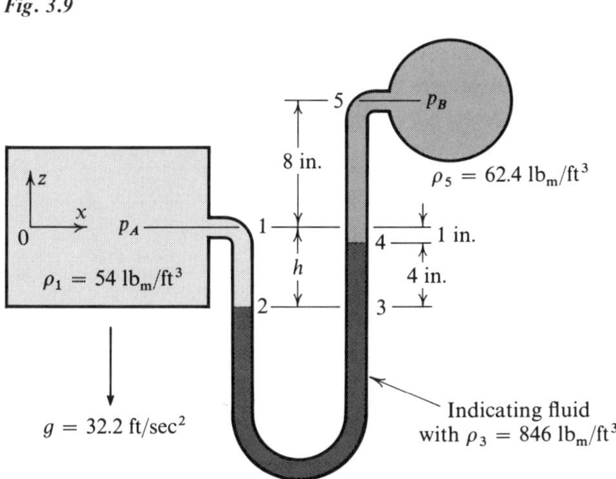

is 32.2 ft/sec.² All three fluids are at rest and have uniform and different densities. Several points within the fluids have been numbered for reference purposes.

Solution. Since points A and 1 are at the same elevation, $p_A = p_1$. Considering the column of fluid between 1 and 2, with $z = 0$ at point A, one applies

$$dp = -\rho_1 g\, dz$$

$$\int_1^2 dp = -\rho_1 g \int_0^{-h} dz = \rho_1 g h$$

or

$$p_2 = p_1 + \rho_1 g h = p_A + \rho_1 g h = p_A + \rho_1 g \frac{5}{12}$$

Examination of the fluid between 2 and 3 reveals that $p_2 = p_3$ since 2 and 3 are at the same elevation. This results in p_3 being expressed in terms of p_A and given quantities.

It is desirable to find p_B in terms of p_3. Noting that $p_B = p_5$, and considering the fluid between 5 and 4,

$$\int_5^4 dp = p_4 - p_5 = -\rho_5 g \int_5^4 dz = \rho_5 g \frac{9}{12}$$

and

$$p_4 = p_5 + \frac{9}{12}\rho_5 g = p_B + \frac{9}{12}\rho_5 g$$

For the fluid between 4 and 3,

$$\int_4^3 dp = p_3 - p_4 = -\rho_3 g \int_4^3 dz = \rho_3 g \frac{4}{12}$$

and

$$p_3 = p_4 + \frac{4}{12}\rho_3 g = p_B + \frac{9}{12}\rho_5 g + \frac{4}{12}\rho_3 g$$

Equating the expressions for p_2 and p_3, one obtains

$$p_A + \frac{5}{12}\rho_1 g = p_B + \frac{9}{12}\rho_5 g + \frac{4}{12}\rho_3 g$$

Solving for $p_A - p_B$ yields

$$p_A - p_B = \frac{9}{12}\rho_5 g + \frac{4}{12}\rho_3 g - \frac{5}{12}\rho_1 g$$

$$= \frac{32.2 \text{ ft/sec}^2}{12 \text{ in./ft } (32.2 \text{ ft-lb}_m/\text{lb}_f\text{-sec}^2)}[9 \text{ in. } (62.4 \text{ lb}_m/\text{ft}^3)$$

$$+ 4 \text{ in. } (846 \text{ lb}_m/\text{ft}^3) - 5 \text{ in. } (54 \text{ lb}_m/\text{ft}^3)]$$

$$= 306 \text{ lb}_f/\text{ft}^2$$

◀

An instrument of the type described in the preceding example is called a *manometer*. The calculation of the pressure difference is a lengthy task, but one will gain some skill by practice provided through the problems at the end of this chapter. No attempt should be made to memorize equations that apply to a specific situation. A better procedure is to start from basic principles and develop one's solution.

Another integration of the equations of motion may be made when the fluid is flowing, but with the assumption that shear stresses are absent. Such a fluid field will be referred to as an *ideal fluid field* throughout the text. With no shear stresses present, Eqs. (3.10), (3.11), and (3.12) reduce to the following:

$$-\frac{1}{\rho}\frac{\partial p}{\partial x} + (\mathbf{f}_B)_x = \frac{DV_x}{Dt}$$

$$-\frac{1}{\rho}\frac{\partial p}{\partial y} + (\mathbf{f}_B)_y = \frac{DV_y}{Dt}$$

$$-\frac{1}{\rho}\frac{\partial p}{\partial z} + (\mathbf{f}_B)_z = \frac{DV_z}{Dt}$$

These three scalar equations can be written as the single vector expression called *Euler's equation*:

$$-\mathbf{grad}\, p + \rho\mathbf{f}_B = \rho\frac{D\mathbf{V}}{Dt} \tag{3.13}$$

This result is obtained by multiplying each scalar equation of motion by the appropriate unit vector **i**, **j**, **k** and then adding the results. If the body force field $\mathbf{f}_B$ is conservative, then it can be defined by the gradient of a scalar body-force potential (designate this potential by $-\Phi_B$), thus $\mathbf{f}_B = -\mathbf{grad}\,\Phi_B$. Equation (3.13) may now be written as

$$-\mathbf{grad}\, p - \rho\,\mathbf{grad}\,\Phi_B = \rho\frac{D\mathbf{V}}{Dt} \tag{3.14}$$

This equation may be integrated under several restrictions. Consider first its integration along a streamline when the velocity field is steady. Let $d\mathbf{l}$ be a small displacement along a streamline shown in Fig. 3.10, and form the scalar product of Eq. (3.14) and $d\mathbf{l}$ to obtain

$$-\mathbf{grad}\, p \cdot d\mathbf{l} - \rho\,\mathbf{grad}\,\Phi_B \cdot d\mathbf{l} = \rho\frac{D\mathbf{V}}{Dt}\cdot d\mathbf{l}$$

As $\mathbf{grad}\, p$ represents the maximum directional change in p, $\mathbf{grad}\, p \cdot d\mathbf{l} = dp$ (the change in p along l, the streamline). Then the term $D\mathbf{V}/Dt \cdot d\mathbf{l}$ may be simplified by noting that $D\mathbf{V}/Dt = V(\partial\mathbf{V}/\partial l)$ because $\mathbf{V}$ is steady. Since the velocity vector in a steady field can only change along the streamline,

$$\frac{D\mathbf{V}}{Dt}\cdot d\mathbf{l} = V\frac{\partial\mathbf{V}}{\partial l}\cdot d\mathbf{l} = V\,dV$$

in which V is the magnitude of the velocity. The equation of motion may now be written as

$$-\frac{dp}{\rho} - d\Phi_B = V\,dV$$

Transferring all terms to one side of the equation and integrating between points 1 and 2 on the streamline shown in Fig. 3.10, one obtains

$$\frac{V_2^2 - V_1^2}{2} + \Phi_{B_2} - \Phi_{B_1} + \int_1^2 \frac{dp}{\rho} = 0 \qquad (3.15)$$

This is one form of *Bernoulli's equation* and it accrued as a consequence of integrating Eq. (3.14) along a streamline. The body force commonly considered in elementary fluid mechanics is gravity. If the negative z axis is arbitrarily selected in the direction of g, the local acceleration due to gravity, then Φ_B becomes gz. (Note: $-\mathbf{grad}\; gz = -g\mathbf{k}$, which is the body force per unit mass for g in the negative z direction.) This leads to the specialization

$$\frac{V_2^2 - V_1^2}{2} + g(z_2 - z_1) + \int_1^2 \frac{dp}{\rho} = 0 \qquad (3.16)$$

If in addition the density is assumed uniform and steady, then it may be removed from the integrand, and a further specialization accrues as

$$\frac{V_2^2 - V_1^2}{2} + g(z_2 - z_1) + \frac{p_2 - p_1}{\rho} = 0 \qquad (3.17)$$

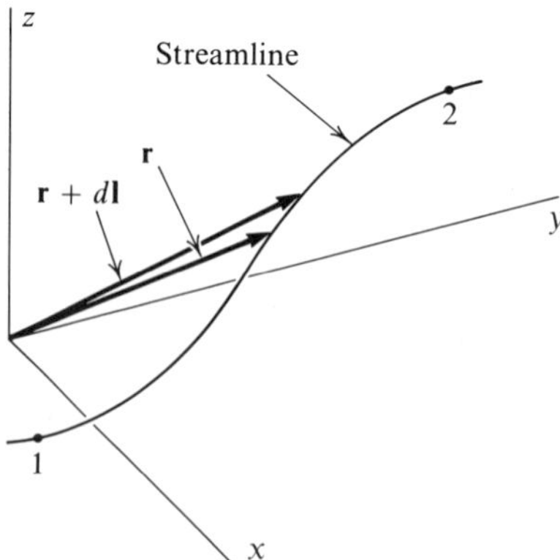

Fig. 3.10 Path of integration taken along the streamline.

When ρ is not constant, one must know how ρ varies with p. A relation of the form $\rho = \rho(p)$ is referred to as a *barotropic relation*.

Equations (3.16) and (3.17) were obtained by integration along a streamline (any streamline) in a steady flow field. Upon rearrangement, Eq. (3.17) may be written as

$$\frac{V_1^2}{2} + gz_1 + \frac{p_1}{\rho} = \frac{V_2^2}{2} + gz_2 + \frac{p_2}{\rho}$$

or

$$\frac{V^2}{2} + gz + \frac{p}{\rho} = C \tag{3.18}$$

in which C is a constant along a given streamline; however, C can vary for different streamlines. The sum of the three terms in Eq. (3.18) is called the *Bernoulli constant*.

Example 3.6

The nozzle shown in Fig. 3.11 has a fluid of uniform and steady density passing through it in steady flow. The velocity field is one dimensional at sections 1 and 2 and shear stresses are negligible. Show that the Bernoulli constant is the same for all streamlines at section 1, and then calculate the mass rate of flow.

Solution. The given information concerning the density and the continuity equation requires that $\partial V_x/\partial x = 0$ at section 1 where V_y and V_z are both zero. This leads to the conclusion that $D\mathbf{V}/Dt$ is zero, and Euler's equation (3.13) may be written as

$$-\mathbf{grad}\, p + \rho(-g\mathbf{k}) = 0 \qquad \text{at any point on section 1}$$

From this equation,

$$\frac{\partial p}{\partial x} = \frac{\partial p}{\partial y} = 0 \qquad \text{and} \qquad \frac{\partial p}{\partial z} = -\rho g$$

Fig. 3.11

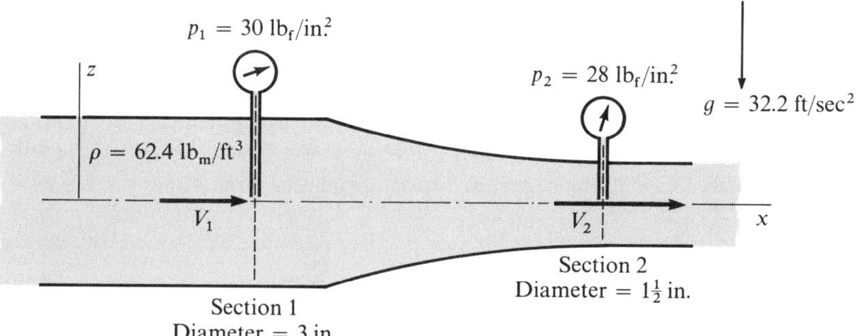

which implies that p is a function of z only, and

$$\frac{dp}{dz} = -\rho g$$

This last equation may be integrated to give

$$p = -\rho g z + \text{a constant}$$

so that

$$\frac{p}{\rho} = -gz + \text{another constant}$$

The above indicates that the sum of p/ρ and gz is constant over section 1. Since V_1 is uniform over this section, the Bernoulli constant is the same for all streamlines passing through section 1. Similar reasoning may be applied at section 2 to obtain the same conclusion at that location. As a consequence, any streamline may be considered typical; and the one coincident with the centerline of the nozzle will be used since the pressure is known at two points on this streamline.

The conservation of matter expressed in the form of Eq. (2.1) may be applied to a control volume defined by sections 1 and 2 and the conduit walls. This yields $A_1 V_1 = A_2 V_2$. Using this with the Bernoulli equation (3.17) to eliminate V_1, and solving for V_2, gives

$$V_2 = \left[\frac{2}{\rho}\,\frac{p_1 - p_2}{1 - (A_2/A_1)^2}\right]^{1/2}$$

$$= \left[\frac{2}{62.4\ \text{lb}_m/\text{ft}^3}\left(\frac{30\ \text{lb}_f/\text{in.}^2 - 28\ \text{lb}_f/\text{in.}^2}{1 - 1/16}\right)(144\ \text{in.}^2/\text{ft}^2)\left(32.2\,\frac{\text{ft-lb}_m}{\text{lb}_f\text{-sec}^2}\right)\right]^{1/2}$$

$$= 17.8\ \text{ft/sec}$$

The mass rate of flow is given by

$$\dot{m} = \int_{A_2} \rho \mathbf{V} \cdot d\mathbf{S} = \rho_2 A_2 V_2 = \left(62.4\,\frac{\text{lb}_m}{\text{ft}^3}\right)\left(\frac{\pi}{4}\right)\left(\frac{1.5}{12}\ \text{ft}\right)^2\left(17.8\,\frac{\text{ft}}{\text{sec}}\right)$$

$$= 13.6\ \text{lb}_m/\text{sec}$$

◀

Example 3.7

The nozzle in Example 3.6 has air flowing through it, and this fluid is assumed to be a perfect gas with a specific gas constant $R = 53.3$ ft-lb$_f$/lb$_m$-°R, and a specific heat ratio of $\gamma = 1.4$. Assume that the entropy of the air is uniform and steady for the flow, and that $p_1 = 100$ lb$_f$/in.^{2} abs, $T_1 = 200$°F, and $p_2 = 60$ lb$_f$/in.^{2} abs. Calculate the velocity V_2 using Eq. (3.16) and integrating along the central streamline.

Solution.

$$\frac{V_2^2 - V_1^2}{2} + g(z_2 - z_1) + \int_1^2 \frac{dp}{\rho} = 0 \qquad (3.16)$$

In this example, ρ is not removed from the integrand as one would expect the density to change as the pressure changes from 100 $lb_f/in.^2$ to 60 $lb_f/in.^2$. To integrate the third term above, there is a need for a barotropic relation. Thermodynamics supplies a pressure-density relation for the isentropic change of state of a perfect gas, which is given by $p = \rho^\gamma C$ where C is a constant. Substitution for ρ yields

$$\int_1^2 \frac{dp}{\rho} = C^{1/\gamma} \int_1^2 \frac{dp}{p^{1/\gamma}}$$

Carrying out the integration and using

$$C^{1/\gamma} = \frac{p_1^{\ 1/\gamma}}{\rho_1} = \frac{p_2^{\ 1/\gamma}}{\rho_2}$$

gives

$$\int_1^2 \frac{dp}{\rho} = \frac{\gamma}{\gamma - 1}\left(\frac{p_2}{\rho_2} - \frac{p_1}{\rho_1}\right)$$

One of the velocities may be eliminated by use of conservation of matter: $\rho_1 A_1 V_1 = \rho_2 A_2 V_2$. Then solving for V_1 and squaring the result gives

$$V_1^{\ 2} = \left(\frac{\rho_2}{\rho_1}\right)^2 \left(\frac{A_2}{A_1}\right)^2 V_2^{\ 2}$$

For the configuration shown in Fig. 3.11, note that $g(z_2 - z_1) = 0$. By substituting for $V_1^{\ 2}$ and $\int_1^2 dp/\rho$, the integrated equation of motion becomes

$$\frac{V_2^{\ 2} - V_2^{\ 2}(\rho_2/\rho_1)^2(A_2/A_1)^2}{2} + \frac{\gamma}{\gamma - 1}\left(\frac{p_2}{\rho_2} - \frac{p_1}{\rho_1}\right) = 0$$

and

$$V_2 = \left[\frac{2\gamma(p_1/\rho_1 - p_2/\rho_2)/(\gamma - 1)}{1 - (\rho_2/\rho_1)^2(A_2/A_1)^2}\right]^{1/2}$$

By use of $p = \rho RT$ and $p_2/\rho_2^{\ \gamma} = p_1/\rho_1^{\ \gamma}$, the last expression can be written as

$$V_2 = \left\{\frac{[2\gamma RT_1/(\gamma - 1)][1 - (p_2/p_1)^{(\gamma-1)/\gamma}]}{1 - (p_2/p_1)^{2/\gamma}(A_2/A_1)^2}\right\}^{1/2}$$

which gives the desired velocity in terms of the given data.

$$V_2 = \left\{\frac{\left(\frac{2.8}{0.4}\right)\left(53.3\,\frac{\text{ft-lb}_f}{\text{lb}_m\,°\text{R}}\right)(660°\text{R})\left[1 - \left(\frac{60\ \text{lb}_f/\text{in.}^2}{100\ \text{lb}_f/\text{in.}^2}\right)^{0.4/1.4}\right]\left(\frac{32.2\ \text{ft-lb}_m}{\text{lb}_f\text{-sec}^2}\right)}{1 - \left(\frac{60\ \text{lb}_f/\text{in.}^2}{100\ \text{lb}_f/\text{in.}^2}\right)^{2/1.4}\left(\frac{1}{16}\right)}\right\}^{1/2}$$

$$= 1055\ \text{ft/sec}$$

◀

3.6 Closing Remarks

The two preceding examples applied to fluids moving with no shear stresses present. The first treated a fluid of uniform and steady density and the other involved a compressible fluid with a given barotropic relation. Some question naturally arises concerning the handling of Eqs. (3.10), (3.11), and (3.12) when shear stresses are present. One must thus consider how to relate the stresses to rates of strain. Such relations are referred to as constitutive equations and are treated in the next chapter.

Self-Study Questions

3.1 In the definition of normal stress, is the limit of ΔS taken to zero in the continuum sense as in the definition of density?

3.2 State a difference between surface and body forces.

3.3 Does a nonconservative force only depend on a spatial position in a field?

3.4 What is an expression for the gravity force acting on a fluid particle on a per unit volume basis?

3.5 A force of $\mathbf{V} \times \mathbf{B}$ acts on an electrically charged particle moving at a velocity $\mathbf{V}$ in a magnetic field of magnetic induction $\mathbf{B}$. Write an integral expression for the force acting on a system of such fluid particles using a charge density ρ_e.

3.6 In defining a sign convention for stresses,
a. What is the assigned positive direction for a surface area?
b. What is the assigned direction for the surface stress force?
c. What combination(s) of area and surface force directions yield a negative stress?

3.7 Is a positive normal stress tensile or compressive in character?

3.8 What quantities provide a description of the state of stress in a fluid?

3.9 Discuss the following with respect to Eqs. (3.4), (3.5), and (3.6):
a. What do σ and τ represent?
b. From what basic law are these equations derived?
c. In these expressions relating σ and τ to normal and shear stresses on surfaces oriented in the direction of coordinate axes, does the result apply to a fluid that is accelerating? To a fluid with a body force?

3.10 Establish the validity of Eq. (3.8) by taking the scalar product of two orthogonal unit vectors.

3.11 Is it possible for $\sigma_{xx} = \sigma_{yy} = \sigma_{zz}$ in a fluid
a. In steady flow with nonzero shear stresses?
b. In accelerating flow with nonzero shear stresses?
c. In steady flow with zero shear stresses everywhere?
d. In unsteady flow with zero shear stresses everywhere?
e. In accelerating flow with zero shear stresses everywhere, but in a gravity field?
f. Not in motion?

3.12 Consider an infinitesimal element of volume $\Delta x\ \Delta y\ \Delta z$ in the positive xyz octant.
a. Show and label all surface and body forces.
b. Explain why the labeled stresses on the surface represent "approximate" mean values.

3.13 *a.* Write the equation of motion (in the x direction only) of a fluid particle in any fluid flow field.
b. What are the basic dimensions of each term in the answer to (*a*)?
c. State a physical meaning of each term in the answer to (*a*).

3.14 What term(s) in the equation of motion for a fluid particle are zero everywhere in a fluid field when
a. The flow is steady?
b. The density is uniform and steady in value?
c. The flow is parallel and in the x direction?
d. There is no flow and the force of gravity is in the negative z direction? No other body forces exist.
e. Shear stress forces are negligible compared to other forces?
f. The velocity field is steady, in the x direction, and one dimensional in the y direction?
g. The flow field is one dimensional (in x) and motion is in the x direction only?
h. The velocity and stress field is two dimensional (in x and y) and body forces are negligible?

3.15 Describe a stress and velocity field to which the following reduced equations of motion apply:

a.
$$\frac{\partial \sigma_{xx}}{\partial x} = \rho\left(V_x\frac{\partial V_x}{\partial x} + V_y\frac{\partial V_x}{\partial y}\right)$$
$$\frac{\partial \sigma_{yy}}{\partial y} = \rho\left(V_x\frac{\partial V_y}{\partial x} + V_y\frac{\partial V_y}{\partial y}\right)$$
$$\frac{\partial \sigma_{zz}}{\partial z} = 0$$

b.
$$\frac{\partial \sigma_{xx}}{\partial x} + \frac{\partial \tau_{yx}}{\partial y} = \rho V_y\frac{\partial V_x}{\partial y}$$
$$\frac{\partial \sigma_{yy}}{\partial y} + \frac{\partial \tau_{xy}}{\partial x} - \rho g = \rho V_y\frac{\partial V_y}{\partial y}$$
$$\frac{\partial \tau_{xz}}{\partial x} + \frac{\partial \tau_{yz}}{\partial y} = 0$$

c.
$$\frac{\partial \sigma_{xx}}{\partial x} + \frac{\partial \tau_{yx}}{\partial y} = \rho\left(V_x\frac{\partial V_x}{\partial x} + \frac{\partial V_x}{\partial t}\right)$$
$$\frac{\partial \tau_{xy}}{\partial x} + \frac{\partial \sigma_{yy}}{\partial y} = 0$$
$$\frac{\partial \tau_{xz}}{\partial x} + \frac{\partial \tau_{yz}}{\partial y} = 0$$

d.
$$\frac{dp}{dz} + \rho g = 0$$

e.
$$-\frac{1}{\rho}\,\mathbf{grad}\,p - g\mathbf{k} = \frac{D\mathbf{V}}{Dt}$$

3.16 *a.* State any restrictions concerning the application of Euler's equation as opposed to Eqs. (3.10), (3.11), and (3.12).
b. Discuss in words how the Bernoulli equation (3.15) is obtained.
c. Does the Bernoulli equation apply to an unsteady flow field?
d. Describe a flow field in which the Bernoulli constant does not vary spatially.
e. If any part of a steady flow field (containing all of the streamlines) has the same value as the Bernoulli constant, what may be concluded about the variation of the Bernoulli constant in the remainder of the field?

3.17 If a given ideal fluid is in a gaseous form, does this convey the same meaning as a definition of a perfect (sometimes called ideal) gas in thermodynamics?

Problems

3.1 Sketch the following stresses on the appropriate faces of an infinitesimal cube located in the positive octant of the x, y, z axes.

$\sigma_{xx} = -2000 \quad \tau_{yx} = 5$

$\sigma_{yy} = -2005 \quad \tau_{xz} = -5$

3.2 Sketch the following stresses on an infinitesimal cube located in the positive octant of the x, y, z axes.

$\sigma_{xx} = -2000 \quad \tau_{yx} = -3$

$\sigma_{yy} = -2005 \quad \tau_{zx} = 5$

3.3 Calculate the value of the stress normal to a plane with direction cosines of $l = 0$, $m = \frac{1}{2}$, and $n = (3)^{1/2}/2$ for a flow field where

$\sigma_{xx} = \sigma_{yy} = \sigma_{zz} = -2100 \quad \tau_{yz} = \tau_{zy} = 0$

$\tau_{xz} = \tau_{zx} = 0 \quad \tau_{xy} = \tau_{yx} = 2$

3.4 Calculate the normal stress acting on a plane oriented such that its direction cosines have values of $l = (1/2)^{1/2}$ and $m = (1/2)^{1/2}$ at a point in the flow field where

$\sigma_{xx} = -398 \quad \tau_{zx} = \tau_{xz} = 3$

$\sigma_{yy} = -400 \quad \tau_{yz} = \tau_{zy} = 0$

$\sigma_{zz} = -402 \quad \tau_{xy} = \tau_{yx} = 3$

3.5 Calculate the normal stresses which act on surfaces with zero shear stresses if

$\sigma_{xx} = -198 \quad \tau_{zx} = \tau_{xz} = 0$

$\sigma_{yy} = -198 \quad \tau_{yz} = \tau_{zy} = 0$

$\sigma_{zz} = -200 \quad \tau_{xy} = \tau_{yx} = 1$

3.6 Determine the direction cosines for the normal stresses that act on surfaces with zero shear stresses if

$\sigma_{xx} = -1000 \quad \tau_{zx} = \tau_{xz} = 0$

$\sigma_{yy} = -1000 \quad \tau_{yz} = \tau_{zy} = 0$

$\sigma_{zz} = -1000 \quad \tau_{xy} = \tau_{yx} = 1$

3.7 Obtain an algebraic expression for the pressure at 30,000 ft above the earth's surface assuming that the temperature variation is linear at a rate of $-3.5°$F/1000 ft. Assume perfect-gas behavior for the static atmosphere.

3.8 With reference to Fig. 3.12, calculate the net force, due to the water and atmosphere, acting on the right triangular gate. The density of the water is 62.4 lb_m/ft^3.

Fig. 3.12

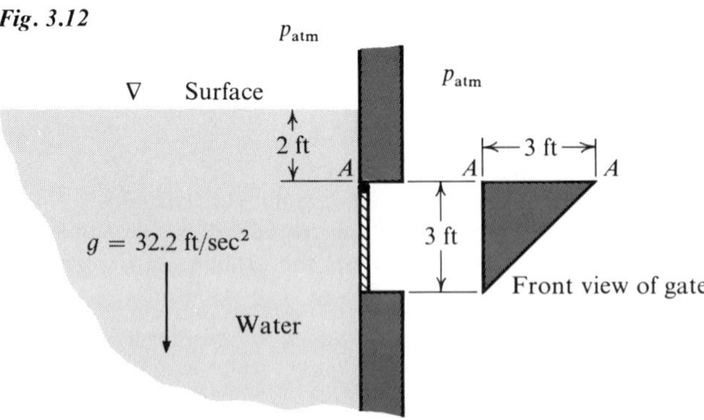

3.9 A hypothetical vertical wall extends 10,000 ft into the earth's atmosphere. The atmospheric temperature of 40° F is uniform. Estimate the total force exerted by the atmosphere on one side of the wall. The width is 10 ft. Assume perfect-gas behavior.

3.10 Calculate the minimum force F required to open the gate shown in Fig. 3.13, where the hinge axis is at point A,

Fig. 3.13

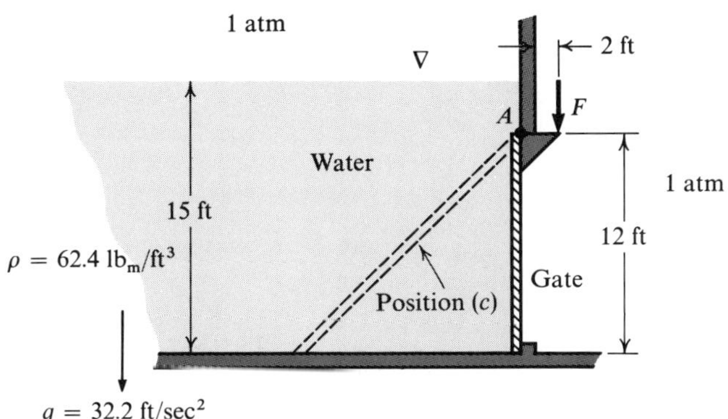

a. If the vertical gate is square.
b. If the vertical gate is an equilateral triangle hinged on one side.
c. If the rectangular gate is inclined at 45° to the horizontal. Use a 12-ft width.

3.11 A gate is placed in a wall at a submerged location in the static fluid as illustrated in Fig. 3.14. The fluid density is uniform and steady. The axis of rotation for the gate is horizontally oriented.

Fig. 3.14

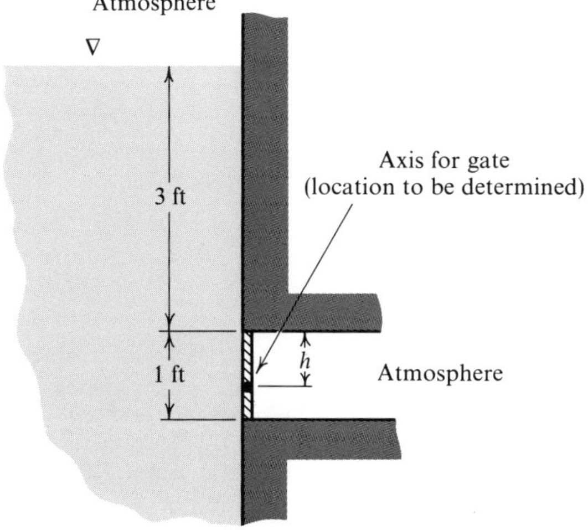

a. Locate the axis for minimum external torque to keep the gate closed (measured from the top of the gate) if the gate is square, 1 ft on a side.

b. Would the answer for (*a*) change if the fluid level changed?

3.12 A cube of solid material (1 ft on a side) is partially immersed in two fluids, labeled *A* and *B* in Fig. 3.15. The cube weighs 100 lb_f. Calculate the depth that it is immersed in fluid *B*.

Fig. 3.15

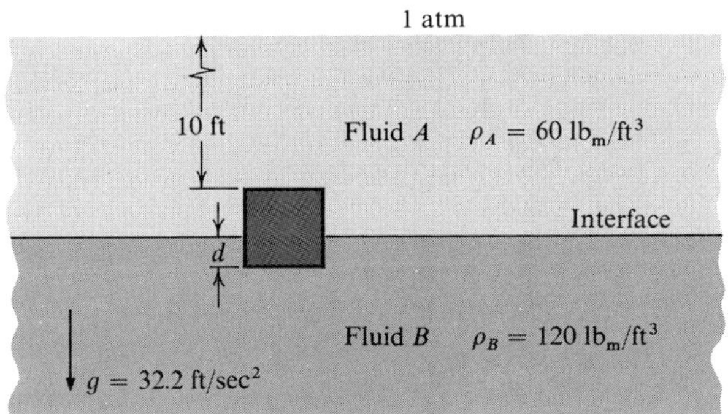

3.13 Calculate the pressure of the gas contained as shown in Fig. 3.16. The vapor pressure of the liquids is negligibly small.

Fig. 3.16

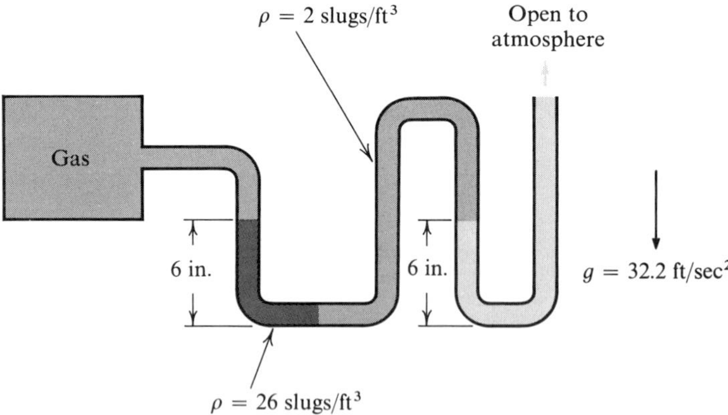

3.14 A tube is vertically oriented and completely filled with water. This tube is shown closed at one end and open at the other in Fig. 3.17. Find the value of acceleration in the x direction to make the pressure at *A* equal to zero gage pressure (1 atm) if the local acceleration of gravity is 28 ft/sec^2.

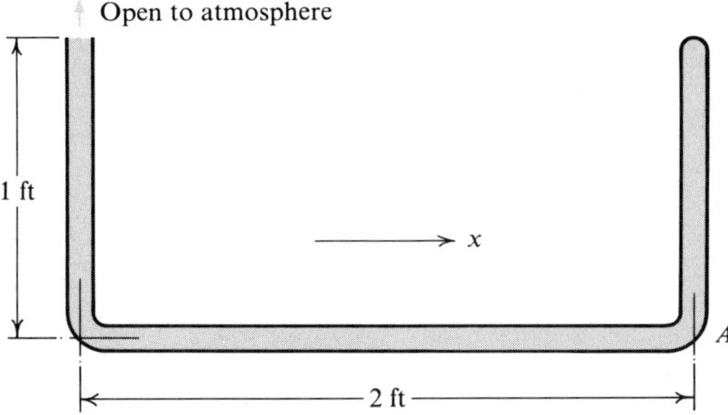

Fig. 3.17

3.15 A vertical glass column contains miscible fluids of different densities. The density of the fluid mixture is 40 lb_m/ft^3 at the top and increases linearly at a rate of 2 lb_m/ft^4. Calculate the density (assumed uniform) of a straight rod vertically oriented and immersed in such a fluid when three-fourths of the 1 ft length is submerged.

3.16 A tank is filled with a fluid (ρ = a constant) and the level is maintained at a fixed point. A number of small holes are punched in the wall in a vertical line. Which vertical variation in Fig. 3.18 represents the variation of the fluid exit velocity at the wall? Justify your answer.

Fig. 3.18

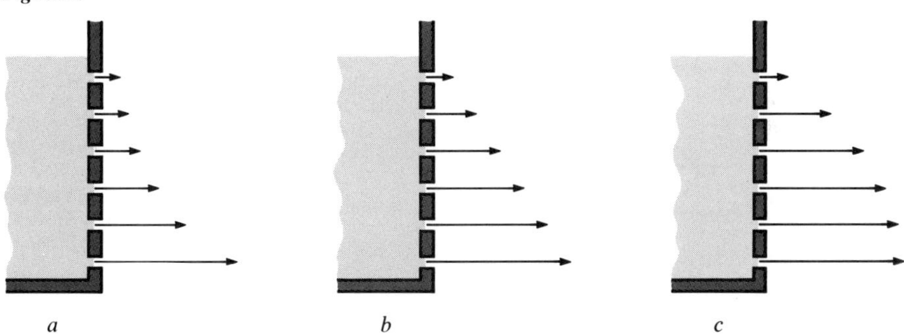

3.17 A fluid flows through a vertical pipe with changing cross-sectional area. The given information at two cross sections is

$p_1 = 5$ $lb_f/in.^2$ gage $\quad z_1 = 0$ ft $\quad V_1 = 10$ ft/sec

$p_2 = 1$ atm $\quad z_2 = 10$ ft $\quad g = 32.2$ ft/sec^2

a. Calculate the velocity at section 2 for a fluid with a density of 1.94 slug/ft^3 and negligible shear stress effects.

b. Calculate the velocity at section 2 for an ideal fluid with a barotropic relation of p/ρ = constant. The density at section 1 is 0.10 lb_m/ft^3.

3.18 Calculate the steady volume flow rate (ft^3/sec) for an ideal fluid passing through the reducer shown in Fig. 3.19 if the gage pressure difference is 3 lb_f/in^2. The fluid density is 1.94 slug/ft^3.

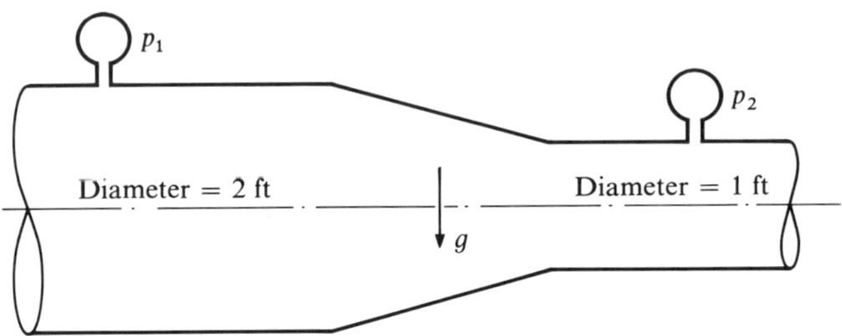

Fig. 3.19

3.19 An ideal fluid flows steadily through a branching pipe as shown in Fig. 3.20. Fluid properties are assumed uniform at each designated cross section (each at the same horizontal location). The fluid density is 1.9 slug/ft^3. Calculate the pressure at section 3.

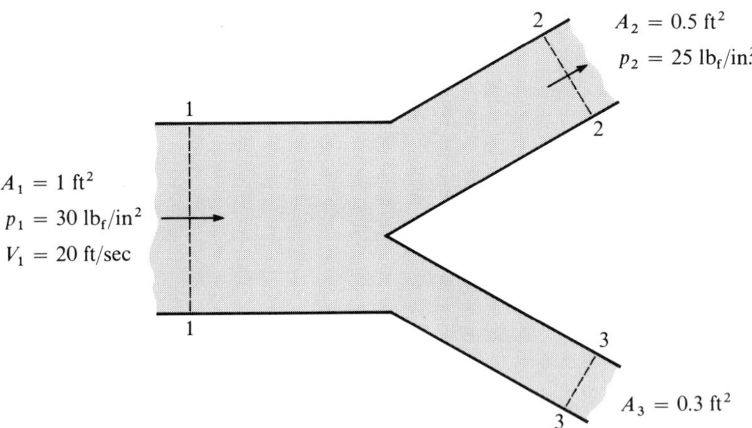

Fig. 3.20

3.20 Refer to Fig. 3.21 and estimate the velocity of the water leaving the nozzle. Assume the pressure at the exit is atmospheric.

3.21 Calculate the total vertical force (due to water and air) acting on the hemispherical surface located in the bottom of the tank as shown in Fig. 3.22.

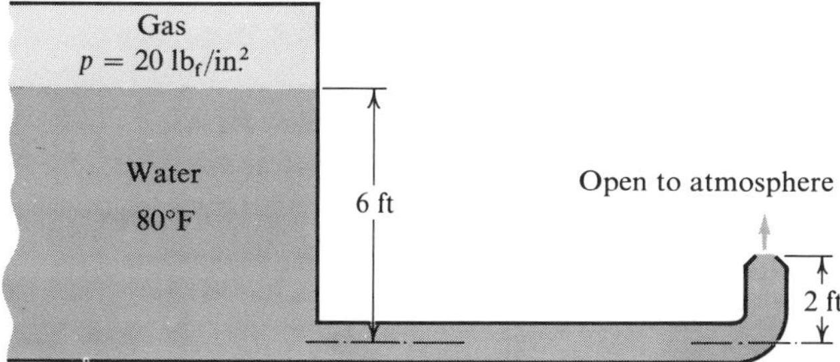

Fig. 3.21

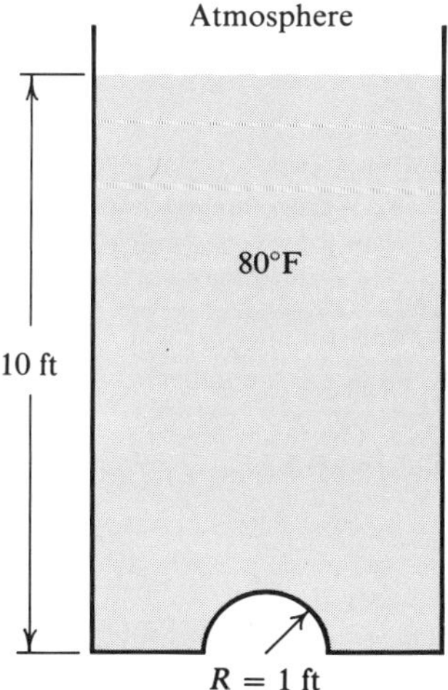

Fig. 3.22

3.22 Estimate the mass flow rate through the piping system shown for liquid water in Fig. 3.23.

3.23 Initially, air at a pressure of 30 lb_f/in^2 and at a temperature of 70°F is contained in a cubical chamber (3 ft on a side) with liquid water which fills two-thirds of the chamber. Assuming that the temperature of the air does not change, calculate the water level at which the flow could stop when a valve located at the bottom of the tank is open to the atmosphere.

3.24 Oil, with a density of 50 lb_m/ft^3 flows steadily through a pipe of changing area as indicated in Fig. 3.24. A mercury manometer is used to evaluate the flow. The density of mercury is 850 lb_m/ft^3 Estimate the mass flow rate in lb_m/sec.

3.25 Calculate the density of oil in a container with water ($\rho = 62.4\ lb_m/ft^3$) if a metal sphere with a radius of 1 in. and weighing 0.14 lb_f is observed to be half submerged in the water and oil.

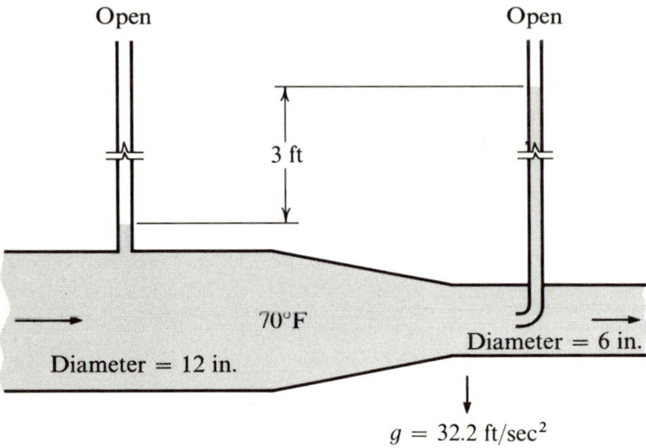

Fig. 3.23

3.26 A crop-dusting aircraft has "booms" that serve as manifolds along the wing, with 10 nozzles spaced on each side of the fuselage. It is desired that the pressure in the liquid be the same before each nozzle. Estimate the variation in area needed along the boom.

3.27 Estimate the maximum error in the answer to Prob. 3.24 if air were unknowingly trapped in the manometer where the oil is shown. Use air density of 0.1 lb_m/ft^3.

3.28 The water level is maintained as two streams exit from the side of the tank shown in Fig. 3.25. Estimate the horizontal distance from the origin where each stream of fluid will pass the x axis.

Fig. 3.24

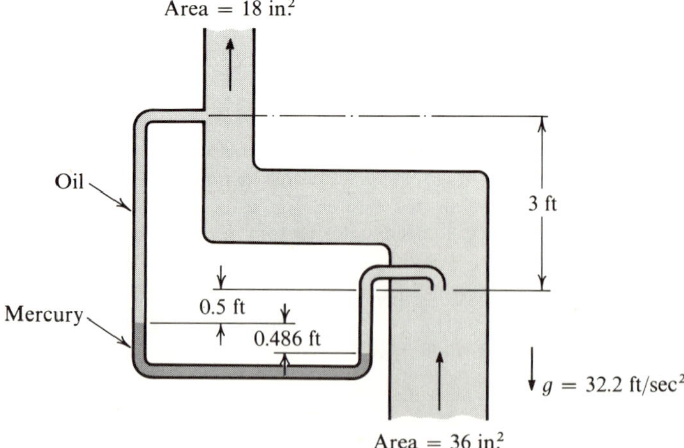

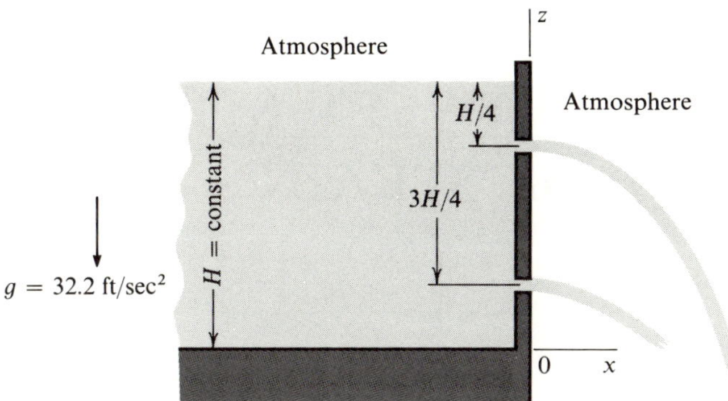

Fig. 3.25

3.29 Show that the net force of a static fluid on a submerged object is equal to the "weight" of the fluid displaced.

3.30 A two-dimensional "air-cushion" device illustrated in Fig. 3.26 causes the flow of air downward as shown at A. A pressure greater than 1 atm exists at B causing the flow to turn outward. The fluid is assumed ideal.

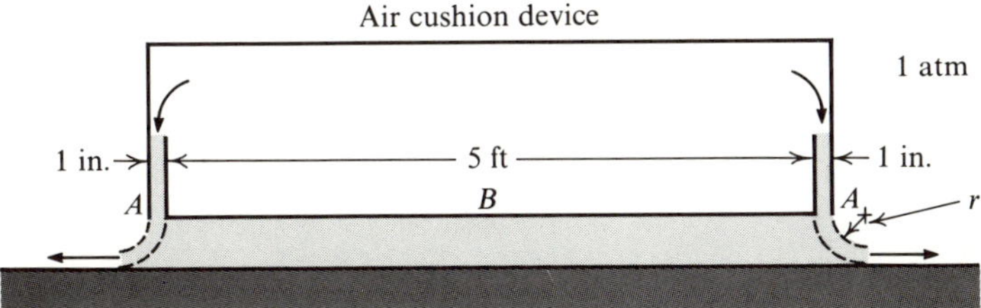

Fig. 3.26

a. Write Euler's equation in r, θ coordinates.

b. Assume that a uniform and steady velocity of 100 ft/sec exists at two locations A and that the density of the fluid is 0.1 lb_m/ft^3. Estimate the pressure along the bottom of the plate B if the flow turns in a circular path and this plate is 3 in. above the bottom solid surface.

c. Calculate the force exerted by the air along B on a per unit depth basis.

3.31 A fluid in an open container is given a constant acceleration in a horizontal direction. Obtain an expression for the slope of the free surface after the fluid becomes stationary relative to the container.

3.32 Fluid in a cylinder is rotated about its axis of symmetry with a constant angular velocity. Obtain an equation for the surface of the liquid.

4

RELATION BETWEEN STRESS AND RATE OF STRAIN

4.1 Introduction

Attention is now directed to the problem of finding the stress distribution in a moving fluid with shear stresses present. Equations (3.10), (3.11), (3.12), and (2.3) provide four equations. In rectangular cartesian coordinates they are

$$\frac{\partial \sigma_{xx}}{\partial x} + \frac{\partial \tau_{yx}}{\partial y} + \frac{\partial \tau_{zx}}{\partial z} + \rho(\mathbf{f}_B)_x = \rho \frac{DV_x}{Dt} \tag{3.10}$$

$$\frac{\partial \tau_{xy}}{\partial x} + \frac{\partial \sigma_{yy}}{\partial y} + \frac{\partial \tau_{zy}}{\partial z} + \rho(\mathbf{f}_B)_y = \rho \frac{DV_y}{Dt} \tag{3.11}$$

$$\frac{\partial \tau_{xz}}{\partial x} + \frac{\partial \tau_{yz}}{\partial y} + \frac{\partial \sigma_{zz}}{\partial z} + \rho(\mathbf{f}_B)_z = \rho \frac{DV_z}{Dt} \tag{3.12}$$

$$\frac{\partial \rho}{\partial t} + \frac{\partial(\rho V_x)}{\partial x} + \frac{\partial(\rho V_y)}{\partial y} + \frac{\partial(\rho V_z)}{\partial z} = 0 \tag{2.3}$$

An inventory of the dependent variables (unknowns) indicates that in general there are nine unknown stresses, three unknown velocity components, the density, and the body force field of three scalar components. Thus there are a total of 16 unknown scalar quantities, and the above four equations are obviously inadequate for determination of these quantities. The problem can be alleviated in two ways: a reduction in the number of unknowns by virtue of the given information in a problem and/or an introduction of more equations from basic laws and experimental evidence that relate the dependent variables.

The first task will be to show that there are only six distinct components required to define the state of stress. Next, constitutive relations will be introduced that relate stress to the space derivatives of velocity components. These derivatives are referred to as rates of strain, and their relation to stresses provides a set of constitutive relations. Studies in solid mechanics entail relating stress to strain. The constitutive law involved there is referred to as *Hooke's law*, which simply states that *stress is proportional to strain for the deformation of an elastic solid.* Without such a relation, it would not be possible to solve problems in solid mechanics. Brief reference to basic electrical science also indicates that a constitutive relation known as *Ohm's law* is used to relate electrical field intensity to current density. To summarize: physical phenomena associated with the behavior of material systems require some "cause and effect" laws or relations to describe the desired behavior. Fluid flow fields are no exception in this respect. It will be found that constitutive relations provide six equations between stresses and strain rates.

The latter portion of the chapter will be devoted to recognition of certain boundary conditions imposed on fluid flow fields, and the solution of some simple fluid flow examples with shear stresses considered.

4.2 Symmetry of the Stress Tensor

Consider a fluid element of volume $\Delta x\ \Delta y\ \Delta z$ in a flow field in which shear and normal stresses are present. If moments are taken about an axis parallel to the z axis and passing through the center of mass of the element, one may formalize[1] the sum of these as

$$M_z = I_{zz}\dot{\omega}_z + (I_{yy} - I_{xx})\,\omega_x\omega_y$$

in which I_{xx}, I_{yy}, and I_{zz} are the principal mass moments of inertia at the center of mass of the element; ω_x and ω_y are the angular velocity components about the x and y axis, respectively; $\dot{\omega}_z$ is the angular acceleration component

[1]Irving H. Shames, "Engineering Mechanics—Dynamics," 2d ed., p. 571, Prentice-Hall, Inc., Englewood Cliffs, N.J., 1966.

about the z axis; and M_z is the moment of all the forces about the axis parallel to the z axis and passing through the center of mass of the element. In terms of the dimensions of the element and the approximate values of surface forces acting on a finite-sized element, the preceding equation may be written as

$$\left(\tau_{xy} + \frac{\partial \tau_{xy}}{\partial x}\frac{\Delta x}{2}\right)\Delta y\,\Delta z\,\frac{\Delta x}{2} + \left(\tau_{xy} - \frac{\partial \tau_{xy}}{\partial x}\frac{\Delta x}{2}\right)\Delta y\,\Delta z\,\frac{\Delta x}{2}$$

$$-\left(\tau_{yx} + \frac{\partial \tau_{yx}}{\partial y}\frac{\Delta y}{2}\right)\Delta x\,\Delta z\,\frac{\Delta y}{2} - \left(\tau_{yx} - \frac{\partial \tau_{yx}}{\partial y}\frac{\Delta y}{2}\right)\Delta x\,\Delta z\,\frac{\Delta y}{2}$$

$$= \rho\,\Delta x\,\Delta y\,\Delta z\,\frac{(\Delta x)^2 + (\Delta y)^2}{12}\,\dot{\omega}_z + \left[\rho\,\Delta x\,\Delta y\,\Delta z\,\frac{(\Delta x)^2 + (\Delta z)^2}{12} - \rho\,\Delta x\,\Delta y\,\Delta z\,\frac{(\Delta y)^2 + (\Delta z)^2}{12}\right]\omega_x\omega_y$$

If this expression is simplified and put on a per unit volume basis by dividing through by $\Delta x\ \Delta y\ \Delta z$, one obtains

$$\tau_{xy} - \tau_{yx} = \rho\,\frac{(\Delta x)^2 + (\Delta y)^2}{12}\,\dot{\omega}_z + \rho\omega_x\omega_y\,\frac{(\Delta x)^2 - (\Delta y)^2}{12}$$

Taking the limit of both sides of this equation as Δx, Δy, and Δz approach zero makes the expression exact and yields the following result:

$$\tau_{xy} = \tau_{yx}$$

and similarly for the other axes

$$\tau_{yz} = \tau_{zy} \qquad (4.1)$$

$$\tau_{zx} = \tau_{xz}$$

It is now apparent that only six of the nine components of the stress tensor are distinct. Equations (4.1) establish the symmetry alluded to in the title of this section.

4.3 One Invariant of the Stress Tensor

There are certain invariants inherent in the state of stress in a continuum. The term *invariant* is employed here in the following sense: A quantity is said to be an invariant at a point in the field if this quantity is the same when evaluated with respect to any and all cartesian coordinate systems having their origins at the point. In brief, *invariance means "invariance to rotation of axes" at any point in the field.* It is not difficult to show that the sum of the three normal stresses at a point is invariant with respect to the rotation of axes. If the forces associated with the stresses shown in Fig. 3.4 are summed in a direction normal to the oblique face, the result is, after limiting the equation as $\Delta x \to 0$, $\Delta y \to 0$, and $\Delta z \to 0$,

$$\sigma = \sigma_{xx} l^2 + \sigma_{yy} m^2 + \sigma_{zz} n^2 + \tau_{xy} ml + \tau_{yx} ml + \tau_{xz} nl + \tau_{zx} nl + \tau_{yz} mn + \tau_{zy} mn$$

in which l, m, and n are the direction cosines of the normal to the oblique face and σ is the resultant normal stress on this oblique face. As a result of Eqs. (4.1), one may write the preceding equation as

$$\sigma = \sigma_{xx} l^2 + \sigma_{yy} m^2 + \sigma_{zz} n^2 + 2\,\tau_{xy} ml + 2\,\tau_{xz} nl + 2\,\tau_{yz} mn \tag{4.2}$$

Equation (4.2) may now be applied to three different planes passing through a given point in the fluid flow field. Consider the x,y,z coordinate system shown in Fig. 4.1. The primed coordinate system is referred to the x,y,z frame as follows:

let l_1, m_1, n_1 = respective direction cosines of x' axis
l_2, m_2, n_2 = respective direction cosines of y' axis
l_3, m_3, n_3 = respective direction cosines of z' axis

Then it follows that the respective direction cosines of the x, y, and z axes referred to the x',y',z' frame of reference are l_1, l_2, l_3; m_1, m_2, m_3; and n_1, n_2, n_3, respectively.

Now if σ is formulated as $\sigma_{x'x'}$ from Eq. (4.2), then $l = l_1$, $m = m_1$, $n = n_1$, and

$$\sigma_{x'x'} = \sigma_{xx} l_1{}^2 + \sigma_{yy} m_1{}^2 + \sigma_{zz} n_1{}^2 + 2\tau_{xy} l_1 m_1 + 2\tau_{xz} l_1 n_1 + 2\tau_{yz} m_1 n_1 \tag{4.3a}$$

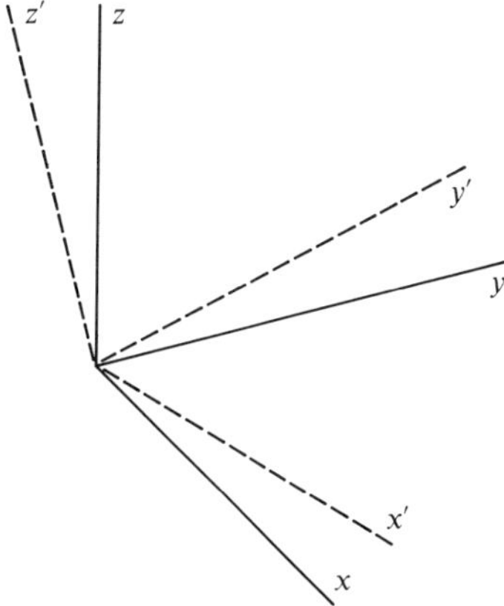

Fig. 4.1 The primed coordinate system x',y',z' rotated with respect to the x,y,z system.

Formulating σ as $\sigma_{y'y'}$ gives

$$\sigma_{y'y'} = \sigma_{xx}l_2^{\,2} + \sigma_{yy}m_2^{\,2} + \sigma_{zz}n_2^{\,2} + 2\tau_{xy}l_2m_2 + 2\tau_{xz}l_2n_2 + 2\tau_{yz}m_2n_2 \tag{4.3b}$$

Finally, considering σ as $\sigma_{z'z'}$ yields

$$\sigma_{z'z'} = \sigma_{xx}l_3^{\,2} + \sigma_{yy}m_3^{\,2} + \sigma_{zz}n_3^{\,2} + 2\tau_{xy}l_3m_3 + 2\tau_{xz}l_3n_3 + 2\tau_{yz}m_3n_3 \tag{4.3c}$$

Equations (4.3*a*,*b*,*c*) are now added to obtain

$$\begin{aligned}\sigma_{x'x'} + \sigma_{y'y'} + \sigma_{z'z'} &= (l_1^{\,2} + l_2^{\,2} + l_3^{\,2})\sigma_{xx} + (m_1^{\,2} + m_2^{\,2} + m_3^{\,2})\sigma_{yy} \\ &+ (n_1^{\,2} + n_2^{\,2} + n_3^{\,2})\sigma_{zz} + 2(l_1m_1 + l_2m_2 + l_3m_3)\,\tau_{xy} \\ &+ 2(l_1n_1 + l_2n_2 + l_3n_3)\tau_{xz} + 2(m_1n_1 + m_2n_2 + m_3n_3)\tau_{yz}\end{aligned}$$

Since l_1, l_2, and l_3 are the direction cosines of a given line, the coefficient of σ_{xx} is unity in the preceding expression. This is also true for the coefficients of σ_{yy} and σ_{zz}. The coefficients of the terms containing shear stresses are zero as was noted in Eq. (3.8). This establishes that

$$\sigma_{x'x'} + \sigma_{y'y'} + \sigma_{z'z'} = \sigma_{xx} + \sigma_{yy} + \sigma_{zz} \tag{4.4}$$

Equation (4.4) demonstrates the *invariance of the sum of the normal stresses at a point with respect to rotation of axes.*

One-third of this invariant is referred to as the *mean bulk stress* and is noted by

$$\bar{\sigma} = \frac{1}{3}(\sigma_{xx} + \sigma_{yy} + \sigma_{zz}) \tag{4.5}$$

The significance of $\bar{\sigma}$ is its relation to pressure and rate of unit volume strain. A discussion of this follows the development of the constitutive equations in Section 4.7.

4.4 Rate of Strain

The *rate of strain* at a point in a fluid may be studied by examining the relative velocity of two closely neighboring fluid particles in the flow field at a given instant of time. Figure 4.2 indicates these two particles by position vectors that are functions of the material coordinates and time. The x component of the relative displacement of the two particles $\Delta\mathbf{r}$, considered as a function of the material coordinates, is

$$\Delta x = \frac{\partial x}{\partial x_0}dx_0 + \frac{\partial x}{\partial y_0}dy_0 + \frac{\partial x}{\partial z_0}dz_0$$

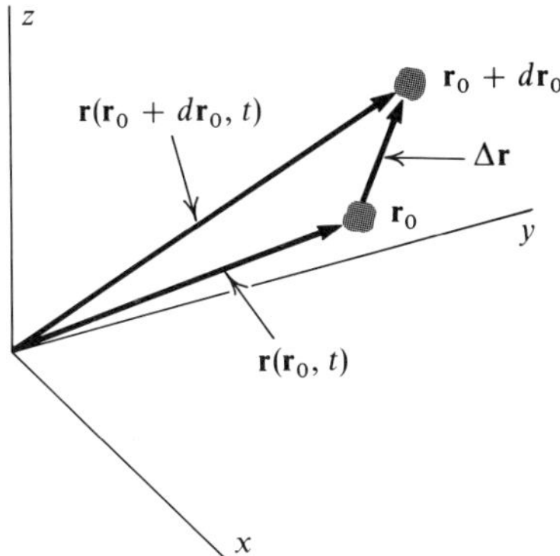

Fig. 4.2 *Neighboring fluid particles identified as* $\mathbf{r}_0$ *and* $\mathbf{r}_0 + d\mathbf{r}_0$ *with relative displacement* $\Delta\mathbf{r}$ *at time* t.

neglecting higher-order terms in dx_0, dy_0, and dz_0. The material derivative of Δx is

$$\left(\frac{\partial\, \Delta x}{\partial t}\right)_{\mathbf{r}_0} = \frac{\partial}{\partial t}\left(\frac{\partial x}{\partial x_0}\, dx_0 + \frac{\partial x}{\partial y_0}\, dy_0 + \frac{\partial x}{\partial z_0}\, dz_0\right)_{\mathbf{r}_0}$$

and the order of differentiation may be interchanged on the right-hand side since t and $\mathbf{r}_0$ are independent variables and the indicated derivatives have been assumed to be continuous. This interchange yields

$$\left(\frac{\partial\, \Delta x}{\partial t}\right)_{\mathbf{r}_0} = \frac{\partial}{\partial x_0}\left(\frac{\partial x}{\partial t}\right)_{\mathbf{r}_0} dx_0 + \frac{\partial}{\partial y_0}\left(\frac{\partial x}{\partial t}\right)_{\mathbf{r}_0} dy_0 + \frac{\partial}{\partial z_0}\left(\frac{\partial x}{\partial t}\right)_{\mathbf{r}_0} dz_0$$

The left-hand side of the preceding equation is identified as dV_x, the rate at which $\Delta\mathbf{r}$ is changing with respect to time in the x direction. Equation (1.9) permits the substitution of V_x for $[\partial x/\partial t]_{\mathbf{r}_0}$; $(\partial x_0/\partial x)\, dx$, $(\partial y_0/\partial y)\, dy$, and $(\partial z_0/\partial z)\, dz$ may also be substituted for dx_0, dy_0, and dz_0, respectively. Carrying out these substitutions gives

$$dV_x = \frac{\partial V_x}{\partial x_0}\frac{\partial x_0}{\partial x}\, dx + \frac{\partial V_x}{\partial y_0}\frac{\partial y_0}{\partial y}\, dy + \frac{\partial V_x}{\partial z_0}\frac{\partial z_0}{\partial z}\, dz$$

which is equivalent to[1]

$$dV_x = \frac{\partial V_x}{\partial x}\, dx + \frac{\partial V_x}{\partial y}\, dy + \frac{\partial V_x}{\partial z}\, dz$$

By similar formulation, one can obtain

$$dV_y = \frac{\partial V_y}{\partial x}\, dx + \frac{\partial V_y}{\partial y}\, dy + \frac{\partial V_y}{\partial z}\, dz$$

[1]The following expressions for the components of the relative velocity of neighboring particles are obvious; however, these were obtained using material coordinates to emphasize the rate of strain of the material.

and

$$dV_z = \frac{\partial V_z}{\partial x}dx + \frac{\partial V_z}{\partial y}dy + \frac{\partial V_z}{\partial z}dz$$

The last three expressions may be written as a single matrix equation of the form

$$\begin{bmatrix} dV_x \\ dV_y \\ dV_z \end{bmatrix} = \begin{bmatrix} \dfrac{\partial V_x}{\partial x} & \dfrac{\partial V_x}{\partial y} & \dfrac{\partial V_x}{\partial z} \\ \dfrac{\partial V_y}{\partial x} & \dfrac{\partial V_y}{\partial y} & \dfrac{\partial V_y}{\partial z} \\ \dfrac{\partial V_z}{\partial x} & \dfrac{\partial V_z}{\partial y} & \dfrac{\partial V_z}{\partial z} \end{bmatrix} \begin{bmatrix} dx \\ dy \\ dz \end{bmatrix} \tag{4.6}$$

which indicates that the relative velocity of two closely neighboring fluid particles is related to the relative spatial displacements at a given instant of time through an array of nine quantities which comprise the components of the velocity gradient tensor.

To facilitate interpretation of the matrix equation (4.6), it may be arranged[1] in the following form:

$$\begin{bmatrix} dV_x \\ dV_y \\ dV_z \end{bmatrix} = \begin{bmatrix} \dfrac{\partial V_x}{\partial x} & \dfrac{1}{2}\left(\dfrac{\partial V_x}{\partial y} + \dfrac{\partial V_y}{\partial x}\right) & \dfrac{1}{2}\left(\dfrac{\partial V_x}{\partial z} + \dfrac{\partial V_z}{\partial x}\right) \\ \dfrac{1}{2}\left(\dfrac{\partial V_y}{\partial x} + \dfrac{\partial V_x}{\partial y}\right) & \dfrac{\partial V_y}{\partial y} & \dfrac{1}{2}\left(\dfrac{\partial V_y}{\partial z} + \dfrac{\partial V_z}{\partial y}\right) \\ \dfrac{1}{2}\left(\dfrac{\partial V_z}{\partial x} + \dfrac{\partial V_x}{\partial z}\right) & \dfrac{1}{2}\left(\dfrac{\partial V_z}{\partial y} + \dfrac{\partial V_y}{\partial z}\right) & \dfrac{\partial V_z}{\partial z} \end{bmatrix} \begin{bmatrix} dx \\ dy \\ dz \end{bmatrix}$$

$$+ \begin{bmatrix} 0 & \dfrac{1}{2}\left(\dfrac{\partial V_x}{\partial y} - \dfrac{\partial V_y}{\partial x}\right) & \dfrac{1}{2}\left(\dfrac{\partial V_x}{\partial z} - \dfrac{\partial V_z}{\partial x}\right) \\ \dfrac{1}{2}\left(\dfrac{\partial V_y}{\partial x} - \dfrac{\partial V_x}{\partial y}\right) & 0 & \dfrac{1}{2}\left(\dfrac{\partial V_y}{\partial z} - \dfrac{\partial V_z}{\partial y}\right) \\ \dfrac{1}{2}\left(\dfrac{\partial V_z}{\partial x} - \dfrac{\partial V_x}{\partial z}\right) & \dfrac{1}{2}\left(\dfrac{\partial V_z}{\partial y} - \dfrac{\partial V_y}{\partial z}\right) & 0 \end{bmatrix} \begin{bmatrix} dx \\ dy \\ dz \end{bmatrix} \tag{4.7}$$

[1]This may be accomplished by use of the identity

$$a_{ij} = \frac{1}{2}(a_{ij} + a_{ji}) + \frac{1}{2}(a_{ij} - a_{ji})$$

in which the first and second subscripts, respectively, call out the row and column in which the element of the matrix is located.

The reason for placing Eq. (4.6) in the above form is for identification of the rigid body rotation associated with the relative velocity components. This is readily seen by carrying out the matrix multiplication in the last term of Eq. (4.7), which results in the following column matrix:

$$\begin{bmatrix} \frac{1}{2}\left(\frac{\partial V_x}{\partial y} - \frac{\partial V_y}{\partial x}\right) dy + \frac{1}{2}\left(\frac{\partial V_x}{\partial z} - \frac{\partial V_z}{\partial x}\right) dz \\ \frac{1}{2}\left(\frac{\partial V_y}{\partial x} - \frac{\partial V_x}{\partial y}\right) dx + \frac{1}{2}\left(\frac{\partial V_y}{\partial z} - \frac{\partial V_z}{\partial y}\right) dz \\ \frac{1}{2}\left(\frac{\partial V_z}{\partial x} - \frac{\partial V_x}{\partial z}\right) dx + \frac{1}{2}\left(\frac{\partial V_z}{\partial y} - \frac{\partial V_y}{\partial z}\right) dy \end{bmatrix}$$

and in terms of the components of the rotational rate given by Eq. (1.27), the above array may be noted by

$$\begin{bmatrix} -\omega_z\, dy + \omega_y\, dz \\ \omega_z\, dx - \omega_x\, dz \\ -\omega_y\, dx + \omega_x\, dy \end{bmatrix}$$

The three elements of this matrix are clearly the components of $\boldsymbol{\omega} \times d\mathbf{r}$ representing the part of the relative velocity due to rigid body rotation; hence these elements do not contribute to the rate of strain of the fluid particle. However, the first factor in the middle term of Eq. (4.7) represents the nine components of the *rate-of-strain tensor*; if the generic element of this matrix is noted by the symbol e with a double subscript, then

$$e_{xx} = \frac{\partial V_x}{\partial x} \tag{4.8a}$$

$$e_{yy} = \frac{\partial V_y}{\partial y} \tag{4.8b}$$

$$e_{zz} = \frac{\partial V_z}{\partial z} \tag{4.8c}$$

$$e_{xy} = e_{yx} = \frac{1}{2}\left(\frac{\partial V_x}{\partial y} + \frac{\partial V_y}{\partial x}\right) = \frac{1}{2}\gamma_{xy} = \frac{1}{2}\gamma_{yx} \tag{4.8d}$$

$$e_{yz} = e_{zy} = \frac{1}{2}\left(\frac{\partial V_y}{\partial z} + \frac{\partial V_z}{\partial y}\right) = \frac{1}{2}\gamma_{yz} = \frac{1}{2}\gamma_{zy} \tag{4.8e}$$

$$e_{zx} = e_{xz} = \frac{1}{2}\left(\frac{\partial V_z}{\partial x} + \frac{\partial V_x}{\partial z}\right) = \frac{1}{2}\gamma_{zx} = \frac{1}{2}\gamma_{xz} \tag{4.8f}$$

The γ's represent the rates of shear strains. The following example will serve to amplify the physical nature of the rate of strain.

Example 4.1

Use Fig. 4.3 to show that γ_{xy} and e_{xx} represent shear and longitudinal rates of strain, respectively.

Solution. The shear strain rate is formulated by noting that the angle AOB in Fig. 4.3 is initially 90°, and that at a time Δt later, this angle has decreased by $\alpha + \beta$. These angles may be approximated by

$$\alpha \simeq \left(\frac{\partial V_x}{\partial y}\,\Delta y\,\Delta t\right)\frac{1}{\Delta y} = \frac{\partial V_x}{\partial y}\,\Delta t$$

$$\beta \simeq \left(\frac{\partial V_y}{\partial x}\,\Delta x\,\Delta t\right)\frac{1}{\Delta x} = \frac{\partial V_y}{\partial x}\,\Delta t$$

The rate of decrease of the angle AOB is given by

$$\left(\frac{\partial V_x}{\partial y}\,\Delta t + \frac{\partial V_y}{\partial x}\,\Delta t\right)\frac{1}{\Delta t}$$

which, on limiting with $\Delta t \to 0$,

$$\lim_{\Delta t \to 0}\left(\frac{\partial V_x}{\partial y}\,\Delta t + \frac{\partial V_y}{\partial x}\,\Delta t\right)\frac{1}{\Delta t} = \frac{\partial V_x}{\partial y} + \frac{\partial V_y}{\partial x} = \gamma_{xy}$$ ◀

The longitudinal strain in the x direction is the change of the length of OA in this direction and is approximately given by

$$\left(\Delta x + \frac{\partial V_x}{\partial x}\,\Delta x\,\Delta t\right) - \Delta x = \frac{\partial V_x}{\partial x}\,\Delta x\,\Delta t$$

Fig. 4.3 *Initial and final locations of fluid element indicated by solid and dashed outlines, respectively.*

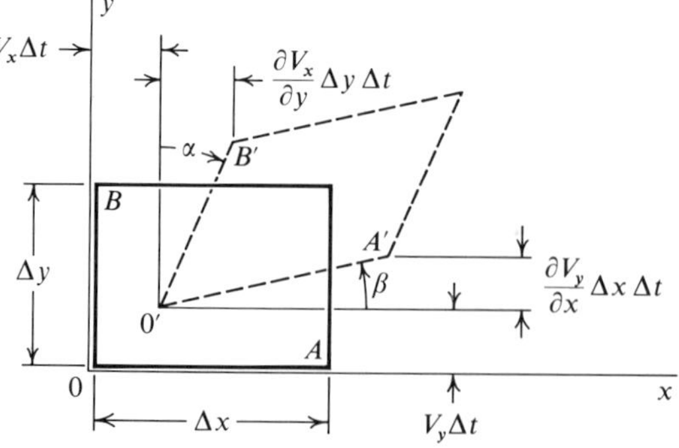

The unit rate of strain may be obtained by dividing by Δt and the initial length of OA, and by limiting:

$$\lim_{\Delta t \to 0} \frac{(\partial V_x/\partial x)\,\Delta x\,\Delta t}{\Delta t\,\Delta x} = \frac{\partial V_x}{\partial x} = e_{xx}$$ ◀

Interpretations may be attributed to γ_{yz}, γ_{zx}, e_{yy}, and e_{zz} similar to those for γ_{xy} and e_{xx} in Example 4.1. Note that the dimension of each of the components of rate of strain is reciprocal time. The set f Eqs. (4.8) must now be related to the six distinct components representing the state of stress.

4.5 Transformation of Rates of Strain

Equations (4.3*a*, *b*, and *c*) provide a set of transformation equations expressing the normal stress referred to a set of axes that have been rotated, and these stresses are given as functions of the stress components referred to the initial orientation of the axes. The transformation of rates of strain to be developed next may be used in obtaining relations between stress and these rates.

The transformation of strains and their associated rates is an easy task since these only require the forming of partial derivatives of velocity components. The transformation of velocity components is a familiar operation employed in engineering mechanics, and calculus provides the method whereby space derivatives of the components may be transformed. Figure 4.4 shows the two coordinate systems, one rotated with respect to

Fig. 4.4 *Rotation of axes showing point fixed in either set of coordinates by position vector* **r**.

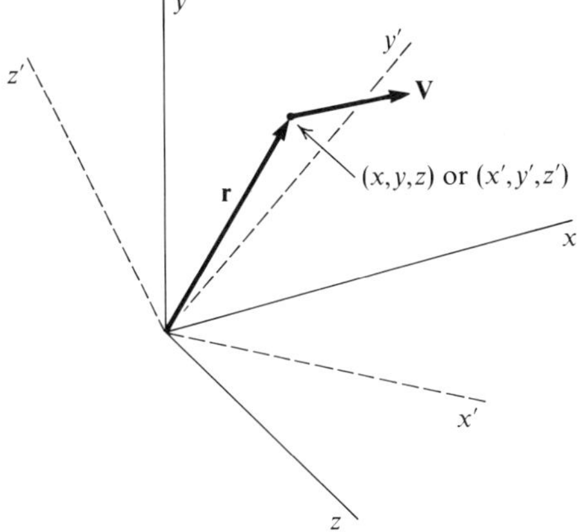

the other. As noted in Section 4.3: l_1, m_1, and n_1 are the respective direction cosines of the x' axis, while l_2, m_2, and n_2 and l_3, m_3, and n_3 are the direction cosines of the y' and z' axes, respectively. The transformations for the coordinates are given by

$$x' = l_1 x + m_1 y + n_1 z \tag{4.9a}$$

$$y' = l_2 x + m_2 y + n_2 z \tag{4.9b}$$

$$z' = l_3 x + m_3 y + n_3 z \tag{4.9c}$$

and

$$x = l_1 x' + l_2 y' + l_3 z' \tag{4.9d}$$

$$y = m_1 x' + m_2 y' + m_3 z' \tag{4.9e}$$

$$z = n_1 x' + n_2 y' + n_3 z' \tag{4.9f}$$

These may be obtained by projection of the position vector **r** on the coordinate axes. This same procedure may be employed by projecting the velocity vector **V** of the fluid particle on the coordinate axes to obtain

$$V_{x'} = l_1 V_x + m_1 V_y + n_1 V_z \tag{4.10a}$$

$$V_{y'} = l_2 V_x + m_2 V_y + n_2 V_z \tag{4.10b}$$

$$V_{z'} = l_3 V_x + m_3 V_y + n_3 V_z \tag{4.10c}$$

and similarly for V_x, V_y, and V_z. This should be an expected result since the velocity components are time derivatives of the coordinates of the fluid particle at the point at any given instant of time.

To form the rates of strain in the primed coordinate system, the chain rule of calculus[1] is used as follows:

$$\frac{\partial V_{x'}}{\partial x'} = \frac{\partial V_{x'}}{\partial x}\frac{\partial x}{\partial x'} + \frac{\partial V_{x'}}{\partial y}\frac{\partial y}{\partial x'} + \frac{\partial V_{x'}}{\partial z}\frac{\partial z}{\partial x'}$$

From Eqs. (4.9),

$$\frac{\partial x}{\partial x'} = l_1 \qquad \frac{\partial y}{\partial x'} = m_1 \qquad \frac{\partial z}{\partial x'} = n_1$$

and hence

$$\frac{\partial V_{x'}}{\partial x'} = l_1 \frac{\partial V_{x'}}{\partial x} + m_1 \frac{\partial V_{x'}}{\partial y} + n_1 \frac{\partial V_{x'}}{\partial z} \tag{4.11}$$

[1]Wilfred Kaplan, "Advanced Calculus," p. 86, Addison-Wesley Publishing Company, Inc., Reading, Mass., 1952.

Partial differentiation of Eq. (4.10*a*) yields

$$\frac{\partial V_{x'}}{\partial x} = l_1 \frac{\partial V_x}{\partial x} + m_1 \frac{\partial V_y}{\partial x} + n_1 \frac{\partial V_z}{\partial x}$$

$$\frac{\partial V_{x'}}{\partial y} = l_1 \frac{\partial V_x}{\partial y} + m_1 \frac{\partial V_y}{\partial y} + n_1 \frac{\partial V_z}{\partial y}$$

$$\frac{\partial V_{x'}}{\partial z} = l_1 \frac{\partial V_x}{\partial z} + m_1 \frac{\partial V_y}{\partial z} + n_1 \frac{\partial V_z}{\partial z}$$

Substituting these three derivatives into Eq. (4.11), utilizing Eqs. (4.8), and simplifying, gives

$$e_{x'x'} = \frac{\partial V_{x'}}{\partial x'} = l_1^2 e_{xx} + m_1^2 e_{yy} + n_1^2 e_{zz} + l_1 m_1 \gamma_{xy} + l_1 n_1 \gamma_{xz} + m_1 n_1 \gamma_{yz} \qquad (4.12a)$$

In a similar manner,

$$\frac{\partial V_{y'}}{\partial y'}, \frac{\partial V_{z'}}{\partial z'}, \frac{\partial V_{x'}}{\partial y'}, \frac{\partial V_{x'}}{\partial z'}, \frac{\partial V_{y'}}{\partial x'}, \frac{\partial V_{y'}}{\partial z'}, \frac{\partial V_{z'}}{\partial x'}, \text{ and } \frac{\partial V_{z'}}{\partial y'}$$

may be generated, with the formulation of the remaining five rates of strain transformations. These are

$$e_{y'y'} = l_2^2 e_{xx} + m_2^2 e_{yy} + n_2^2 e_{zz} + l_2 m_2 \gamma_{xy} + l_2 n_2 \gamma_{xz} + m_2 n_2 \gamma_{yz} \qquad (4.12b)$$

$$e_{z'z'} = l_3^2 e_{xx} + m_3^2 e_{yy} + n_3^2 e_{zz} + l_3 m_3 \gamma_{xy} + l_3 n_3 \gamma_{xz} + m_3 n_3 \gamma_{yz} \qquad (4.12c)$$

$$\gamma_{x'y'} = 2\, l_1 l_2 e_{xx} + 2\, m_1 m_2 e_{yy} + 2\, n_1 n_2 e_{zz} + (l_1 m_2 + l_2 m_1)\gamma_{xy} + (l_1 n_2 + l_2 n_1)\gamma_{xz} + (m_1 n_2 + m_2 n_1)\gamma_{yz} \qquad (4.12d)$$

$$\gamma_{y'z'} = 2\, l_2 l_3 e_{xx} + 2\, m_2 m_3 e_{yy} + 2\, n_2 n_3 e_{zz} + (l_2 m_3 + l_3 m_2)\gamma_{xy} + (l_2 n_3 + l_3 n_2)\gamma_{xz} + (m_2 n_3 + m_3 n_2)\gamma_{yz} \qquad (4.12e)$$

$$\gamma_{z'x'} = 2\, l_1 l_3 e_{xx} + 2\, m_1 m_3 e_{yy} + 2\, n_1 n_3 e_{zz} + (l_1 m_3 + l_3 m_1)\gamma_{xy} + (l_1 n_3 + l_3 n_1)\gamma_{xz} + (m_1 n_3 + m_3 n_1)\gamma_{yz} \qquad (4.12f)$$

Before relating the six components specifying the state of stress to the rates of strain, it will be convenient to examine a simple relation between one stress component and a single rate of strain.

4.6 Newton's Hypothesis and Viscosity

The earliest relation between stress and rate of strain may be attributed to Newton who advanced the following hypothesis: "The resistance arising from the want of lubricity in the parts of a fluid, is, other things being equal, proportional to the velocity with which the parts of the fluid are separated from one another."[1] Applied to the flow between the two parallel plates in Fig. 4.5, this hypothesis may be formulated as

$$\tau = \frac{F}{A} = C\frac{dV}{dy} \qquad d \geq y \geq 0 \tag{4.13}$$

in which F is the force that must be applied to the moving plate of area A. Both plates are very large compared to the distance of separation d, which tends to minimize the edge effects. Note that the fluid adjacent to each of the plates adheres to the plates. This is called the "no slip" condition. The proportionality factor C in Eq. (4.13) is usually denoted by μ, which gives an equation of the form

$$\tau = \mu\frac{dV}{dy} \tag{4.14}$$

This factor is called the *coefficient of viscosity*, the dynamic viscosity, or simply the viscosity; it has the dimensions $ML^{-1}T^{-1}$ or FTL^{-2} in the mass-length-time or force-length-time system of dimensions, respectively. The viscosity of a given fluid depends on temperature but is only a weak function of pressure.[2] The variation of viscosity with temperature is shown for several

[1]Andrew Motte (tr.), "Sir Isaac Newton's Mathematical Principles of Natural Philosophy and His System of the World," p. 385, University of California Press, Berkeley, Calif., 1934. By permission of the publisher.

[2]James G. Knudsen and Donald L. Katz, "Fluid Dynamics and Heat Transfer," p. 15, McGraw-Hill Book Company, New York, 1958.

Fig. 4.5 *Velocity profile for fluid motion between two parallel plates.*

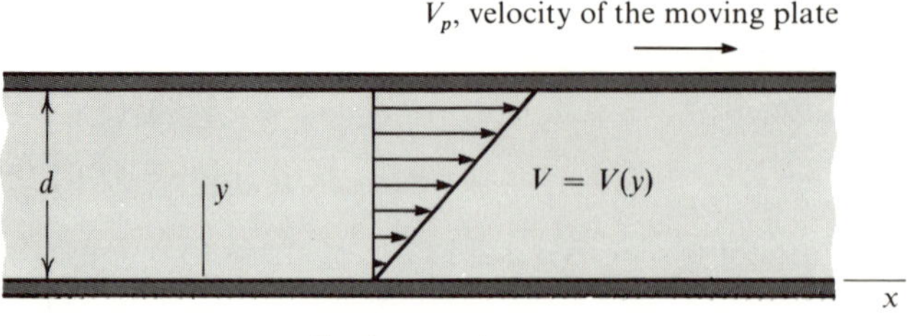

liquids and gases in Fig. 4.6. These curves indicate that the viscosity of a liquid decreases with an increase in temperature while the opposite trend is true for a gas. The difference between viscosity-temperature variations for gases and for liquids stems from the prevalent forces among the molecules in each fluid. In the gas the predominant forces among molecules are those due to collisions; while in a liquid, the attraction of the molecules for one another (cohesive forces) prevails over collision effects. This cohesion provides the forces that assist in giving liquids an observable "form," as can be seen by filling a glass with water.

The viscosity of a fluid is a consequence of the forces among molecules and the transfer of momentum of the molecules associated with these forces. Equation (4.14) provides a simple relation between these effects through a coefficient of viscosity. The transfer of momentum at a molecular level is brought out here to emphasize that the relation between stress and the velocity derivative in Eq. (4.14) is restricted to those flows where transfer of momentum in the fluid is among molecules. This regime of flow is referred to as a *well-ordered flow* or *laminar flow* in contrast to *turbulent flow* in which transfers of mass, momentum, and energy are manifest at a macroscopic level. A well-ordered flow is one free of macroscopic velocity fluctuations. Chapter 9 gives a more detailed treatment of laminar and turbulent flows.

The kinematic viscosity ν is defined by

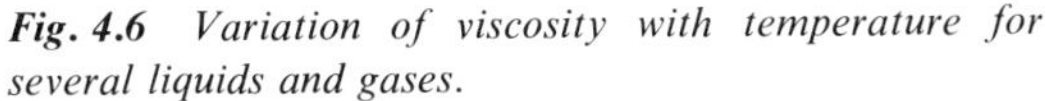

$$\nu = \frac{\mu}{\rho} \tag{4.15}$$

and its dimensions are L^2T^{-1}. This quantity arises naturally in some problems in fluid mechanics.

Fig. 4.6 Variation of viscosity with temperature for several liquids and gases.

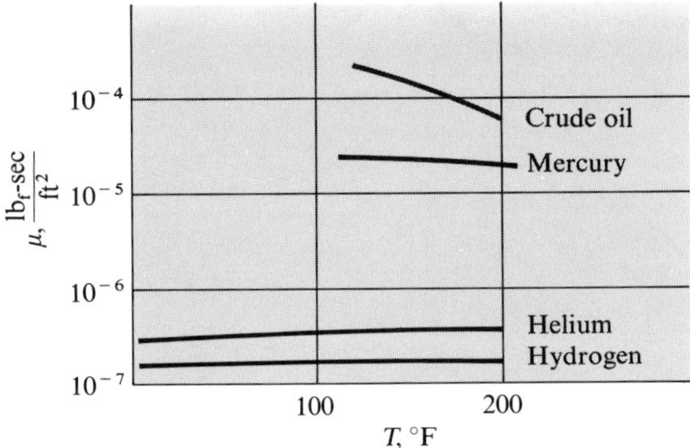

Example 4.2

The two cylinders shown in Fig. 4.7 are coaxial; they are long enough so that end effects may be ignored. The space between them is filled with a fluid of viscosity μ. A constant torque T is applied to the outside cylinder. The inside cylinder is stationary. Using the applicable form of Eq. (4.14), obtain an expression for the steady-state angular speed of the outside cylinder under the action of fluid shear and the applied torque.

Solution. The velocity profile in the fluid is indicated on the sketch; however, it is not known at the outset. The fluid adjacent to the outer surface of the inside cylinder has zero velocity, while that adjacent to the inner surface of the outside cylinder has the velocity of a point on this surface, which is ωR_2. This is the no-slip condition referred to previously. An applicable form[1] of Eq. (4.14) from Newton's statement concerning lubricity is

$$\tau = \frac{F}{A} = \mu \frac{dV}{dr}$$

This may be applied to a typical area through which the torque is transmitted by shear action. This area is taken as $2\pi rL$. The force transmitted across this area is T/r, so that

$$\frac{T}{r\,(2\pi rL)} = \mu \frac{dV}{dr}$$

[1]In this problem the correct expression for $\tau_{r\theta}$ is $\mu(dV/dr - V/r)$; however, V/r is generally small compared to dV/dr.

Fig. 4.7

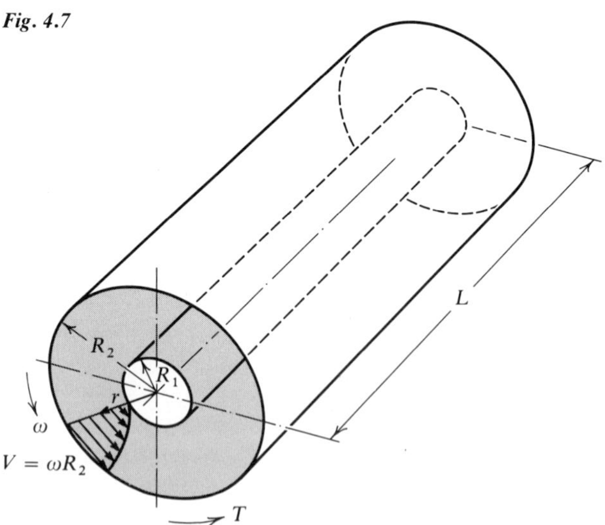

Separation of the variables yields

$$\frac{T}{2\pi L}\int_{R_1}^{R_2}\frac{dr}{r^2} = \mu\int_0^{\omega R_2} dV$$

The torque has been taken outside the integral, since it is independent of r when the outside cylinder has attained a steady-state angular speed. Integration, substitution of limits, and solution for the angular speed gives the required result

$$\omega = \frac{T(1/R_1 - 1/R_2)}{2\pi L\mu R_2}$$ ◀

If in addition to this result, the velocity profile is desired, the differential equation may be integrated with variable upper limits, that is,

$$\frac{T}{2\pi L}\int_{R_1}^{r}\frac{dr}{r^2} = \mu\int_0^{V} dV$$

and

$$V = \frac{T(1/R_1 - 1/r)}{2\pi L\mu}$$

4.7 Constitutive Equations

The *constitutive equations*, which relate the stress components to the strain rates, are based on the following assumptions:

1. Each of the stress components is linearly proportional to the strain rates.
2. The fluid is isotropic and hence there are no preferred directions.
3. In the absence of strain rates, the normal stresses must reduce to the pressure, and the shear stresses must vanish.

The most general relation between each of the six distinct stress components and the six rates of strain results in six equations including 42 constants[1] according to assumption 1. From assumption 2, the number of constants is reduced to five, and three of these are related to the pressure p through assumption 3. Finally, application of Eq. (4.14) to the flow through the parallel plates shown in Fig. 4.5 yields a relationship among the two remaining constants and the coefficient of viscosity μ. The assumed linear relations may now be expressed as

[1]Details of this development to obtain Eq. (4.16) are given in Appendix A.

$$\begin{aligned}
\sigma_{xx} &= 2\mu e_{xx} + C_{12}(e_{xx} + e_{yy} + e_{zz}) - p \\
\sigma_{yy} &= 2\mu e_{yy} + C_{12}(e_{xx} + e_{yy} + e_{zz}) - p \\
\sigma_{zz} &= 2\mu e_{zz} + C_{12}(e_{xx} + e_{yy} + e_{zz}) - p \\
\tau_{xy} &= \mu\gamma_{xy} \\
\tau_{xz} &= \mu\gamma_{xz} \\
\tau_{yz} &= \mu\gamma_{yz}
\end{aligned} \tag{4.16}$$

Addition of the first three equations of set (4.16) provides

$$\sigma_{xx} + \sigma_{yy} + \sigma_{zz} = 2\mu(e_{xx} + e_{yy} + e_{zz}) + 3C_{12}(e_{xx} + e_{yy} + e_{zz}) - 3p$$

$$\bar{\sigma} = \left(\frac{2}{3}\mu + C_{12}\right) \text{div } \mathbf{V} - p \tag{4.17}$$

using Eqs. (4.8). Since the div **V** is the rate of change of volume per unit volume, the quantity $(2/3)\mu + C_{12}$ is called the *volume viscosity*. When the volume viscosity is zero, Eqs. (4.16) reduce to

$$\sigma_{xx} = -p - \frac{2}{3}\mu \text{ div } \mathbf{V} + 2\mu \frac{\partial V_x}{\partial x} \tag{4.18a}$$

$$\sigma_{yy} = -p - \frac{2}{3}\mu \text{ div } \mathbf{V} + 2\mu \frac{\partial V_y}{\partial y} \tag{4.18b}$$

$$\sigma_{zz} = -p - \frac{2}{3}\mu \text{ div } \mathbf{V} + 2\mu \frac{\partial V_z}{\partial z} \tag{4.18c}$$

$$\tau_{xy} = \mu\left(\frac{\partial V_y}{\partial x} + \frac{\partial V_x}{\partial y}\right) \tag{4.18d}$$

$$\tau_{xz} = \mu\left(\frac{\partial V_z}{\partial x} + \frac{\partial V_x}{\partial z}\right) \tag{4.18e}$$

$$\tau_{yz} = \mu\left(\frac{\partial V_z}{\partial y} + \frac{\partial V_y}{\partial z}\right) \tag{4.18f}$$

These are the *constitutive equations* usually credited to Stokes.[1]

Some comments are necessary concerning the quantity $(2/3)\mu + C_{12}$ in Eq. (4.17). If the fluid is incompressible, then the value of the volume viscosity does not matter since div **V** vanishes identically from continuity requirements, and the pressure is the negative of the mean bulk stress.

[1]G. G. Stokes, On the Theories of the Internal Friction of Fluids in Motion, and of the Equilibrium and Motion of Elastic Solids, *Proc. Camb. Phil. Soc.*, vol. 1, no. 2, pp. 16–18, 1845.

Experimental investigation[1] of some liquids indicates that $(2/3)\mu + C_{12}$ is greater than zero; however, most liquids are not very compressible, and $\bar{\sigma} = -p$ is a very good approximation. With regard to gases, which are normally considered compressible, it has been shown[2] that $(2/3)\mu + C_{12} = 0$ for an ideal monatomic gas, and there is some evidence of its validity for the flow of ideal gases where there is very little energy interchange among the various degrees of freedom of the atoms comprising the molecules of the gas. For these reasons, *pressure will be taken as the negative of the mean bulk stress* throughout this text.

Fluids exhibiting a linear stress rate of deformation relation are referred to as *Newtonian fluids*. Some common examples of Newtonian fluids are monatomic gases, water, glycerin, and most liquids of low molecular weight under ordinary conditions of temperature and pressure. Diatomic gases or mixtures of these may be considered newtonian at temperatures low enough to exclude an appreciable number of molecules with rotational and vibrational energies.

Example 4.3

A Newtonian fluid flow field is given by the velocity vector $\mathbf{V} = 4xyz\mathbf{i} + z^2\mathbf{j} - 2yz^2\mathbf{k}$ in ft/sec, when x, y, and z are in feet. Calculate all the normal and shear stresses at the point $x = 2$ ft, $y = 1$ ft, $z = 1$ ft, if the fluid viscosity μ is 10^{-4} $\text{lb}_f\text{-sec/ft}^2$ and the pressure p is 216.000 lb_f/ft^2

Solution. From Eqs. (4.18*a*,*b*,*c*), the normal stresses are

$$\sigma_{xx} = -p - \frac{2}{3}\mu \operatorname{div} \mathbf{V} + 2\mu \frac{\partial V_x}{\partial x}$$

$$\sigma_{yy} = -p - \frac{2}{3}\mu \operatorname{div} \mathbf{V} + 2\mu \frac{\partial V_y}{\partial y}$$

$$\sigma_{zz} = -p - \frac{2}{3}\mu \operatorname{div} \mathbf{V} + 2\mu \frac{\partial V_z}{\partial z}$$

and

$$\operatorname{div} \mathbf{V} = e_{xx} + e_{yy} + e_{zz} = 4yz + 0 - 4yz = 0$$

Then the normal stress distribution is given by

$$\sigma_{xx} = -p + 2\mu(4yz) = -p + 8\mu yz$$

$$\sigma_{yy} = -p + 2\mu(0) = -p$$

$$\sigma_{zz} = -p + 2\mu(-4yz) = -p - 8\mu yz$$

[1]S. M. Karim and L. Rosenhead, The Second Coefficient of Viscosity of Liquids and Gases, *Rev. Mod. Phys.*, vol. 24, no. 2, pp. 108–116, April, 1952.

[2]James Jeans, "An Introduction to the Kinetic Theory of Gases," p. 248, Cambridge University Press, New York, 1960.

At the given point 2, 1, 1, the normal stresses have the following values:

$$\sigma_{xx} = -216.000 + 8(10^{-4})(1)(1) = -215.9992 \text{ lb}_f/\text{ft}^2 \quad \blacktriangleleft$$

$$\sigma_{yy} = -216.000 \text{ lb}_f/\text{ft}^2 \quad \blacktriangleleft$$

$$\sigma_{zz} = -216.000 - 8(10^{-4})(1)(1) = -216.0008 \text{ lb}_f/\text{ft}^2 \quad \blacktriangleleft$$

The shear stresses are given by Eqs. (4.18*d*, *e*, and *f*)

$$\tau_{xy} = \mu\left(\frac{\partial V_y}{\partial x} + \frac{\partial V_x}{\partial y}\right)$$

$$\tau_{yz} = \mu\left(\frac{\partial V_z}{\partial y} + \frac{\partial V_y}{\partial z}\right)$$

$$\tau_{zx} = \mu\left(\frac{\partial V_x}{\partial z} + \frac{\partial V_z}{\partial x}\right)$$

Accordingly, the shear stress distribution is obtained as

$$\tau_{xy} = \mu(0 + 4xz) = \mu(4xz)$$

$$\tau_{yz} = \mu(-2z^2 + 2z)$$

$$\tau_{zx} = \mu(4xy + 0) = \mu(4xy)$$

At the given point 2, 1, 1, the shear stresses have the following values:

$$\tau_{xy} = \mu(4)(2)(1) = 8\mu = 8(10^{-4}) \text{ lb}_f/\text{ft}^2 \quad \blacktriangleleft$$

$$\tau_{yz} = \mu(-2 + 2) = 0 \quad \blacktriangleleft$$

$$\tau_{zx} = \mu(4)(2)(1) = 8\mu = 8(10^{-4}) \text{ lb}_f/\text{ft}^2 \quad \blacktriangleleft$$

It now becomes obvious that once the velocity field is known, one can find the stress field from the constitutive equations if, in addition, the pressure and the viscosity are known.

4.8 Constituted Equations of Motion

The next task is *to constitute* the equations of motion. This is done by substituting the Eqs. (4.18) into Eqs. (3.10), (3.11), and (3.12). This will be carried out in detail for Eq. (3.10), which is

$$\rho\frac{DV_x}{Dt} = \frac{\partial \sigma_{xx}}{\partial x} + \frac{\partial \tau_{yx}}{\partial y} + \frac{\partial \tau_{zx}}{\partial z} + \rho(\mathbf{f}_B)_x \tag{3.10}$$

Substituting for σ_{xx}, τ_{xy}, and τ_{zx} by using Eqs. (4.18*a*), (4.18*d*), and (4.18*e*), respectively, one obtains

$$\rho \frac{DV_x}{Dt} = -\frac{\partial p}{\partial x} - \frac{\partial}{\partial x}\left(\frac{2}{3}\mu \operatorname{div} \mathbf{V} - 2\mu \frac{\partial V_x}{\partial x}\right) + \frac{\partial}{\partial y}\left[\mu\left(\frac{\partial V_y}{\partial x} + \frac{\partial V_x}{\partial y}\right)\right] + \frac{\partial}{\partial z}\left[\mu\left(\frac{\partial V_x}{\partial z} + \frac{\partial V_z}{\partial x}\right)\right] + \rho(\mathbf{f}_B)_x \tag{4.19a}$$

Similar substitutions can be made in Eqs. (3.11) and (3.12), which yield

$$\rho \frac{DV_y}{Dt} = -\frac{\partial p}{\partial y} - \frac{\partial}{\partial y}\left(\frac{2}{3}\mu \operatorname{div} \mathbf{V} - 2\mu \frac{\partial V_y}{\partial y}\right) + \frac{\partial}{\partial x}\left[\mu\left(\frac{\partial V_x}{\partial y} + \frac{\partial V_y}{\partial x}\right)\right] + \frac{\partial}{\partial z}\left[\mu\left(\frac{\partial V_y}{\partial z} + \frac{\partial V_z}{\partial y}\right)\right] + \rho(\mathbf{f}_B)_y \tag{4.19b}$$

and

$$\rho \frac{DV_z}{Dt} = -\frac{\partial p}{\partial z} - \frac{\partial}{\partial z}\left(\frac{2}{3}\mu \operatorname{div} \mathbf{V} - 2\mu \frac{\partial V_z}{\partial z}\right) + \frac{\partial}{\partial y}\left[\mu\left(\frac{\partial V_y}{\partial z} + \frac{\partial V_z}{\partial y}\right)\right] + \frac{\partial}{\partial x}\left[\mu\left(\frac{\partial V_x}{\partial z} + \frac{\partial V_z}{\partial x}\right)\right] + \rho(\mathbf{f}_B)_z \tag{4.19c}$$

The above set is referred to as the *Navier-Stokes equations of motion.*

A considerable simplification occurs in problems where viscosity can be considered constant. In this case, the set of Eqs. (4.19) reduces to the following:

$$\rho \frac{DV_x}{Dt} = -\frac{\partial p}{\partial x} + \frac{1}{3}\mu \frac{\partial}{\partial x} \operatorname{div} \mathbf{V} + \mu \operatorname{div} \mathbf{grad}\, V_x + \rho(\mathbf{f}_B)_x \tag{4.20a}$$

$$\rho \frac{DV_y}{Dt} = -\frac{\partial p}{\partial y} + \frac{1}{3}\mu \frac{\partial}{\partial y} \operatorname{div} \mathbf{V} + \mu \operatorname{div} \mathbf{grad}\, V_y + \rho(\mathbf{f}_B)_y \tag{4.20b}$$

$$\rho \frac{DV_z}{Dt} = -\frac{\partial p}{\partial z} + \frac{1}{3}\mu \frac{\partial}{\partial z} \operatorname{div} \mathbf{V} + \mu \operatorname{div} \mathbf{grad}\, V_z + \rho(\mathbf{f}_B)_z \tag{4.20c}$$

If the fluid is incompressible, then each of the second terms on the right-hand side becomes zero, and a further reduction is possible. It should be observed that Eqs. (4.20) may be reduced to the single vector equation,[1]

[1]The terms **grad** div $\mathbf{V}$ − **curl** (**curl** $\mathbf{V}$) are replaced by $\nabla^2\mathbf{V}$ by some authors. It is referred to as the *vector laplacian.* The symbol ∇, pronounced "del," is not used throughout this text since its use requires some special considerations where vector identities are involved.

$$\rho \frac{D\mathbf{V}}{Dt} = -\mathbf{grad}\, p + \frac{1}{3}\mu\, \mathbf{grad}\, \mathrm{div}\, \mathbf{V} + \mu\, [\mathbf{grad}\, \mathrm{div}\, \mathbf{V} - \mathbf{curl}\, (\mathbf{curl}\, \mathbf{V})] + \rho \mathbf{f}_B \qquad (4.21)$$

or

$$\rho \frac{D\mathbf{V}}{Dt} = -\mathbf{grad}\, p + \frac{4}{3}\mu\, \mathbf{grad}\, \mathrm{div}\, \mathbf{V} - \mu\, \mathbf{curl}\, (\mathbf{curl}\, \mathbf{V}) + \rho \mathbf{f}_B \qquad (4.22)$$

This equation may be expressed in a coordinate system of the analyst's choice by formulation of the gradient, divergence, curl, and material derivative in the desired system. Appendix B lists these for some of the more common coordinate systems.

It is now desirable to return to a discussion of the number of dependent variables and independent equations introduced in Section 4.1. Equations (4.20) and (2.3) provide four independent scalar equations containing the following unknowns: three velocity components, pressure, density, and three body-force components if μ is assumed constant. Usually the body-force field is a known function of coordinates or is considered a known constant (such as a gravitational force). Then there are five unknowns, and this requires another equation, which is provided by the equation of state in a form of $p = p(\rho, T)$. The equation of state introduces another unknown, the absolute temperature T, and so still another equation is needed. This is provided by an energy equation, which is a statement of the first law of thermodynamics and specific heat data, relating energy to temperature and pressure.

In summary: the dependent variables or unknowns are three velocity components, pressure, density, and temperature; and the equations relating these variables are three constituted equations of motion, conservation of matter, a p-ρ-T equation of state, an energy equation, and specific heat data. Obviously the independent variables are three space coordinates and time. In problems treating the flow of a fluid with steady and uniform density, the four scalar equations (4.20) and (2.3) are adequate for determination of the three velocity components and the pressure. The second law of thermodynamics must be satisfied also. Since entropy is a function of the thermodynamic variables, it did not enter explicitly into the discussion of the equations relating the variables of the flow. However, the satisfaction of the second law with respect to particular flows is discussed in Chapters 6 and 8.

There exists no general method for solving the Navier-Stokes equations. Their nonlinear nature, which arises from the convective terms in the acceleration, renders these equations difficult to solve. However, solutions have been obtained for some simple, special cases. A few of these are presented

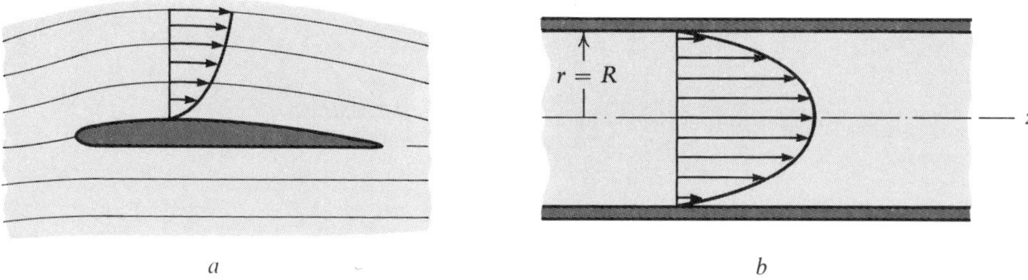

Fig. 4.8 *Fluids bounded by a solid surface with velocity profiles shown at a given locality.*

as examples and assigned problems after a brief discussion of boundary conditions.

4.9 Some General Boundary Conditions

Only two of many boundary conditions will be treated. These have been selected because of their frequent occurrence. Generally a fluid is in contact with or confined by a solid surface, another fluid, or a solid surface and another fluid.

When a fluid is bounded by a solid surface such as an airfoil or a pipe as shown in Fig. 4.8, one might find that the velocity profile would be as indicated. In each case the velocity of the fluid relative to the solid surface is zero at all points on the surface. This is a general boundary condition, the no-slip condition mentioned in Section 4.6. Physically this means that the fluid adjacent to the solid wall sticks or adheres to the wall. The mathematical formulation in cylindrical coordinates of the boundary condition for the flow in Fig. 4.8*b* is

$$V_z(R,\theta,z) = 0$$

When two fluids are in contact so that there exists an interface between them as shown in Fig. 4.9, one fluid is said to be bounded by the other. The appropriate boundary condition is that the stress and velocity distribution must be continuous at a plane fluid interface. To gain a physical interpretation of this condition, imagine a small fluid element in the neighborhood of the interface and consider this small element to extend into both fluids. Such an element is shown in Fig. 4.9. If a stress discontinuity is assumed at the interface and the equation of motion is formulated for the element of fluid and then the element is shrunk to a point at the interface, one finds that this point in the fluid flow field would have infinite acceleration along the interface

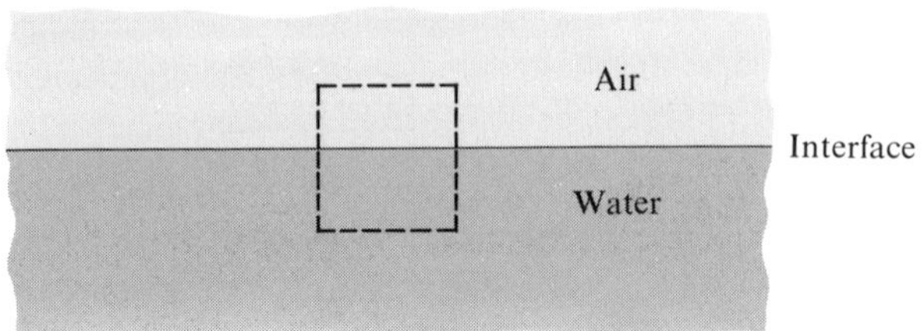

Fig. 4.9 *Two fluids and their interface.*

since the surface forces acting on the element are infinitesimals of a second order while the mass of the element is an infinitesimal of the third order. Therefore the stress distribution is continuous across the plane fluid interface. The details of this demonstration are left as an exercise at the end of the chapter.

Example 4.4

An incompressible Newtonian fluid with constant viscosity flows between two parallel plates. One plate is fixed and the other moves in the positive x direction with a velocity V_p as indicated in Fig. 4.10. The motion is steady and laminar, and none of the variables depend on the z coordinate. There is a pressure gradient maintained in the x direction, and there is no y or z component of velocity at any point in the field. Obtain a solution for V_x by neglecting body forces and discuss the influence of several pressure gradients.

Solution. With body forces neglected, and div $\mathbf{V} = 0$ for the incompressible fluid, the constituted equations of motion reduce to

$$\rho\left(V_x \frac{\partial V_x}{\partial x} + V_y \frac{\partial V_x}{\partial y} + V_z \frac{\partial V_x}{\partial z} + \frac{\partial V_x}{\partial t}\right) = -\frac{\partial p}{\partial x} + \mu\left(\frac{\partial^2 V_x}{\partial x^2} + \frac{\partial^2 V_x}{\partial y^2} + \frac{\partial^2 V_x}{\partial z^2}\right)$$

$$\rho\left(V_x \frac{\partial V_y}{\partial x} + V_y \frac{\partial V_y}{\partial y} + V_z \frac{\partial V_y}{\partial z} + \frac{\partial V_y}{\partial t}\right) = -\frac{\partial p}{\partial y} + \mu\left(\frac{\partial^2 V_y}{\partial x^2} + \frac{\partial^2 V_y}{\partial y^2} + \frac{\partial^2 V_y}{\partial z^2}\right)$$

$$\rho\left(V_x \frac{\partial V_z}{\partial x} + V_y \frac{\partial V_z}{\partial y} + V_z \frac{\partial V_z}{\partial z} + \frac{\partial V_z}{\partial t}\right) = -\frac{\partial p}{\partial z} + \mu\left(\frac{\partial^2 V_z}{\partial x^2} + \frac{\partial^2 V_z}{\partial y^2} + \frac{\partial^2 V_z}{\partial z^2}\right)$$

and the continuity equation, of course, requires that

$$\frac{\partial V_x}{\partial x} + \frac{\partial V_y}{\partial y} + \frac{\partial V_z}{\partial z} = 0$$

Other given conditions require that $\partial V_x/\partial t = 0$ since the motion is steady, and that all of the derivatives of V_y and V_z are zero since $V_y = 0$ and $V_z = 0$.

The continuity equation thus reduces to $\partial V_x/\partial x = 0$. Carrying out the operation of $\int (\partial V_x/\partial x)\,dx$ yields $V_x = V_x(y,z,t)$, which means $V_x = V_x(y)$ only since V_x is not a

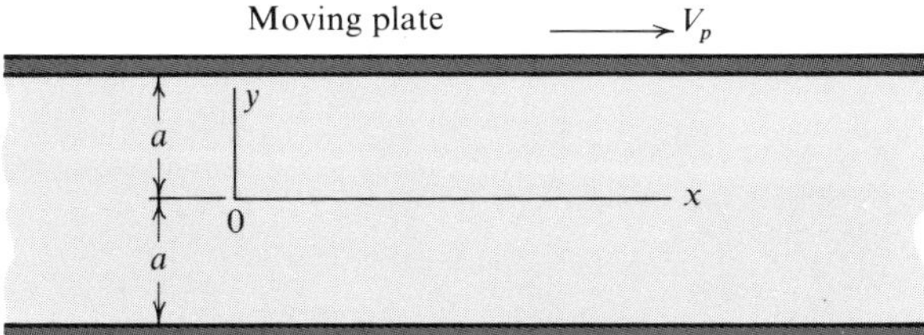

Fig. 4.10

function of z and time according to the given data. Since $\partial V_x/\partial x = 0$ and $\partial V_x/\partial t = 0$, the second derivatives of V_x with respect to the x or z also vanish. The three constituted equations of motion may now be written respectively as

$$0 = -\frac{\partial p}{\partial x} + \mu \frac{\partial^2 V_x}{\partial y^2}$$

$$0 = -\frac{\partial p}{\partial y}$$

$$0 = -\frac{\partial p}{\partial z}$$

The second and third of these imply that pressure is a function of x only, so that the first equation may now be written in terms of ordinary derivatives as

$$\frac{d^2 V_x}{dy^2} = \frac{1}{\mu} \frac{dp}{dx}$$

Two integrations of the above equation yield

$$\frac{dV_x}{dy} = \frac{1}{\mu} \frac{dp}{dx} y + K_1$$

and

$$V_x = \frac{1}{2\mu} \frac{dp}{dx} y^2 + K_1 y + K_2$$

in which K_1 and K_2 are constants to be specified by the conditions to which the flow is subjected. These conditions or constraints are $V_x(a) = V_p$ and $V_x(-a) = 0$, using the previously chosen fixed reference axes as shown in Fig. 4.10. These conditions are substituted into the expression for V_x to obtain

$$V_p = \frac{1}{2\mu} \frac{dp}{dx} a^2 + K_1 a + K_2$$

$$0 = \frac{1}{2\mu} \frac{dp}{dx} a^2 - K_1 a + K_2$$

Solving for these constants gives

$$K_1 = \frac{V_p}{2a}$$

and

$$K_2 = \frac{V_p}{2} - \frac{1}{2\mu}\frac{dp}{dx}a^2$$

Substituting K_1 and K_2 into the expression for V_x, one obtains the following solution:

$$V_x = \frac{V_p}{2}\left(1 + \frac{y}{a}\right) - \frac{a^2}{2\mu}\frac{dp}{dx}\left(1 - \frac{y^2}{a^2}\right) \quad ◀$$

A discussion of this solution will provide better appreciation of the flow under consideration. First, if the flow were maintained by the motion of one plate with zero pressure gradient, the velocity profile would be linear because of the effects of viscosity. For motion of the plate to the right, the established flow between the plates would be from left to right. Second, if both plates are fixed, that is, $V_p = 0$, and the flow between the plates is maintained by the applied pressure gradient dp/dx, then the velocity profile is a parabola (or a parabolic wedge since there are no variations of the dependent variables with the z coordinate). The flow is to the right if $dp/dx < 0$; and it is to the left if $dp/dx > 0$. Third, if the fluid is flowing under the action of the moving plate and a pressure gradient, then the velocity profile may take on a number of forms as indicated by the solution of this problem. A few interesting cases are illustrated in Fig. 4.11.

Note that it is possible at a cross section to have some flow to the right and to the left when a suitably large positive pressure gradient exists. It is also possible to maintain a particular value of the pressure gradient, $dp/dx = \mu V_p/2a^2$, such that the velocity profile is perpendicular to the stationary plate at this surface. In this case the constitutive equation for τ_{yx} indicates that the shear stress would be zero at this plate.

Fig. 4.11

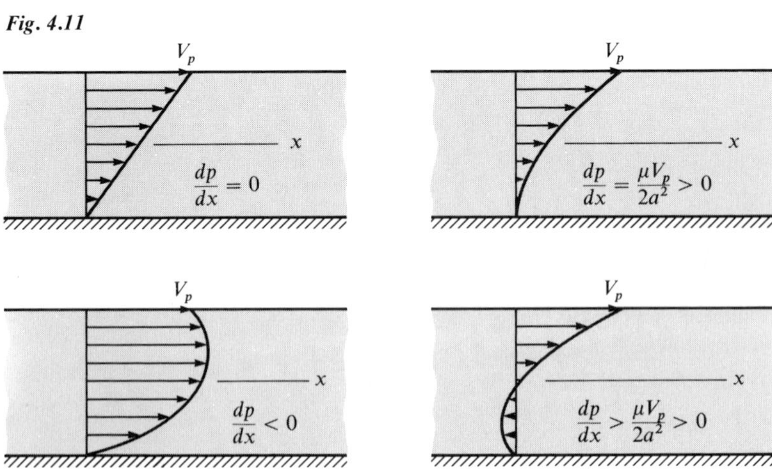

Example 4.5

Consider a fluid field initially at rest and neglect body forces. Figure 4.12 shows a very thin flat plate with a large area immersed in this fluid. The plate undergoes a sudden acceleration such that, for all times $t > 0$, the plate has a velocity of magnitude V_p in the plane of the plate. This produces a motion in the surrounding fluid which is characterized by the fluid having a velocity component in the direction of V_p only. The density is uniform and steady, and the Newtonian fluid has a constant viscosity. Find $V_x = V_x(y,t)$ for laminar flow.

Solution. Since the motion of the fluid is the same above and below the x,z coordinate plane, the motion is analyzed for $y \geq 0$ only. From given information and a sketch of the plate relative to the chosen coordinate system, $V_y = V_z = 0$ at all times and in all positions in space. Equation (2.3), the continuity equation, reduces to $\partial V_x/\partial x = 0$; and this implies that $V_x = V_x(y,t)$, a function of y and t only if it is assumed that V_x is not a function of z which seems plausible in view of the given conditions (the fluid is in no way confined in the z direction). Equations (4.20), the constituted equations of motion, now reduce to

$$\rho \frac{\partial V_x}{\partial t} = -\frac{\partial p}{\partial x} + \mu \frac{\partial^2 V_x}{\partial y^2}$$

$$\frac{\partial p}{\partial y} = 0$$

$$\frac{\partial p}{\partial z} = 0$$

In the absence of any body forces or confinement in the x direction, it is further assumed $\partial p/\partial x = 0$, leading to a final reduced equation

$$\frac{\partial V_x}{\partial t} = \nu \frac{\partial^2 V_x}{\partial y^2}$$

The variable V_x in this equation is subject to the following boundary/initial conditions:

1. $V_x = V_p$ for $y = 0$ and $t > 0$
2. $\lim_{y \to \infty} V_x = 0$ for all values of t
3. $V_x = 0$ for $t = 0$ and all values of y

Fig. 4.12 The plate for time $t > 0$.

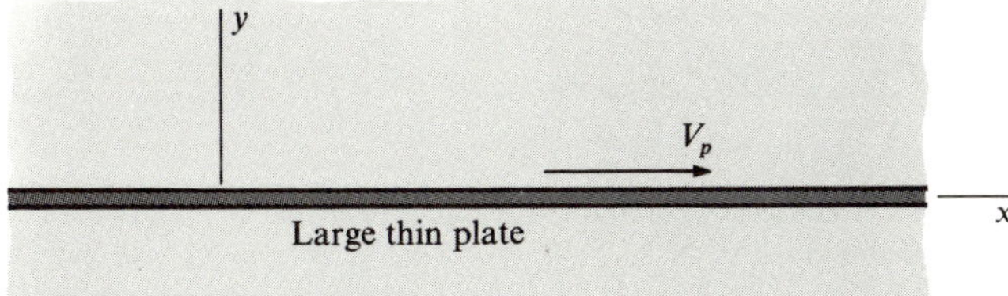

Some partial differential equations may be transformed to ordinary differential equations which are more easily solved.[1] In this problem it is desirable to express V_x as a function of a single variable in such a manner that an ordinary differential equation results after a change in variable is effected. A solution of the form $V_x = V_p f(\eta)$ is assumed on a trial basis. The single independent variable η should be a function of y and t in the physical problem, and it is logical that η should be a dimensionless combination of y and t. One possible choice is $\eta = y/2(\nu t)^{1/2}$, adding the factor of $\frac{1}{2}$ for convenience.

According to rules of partial differentiation,

$$\frac{\partial V_x}{\partial y} = \frac{\partial V_x}{\partial \eta}\frac{\partial \eta}{\partial y} = V_p \frac{df}{d\eta}\frac{1}{2(\nu t)^{1/2}}$$

similarly

$$\frac{\partial^2 V_x}{\partial y^2} = \frac{V_p}{4\nu t}\frac{d^2 f}{d\eta^2}$$

and

$$\frac{\partial V_x}{\partial t} = \frac{\partial V_x}{\partial \eta}\frac{\partial \eta}{\partial t} = V_p \frac{df}{d\eta}\left(-\frac{\eta}{2t}\right) = -\frac{\eta V_p}{2t}\frac{df}{d\eta}$$

Substitution of these into the partial differential equation yields

$$\frac{d^2 f}{d\eta^2} = -2\eta \frac{df}{d\eta}$$

which is an ordinary differential equation. The previously mentioned boundary/initial conditions in the new variable are, respectively,

1. $f(\eta) = 1 \qquad \text{for } \eta = 0$

2 and 3. $\lim_{\eta \to \infty} f(\eta) = 0$

The first integration of the ordinary differential equation gives

$$\ln \frac{df}{d\eta} = -\eta^2 + \ln K \qquad \text{or} \qquad \frac{df}{d\eta} = Ke^{-\eta^2}$$

where K is a constant that arises from the integration.

The first transformed boundary condition may be applied as follows:

$$\int_1^f df = K \int_0^\eta e^{-\eta^2}\, d\eta$$

$$f = 1 + K \int_0^\eta e^{-\eta^2}\, d\eta$$

[1]Arthur G. Hansen, "Similarity Analyses of Boundary Value Problems in Engineering," pp. 1–5, Prentice-Hall, Inc., Englewood Cliffs, N.J., 1964.

The second transformed boundary condition may be used to ascertain the constant K, as follows:

$$0 = 1 + K \int_0^\infty e^{-\eta^2}\, d\eta$$

Solving for K, one obtains

$$K = \left(- \int_0^\infty e^{-\eta^2}\, d\eta \right)^{-1}$$

The definite integral in the above expression is given in most integral tables[1] or it can be evaluated by techniques of the calculus,[2] and its value is $\pi^{1/2}/2$, so that $K = -2/\pi^{1/2}$. The function $f(\eta)$ is then

$$f(\eta) = 1 - \frac{2}{\pi^{1/2}} \int_0^\eta e^{-\eta^2}\, d\eta$$

and

$$V_x = V_p f(\eta) = V_p \left(1 - \frac{2}{\pi^{1/2}} \int_0^\eta e^{-\eta^2}\, d\eta \right)$$

The term containing the integral is called the *error function* (abbreviated erf) and is tabulated in numerous handbooks (values of the error function may be found in Appendix D). The solution is written as

$$V_x = V_p (1 - \operatorname{erf} \eta) \qquad \text{where } \eta = \frac{y}{2(\nu t)^{1/2}}$$ ◀

To use the solution, η is calculated from ν and given values of y and t. The value of the error function for the given η is obtained. Then V_x may be calculated for a given V_p. A velocity profile for a given time ($t > 0$) is shown in Fig. 4.13.

A general comment should be made with regard to this profile as a function of time. With an increase in time, the value of η decreases for a given finite value of y; and erf η decreases as η decreases, so that the velocity V_x increases at a given y as time increases. Of course the value of V_x never exceeds the value of V_p at any value of y since erf $0 = 0$.

Example 4.6

Consider the steady and laminar flow of an incompressible Newtonian fluid in a pipe in the absence of body forces. The flow is assumed to have a velocity component in the direction of the axis of the pipe, which is shown as the z axis in Fig. 4.14. This velocity

[1] Milton Abramowitz and Irene Stegun (eds.), "Handbook of Mathematical Functions," p. 302 (entry 7.4.1), Dover Publications, Inc., New York, 1965.

[2] Wilfred Kaplan, "Advanced Calculus," p. 218, Addison-Wesley Publishing Company, Inc., Reading, Mass., 1952.

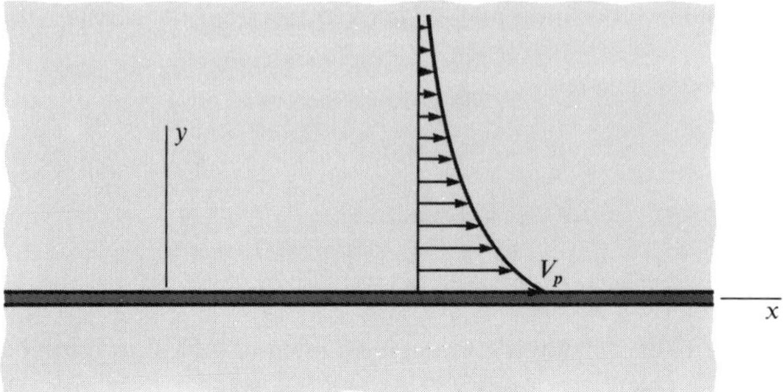

Fig. 4.13

component is considered to be a function of r, θ, and z (cylindrical coordinates). A solution of the velocity distribution in the fluid as a function of space coordinates, and the space average of the velocity of the fluid at any cross section in the pipe are desired.

Solution. A reference set of axes is chosen as shown in Fig. 4.14. The continuity equation, Eq. (2.3), demands that the divergence of the velocity vector vanish for the flow of an incompressible fluid. In cylindrical coordinates, this requires

$$\frac{1}{r}\frac{\partial(rV_r)}{\partial r} + \frac{1}{r}\frac{\partial V_\theta}{\partial \theta} + \frac{\partial V_z}{\partial z} = 0$$

Since there is only a z component of velocity, that is, $V_r = V_\theta = 0$, only the last partial derivative is considered, and $\partial V_z/\partial z = 0$, which on integration yields $V_z = V_z(r,\theta)$. The dependence of V_z on θ is ruled out on further assuming symmetry of the flow about the z axis, and so V_z is established, from conservation of matter and symmetry consideration, as a function of r only.

The Navier-Stokes equation in vector form, Eq. (4.22), may be expressed in cylindrical coordinates by use of the expressions for **grad**, div, and **curl** in Appendix B. Carrying this out, expressing the results in component form, and dropping the terms known or assumed to be zero, gives

$$\frac{\partial p}{\partial r} = 0 \qquad \text{for the } r \text{ component}$$

Fig. 4.14

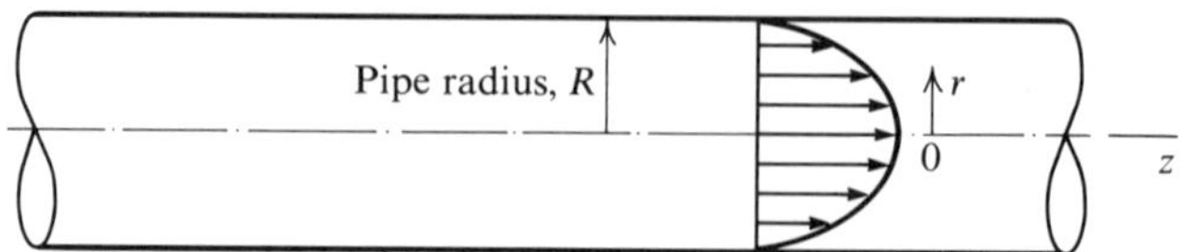

$$\frac{\partial p}{\partial \theta} = 0 \qquad \text{for the } \theta \text{ component}$$

$$-\frac{\partial p}{\partial z} + \frac{\mu}{r}\frac{\partial}{\partial r}\left(r\frac{\partial V_z}{\partial r}\right) = 0 \qquad \text{for the } z \text{ component}$$

The first two of the preceding equations indicate that the pressure is a function of z only, so $\partial p/\partial z = dp/dz$; also $\partial V_z/\partial r = dV_z/dr$ since $V_z = V_z(r)$ only. The problem may then be stated as

$$\frac{d}{dr}\left(r\frac{dV_z}{dr}\right) = \frac{r}{\mu}\frac{dp}{dz}$$

$$V_z(R,\theta,z) = 0 \qquad \text{the no-slip condition}$$

$$\frac{d}{dr}[V_z(0,\theta,z)] = 0 \qquad \text{from axial symmetry}$$

Integrating the first of the above three equations results in

$$\frac{dV_z}{dr} = \frac{1}{2\mu}\frac{dp}{dz}r + \frac{K_1}{r}$$

and

$$V_z = \frac{1}{4\mu}\frac{dp}{dz}r^2 + K_1 \ln r + K_2$$

The condition from axial symmetry requires that $K_1 = 0$, and the no-slip condition yields

$$K_2 = -\frac{1}{4\mu}\frac{dp}{dz}R^2$$

The solution is

$$V_z = \frac{1}{4\mu}\frac{dp}{dz}R^2\left(\frac{r^2}{R^2} - 1\right)$$ ◀

The profile as shown in Fig. 4.14 is a paraboloid of revolution.

In the study of fluid mechanics, a *space-average velocity* is frequently used (e.g., Chapter 8 concerning simplified frictional flows). This velocity is designated by V_{SA} and is defined by

$$V_{SA} = \frac{\int_A \rho \mathbf{V} \cdot d\mathbf{S}}{\rho A}$$

It is the velocity imagined to prevail uniformly over the surface A for the purpose of calculating the mass flow rate through the area. If one knows the velocity profile, the space-average velocity can be obtained by performing the integration indicated by the definition. For this example, the space-average velocity will be found over an area transverse to the pipe. The general element of area is shown in Fig. 4.15. From the solution for the velocity profile, the space-average velocity is (on further assuming that the density is uniform)

$$V_{SA} = \frac{\rho \int_A \{[(1/4\mu)(dp/dz)\, R^2](r^2/R^2 - 1)\}\, \mathbf{k} \cdot 2\pi r\, dr\, \mathbf{k}}{\rho \pi R^2}$$

$$= \frac{(2\pi\rho\, R^2/4\mu)(dp/dz) \displaystyle\int_{r=0}^{r=R} (r^2/R^2 - 1)\, r\, dr}{\rho \pi R^2}$$

$$= -\frac{R^2}{8\mu}\frac{dp}{dz}$$

◀

This velocity, when multiplied by the cross-sectional area of the pipe and the density of the fluid, will yield the mass flow rate through the pipe. Of course one notes that the use of the density is superfluous in defining the space-average velocity for a fluid of uniform density.

Example 4.7

For the previous example, sketch a "typical element" of fluid in the pipe and formulate Newton's law of motion for this element.

Solution. Since the motion is characterized by axial symmetry, an annulus symmetric with respect to the z axis is selected as a typical element. If the fluid is not accelerating, then the shear forces acting on this element are balanced by the forces due to the pressure gradient dp/dz. These forces are indicated in Fig. 4.16. Newton's law of motion requires that the sum of the forces in an axial direction be zero, so

$$p2\pi\left(r + \frac{\Delta r}{2}\right)\Delta r - \left(p + \frac{\partial p}{\partial z}\Delta z\right)2\pi\left(r + \frac{\Delta r}{2}\right)\Delta r - \tau_{rz}(2\pi r\, \Delta z)$$
$$+ \left(\tau_{rz} + \frac{\partial \tau_{rz}}{\partial r}\Delta r\right)[2\pi\,(r + \Delta r)]\,\Delta z = 0$$

Fig. 4.15

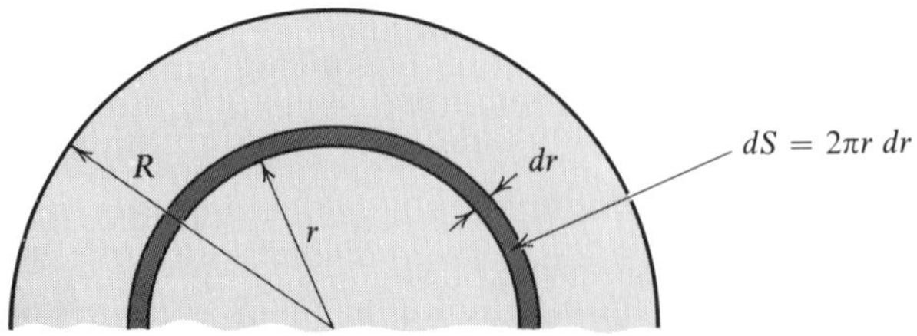

Simplifying the preceding expression and limiting it as Δr and Δz each approach zero gives

$$-\frac{\partial p}{\partial z} + \frac{\tau_{rz}}{r} + \frac{\partial \tau_{rz}}{\partial r} = 0$$

This equation would apply for "any fluid flowing" in a pipe under the given conditions of this problem. If the fluid flowing is newtonian, then $\tau_{rz} = \mu(\partial V_z/\partial r)$, and the equation of motion reduces to

$$-\frac{\partial p}{\partial z} + \frac{\mu}{r}\frac{\partial V_z}{\partial r} + \mu\frac{\partial^2 V_z}{\partial r^2} = 0$$

This is the same as the z component of the Navier-Stokes equation in Example 4.6. To demonstrate that p is a function of z only will require the consideration of a small sector-shaped element rather than a complete annular cylinder. Subsequent summing of the forces in the radial direction, equating this sum to zero, and recognizing that axial symmetry requires no change of p in the θ direction, yields

$$\frac{\partial p}{\partial r} = 0$$

and it is concluded that p is a function of z only. The governing differential equation then becomes

$$-\frac{dp}{dz} + \frac{\mu}{r}\frac{d}{dr}\left(r\frac{dV_z}{dr}\right) = 0$$ ◀

Fig. 4.16

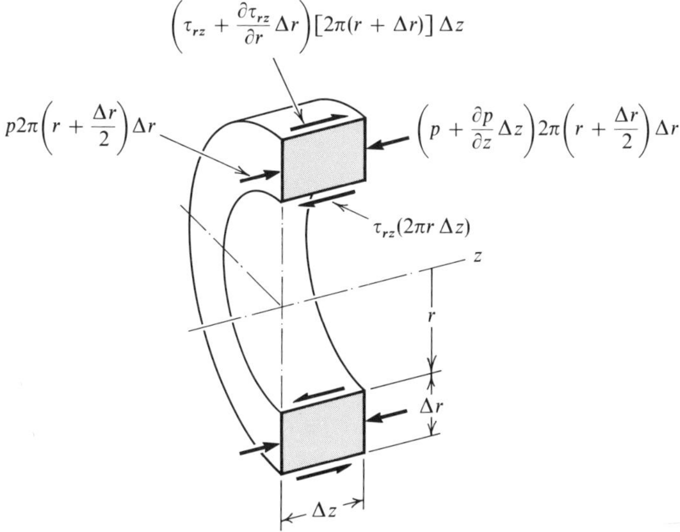

This exercise should illustrate that the simpler physical problems may be handled more easily by selecting a typical element and applying the basic physical laws to obtain a simple equation rather than simplifying the constituted equations of motion. However, this simplicity is small compensation if the recognition of the aspects of more general problems is lost.

Example 4.8

The inside cylinder shown in Fig. 4.17 rotates at a constant angular velocity ω and is coaxial with the stationary outside cylinder. The space between the two cylinders is partially filled with a fluid of uniform and steady density and viscosity. The velocity of the Newtonian fluid has only a transverse component V_θ; and this is a function of r only. For the orientation of the gravity body force indicated, obtain an expression for the velocity field V_θ, the pressure field $p = p(r,z)$, and the equation of the free surface.

Solution. Simplifying the Navier-Stokes equations in cylindrical coordinates by using the given information, the following equations accrue:

$$\frac{\rho V_\theta^2}{r} = \frac{\partial p}{\partial r} \qquad \text{for the } r \text{ component}$$

$$\mu \frac{d}{dr}\left[\frac{1}{r}\frac{d}{dr}(rV_\theta)\right] = 0 \qquad \text{for the } \theta \text{ component}$$

$$\frac{\partial p}{\partial z} + \rho g = 0 \qquad \text{for the } z \text{ component}$$

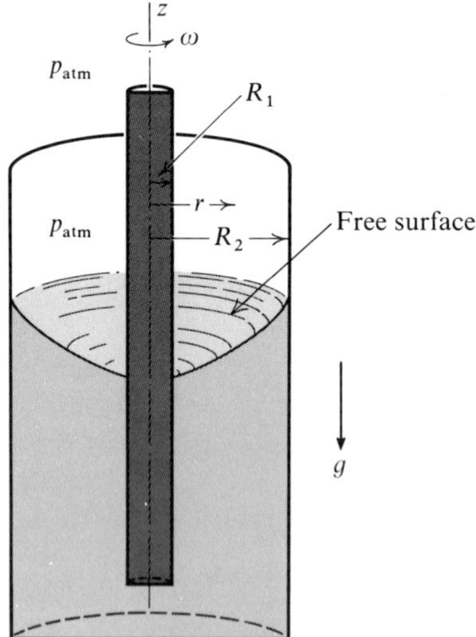

Fig. 4.17

The second equation may be integrated twice to give $V_\theta = C_1 r/2 + C_2/r$; and the constants C_1 and C_2 are found from the no-slip conditions:

$$V_\theta = \omega R_1 \qquad \text{for } r = R_1$$

and

$$V_\theta = 0 \qquad \text{for } r = R_2$$

These constants are

$$C_1 = \frac{2\omega R_1^2}{R_1^2 - R_2^2} \qquad \text{and} \qquad C_2 = \frac{-\omega R_1^2 R_2^2}{R_1^2 - R_2^2}$$

The velocity profile is given by

$$V_\theta = \frac{\omega R_1^2(r^2 - R_2^2)}{r(R_1^2 - R_2^2)}$$ ◀

and is shown in Fig. 4.18. This result is valid only if the "end effects" due to the finite length of the inside cylinder are neglected. One would not expect the solution to be valid in the region near the bottom of the inside cylinder.

The pressure distribution may be found from the r and z components of the Navier-Stokes equations by substitution of V_θ into $\rho V_\theta^2/r = \partial p/\partial r$. This gives, after simplification,

$$\frac{\partial p}{\partial r} = \frac{\rho\omega^2 R_1^4}{(R_1^2 - R_2^2)^2}\left(r - \frac{2R_2^2}{r} + \frac{R_2^4}{r^3}\right)$$

Integrating with respect to r (holding z constant),

$$p = \frac{\rho\omega^2 R_1^4}{(R_1^2 - R_2^2)^2}\left(\frac{r^2}{2} - 2R_2^2 \ln r - \frac{R_2^4}{2r^2}\right) + f(z)$$

in which $f(z)$ is an arbitrary function of z at this stage of the analysis. However, $f(z)$ may be determined as follows: from the preceding expression for p

$$\frac{\partial p}{\partial z} = \frac{d\,f(z)}{dz}$$

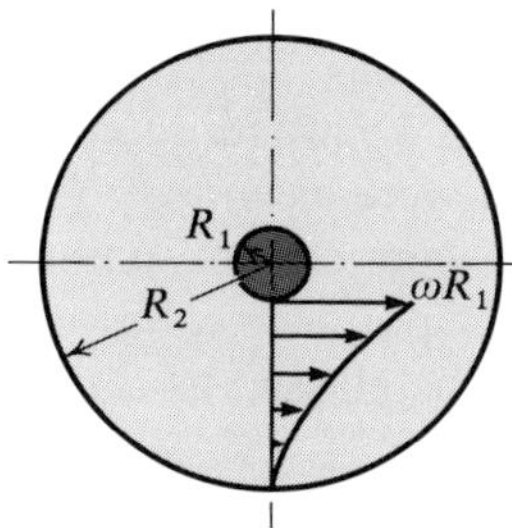

Fig. 4.18

substitute this into the expression for the z component of the Navier-Stokes equations to obtain

$$\frac{df(z)}{dz} + \rho g = 0$$

from which

$$f(z) = -\rho gz + C_3$$

The pressure distribution (or field) is then given by

$$p = \frac{\rho\omega^2 R_1^4}{(R_1^2 - R_2^2)^2}\left(\frac{r^2}{2} - 2R_2^2 \ln r - \frac{R_2^4}{2r^2}\right) - \rho gz + C_3$$

The constant C_3 requires specification of the pressure at a given value of r and z. If the origin of the coordinate system is selected so that $p = p_{\text{atm}}$ where $r = R_1$ and $z = 0$, then

$$p = \frac{\rho\omega^2 R_1^4}{(R_1^2 - R_2^2)^2}\left[\frac{r^2 - R_1^2}{2} + 2R_2^2 \ln \frac{R_1}{r} - \frac{R_2^4}{2}\left(\frac{1}{r^2} - \frac{1}{R_1^2}\right)\right] - \rho gz + p_{\text{atm}} \quad ◀$$

The equation of the free surface is given by the locus of points such that $p = p_{\text{atm}}$, a constant; and substitution of this condition into the solution gives

$$z = \frac{\omega^2 R_1^4}{g(R_1^2 - R_2^2)^2}\left[\frac{r^2 - R_1^2}{2} + 2R_2^2 \ln \frac{R_1}{r} - \frac{R_2^4}{2}\left(\frac{1}{r^2} - \frac{1}{R_1^2}\right)\right] \quad ◀$$

which is indicated by the solid curved line on Fig. 4.17.

Examples 4.4, 4.5, 4.6, and 4.8 could be solved by the approach taken in Example 4.7. This method involved selecting a general infinitesimal element in a natural coordinate system, applying Newton's law of motion, and formulating the stresses in terms of rates of strain. Then, Newton's simple statement concerning fluid shear was used to obtain the differential equations of motion.

4.10 Non-Newtonian Fluids

The application of the constitutive equations in Section 4.9 was limited to the *Newtonian fluid.* The stress–rate-of-strain relations were assumed to be linear, which is not true for all fluids. Some notable exceptions are paints, asphalts, some glues and suspensions, and a number of polymer solutions. A fluid that does not conform to the linear hypothesis is classed as a *non-Newtonian fluid.*

No attempt will be made to go into the general theory of constitutive equations, as this falls within the scope of more advanced texts concerning the general aspects of stress–rate-of-strain relations, a subject referred to as rheology. However, by considering the stress–rate-of-strain relations for a few non-Newtonian fluids, and discussing the solution of several problems, it is hoped that the fact is conveyed that all fluids are not Newtonian fluids.

A few examples will serve to point out differences and some similarities of a Newtonian fluid and several non-Newtonian fluids.

Example 4.9

A non-Newtonian fluid, which is a plastic, passes through a pipe similar to that of Example 4.7. The fluid is assumed to deform at a rate that varies linearly with the shear stress once the yield stress is exceeded, and this is expressed as

$$\tau_{rz} = \tau_0 + \eta_p \frac{dV_z}{dr} \qquad \text{for } \tau_{rz} \geq \tau_0$$

The constant τ_0 is the *yield stress* in *shear*, and η_p is called the *plastic viscosity* or *stiffness*. Obtain the velocity profile for a given pressure[1] gradient and discuss the resultant flow.

Solution. If the pressure is a function of z only, then the applicable equation of motion is

$$-\frac{dp}{dz} + \frac{\tau_{rz}}{r} + \frac{d\tau_{rz}}{dr} = 0$$

or

$$-\frac{dp}{dz} + \frac{1}{r}\frac{d(r\tau_{rz})}{dr} = 0$$

An integration with respect to r (at a given z location) between the limits of 0 to r yields

$$\tau_{rz} = \frac{r}{2}\frac{dp}{dz}$$

Constituting this equation for the given non-Newtonian fluid yields

$$\tau_0 + \eta_p \frac{dV_z}{dr} = \frac{r}{2}\frac{dp}{dz} \qquad \text{for } \tau_{rz} \geq \tau_0$$

Integration with respect to r yields

$$V_z = \frac{r^2}{4\eta_p}\frac{dp}{dz} - \frac{\tau_0 r}{\eta_p} + C_1$$

[1]The pressure that appears in the following is taken as the normal compressive stress in the z direction, $-\sigma_{zz}$. This simple interpretation is offered in lieu of a more precise and general definition of pressure.

and for the no-slip condition of $V_z = 0$ at $r = R$,

$$C_1 = \frac{\tau_0 R}{\eta_p} - \frac{R^2}{4\eta_p}\frac{dp}{dz}$$

The equation of the velocity profile is

$$V_z = -\frac{R^2}{4\eta_p}\frac{dp}{dz}\left(1 - \frac{r^2}{R^2}\right) + \frac{\tau_0 R}{\eta_p}\left(1 - \frac{r}{R}\right) \qquad \text{for } \tau_{rz} \geq \tau_0 \qquad ◀$$

The restriction provided by the inequality is necessary since the plastic material does not "exhibit fluid behavior" until it begins to flow. In the establishment of this flow, the material would begin to deform where the shear stress is the greatest.

Since $\tau_{rz} = (r/2)(dp/dz)$, the shear stress is of maximum magnitude at the wall, where $r = R$, and this magnitude decreases with r so that in a region near the axis of the pipe, τ_{rz} would not be sufficient to cause plastic deformation or flow. The velocity profile is indicated in Fig. 4.19 for $dp/dz < 0$. There is a portion of the fluid near the z axis that moves as a solid body and without shear rate of strain. To find the radius r_p of this "plug," the shear stress is equated to yield stress τ_0 to give $\tau_0 = (r/2)(dp/dz)$ at $r = r_p$, from which $r_p = 2\,\tau_0/(dp/dz)$. The velocity of the plug may be obtained by substitution of $(r_p/2)(dp/dz)$ and r_p for τ_0 and r, respectively, in the equation for V_z. This gives

$$V_p = -\frac{1}{4\eta_p}\frac{dp}{dz}(R - r_p)^2 \qquad \text{for } 0 \leq r \leq r_p \qquad ◀$$

Substances closely characterized by $\tau_{rz} = \tau_0 + \eta_p\,(dV_z/dr)$ are referred to as *Bingham plastics*. Examples of this are some paints, slurries, and some polymer solutions. The fact that a layer of fresh paint applied to a vertical wall will adhere to the wall in a thin coat while that portion of the layer adjacent to the air falls more rapidly is partially explained by the plastic characteristics of the paint.

If the equations for V_z (the two corresponding to $r_p \leq r \leq R$ and $0 \leq r \leq r_p$) are integrated over their respective areas, the volume flow rate of a plastic material extruded[1] from a tube may be estimated. Note that the velocity profile for this flow does not violate the requirement of the shear stress being continuous at $r = r_p$. With reference to the velocity-profile equation for V_z, note that τ_0 is negative (or positive)

[1] For some excellent photographs of plastic flow in a tube, see Arnold G. Fredrickson, "Principles and Applications of Rheology," p. 178, Prentice-Hall, Inc., Englewood Cliffs, N.J., 1964.

Fig. 4.19

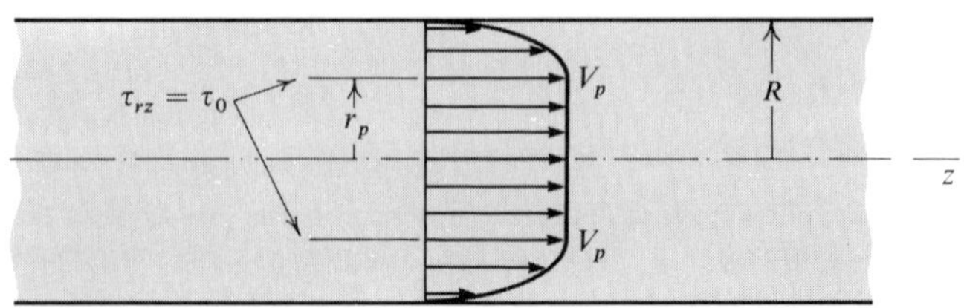

when dp/dz is negative (or positive), which accrues from the manner in which this problem was formulated (see Example 4.7). The last statement emphasizes that the forces due to the pressure gradient are in balance with the forces due to shear stresses for steady flow.

Example 4.10

Consider a non-Newtonian fluid steadily flowing through a pipe similar to that of Example 4.9. The fluid shear stress is related to a strain rate according to a power-law model, expressed as

$$\tau_{rz} = K\left(\frac{dV_z}{dr}\right)^n$$

where K is called a *consistency index* and n is a *flow-behavior index*. The coefficient K should not be thought of as a coefficient of viscosity, but of course if the value of $K = \mu$, and n were unity, the relation would describe the behavior of a Newtonian fluid. Obtain an expression for the velocity profile and discuss these results for various values of n.

Solution. Integration of the applicable form of the equation of motion

$$\frac{dp}{dz} = \frac{1}{r}\frac{d(r\tau_{rz})}{dr}$$

yields

$$\tau_{rz} = \frac{r}{2}\frac{dp}{dz}$$

Constituting this result for the given fluid results in

$$K\left(\frac{dV_z}{dr}\right)^n = \frac{r}{2}\frac{dp}{dz}$$

The solution of this differential equation, subject to the no-slip condition at $r = R$, is

$$V_z = \left|\frac{1}{2K}\frac{dp}{dz}\right|^{1/n}\frac{nR^{(n+1)/n}}{n+1}\left[1 - \left(\frac{r}{R}\right)^{(n+1)/n}\right]$$

The velocity profiles for several values of n are shown in Fig. 4.20. The somewhat flat velocity profile for $n = \frac{1}{2}$ appears much like that for a plastic with plug flow near the axis of the pipe. Fluids with a flow-behavior index of $n < 1$ are referred to as *pseudoplastic fluids*. A few examples are paper pulp, some polymer solutions and soaps, greases, and some paints; and the shear-stress–rate-of-strain relation is closely approximated by the power-law equation. Several disadvantages of this relation are that values of K and n determined from experiment in a pipe or tube cannot be used in another experiment with a different geometry and that the cumbersome dimensions of K depend on the value of n.

Fluids with a flow-behavior index greater than one are referred to as *dilatant fluids*. Examples of these are some dispersions containing high concentrations of solids

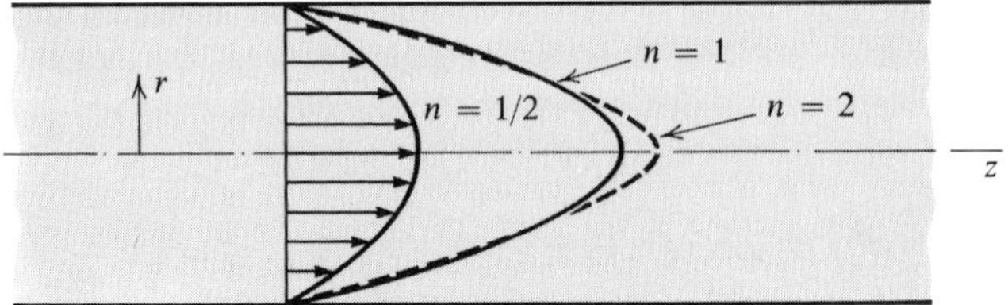

Fig. 4.20

in suspension, quicksand, laundry starch solutions, and some syrups. A rheological diagram is shown in Fig. 4.21, which permits quick comparison of the fluids discussed in this chapter.

4.11 Closing Remarks

With some knowledge concerning the relation of stress to rate of strain, it is possible to solve a few simple problems involving the flow of a fluid with shear stresses present. These problems require the solving of differential equations, and the solutions represent facts that are valid at points within a region. It is also possible to pose a problem in the integral form; this was done in Chapter 2 where a statement of the conservation of matter was formulated for a control volume. In the next chapter the forces acting on a fluid in a given region and the relation of these forces to the change in momentum of the fluid will be considered.

Self-Study Questions

4.1 Upon inspection of Eqs. (2.3), (3.9), (3.10), (3.11), (3.12), and (4.1), name and determine the number of dependent and independent variables for the following cases:
a. Any ideal fluid field

Fig. 4.21

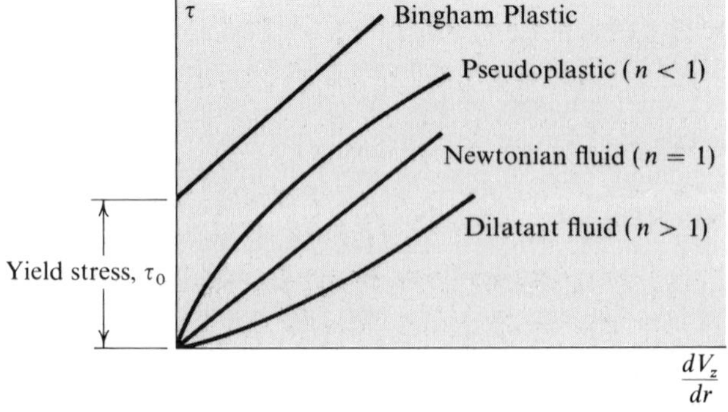

b. A steady and incompressible density fluid field in steady flow
c. An ideal fluid in steady flow in a gravity force field
d. A fluid with negligible body forces in two-dimensional flow
e. Same as (*d*), except that the density is uniform and steady
f. A one-dimensional and steady flow in a gravity force field
g. A one-dimensional flow with a fluid density field that is everywhere constant

4.2 In planes of different orientations through a point, how does mean bulk stress $\bar{\sigma}$ vary?

4.3 State an important difference in the character of σ_{nn} and $\bar{\sigma}$.

4.4 Briefly discuss the meaning of pressure in fluid mechanics and in the thermodynamic equation of state (relating p, ρ, and T).

4.5 State two different conditions concerning the velocity that would yield a zero shear strain rate such that $\gamma_{xy} = 0$.

4.6. Derive expressions for shear strain rates in cylindrical coordinates.

4.7 *a.* If a viscous fluid moves in translation as a solid body, what are the values of rates of shear and longitudinal strain?

b. If a viscous liquid moves in rotation as a solid body (such as a right-circular cylinder spinning on its axis), what are the values of rates of shear and longitudinal strain?

4.8 Obtain an expression for the longitudinal strain rates in cylindrical coordinates.

4.9 Express the different conditions on velocity that would yield a zero rate of longitudinal strain (in x, y, z coordinates).

4.10 Prepare a word statement of the meaning of the set of Eqs. (4.12).

4.11 *a.* Make a word statement of Newton's hypothesis relating stress and rate of strain for a fluid.
b. Obtain the basic dimensions of the coefficient of viscosity μ from the definition. Why is it called dynamic viscosity?
c. Obtain the basic dimensions of the kinematic viscosity ν. Why is it called kinematic viscosity?

4.12 As a fluid changes to a much higher velocity and a corresponding lower temperature, would an increase of viscous effects be expected for a gas? For a liquid?

4.13 Does the viscosity definition apply to both laminar and turbulent flow?

4.14 If the formulation of Newton's hypothesis were taken to apply to any point in the fluid shown in Fig. 4.5, what subscript notation could be added to τ and V?

4.15 What were three assumptions made by Stokes to describe or constitute a fluid in relating stress components and rates of strain?

4.16 *a.* Check the homogeneity of dimensions in the constitutive relations given by Eqs. (4.18).
b. Can shear stresses be other than zero for one-dimensional flow?
c. State the various conditions in a flow field that yield zero shear stress.
d. State the various conditions necessary for an incompressible flow field that would yield each normal stress component equal to the negative of the pressure.
e. State the various conditions necessary in a flow field that would yield each normal stress component equal to the negative of the pressure for fluid fields other than the incompressible type.

4.17 What stress–rate-of-strain relations apply to a Newtonian fluid?

4.18 With reference to Example 4.3, briefly discuss how one could proceed to find the normal stress at a given point on a surface oriented such that the shear stress is zero.

4.19 *a*. The Navier-Stokes equations may be applied to the motion of what type of fluid?
b. Check the dimensional homogeneity of Eq. (4.20*a*).
c. Must the value of viscosity be zero in order for the Navier-Stokes equations of motion to reduce to Euler's equation of motion?

4.20 Write the reduced form of the Navier-Stokes equations of motion for an incompressible and irrotational fluid field.

4.21 Give the reduced form of the constituted equations of motion and the continuity equation, and point out the names and number of independent variables and the number of independent equations for the following situations:
a. A Newtonian fluid (μ = constant) where ρ = constant and the body-force field is known.
b. A Newtonian fluid (μ = constant) in a two-dimensional (x, z) steady flow field where $V_y = 0$ and $V_z = 0$ in a gravity force field only.
c. An incompressible Newtonian fluid (μ = constant) in parallel flow in the x direction and in a gravity force field only.
d. Same as (*c*) but add steady flow.

4.22 Make a mathematical statement concerning the boundary condition of the fluid for each case in Fig. 4.8 if the wing has a velocity V_{wing} and the pipe is stationary relative to an observer.

4.23 Sketch a velocity profile for laminar steady flow of a Newtonian fluid
a. Between two flat parallel plates with one plate moving and a zero pressure gradient.
b. Same as (*a*) except that a negative pressure gradient exists and diminishes the viscosity effect.
c. In a circular pipe.
d. Along a large flat wall with a negative pressure gradient in the fluid.
e. Same as (*d*) but the shear stress is zero at some point at the wall.

4.24 Does the use of space-average values allow the possibility of treating fluid flow problems as one-dimensional in the direction of flow?

4.25 State the difference or similarity of approach to obtain applicable reduced equations of motion in Examples 4.6 and 4.7.

4.26 Give a definition of a non-Newtonian fluid.

4.27 Show that the space-average velocity for fully developed laminar flow of an incompressible fluid in a pipe of circular cross section is one-half the maximum velocity.

Problems

4.1 *a*. Derive the invariance of the sum of normal stresses in two dimensions (e.g., about the z axis).
b. For a given state of stress, obtain an expression for the difference between a normal stress on a plane of arbitrary orientation and the mean bulk stress at a point.

4.2 Obtain expressions for time rates of strain for a fluid flow field with *a.* $\mathbf{V} = \left[A\left(1 - \dfrac{y}{a}\right) + B\left(1 - \dfrac{y^2}{b^2}\right)\right]\mathbf{i}$

where A, a, B, and b are dimensional constants.

b. $\mathbf{V} = \left[V_{\max}\left(\dfrac{y}{y_{\max}}\right)^{1/7}\right]\mathbf{i}$

where $V_{\max}$ is a constant and $y_{\max} = f(x)$.

4.3 Calculate the time rates of shear strain for a fluid flow field with

$$\mathbf{V} = \frac{A}{r^2}\,\boldsymbol{\varepsilon}_r + 0\,\boldsymbol{\varepsilon}_\theta + 0\,\boldsymbol{\varepsilon}_z$$

and a density field of $\rho = Br$ where A and B are dimensional constants. The flow field excludes the origin.

4.4 Calculate the time rate of change of volume per unit volume at any point in the flow field where

$$\mathbf{V} = A\left(\frac{r^2}{R^2} - 1\right)\boldsymbol{\varepsilon}_z$$

A is a dimensional constant and R is the maximum value of r.

4.5 Calculate the rates of shear strain for the flow field given in Prob. 4.4. Locate any points in the fluid where the rate of shear strain is zero.

4.6 Obtain relations for the variation of the normal and shear stresses for Newtonian fluids in each flow field of Prob. 4.2. The pressure at any point in the flow is p.

4.7 Consider a flow field with negligible body forces acting on a Newtonian fluid in the region of $r > 1$ and between $\theta = \pm\pi/4$ and with the fields as given in Prob. 4.3.
a. Show that $\tau_{rz} = \tau_{r\theta} = \tau_{\theta z} = 0$.
b. Establish that $p = p(r)$ for the flow field.
c. If $A = 500$ ft^3/sec and $\mu = 10^{-4}$ lb$_f$-sec/ft^2, evaluate σ_{xx} for $p = 15$ lb$_f$/in.^{2} and $r = 2$ ft at $x = 1.41$ ft.
d. Comment on the significance of the variation of the normal stress σ_{xx} from the given value of pressure.

4.8 In which of the following situations will $\sigma_{xx} = \sigma_{yy} = \sigma_{zz} = -p$ for a Newtonian fluid?
a. $V_x \neq f(x)$, $V_y \neq f(y)$, and $V_z \neq f(z)$.
b. A field of uniform velocity with flow in the x direction only, and where the density is one value everywhere in the field.
c. A flow field in the x direction only, where $V_x = V_x(y)$.
d. A flow field in the z direction only, where $V_r = 0$, $V_\theta = 0$, and $V_z = V_z(r)$.
e. An unsteady incompressible flow in the x direction only.
f. A flow field where the density is everywhere the same value.

4.9 Cite two velocity fields for a Newtonian fluid in which the shear stress is constant.

4.10 Demonstrate the validity of the second boundary condition that concerns the stress continuity at a fluid interface; i.e., show that an infinite acceleration results if a stress discontinuity is assumed.

4.11 With reference to Example 4.4,
a. Obtain expressions for σ_{xx}, σ_{yy}, σ_{zz}, τ_{xy}, τ_{yz}, and τ_{xz}.

b. Show that the normal stress at any point (in any direction in the x, y plane) can be obtained from

$$\sigma = -p + \left(\frac{\mu V_p}{2a} + y\frac{dp}{dx}\right)2lm$$

c. Obtain an expression for the maximum value of $\sigma - \bar{\sigma}$ in the given flow for a positive pressure gradient. What is the location and the values of l and m for this maximum?
d. What is the fractional variation of normal stress from the mean bulk stress at a point $y = a$ and a direction of $l = m = 1/2^{1/2}$ if

$\mu = 10^{-4}\ \text{lb}_f\text{-sec/ft}^2 \qquad a = 0.01\ \text{ft}$

$V_p = 50\ \text{ft/sec} \qquad \dfrac{dp}{dx} = 0.694\ \text{lb}_f/\text{in.}^2\text{-ft}$

$p = 4000\ \text{lb}_f/\text{ft}^2$

e. Calculate the external force needed to move the plate for the conditions given in (*d*) on a per unit length and width basis.

4.12 With reference to Example 4.4,
a. If both plates were stationary, determine any location in the fluid flow field where the shear stress would be zero. Would this be true at any x location? What is the direction of the shear stress at points in the fluid that are adjacent to the wall (at $y = a$ and $y = -a$) for $dp/dx < 0$ and $dp/dx > 0$?
b. Obtain an expression for the pressure gradient that
i. Yields a zero shear stress at the stationary surface.
ii. Yields a zero shear stress at the moving surface.
iii. Will produce zero net mass flow between the two plates.

4.13 With reference to Example 4.5, and given

$V_p = 20\ \text{ft/sec} \qquad \mu = 2(10^{-5})\ \text{lb}_f\text{-sec/ft}^2 \qquad \text{and} \qquad \rho = 1.93\ \text{slug/ft}^3$

a. Plot a velocity profile for $t = 0$, $t = 1$ sec, and $t = 10$ sec.
b. Evaluate the shear stress in the fluid at the moving plate when $t = 10$ sec.

4.14 With reference to Example 4.6, and given $dp/dz = -6(10^{-2})\ \text{lb}_f/\text{in.}^2\text{-ft}$, calculate the force exerted by the fluid on the pipe wall per unit length of 1-in. diameter pipe.

4.15 Select a fluid element in the given flow of Example 4.4, show the forces acting on the element, and formulate the reduced equations of motion consistent with the assumptions used in the solution.

4.16 A Newtonian fluid in liquid form with constant viscosity flows down a vertical wall because of gravity effects. The free surface is exposed to air at 1 atm. Assuming that the flow is steady and of constant thickness from the wall and that the density value is the same everywhere in the flow field,
a. Obtain an expression for the velocity in terms of the thickness t.
b. Sketch the velocity profile.
c. Estimate the free surface velocity for a thickness of $\frac{1}{8}$ in. if $\rho = 1.94\ \text{slug/ft}^3$ and $\mu = 10^{-4}\ \text{lb}_f\text{-sec/ft}^2$.

4.17 An incompressible Newtonian fluid with steady properties passes through the annular space between two straight pipes. The inner and outer radius of the annulus are R_1 and

R_2, respectively. Assuming that V_r and V_θ are zero everywhere, obtain an expression for the velocity profile in terms of r, R_1, R_2, dp/dz, and μ, which is assumed constant.

4.18 A straight rod of radius R_1, immersed in an incompressible and Newtonian fluid, is centrally located in a circular sleeve of radius R_2 as shown in Fig. 4.22. The rod moves to the right at a constant velocity V_{rod}. Obtain an expression for the force per unit length that must be exerted on the rod, assuming steady laminar flow, negligible end effects, and a constant pressure gradient.

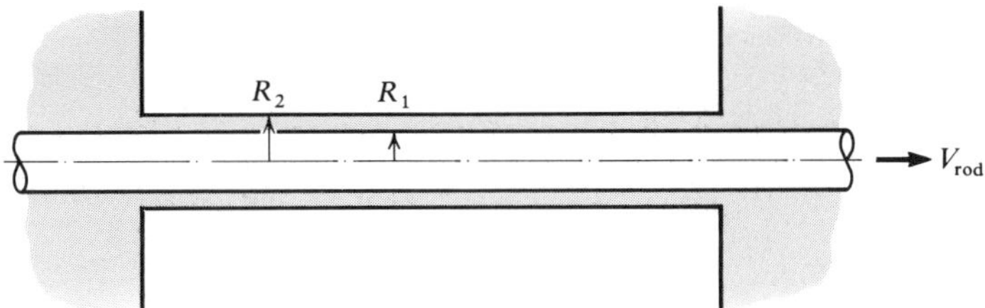

Fig. 4.22

4.19 A flat plate is slowly moved toward a fixed wall as illustrated in Fig. 4.23. The Newtonian fluid has a constant density.

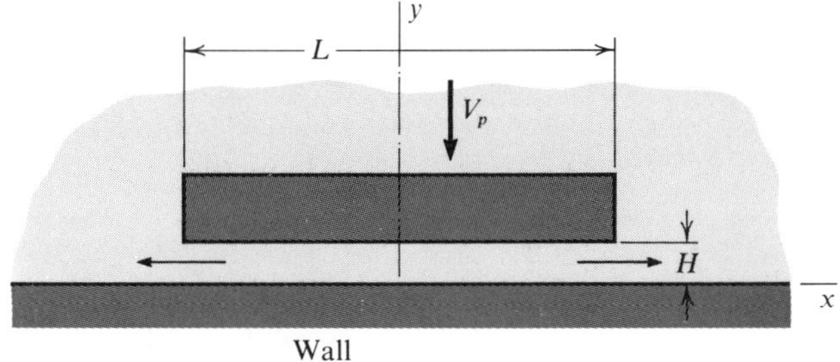

Fig. 4.23

a. Assuming two-dimensional flow, simplify the Navier-Stokes equations for this flow.
b. List the boundary/initial conditions necessary to solve the problem.
c. Do not solve the problem, but sketch a V_x profile at $x = L/2$.

4.20 Consider the flow situation discussed in Example 4.2. Assume a Newtonian fluid of constant density and $V_\theta = V_\theta(r)$ while $V_r = V_z = 0$. Use Eq. (4.22) in cylindrical coordinates to determine the velocity and pressure-profile equations.

4.21 A Bingham plastic material flows along a vertical wall due to gravity effects. One surface is exposed to the atmosphere. Assume that the flow is steady and of constant thickness from the wall. The density and the plastic viscosity η_p are constant.
a. Obtain an expression for the velocity profile at a cross section for a thickness t.
b. Sketch the velocity profile.
c. Obtain an expression for the velocity at the free surface.

5

INTEGRAL FORMS OF THE MOMENTUM EQUATION

5.1 Introduction

The methods presented in Chapter 4 offered a means of determining the velocity and pressure field by solving differential equations subject to certain boundary and initial conditions. By using the velocity field, the stress field can be ascertained from constitutive equations and a knowledge of the pressure field, and these stresses may be integrated over surfaces that partially bound the fluid to find the force the fluid exerts on these surfaces. A prediction of such forces is necessary in many problems of engineering interest, namely: The force or torque a fluid transmits to the rotor of a turbine is important if the power output is to be predicted, the thrust a rocket can develop depends on the force exerted by the hot gases on the thrust chamber, and the power required to propel an airplane is dependent in part on the drag due to fluid friction. These are just a few of many problems in which the engineer must be able to predict forces transmitted between a flowing fluid and a bounding surface.

Since the differential equations of motion are frequently difficult to solve, it appears desirable to formulate Newton's second law of motion in a manner more easily applied to a flowing fluid. The integral forms of the laws of linear and angular momenta offer an alternative, and these forms may be applied to situations where the velocity profiles can be measured or reasonably estimated at certain sections in the flow field. This method has been employed extensively and successfully for estimating the shear stress in boundary-layer flows. The integral form of the boundary-layer equations is treated in Chapter 9.

5.2 Linear Momentum Principle for an Inertial Control Volume

Equation (1.37) may be applied to any control volume in which matter is conserved, with the understanding that the vector **B** is an extensive property and that the velocity **V** is relative to the control surface. If the control volume is fixed with respect to an inertial[1] frame of reference, then all quantities may be referred to a frame of reference that is fixed in inertial space; and the time change of vector quantities at a point in this frame of reference presents no problem. The equation relating the time rate of change of **B** to control volume and surface integrals may be written as

$$\frac{D\mathbf{B}}{Dt} = \int_{V} \frac{\partial(\mathbf{b}\rho)}{\partial t}\, dV + \oint_{S} \rho\mathbf{b}\mathbf{V}\cdot d\mathbf{S} \tag{5.1}$$

which is Eq. (1.37) slightly modified in symbolism only.[2] Since the linear momentum **P** of a small fluid element is the product of its mass and velocity, **P** is an extensive quantity, and specific linear momentum is the linear momentum per unit mass which is the velocity vector **V**. Equation (5.1) may be written for **P** as

$$\frac{D\mathbf{P}}{Dt} = \int_{V} \frac{\partial(\rho\mathbf{V})}{\partial t}\, dV + \oint_{S} \rho\mathbf{V}\,(\mathbf{V}\cdot d\mathbf{S}) \tag{5.2}$$

The left-hand side of Eq. (5.2) is the time rate of change of linear momentum of the system that instantaneously occupies the control volume, and this must be equal to the sum of the external forces acting on the system. These external forces may be surface forces $\mathbf{F}_S$, or body forces $\mathbf{F}_B$. Note that the boundaries of the system are coincident with the control surface at the instant of observation, and the mass comprising the system is the same as the mass inside of the control volume at the instant of observation, so that the left-hand side of Eq. (5.2) may be equated to the vector sum of the surface

[1]The term *inertial control volume* will be used to denote a control volume that is fixed relative to an inertial frame of reference.

[2]The modifications consist of the addition of three component equations, each of the form of Eq. (1.37).

forces acting on the control surface and the body forces acting on the material inside of the control volume. The *linear momentum principle for a control volume fixed with respect to an inertial frame of reference* may then be written as

$$\sum \mathbf{F}_S + \sum \mathbf{F}_B = \int_{\forall} \frac{\partial(\rho \mathbf{V})}{\partial t}\, d\forall + \oint_S \rho \mathbf{V}\, (\mathbf{V} \cdot d\mathbf{S}) \tag{5.3}$$

The first term on the right-hand side of the preceding equation is the time rate of change of linear momentum within the control volume, while the second term is the net efflux of linear momentum through the control surface. Equation (5.3) has many useful applications in solving engineering problems since a careful choice of a control surface can simplify the evaluation of the surface forces.

Example 5.1

Water with a constant density of 1.94 slug/ft^3 flows steadily through a 12-in. diameter pipe that terminates in a nozzle at one end. The exit diameter of the nozzle is 3 in. The uniform velocity of the fluid at the nozzle exit is 112 ft/sec. Pressure information is given in Fig. 5.1. What is the force the water exerts on the interior of the nozzle wall?

Solution. A control volume is selected with a surface inside the nozzle and next to the nozzle wall. This portion of the control surface is indicated by dashed lines on Fig. 5.1. The control surface is then composed of this lateral area of the interior of the nozzle wall, a transverse area of 1-ft diameter at nozzle entrance, and a transverse area of 3-in. diameter at nozzle exit. The forces acting on the fluid in the control volume are shown in Fig. 5.2 where $p_1 A_1$ is the force the water (upstream of the nozzle entrance area) exerts on the material in the control volume, $p_2 A_2$ is the force the water (downstream of the nozzle exit) exerts on the material in the control volume, and the force $\mathbf{F}_w$ (due to p_w and τ_w) is the resultant force exerted by the nozzle on the fluid in the control volume. All these forces are surface forces. The only body force is the weight (F_{wt}) of the fluid inside the control volume.

Fig. 5.1

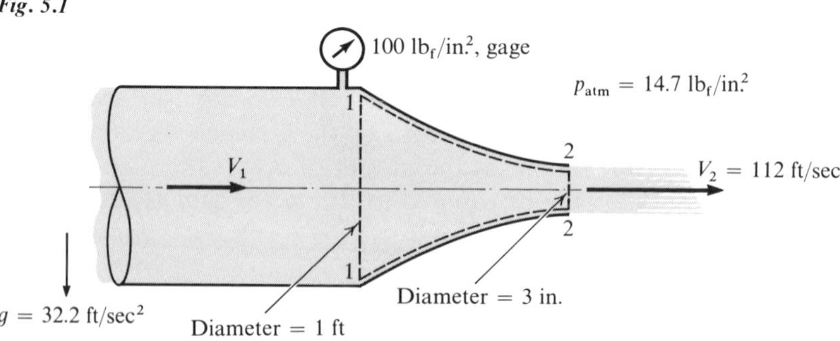

Reference to the control volume indicates that the pressure has been used to obtain surface forces normal to the control surfaces, when the normal stress should have been used, in a strict sense. It has been assumed that the value of the pressure does not deviate appreciably from the value of the normal stress (refer to Section 4.7). It has been further assumed that the pressure at sections 1 and 2 is uniformly distributed over these sections. The uniformly distributed characteristic will also be assumed for the velocities V_1 and V_2.

Applying Eq. (5.3) to this control volume, two scalar equations result

$$p_1 A_1 - p_2 A_2 - (\mathbf{F}_w)_x = \int_{V} \overset{0}{\frac{\partial(\rho V_x)}{\partial t}} \, dV + \oint_S \rho V_x (\mathbf{V} \cdot d\mathbf{S})$$

and

$$-F_{wt} + (\mathbf{F}_w)_y = \int_{V} \overset{0}{\frac{\partial(\rho V_y)}{\partial t}} \, dV + \overset{0}{\oint_S \rho V_y (\mathbf{V} \cdot d\mathbf{S})}$$

As indicated, several of these integrals are zero since the flow is steady and there is no y component of velocity at sections 1 and 2. The equations can be simplified to

$$(\mathbf{F}_w)_x = p_1 A_1 - p_2 A_2 - \oint_S \rho V_x (\mathbf{V} \cdot d\mathbf{S})$$

$$(\mathbf{F}_w)_y = F_{wt}$$ ◀

If one knows the weight of water in the nozzle, then $(\mathbf{F}_w)_y$ is determined. This is a logical result when one looks at the problem. The equation for $(\mathbf{F}_w)_x$ can be written as

$$(\mathbf{F}_w)_x = p_1 A_1 - p_2 A_2 - (\rho A_2 V_2 V_{2x} - \rho A_1 V_1 V_{1x})$$

since the velocities have been assumed uniform over A_1 and A_2 and there is no momentum flux passing through the nozzle wall. The terms $\rho A_2 V_2$ and $\rho A_1 V_1$ are equal and each represents the steady mass flow passed by the nozzle. This follows from the conservation of matter, and $V_1 = A_2 V_2 / A_1 = 7$ ft/sec. The minus sign appears in front of the term $\rho A_1 V_1 V_{1x}$ since it includes the mass flow of material into the control

Fig. 5.2

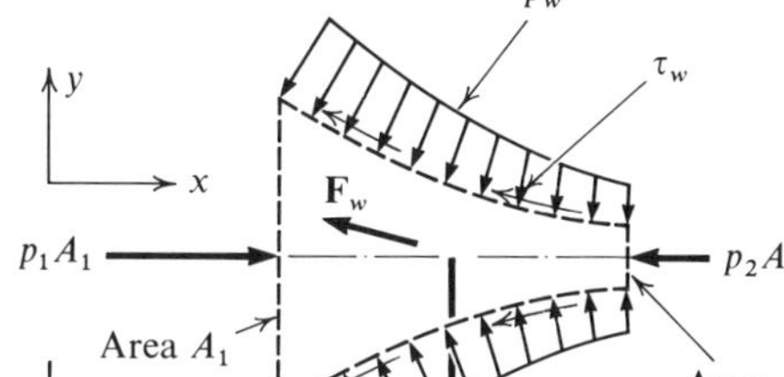

volume, and the cosine of the angle between the velocity vector and the outward-drawn normal to the control surface at section 1 is negative, that is,

$$\int_{A_1} V_{1x}\, \rho \mathbf{V}_1 \cdot d\mathbf{S}$$

is negative until a sign is assigned to V_{1x}. A corresponding reason can be given for the plus sign in front of the term $\rho A_2 V_2 V_{2x}$. Putting in numbers for the symbols in the last equation for $(\mathbf{F}_w)_x$, one obtains

$$\begin{aligned}(\mathbf{F}_w)_x &= \left(100\,\frac{\text{lb}_f}{\text{in.}^2} + 14.7\,\frac{\text{lb}_f}{\text{in.}^2}\right)\frac{\pi}{4}(12\text{ in.})^2 - 14.7\,\frac{\text{lb}_f}{\text{in.}^2}\,\frac{\pi}{4}(3\text{ in.})^2 \\ &\quad - 1.94\,\frac{\text{slugs}}{\text{ft}^3}\,\frac{\pi}{4}\left(\frac{3\text{ in.}}{12\text{ in./ft}}\right)^2 112\,\frac{\text{ft}}{\text{sec}}\left[+112\,\frac{\text{ft}}{\text{sec}} - \left(+7\,\frac{\text{ft}}{\text{sec}}\right)\right] \\ &= 12{,}980 - 104 - 1120 \\ &= 11{,}756\text{ lb}_f\end{aligned}$$

◀

Note that a plus sign was placed in front of both 112 ft/sec and 7 ft/sec because both V_2 and V_1 are in the arbitrarily chosen positive x direction. The plus sign in front of 11,756 lb_f indicates that the force $(\mathbf{F}_w)_x$, which the nozzle walls exert on the water, was assumed in the correct direction. Since the force the water exerts on the nozzle wall is sought, the answer is equal and opposite to $(\mathbf{F}_w)_x$ and $(\mathbf{F}_w)_y$.

It appears that there are a number of assumptions involved in the solution of Example 5.1. It is a good practice to list one's assumptions in the solution of a given problem, and to associate the assumptions with the simplification of the control-volume equations. Also, care must be exercised in assigning the proper signs to the terms in the equations, and this requires being consistent with the sign conventions pertaining to the mass flows in and out of the control volume as well as those pertaining to the velocity components associated with the various momenta.

Some comment should be made concerning the pressure p_2 at the nozzle exit in Example 5.1. One will observe that p_2 was taken to be atmospheric pressure. It was assumed that the pressure in the jet is the same as the surrounding or receiving pressure of the atmosphere. This is reasonable by an indirect argument. For example, if it is assumed that $p_2 > p_{\text{atm}}$, the jet would tend to enlarge or spread. This enlargement would result in a reduced velocity according to the conservation of mass for a constant density fluid, and one would expect that a reduction in velocity would further increase the pressure in the jet. Such an instability has not been observed in nature, so one concludes that the jet pressure cannot be greater than the receiver or surrounding pressure. A similar argument leads to the conclusion that the jet pressure cannot be less than the surrounding pressure, so it is therefore concluded that p_{jet} is equal to the receiver pressure. For the case of a compressible fluid, pressure adjustments are possible through shock waves, and the jet pressure does not have to be equal to the receiver pressure. More will be said about this in Chapter 8.

Example 5.2

Consider the steady flow of an incompressible fluid from a nozzle. The velocity of the fluid relative to the nozzle is 50 ft/sec, and the cross-sectional area of the jet is 0.0462 in^2. The fluid leaving the nozzle is received by a blade translating at 20 ft/sec (relative to the nozzle) in the direction of the positive x axis as shown in Fig. 5.3. Consider that the flow of the fluid through the blade is frictionless and that the density remains uniform at 62.4 lb_m/ft^3. Neglect body forces, and calculate the forces F_x and F_y that must be applied to the blade to restrain the blade velocity to a constant value of 20 ft/sec.

Solution. Before applying the linear-momentum control-volume equation, one must select a control volume; and for convenience it is selected to move to the right at a constant velocity equal to the blade velocity. This is an inertial control volume since it moves with constant velocity. Figure 5.4 shows the control surface with the surface forces acting on it.

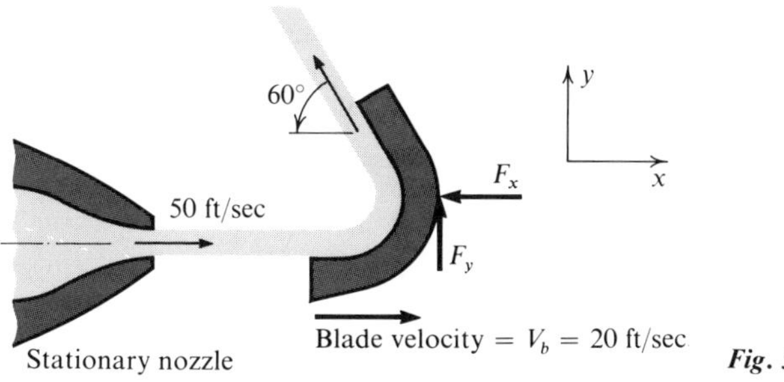

Fig. 5.3

Fig. 5.4

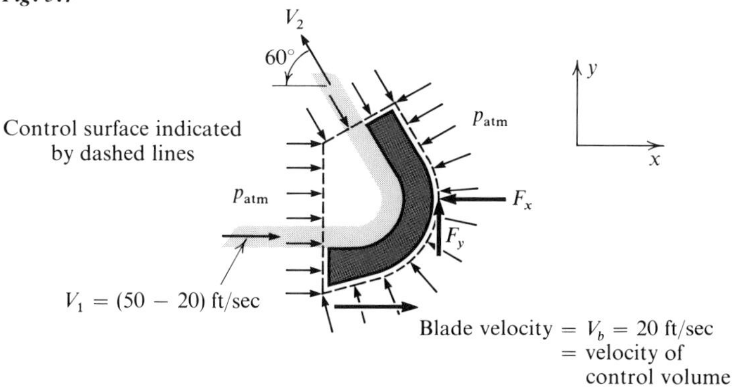

V_2
60°
p_{atm}
y
x
Control surface indicated by dashed lines
p_{atm}
F_x
F_y
$V_1 = (50 - 20)$ ft/sec
Blade velocity = V_b = 20 ft/sec = velocity of control volume

Since body forces are neglected, Eq. (5.3) may be written as

$$\sum \mathbf{F}_S = \int_{\mathcal{V}} \frac{\partial(\rho \mathbf{V})}{\partial t} \, d\mathcal{V} + \oint_S \rho \mathbf{V} \, (\mathbf{V} \cdot d\mathbf{S})$$

and it is understood that velocity **V** in this equation must be relative to the control surface. If it is assumed that velocity of the fluid relative to the blade at entrance (V_1) and exit (V_2) is uniformly distributed over each of these sections, and that the same is true of the jet pressure at each of these sections, then application of Bernoulli's equation with the body-force terms (gz_1 and gz_2) excluded yields $V_1 = V_2 = 30$ ft/sec. It should also be recognized that the mass flow passed by the nozzle is not equal to the mass flow passing through the blade since the blade is "pulling away from the nozzle." The mass flow captured by the blade is

$$\left(62.4 \, \frac{\text{lb}_m}{\text{ft}^3}\right)\left(\frac{0.0462 \text{ in.}^2}{144 \text{ in.}^2/\text{ft}^2}\right)\left(30 \, \frac{\text{ft}}{\text{sec}}\right) = 0.6 \, \frac{\text{lb}_m}{\text{sec}}$$

which is calculated by using the velocity of the fluid relative to the control surface.

The x and y components of the momentum equation in control-volume form are

$$-F_x = \left(\frac{0.6 \text{ lb}_m/\text{sec}}{32.2 \text{ ft-lb}_m/\text{lb}_f\text{-sec}^2}\right)\left[\left(-30 \, \frac{\text{ft}}{\text{sec}}\right) \cos 60^\circ - \left(+30 \, \frac{\text{ft}}{\text{sec}}\right)\right]$$

$$F_y = \left(\frac{0.6 \text{ lb}_m/\text{sec}}{32.2 \text{ ft-lb}_m/\text{lb}_f\text{-sec}^2}\right)\left[\left(30 \, \frac{\text{ft}}{\text{sec}}\right) \sin 60^\circ - (0)\right]$$

and

$$F_x = 0.84 \text{ lb}_f \quad \blacktriangleleft$$

$$F_y = 0.485 \text{ lb}_f \quad \blacktriangleleft$$

The positive answers indicate that the directions of F_x and F_y shown on the control surface were assumed correctly. Atmospheric pressure, which acts over the entire control surface, contributes no net surface force acting on the control surface in any direction.

Example 5.3

The rocket shown in Fig. 5.5 is held stationary by the force F. The velocity of the exit gases relative to the rocket is 5000 ft/sec and the pressure of the gases at exit is 20 $\text{lb}_f/\text{in.}^2$ These values are assumed to be uniform over the exit area of 60 in.^{2} The mass rate of flow is 21 lb_m/sec, and it is held constant by a steady flow of fuel and oxidizer while on a ground test stand. Atmospheric pressure is 15 $\text{lb}_f/\text{in.}^2$ Calculate the force F.

Solution. The control surface is taken around the entire rocket as indicated by the dashed lines in Fig. 5.5. The pressure at the exit has been indicated by the sum of p_{atm} and (20 $\text{lb}_f/\text{in.}^2 - p_{\text{atm}}$). By doing this, it is readily evident that the pressure of the atmosphere, which "acts over the entire control surface," contributes no net force that acts on the control surface; and the resulting effect of all pressures acting on this surface

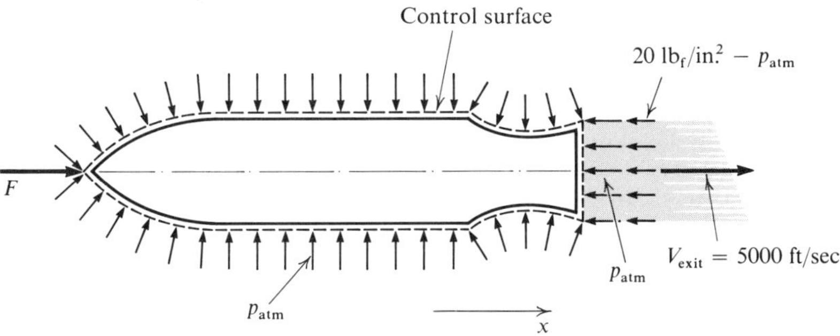

Fig. 5.5

is (20 lb$_f$/in.^{2} − p_{atm}) A_{exit}, which has a direction to the left. The gases leaving the rocket are compressible, hence the jet pressure does not have to equal that of the surrounding atmosphere (recall the argument following Example 5.1). The control volume defined may be considered inertial and Eq. (5.3) applies. The x component of this is

$$F - \left(20\,\frac{\text{lb}_f}{\text{in.}^2} - 15\,\frac{\text{lb}_f}{\text{in.}^2}\right)(60\text{ in.}^2) = \left(\frac{21\text{ lb}_m/\text{sec}}{32.2\text{ ft-lb}_m/\text{lb}_f\text{-sec}^2}\right)\left(5000\,\frac{\text{ft}}{\text{sec}} - 0\right)$$

thus

$$F = 3260\text{ lb}_f + 300\text{ lb}_f = 3560\text{ lb}_f$$ ◀

Example 5.4

An incompressible Newtonian fluid with a density of 50 lb$_m$/ft^3 enters a long tube of 3-in. diameter. The fluid is guided into the tube by a bell-mouth entrance as shown in Fig. 5.6, and the velocity at section 1 is essentially uniform. Because of viscous shear stresses, the fluid velocity decreases toward the inside tube wall as the fluid flows through the pipe. The no-slip condition prevails at the wall. Downstream from the entrance,

Fig. 5.6

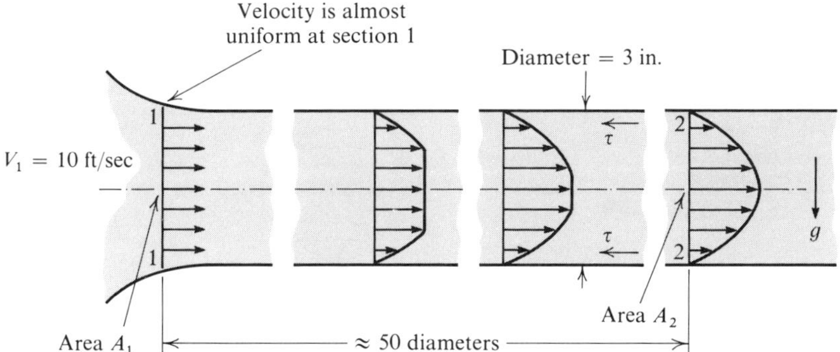

the region called the "core" (in which the velocity is still uniform) begins and continues to decrease as indicated. Finally, at some distance downstream, in the locality of section 2, the velocity profile assumes a form that does not change as the fluid moves farther downstream. The flow at this section is described as *fully developed laminar flow*, assuming that the entering flow at section 1 is laminar. Note that the velocity profile at section 2 is one that was determined in Example 4.6; and the flow is fully developed when the forces due to shear stresses are balanced by the forces due to the pressure gradient. The measured pressure difference $p_1 - p_2$ is 2.61 lb_f/in^2. For a steady velocity field, calculate the force the inside walls exert on the fluid, due to the shear stresses, between sections 1 and 2.

Solution. A control volume bounded by sections 1 and 2 and the inside wall of the pipe is selected; the forces acting on the control surface are indicated in Fig. 5.7 and the only body force is the weight of the fluid in the control volume. This body force is not involved in the calculation of the answer.

The velocity profile at section 2 is a paraboloid of revolution (that ascertained in Example 4.6). The equation of this profile is given by

$$V_z = V_0\,[1 - (r/R)^2]$$

where $R = 1\tfrac{1}{2}$ in., the radius of the tube
V_0 = velocity at axis of the tube

This velocity distribution must be known (V_0 is unknown at this stage of the analysis) before the integral for the momentum flux can be evaluated at this section. Application of Eq. (2.1) for conservation of matter to the control volume of this problem gives $V_0 = 20$ ft/sec. The z component of Eq. (5.3) is

$$p_1 A_1 - p_2 A_2 - F_{\text{shear}} = \int_{V\!\!\!/} \frac{\partial(\rho V_z)}{\partial t}\, dV\!\!\!/ + \oint_S \rho V_z\, (\mathbf{V} \cdot d\mathbf{S})$$

The volume integral is zero since the product ρV_z does not change with respect to time at a given location. Formulating the surface integral and solving for the shear force gives

Fig. 5.7

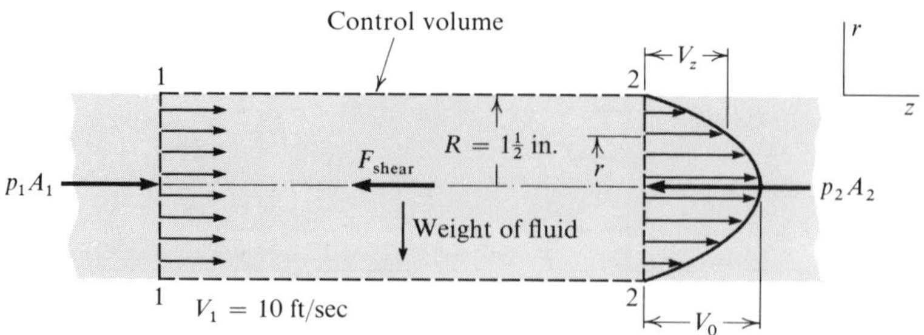

$$F_{\text{shear}} = p_1 A_1 - p_2 A_2 - \left\{ \iint_{A_2} \rho V_0 \left[1 - \left(\frac{r}{R} \right)^2 \right] V_0 \left[1 - \left(\frac{r}{R} \right)^2 \right] \mathbf{k} \cdot 2\pi r \, dr \, \mathbf{k} + \int_{A_1} \rho V_1 [V_1 \mathbf{k} \cdot 2\pi r \, dr (-\mathbf{k})] \right\}$$

$$= p_1 A_1 - p_2 A_2 - 2\pi \rho V_0^2 \int_0^R \left[1 - \left(\frac{r}{R} \right)^2 \right]^2 r \, dr + \rho V_1^2 A_1$$

$$= p_1 A_1 - p_2 A_2 - 2\pi \rho V_0^2 \frac{R^2}{6} + \rho V_1^2 A_1$$

Substitution of the given quantities in the last expression yields

$$F_{\text{shear}} = 2.61 \frac{\text{lb}_f}{\text{in.}^2} \frac{\pi}{4} (3 \text{ in.})^2 - \frac{2\pi(50 \text{ lb}_m/\text{ft}^3)(20 \text{ ft/sec})^2 (1.5 \text{ ft}/12)^2}{6(32.2 \text{ ft-lb}_m/\text{lb}_f\text{-sec}^2)} + \frac{50 \text{ lb}_m/\text{ft}^3 \, (10 \text{ ft/sec})^2 (\pi/4)(3 \text{ ft}/12)^2}{32.2 \text{ ft-lb}_m/\text{lb}_f\text{-sec}^2}$$

$$= 18.5 \text{ lb}_f - 10.2 \text{ lb}_f + 7.6 \text{ lb}_f$$

$$= 15.9 \text{ lb}_f$$ ◀

The fact that the shear force is a positive number indicates that the direction assumed for this force was correct.

Example 5.5

The flat and horizontal plate shown in Fig. 5.8 is in a stream of uniform velocity V_∞ at zero incidence. The density is uniform and steady, and there are no pressure gradients in the flow. There is a layer of low-velocity fluid close to the plate. This is referred to as the

Fig. 5.8

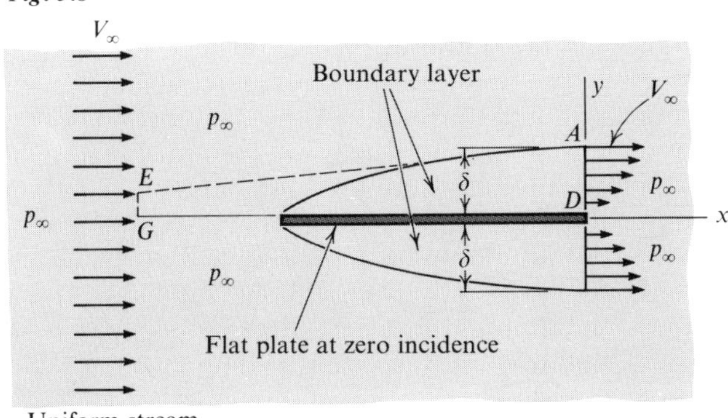

boundary layer, and the flow outside of this layer is considered to be free of shear stress. (Example 5.4 involved a boundary layer on the inside of a cylindrical tube.) The velocity profile at the trailing (downstream) edge of the plate is indicated and measurement of the velocities in this layer provides a profile equation of the form

$$V_x = V_\infty \left[2\frac{y}{\delta} - \left(\frac{y}{\delta}\right)^2 \right] \qquad \text{for } 0 \leq y \leq \delta$$

in which δ is the thickness of the boundary layer at the trailing edge of the plate. More technical terms and definitions are given for the boundary layer in Chapter 9. The velocity field is two dimensional and does not depend on the z coordinate, and the gravity field is in the z direction. For a steady velocity field, obtain an expression for the shear force one side of the plate exerts on the fluid in terms of V_∞, δ, and the density of the fluid.

Solution. A control volume is selected that has a unit length in the z direction and that is bounded by the surfaces shown in Fig. 5.9. The curve AE is the streamline passing through the point A so that there is no flow through the "top portion" of the control surface. The x component of Eq. (5.3) for this control volume is

$$p_\infty \delta - F_{\text{shear}} - p_\infty \delta = \int_0^\delta \rho V_\infty \left[2\frac{y}{\delta} - \left(\frac{y}{\delta}\right)^2 \right] V_\infty \left[2\frac{y}{\delta} - \left(\frac{y}{\delta}\right)^2 \right] \mathbf{i} \cdot (1\, dy\, \mathbf{i})$$

$$+ \int_0^h \rho V_\infty V_\infty \mathbf{i} \cdot (-1\, dy\, \mathbf{i})$$

The term $p_\infty \delta$ is the x component of the surface forces acting on the surfaces $EFGH$ and $EABF$, while $-p_\infty \delta$ is the surface force acting on the surface $ABCD$. There are no

Fig. 5.9

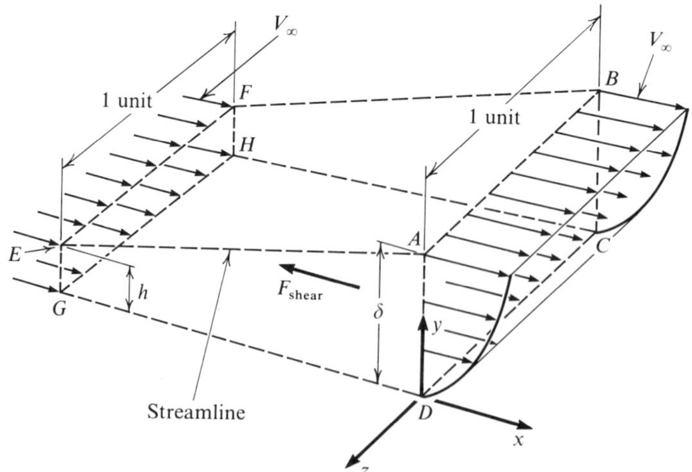

body forces acting in the x direction. Shear forces are zero at the edge or outside of the boundary layer, since $\left.\dfrac{\partial V_x}{\partial y}\right|_{y \geq \delta} = 0$.

The second integral requires some additional information since the upper limit h is unknown; however,

$$\int_0^h \rho V_\infty V_\infty \mathbf{i} \cdot (-1\, dy\, \mathbf{i}) = V_\infty \int_0^h \rho V_\infty \mathbf{i} \cdot (-1\, dy\, \mathbf{i})$$

and

$$\int_0^h \rho V_\infty \mathbf{i} \cdot (-1\, dy\, \mathbf{i}) + \int_0^\delta \rho V_\infty \left[2\frac{y}{\delta} - \left(\frac{y}{\delta}\right)^2 \right] \mathbf{i} \cdot (1\, dy\, \mathbf{i}) = 0$$

by applying the conservation of matter to the control volume. Performing the indicated integration,

$$\int_0^h \rho V_\infty \mathbf{i} \cdot (-1\, dy\, \mathbf{i}) = -\frac{2}{3} \rho V_\infty \delta$$

Substituting this into the momentum equation gives

$$-F_{\text{shear}} = \int_0^\delta \rho V_\infty{}^2 \left[2\frac{y}{\delta} - \left(\frac{y}{\delta}\right)^2 \right]^2 dy - \frac{2}{3} \rho V_\infty{}^2 \delta = \frac{8}{15} \rho V_\infty{}^2 \delta - \frac{2}{3} \rho V_\infty{}^2 \delta$$

$$F_{\text{shear}} = \frac{2}{15} \rho V_\infty{}^2 \delta$$ ◀

This is the shear force that one side of the plate exerts on the fluid for a plate of unit length in the z direction. This force acts to the left as indicated in Fig. 5.9.

The examples indicate that the change of the linear momentum vector of a fluid requires a force or force distribution that can be physically manifested in various forms. In Example 5.1 the designer would have to consider some means of supporting a pipe that terminated in a nozzle. This support would have to bear the dynamic load of the flowing fluid as well as the static load due to the weight of the pipe and the fluid. If a pipeline or duct has elbows or bends, support must be provided in the vicinity where the flow changes direction.

In Example 5.2 the possibility of transmitting a force by changing the linear momentum of a flowing fluid passing over a moving surface becomes apparent. Since a point on the line of action of this force is moving relative to a "fixed frame" of reference, it is possible to extract energy from the moving stream and convert some of this energy into work, which is the function of a system of blades in a steam or gas turbine rotor.

Example 5.3 illustrates a very common means of propulsion, and indicates that no ambient medium or atmosphere is necessary to achieve a propelling force by this method. As a matter of fact, if p_{atm} had been zero in this problem, the force F would have been greater by 900 lb.

Examples 5.4 and 5.5 present methods of estimating "drag forces" due to shear from measured information pertaining to pressure and velocity distributions. Other applications of the linear momentum principle are posed in the assigned problems at the end of this chapter.

5.3 Moment of Momentum Principle for an Inertial Control Volume

The moment of linear momentum **H** of a fluid particle may be obtained by forming the vector product of a position vector **r** and the linear momentum vector **P** of the fluid particle, it being understood that the point about which the moment is taken is considered as the origin from which **r** is drawn to any point on the line of action of the linear momentum vector of the fluid particle. This is illustrated in Fig. 5.10. The moment of **P** about the point 0 is noted by

$$\mathbf{H}_0 = \mathbf{r} \times \mathbf{V}\,\Delta m \tag{5.4}$$

It is apparent that **H** is an extensive quantity, and that the specific moment of linear momentum is simply $\mathbf{r} \times \mathbf{V}$, which is the moment of linear momentum per unit mass. From Eq. (5.1), with **H** substituted for **B**, and $\mathbf{r} \times \mathbf{V}$ for **b**, one obtains

$$\frac{D\mathbf{H}}{Dt} = \int_{\mathcal{V}} \frac{\partial}{\partial t}[(\mathbf{r} \times \mathbf{V})\rho]\, d\mathcal{V} + \oint_S (\mathbf{r} \times \mathbf{V})\rho\mathbf{V} \cdot d\mathbf{S} \tag{5.5}$$

The left-hand side of Eq. (5.5) is the time rate of change of moment of

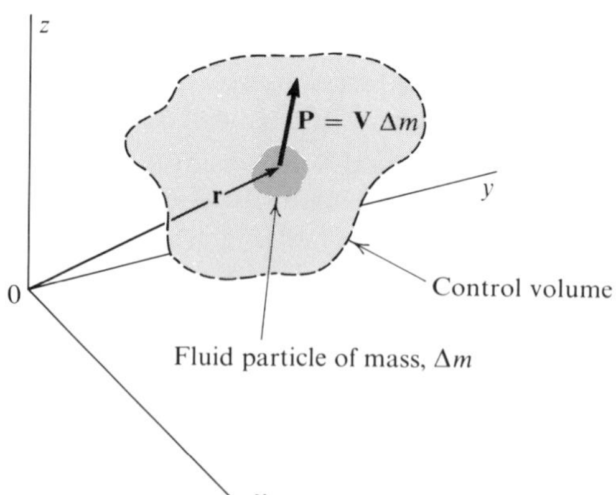

Fig. 5.10 A fluid particle with linear momentum vector **P** *in an aggregate of fluid particles.*

momentum. For a system that is an aggregate of fluid particles, this instantaneous rate of change with time is equal to the sum of the moments of the external forces acting on the system, provided the moments are taken about the point 0. From mechanics it is known that the moment of these forces and the linear momenta may be taken about any one of the following points: a point fixed in an inertial frame of reference, the center of mass of the system, or a point accelerating in a straight line toward or away from the center of mass of the system. If the moments of the external forces and the linear momenta are taken about one of these three points, Eq. (5.5) may be written

$$\sum \mathbf{M}_B + \sum \mathbf{M}_S = \int_{\mathcal{V}} \frac{\partial}{\partial t} [(\mathbf{r} \times \mathbf{V})\rho] \, d\mathcal{V} + \oint_S (\mathbf{r} \times \mathbf{V}) \, \rho \mathbf{V} \cdot d\mathbf{S} \tag{5.6}$$

The terms $\Sigma \mathbf{M}_B$ and $\Sigma \mathbf{M}_S$ represent, respectively, the moments of body forces acting on the material inside of the control volume and the moments of surface forces acting on the control surface, since the system is identical to the material inside of the control volume at the instant of observation. The first term in the right-hand side of Eq. (5.6) is the time rate of change of the moment of momentum of the fluid inside the control volume, and the second term represents the net efflux of the moment of momentum through the control surface. This equation is rather general, and there is considerable detail in analyzing even a simple problem by use of it. The application of the moment of momentum control-volume equation to some simple examples will best illustrate the necessary care and detail involved in its use as an analytical tool in fluid mechanics.

Example 5.6

The rotor of a hydraulic turbine handles 20,000 lb_m/sec. The flow is steady with a constant density of 62.4 lb_m/ft^3. The blade angle and the absolute velocity of the fluid at rotor inlet and exit are shown in Fig. 5.11. The depth of the rotor (its dimension normal to the plane of the figure) is 1 ft. The rotor is turning at a constant speed of 60 rpm. The absolute velocities and the velocities of the fluid relative to coincident points on the rotor are denoted by the subscripts A and R, respectively. The blades are assumed to be of negligible thickness in the reduction of through-flow area. The subscripts r and θ are used to denote radial and tangential components of the various velocities. In addition to these notations, the following nomenclature and data are employed:

r_1 = 2 ft
r_2 = 3 ft
α_1 = 30° (angle between absolute velocity at entrance and the tangent to the rotor at entrance)
α_2 = angle corresponding to α_1, except that location is at exit

β_1 = angle between tangent to blade at rotor entrance and tangent to rotor at entrance

$\beta_2 = 120°$ (angle corresponding to β_1, except that location is at exit of rotor)

The direction of gravity is normal to the plane of Fig. 5.11. Calculate the moment the rotor exerts on the fluid and the horsepower developed.

Solution. The control volume is selected so that the control surfaces are just inside the rotor; and it is noted that this is a control volume fixed in an inertial frame of reference. Assuming that the flow is uniform over the entrance and exit cylindrical surfaces with no z component of velocity, Eq. (5.6) may be applied in a simplified form that will consist of evaluating moments about the axis of the rotor and taking the moments of the momenta about the same axis. This axis pierces the rotor at point 0. There are four distinct portions of the chosen control surface labeled in Fig. 5.12. Surfaces S_1 and S_2 are lateral areas of right and circular cylinders with respective radii r_1 and r_2, while S_3 and S_4 are annular areas with radii r_1 and r_2. Consideration of the surface forces acting on these areas reveals that the moments due to normal stresses that act over the surfaces are zero since these forces are either parallel to the moment axis or pass through this axis. The moments due to shear stresses acting over S_1 and S_2 will be neglected except where S_1 cuts the connections between the rotor and the shaft. Finally, it should be noted that the moment produced by shear stresses acting over S_3 and S_4 are due to portions of the rotor.

The moment due to body forces is zero since the body-force vector is parallel to the moment axis. So Eq. (5.6) is written as

$$\mathbf{M}_S = \oint_S (\mathbf{r} \times \mathbf{V})\, \rho \mathbf{V} \cdot d\mathbf{S}$$

Fig. 5.11

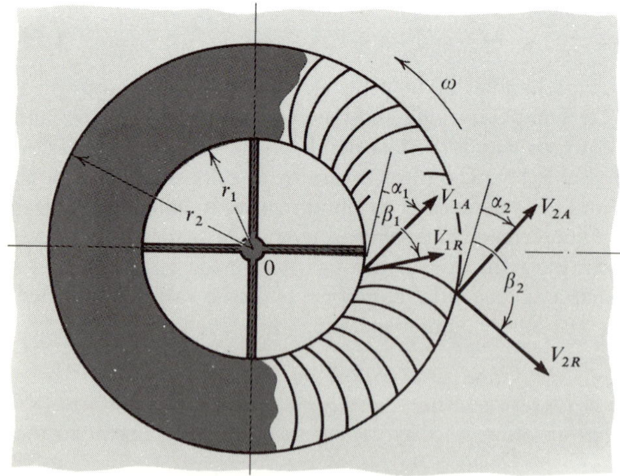

Cutaway of rotor
(looking along axis of rotation)

where the unsteady term represented by the volume integral has been dropped, and $\mathbf{M}_S$ represents the moment (approximately) the rotor exerts on the fluid in the control volume. It is noted that $\mathbf{r}$ and $\mathbf{V}$ in the preceding equation determine a plane perpendicular to the moment axis.

If one assumes that the velocity is uniformly distributed over the entrance surface S_1, and likewise over the exit surface S_2, then

$$\mathbf{M}_S = r_2\boldsymbol{\varepsilon}_r \times (V_{2A_r}\boldsymbol{\varepsilon}_r + V_{2A_\theta}\boldsymbol{\varepsilon}_\theta)\,\rho V_{2A_r} S_2 - r_1\boldsymbol{\varepsilon}_r \times (V_{1A_r}\boldsymbol{\varepsilon}_r + V_{1A_\theta}\boldsymbol{\varepsilon}_\theta)\,\rho V_{1A_r} S_1$$

and

$$\mathbf{M}_S = (r_2 V_{2A_\theta}\mathbf{k} - r_1 V_{1A_\theta}\mathbf{k})\,\rho V_{1A_r} S_1$$

and

$$\rho V_{1A_r} S_1 = \rho V_{2A_r} S_2 = \dot{m}$$

which is the steady mass rate of flow passing through the rotor. The next task is to ascertain the velocity components for use in the preceding equation.

$$V_{1A_r} = \frac{\dot{m}}{\rho S_1} = \frac{20{,}000\ \text{lb}_m/\text{sec}}{(62.4\ \text{lb}_m/\text{ft}^3)(2\pi)(2\ \text{ft})(1\ \text{ft})} = 25.6\ \text{ft/sec}$$

$$V_{1A_\theta} = \frac{V_{1A_r}}{\tan\alpha_1} = \frac{25.6\ \text{ft/sec}}{0.577} = 44.3\ \text{ft/sec}$$

Fig. 5.12

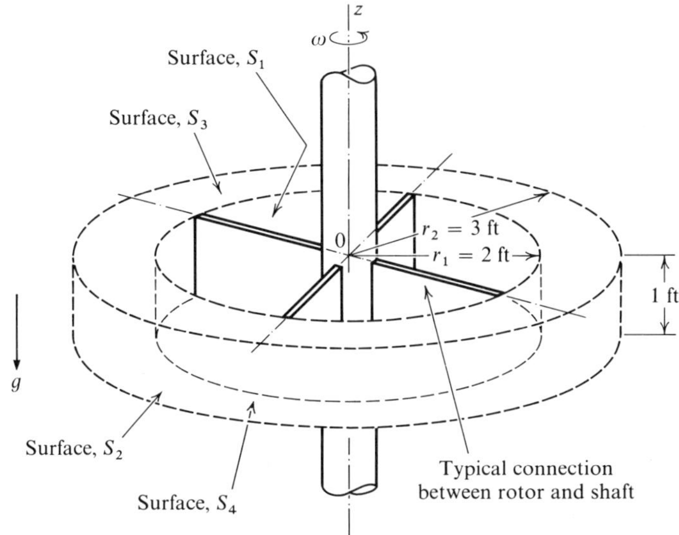

The radial component of velocity V_{2A_r} may be found from the conservation of matter as follows:

$$V_{2A_r} = \frac{\rho V_{1A_r} S_1}{\rho S_2} = V_{1A_r} \frac{(2\pi r_1)(1 \text{ ft})}{(2\pi r_2)(1 \text{ ft})}$$

$$= V_{1A_r} \frac{r_1}{r_2} = \left(25.6 \frac{\text{ft}}{\text{sec}}\right)\left(\frac{2 \text{ ft}}{3 \text{ ft}}\right) = 17.1 \text{ ft/sec}$$

The velocity component V_{2A_θ} can be determined from V_{2A_r} and the blade geometry by assuming that the velocity of the fluid relative to a coincident point on the blade is tangent to the blade at exit. This is commonly assumed to be true at entrance also. It is referred to as the *smooth exit* or *entrance condition*, as the case may be. Figure 5.13 shows this relation. Noting that

$$\mathbf{V}_{2A} = \mathbf{V}_{2R} + \omega r_2 \boldsymbol{\varepsilon}_\theta = \mathbf{V}_{2A_r} + \mathbf{V}_{2A_\theta}$$

and equating the radial components

$$V_{2R_r} = V_{2A_r} = 17.1 \text{ ft/sec}$$

In addition

$$V_{2R_\theta} + \omega r_2 = V_{2A_\theta}$$

so that

$$V_{2A_\theta} = \frac{V_{2R_r}}{\tan \beta_2} + \omega r_2 = \frac{17.1 \text{ ft/sec}}{-1.732} + \left(60 \frac{\text{rev}}{\text{min}}\right)\left(2\pi \frac{\text{rad}}{\text{rev}}\right)\left(\frac{1 \text{ min}}{60 \text{ sec}}\right)(3 \text{ ft})$$

$$= -9.87 \frac{\text{ft}}{\text{sec}} + 18.87 \frac{\text{ft}}{\text{sec}}$$

$$= 9 \text{ ft/sec}$$

Fig. 5.13

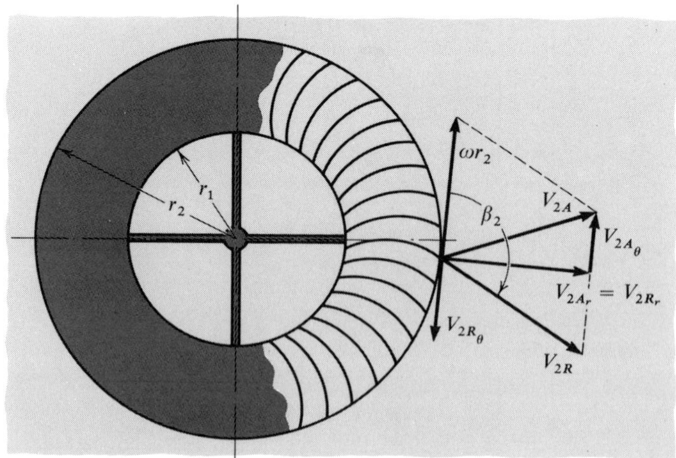

The moment $\mathbf{M}_S$ can now be calculated from

$$\mathbf{M}_S = (r_2 V_{2A_\theta}\mathbf{k} - r_1 V_{1A_\theta}\mathbf{k})\,\rho V_{1A_r} S_1$$

$$= \frac{[(3\text{ ft})(9\text{ ft/sec}) - (2\text{ ft})(44.3\text{ ft/sec})](20{,}000\text{ lb}_m\text{/sec})}{32.2\text{ ft-lb}_m\text{/lb}_f\text{-sec}^2}\mathbf{k}$$

$$= -38{,}300\,\mathbf{k}\text{ ft-lb}_f$$ ◀

This is the moment the rotor exerts on the fluid and it is in the clockwise direction. The moment the fluid exerts on the rotor is equal and opposite to this, or 38,300**k** ft-lb$_f$, and is in the +**k** direction. Since this moment is in the same sense as the indicated rotation of the rotor, power is being developed by the passage of the fluid through the rotor. At the given rotor speed, this power is

$$\frac{dW_S}{dt} = M_S\omega = \frac{(38{,}300\text{ ft-lb}_f)(2\pi\text{ rad/sec})}{550\text{ ft-lb}_f\text{/hp-sec}}$$

$$= 438\text{ hp}$$ ◀

Example 5.7

Water flows steadily through the stationary pipe shown in Fig. 5.14. The exit velocity is 20 ft/sec and may be considered uniform over the exit cross section, which has a diameter of 1 in. This exit section is "small enough" so that the position vector **r** may be taken as a constant for any point on A_2. The fluid has a constant density of 62.4 lb$_m$/ft^3. Atmospheric pressure acts over the outside of the pipe. Neglect any moment due to the weight of the pipe and the fluid in the pipe and calculate the resultant moment at section 1 due to all fluids in contact with the pipe.

Fig. 5.14

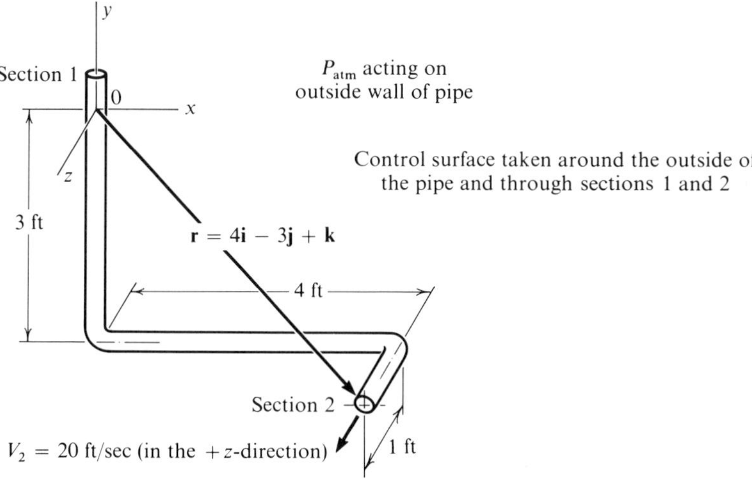

Solution. The control surface chosen encloses an inertial control volume, and the forces due to atmospheric pressure acting over this control surface add to zero, except at the bottom of the bend that joins the 3 ft and 4 ft lengths of pipe. However, the force due to atmospheric pressure at the bottom of this bend and that associated with normal stresses in the fluid at section 1 produce no moment about 0 at this location if the normal stresses are uniform in the fluid passing through A_1. Hence the moments due to surface forces $\Sigma \mathbf{M}_S$ are those associated with the stresses in the pipe wall at section 1, and it is this sum, $\Sigma \mathbf{M}_S$, which is the negative of the desired answer.

From Eq. (5.6),

$$\sum \mathbf{M}_S = \oint_S (\mathbf{r} \times \mathbf{V})\, \rho \mathbf{V} \cdot d\mathbf{S}$$

for the steady velocity field and constant density; and for the given example this may be written as

$$\begin{aligned}\sum \mathbf{M}_S &= \int_{A_2} [(4\mathbf{i} - 3\mathbf{j} + \mathbf{k}) \times 20\mathbf{k}]\, \rho\, 20\mathbf{k} \cdot dS_2\, \mathbf{k} \\ &= 400\rho \int_{A_2} (-4\mathbf{j} - 3\mathbf{i})\, dS_2 \\ &= 400\rho A_2\, (-3\mathbf{i} - 4\mathbf{j})\end{aligned}$$

Supplying the units associated with given quantities gives

$$\begin{aligned}\sum \mathbf{M}_S &= \frac{(400\ \text{ft}^2/\text{sec}^2)(62.4\ \text{lb}_m/\text{ft}^3)(\pi/4)(1\ \text{ft}/12)^2}{32.2\ \text{ft-lb}_m/\text{lb}_f\text{-sec}^2}(-3\,\text{ft}\,\mathbf{i} - 4\,\text{ft}\,\mathbf{j}) \\ &= (-12.7\mathbf{i} - 16.9\mathbf{j})\ \text{ft-lb}_f\end{aligned}$$

The answer is equal and opposite to this, so

$$-\sum \mathbf{M}_S = (12.7\mathbf{i} + 16.9\mathbf{j})\ \text{ft-lb}_f$$ ◀

It is emphasized that the moments due to the weight of the pipe and the fluid should be considered for design purposes. At least these should be calculated to see if they can be safely neglected.

5.4 Linear Momentum Principle for a Noninertial Control Volume

Very often the given or known information concerning a problem or design will require the selection of a control surface such that the control volume enclosed is accelerating and/or rotating with respect to an inertial frame of reference. It is for this reason that the linear momentum control-volume equation must be formulated for a noninertial frame of reference.

In engineering mechanics, expressions were developed for vector functions referred to an inertial (or newtonian) frame of reference in terms of vector quantities referred to a noninertial (or accelerating) frame of

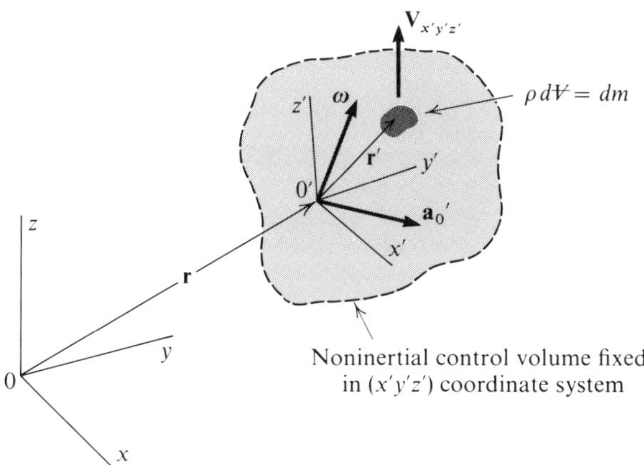

Fig. 5.15 Nomenclature for referencing changes in an accelerating frame of reference to an inertial frame of reference.

reference. Figure 5.15 shows a noninertial control volume, and this control volume is fixed relative to the primed axes. The unprimed coordinate system is an inertial frame of reference.

A differential amount of mass, $dm = \rho d\forall$, is shown in the figure. The forces acting on dm may be related to its acceleration through Newton's law by use of an inertial frame of reference as follows:

$$d\mathbf{F}_B + d\mathbf{F}_S = \mathbf{a}_{xyz}\, dm \tag{5.7}$$

in which the acceleration is referred to the unprimed coordinate system. Reference to Fig. 5.15 and kinematics[1] enables the formulation of $\mathbf{a}_{xyz}$ as

$$\mathbf{a}_{xyz} = \mathbf{a}_{0'} + \left(\frac{d\boldsymbol{\omega}}{dt}\right)_{x'y'z'} \times \mathbf{r}' + 2\boldsymbol{\omega} \times \mathbf{V}_{x'y'z'} + \boldsymbol{\omega} \times (\boldsymbol{\omega} \times \mathbf{r}') + \left(\frac{d^2\mathbf{r}'}{dt^2}\right)_{x'y'z'} \tag{5.8}$$

where $\mathbf{a}_{0'}$ = acceleration of origin of moving coordinate system

$\boldsymbol{\omega}$ = angular velocity of $x'y'z'$ frame of reference

$\left(\frac{d\boldsymbol{\omega}}{dt}\right)_{x'y'z'}$ = angular acceleration viewed from $x'y'z'$ reference; note that $(d\boldsymbol{\omega}/dt)_{x'y'z'} = (d\boldsymbol{\omega}/dt)_{xyz}$

$\mathbf{V}_{x'y'z'}$ = velocity of "center of mass" of dm viewed from $x'y'z'$ reference

[1]Irving Shames, "Engineering Mechanics—Dynamics," 2d ed., pp. 506–507, Prentice-Hall, Inc., Englewood Cliffs, N.J., 1966.

$$\left(\frac{d^2\mathbf{r}'}{dt^2}\right)_{x'y'z'} = \text{acceleration of ``center of mass'' viewed from } x'y'z' \text{ reference}$$

Substitution for $\mathbf{a}_{xyz}$ in Eq. (5.7) by use of Eq. (5.8) yields, after rearranging terms,

$$d\mathbf{F}_S + d\mathbf{F}_B - \left[\left(\frac{d\omega}{dt}\right)_{x'y'z'} \times \mathbf{r}' + 2\omega \times \mathbf{V}_{x'y'z'} + \omega \times (\omega \times \mathbf{r}') + \mathbf{a}_{0'}\right] dm = \left(\frac{d^2\mathbf{r}'}{dt^2}\right)_{x'y'z'} dm$$

Noting that

$$\left(\frac{d^2\mathbf{r}'}{dt^2}\right)_{x'y'z'} dm = \left(\frac{d\,\mathbf{V}_{x'y'z'}}{dt}\right)_{x'y'z'} dm = \frac{D}{Dt}(\mathbf{V}_{x'y'z'}\, dm)_{x'y'z'}$$

since the fluid particle is being followed, Eq. (5.7) may now be written as

$$d\mathbf{F}_S + d\mathbf{F}_B - \left[\left(\frac{d\omega}{dt}\right)_{x'y'z'} \times \mathbf{r}' + 2\omega \times \mathbf{V}_{x'y'z'} + \omega \times (\omega \times \mathbf{r}') + \mathbf{a}_{0'}\right] dm = \frac{D}{Dt}(\mathbf{V}_{x'y'z'}\, dm)_{x'y'z'} \quad (5.9)$$

Equation (5.9) may be integrated over the mass in the control volume at this instant of time to yield

$$\sum \mathbf{F}_S + \sum \mathbf{F}_B - \int_{\mathcal{V}} \left[\left(\frac{d\omega}{dt}\right)_{x'y'z'} \times \mathbf{r}' + 2\omega \times \mathbf{V}_{x'y'z'} + \omega \times (\omega \times \mathbf{r}') + \mathbf{a}_{0'}\right] \rho d\mathcal{V} = \frac{D}{Dt}\left(\int_{\mathcal{V}} \mathbf{V}_{x'y'z'}\, \rho d\mathcal{V}\right)_{x'y'z'} \quad (5.10)$$

Differentiation may be taken outside of the integral sign since this is the time rate of change following the material as viewed from the $x'y'z'$ frame of reference. Inspection of $\int_{\mathcal{V}} \mathbf{V}_{x'y'z'}\, \rho d\mathcal{V}$ reveals that it is the linear momentum of the system that instantaneously occupies the control volume and is passing through it. Since linear momentum is extensive, Eq. (5.1) can be applied to the right-hand side of Eq. (5.10) to give

$$\sum \mathbf{F}_S + \sum \mathbf{F}_B - \int_{\mathcal{V}} \left[\left(\frac{d\omega}{dt}\right)_{x'y'z'} \times \mathbf{r}' + 2\omega \times \mathbf{V}_{x'y'z'} + \omega \times (\omega \times \mathbf{r}') + \mathbf{a}_{0'}\right] \rho d\mathcal{V} = \int_{\mathcal{V}} \left[\frac{\partial}{\partial t}\left(\rho \mathbf{V}\right) d\mathcal{V}\right]_{x'y'z'} + \oint_S \mathbf{V}_{x'y'z'}\, \rho \mathbf{V}_{x'y'z'} \cdot d\mathbf{S} \quad (5.11)$$

The integrand of the volume integral on the left-hand side of Eq. (5.11) may be visualized as the acceleration components that give rise to inertial forces when multiplied by mass ($\rho d\mathcal{V}$ in this case). These accelerations are, respectively: the angular acceleration $d\boldsymbol{\omega}/dt$; Coriolis acceleration $2\boldsymbol{\omega} \times \mathbf{V}_{x'y'z'}$; centripetal acceleration $\boldsymbol{\omega} \times (\boldsymbol{\omega} \times \mathbf{r}')$; and the acceleration of the origin of the moving coordinate system $\mathbf{a}_{0'}$.

Equation (5.11) can be used to solve a large class of problems in fluid mechanics where transmittance of forces between a fluid flow field and solid surfaces is desired. Several applications are given in the following examples.

Example 5.8

The tube shown in Fig. 5.16 has been bent in a quarter circle with radius R_b, which is the radius of the bend. The radius of the tube, R, is *very small* compared to R_b. A fluid whose density is uniform and steady flows through the tube at a steady speed V_f, relative to the tube. The absolute pressures are known at the tube entrance and exit, and these are designated p_1 and p_2. The entire 90° bend is rotating about the x axis with angular velocity and acceleration $\boldsymbol{\omega}$ and $\dot{\boldsymbol{\omega}}$, respectively, as indicated in Fig. 5.16. The point 0 is fixed in inertial space. Find an expression for the x, y, and z components of the force the tube walls exert on the fluid at the instant the bend is in the x, y coordinate plane.

Solution. The control volume selected is bounded by the interior walls of the tube and sections through locations 1 and 2. Consider the primed coordinate axes fixed relative to this control surface; the y' and z' axes each have an angular velocity and acceleration equal to $\boldsymbol{\omega}$ and $\dot{\boldsymbol{\omega}}$, respectively. The unprimed coordinate system is fixed in inertial space, and at the instant of interest it is coincident with the primed coordinate system. Since the control volume selected is rotating, and hence noninertial, Eq. (5.11)

Fig. 5.16

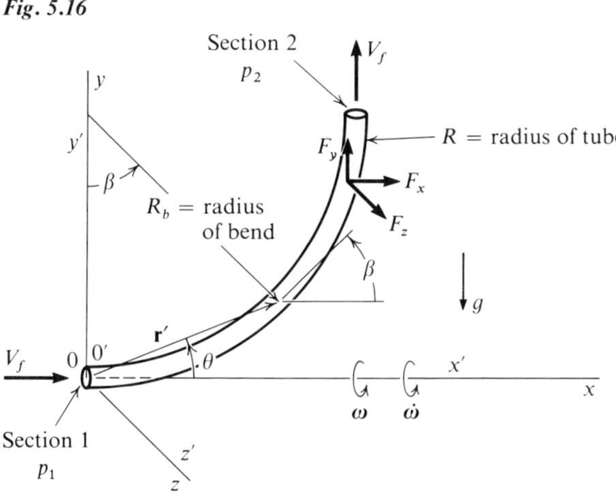

is the applicable form for the linear momentum principle. The primed coordinates in this equation refer to the accelerating coordinate system so that the notation is the same as that selected for this example.

$$\sum \mathbf{F}_S + \sum \mathbf{F}_B - \int_{V} \left[\left(\frac{d\boldsymbol{\omega}}{dt} \right)_{x'y'z'} \times \mathbf{r}' + 2\boldsymbol{\omega} \times \mathbf{V}_{x'y'z'} + \boldsymbol{\omega} \times (\boldsymbol{\omega} \times \mathbf{r}') + \mathbf{a}_{0'} \right] \rho \, dV$$
$$= \int_{V} \left[\frac{\partial(\rho \mathbf{V})}{\partial t} \right]_{x'y'z'} dV + \oint_S \mathbf{V}_{x'y'z'} (\rho \mathbf{V}_{x'y'z'} \cdot d\mathbf{S}) \quad (5.11)$$

It will be instructive to formulate separately each of the terms in this equation and explain how each can be evaluated.

1. Surface forces.

$$\sum \mathbf{F}_S = p_1 \pi R^2 \mathbf{i} - p_2 \pi R^2 \mathbf{j} + F_x \mathbf{i} + F_y \mathbf{j} + F_z \mathbf{k}$$

This is due to pressure acting at each end of the 90° bend and the surface force exerted by the inside walls of the tube on the fluid. The latter has been indicated by F_x, F_y, and F_z, which have been assumed in the positive direction. These are the desired answers.

2. Body forces.

$$\sum \mathbf{F}_B = -\rho \left(\frac{1}{2} \pi R_b \right) (\pi R^2) g \mathbf{j}$$

This is the weight of the fluid inside the bend. The volume inside the bend has been approximated in a form valid for $R \ll R_b$.

3. Inertial forces.

$$a. \quad \int_{V} \left(\frac{d\boldsymbol{\omega}}{dt} \right)_{x'y'z'} \times \mathbf{r}' \rho \, dV = \int_0^{\pi/2} \dot{\omega} \mathbf{i} \times (x' \mathbf{i} + y' \mathbf{j}) \rho \pi R^2 R_b \, d\beta$$

which is the inertial force associated with the angular acceleration $\dot{\omega}$. The position vector $\mathbf{r}'$ has been written as $x'\mathbf{i} + y'\mathbf{j}$ for this instant of time. It can also be written as $\mathbf{r}' = r' \cos \theta \, \mathbf{i} + r' \sin \theta \, \mathbf{j}$, and since the volume integration is to be taken over the entire 90° bend, the tip of $\mathbf{r}'$ is constrained to move along the arc of a circle of radius R_b. Referring to Fig. 5.17, the element of volume (for $R \ll R_b$) is $\pi R^2 R_b \, d\beta$, and integration over the volume of the bend may be accomplished with respect to β by varying β from zero to $\pi/2$ rads. However, geometric considerations give

$$\beta = 2\theta$$

and

$$\mathbf{r}' = R_b \frac{\sin \beta}{\cos \theta} (\cos \theta \, \mathbf{i} + \sin \theta \, \mathbf{j})$$
$$= 2R_b \sin \theta (\cos \theta \, \mathbf{i} + \sin \theta \, \mathbf{j})$$

so that the integral may be expressed as a function of θ by

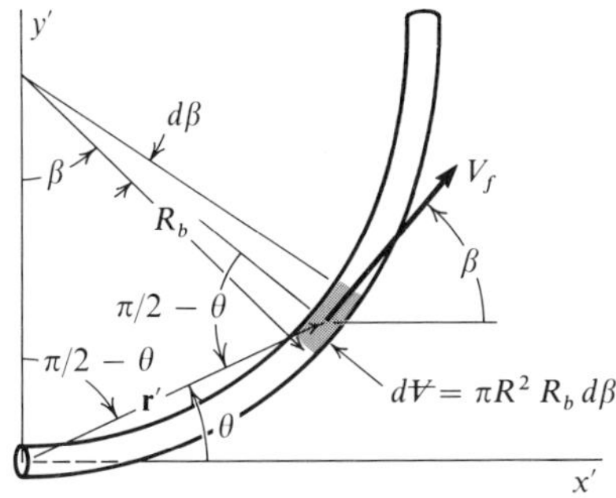

Fig. 5.17

$$\int_{\mathcal{V}} \left(\frac{d\boldsymbol{\omega}}{dt}\right)_{x'y'z'} \times \mathbf{r}' \,\rho\, d\mathcal{V} = \int_0^{\pi/4} \dot{\omega}\mathbf{i} \times 2R_b \sin\theta\,(\cos\theta\,\mathbf{i} + \sin\theta\,\mathbf{j})\,\rho\pi R^2 R_b 2\,d\theta$$

$$= 4R^2 R_b{}^2 \pi\,\rho\,\dot{\omega} \int_0^{\pi/4} \sin^2\theta\,d\theta\,\mathbf{k}$$

$$= 4R^2 R_b{}^2 \pi\,\rho\dot{\omega}\left(\frac{\pi}{8} - \frac{1}{4}\right)\mathbf{k}$$

b. $$\int_{\mathcal{V}} 2\boldsymbol{\omega} \times \mathbf{V}_{x'y'z'}\,\rho\,d\mathcal{V} = \int_{\mathcal{V}} 2\boldsymbol{\omega} \times \mathbf{V}_f\,\rho\,d\mathcal{V}$$

since the velocity of the fluid relative to the bend is the velocity of the fluid as viewed from the primed coordinate system, that is, $\mathbf{V}_f = \mathbf{V}_{x'y'z'}$. Noting that

$$\mathbf{V}_f = V_f\,(\cos\beta\,\mathbf{i} + \sin\beta\,\mathbf{j}) = V_f\,(\cos 2\theta\,\mathbf{i} + \sin 2\theta\,\mathbf{j})$$

and substituting for V_f in the volume integral,

$$\int_{\mathcal{V}} 2\boldsymbol{\omega} \times \mathbf{V}_{x'y'z'}\,\rho\,d\mathcal{V} = \int_0^{\pi/4} 2\omega\mathbf{i} \times V_f\,(\cos 2\theta\,\mathbf{i} + \sin 2\theta\,\mathbf{j})\,\rho\pi R^2 R_b 2\,d\theta$$

$$= 4\,V_f\,\rho\pi R^2 R_b \int_0^{\pi/4} \sin 2\theta\,d\theta\,\mathbf{k}$$

$$= 2\omega V_f \rho\pi R^2 R_b \mathbf{k}$$

c. $$\int_{\mathcal{V}} \boldsymbol{\omega} \times (\boldsymbol{\omega} \times \mathbf{r}')\,\rho\,d\mathcal{V} = \int_0^{\pi/4} \{\omega\mathbf{i} \times [\omega\mathbf{i} \times 2R_b \sin\theta\,(\cos\theta\,\mathbf{i} + \sin\theta\,\mathbf{j})]\}\,\rho\pi R^2 R_b(2)\,d\theta$$

$$= 4\omega^2\,\rho\pi R_b{}^2 R^2 \int_0^{\pi/4} \sin^2\theta\,d\theta\,(-\mathbf{j})$$

$$= -4\omega^2\,\rho\pi R_b{}^2 R^2 \left(\frac{\pi}{8} - \frac{1}{4}\right)\mathbf{j}$$

d. $\int_{V} \mathbf{a}_{0'} \rho \, dV = 0$

since the origin of the primed coordinate system is not accelerating. This concludes the accounting of all inertial forces.

4. Momentum change within the control volume.

$$\int_{V} \left[\frac{\partial(\rho \mathbf{V})}{\partial t} dV \right]_{x'y'z'} = 0$$

since the velocity of the fluid is steady as viewed from the primed coordinate system, and ρ is uniform and steady.

5. The net momentum efflux. The integral $\oint_S \mathbf{V}_{x'y'z'} (\rho \mathbf{V}_{x'y'z'} \cdot d\mathbf{S})$ may be evaluated by observing that sections 1 and 2 are the only portions of the control surface through which fluid (and momentum) is passing. Therefore this integral may be expressed as

$$\int_{A_1} (V_f \mathbf{i})(\rho V_f \mathbf{i}) \cdot (-dS\,\mathbf{i}) + \int_{A_2} (V_f \mathbf{j})(\rho V_f \mathbf{j}) \cdot (dS\,\mathbf{j})$$

and assuming that V_f is uniform over A_1 and A_2,

$$\oint_S \mathbf{V}_{x'y'z'} \, \rho \mathbf{V}_{x'y'z'} \cdot d\mathbf{S} = -\pi R^2 \rho V_f^2 \mathbf{i} + \pi R^2 \rho V_f^2 \mathbf{j}$$

Substitution of these terms into Eq. (5.11) yields

$$p_1 \pi R^2 \mathbf{i} - p_2 \pi R^2 \mathbf{j} + F_x \mathbf{i} + F_y \mathbf{j} + F_z \mathbf{k} - \frac{1}{2}\pi^2 \rho g R^2 R_b \mathbf{j}$$
$$- 4R^2 R_b^2 \pi \rho \dot{\omega} \left(\frac{\pi}{8} - \frac{1}{4} \right) \mathbf{k} - 2\omega V_f \rho \pi R^2 R_b \mathbf{k} + 4\omega^2 \rho \pi R_b^2 R^2 \left(\frac{\pi}{8} - \frac{1}{4} \right) \mathbf{j} - 0$$
$$= 0 + \pi R^2 \rho V_f^2 \mathbf{j} - \pi R^2 \rho V_f^2 \mathbf{i}$$

Equating components of this vector equation gives

$$F_x = -p_1 \pi R^2 - \pi R^2 \rho V_f^2 \quad ◀$$

$$F_y = p_2 \pi R^2 + \frac{1}{2}\pi^2 \rho g R^2 R_b - 4\omega^2 \rho \pi R_b^2 R^2 \left(\frac{\pi}{8} - \frac{1}{4} \right) + \pi R^2 \rho V_f^2 \quad ◀$$

$$F_z = 4R^2 R_b^2 \pi \rho \dot{\omega} \left(\frac{\pi}{8} - \frac{1}{4} \right) + 2\omega V_f \rho \pi R^2 R_b \quad ◀$$

Such a flow configuration as presented in this example may appear strange and unusual; however, the "plumbing hardware" on space vehicles is subjected to motions of the most general nature. Also, rotating-flow bends and loops have been employed as a means of flow-rate measurement,[1] where the torque exerted by the fluid on the bend

[1] W. J. Alspach, C. E. Miller, and T. M. Flynn, Flow Measurement—Part 2: Mass Flowmeters in Cryogenics, *Mech. Eng.*, vol. 89, no. 5, pp. 105–113, 1967.

or loop is measured by a torque cell, and from this measured torque, the flow is inferred. Analysis involves consideration of the moment of linear momentum applied to a noninertial control volume.

Example 5.9

The cart shown in Fig. 5.18 is constrained to move in the horizontal direction and it carries a deflector that receives the flow of a fluid (of uniform and steady density) from a stationary nozzle. The velocity of the jet relative to the nozzle is $\mathbf{V}_j$ and the area of the jet is A_j which does not vary as the jet passes from the nozzle and through the deflector on the cart. The velocity of the fluid relative to the cart does not vary in magnitude as the fluid passes through the deflector, a requirement imposed by the conservation of matter for ρ and for A_j remaining constant. The total mass of the cart and deflector is m_c, and the deflector as viewed is a circular arc of radius R_d which is very large compared to the radial thickness of the jet on the deflector. The air resistance of the atmosphere surrounding the cart and the jet and the rolling resistance is to be neglected. Find the acceleration of the cart as a function of m_c, A_j, R_d, ρ, V_j, θ, and V_c (speed of the cart).

Solution. A control surface is taken around the entire cart and the fluid on the deflector. This is indicated by the dashed lines in Fig. 5.18. The control volume is then moving to the right with the velocity of the cart, and this velocity is changing with respect to time. This control volume is noninertial, and the $x'y'$ axes are fixed in this control volume. The xy axes constitute an inertial frame of reference, and these are fixed in the nozzle.

The x component of Eq. (5.11) is

$$\sum F_{S_x} + \sum F_{B_x} - \int_{V\!\!\!\!-} a_{0'} \rho \, dV\!\!\!\!- = \int_{V\!\!\!\!-} \left[\frac{\partial}{\partial t} (\rho V_{x'}) \, dV\!\!\!\!- \right]_{x'y'} + \oint_S V_{x'} \rho \mathbf{V}_{x'y'} \cdot d\mathbf{S}$$

the terms containing ω and its time derivative are omitted since these are zero for this problem. The sum of the surface forces in the x direction is zero since the force due to

Fig. 5.18

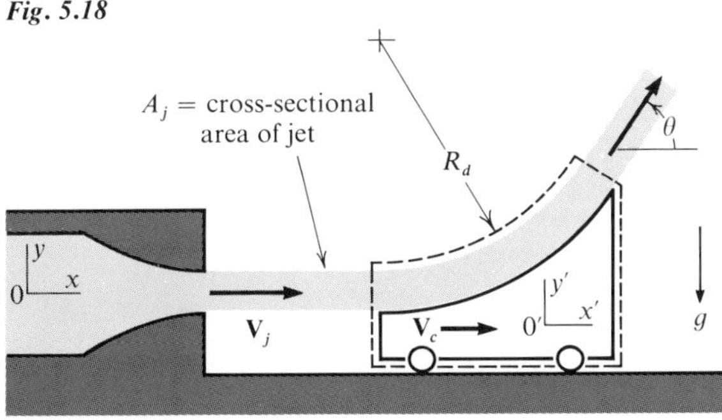

atmospheric pressure acts over all of the control surface except the small portion of line contact between the wheels and the surface on which the cart is rolling. It was assumed that air resistance due to viscous effects is negligible. The sum of the body forces is zero for the x direction with the gravity field shown in Fig. 5.18. The remaining three terms are evaluated as follows:

1. The inertial force.

$$\int_{\mathcal{V}} a_{0'} \rho \, d\mathcal{V} = a_{0'}(m_c + R_d \theta A_j \rho)$$

which is the acceleration of the moving origin (and also the cart) multiplied by the total mass in the control volume—the mass of the cart plus the mass of the fluid on the deflector. The mass has been approximated by using a volume of $R_d A_j \theta$, which is valid if R_d is large compared to the radial thickness of the jet on the deflector. See Fig. 5.19.

2. The change of momentum in the control volume. The momentum of the cart does not change when viewed from the $x'y'$ frame of reference since the velocity of the cart does not change relative to the control surface which moves with the primed coordinate system. However, the velocity of the fluid does change with respect to time when viewed from the $x'y'$ frame. This can be seen in Fig. 5.19, and noting that

$$\begin{aligned}
\int_{\mathcal{V}} \left[\frac{\partial}{\partial t}(\rho V_{x'}) \, d\mathcal{V} \right]_{x'y'} &= \int_0^\theta \frac{\partial}{\partial t} [\rho(V_j - V_c) \cos \beta] R_d A_j \, d\beta \\
&= \rho R_d A_j \int_0^\theta - \frac{\partial V_c}{\partial t} \cos \beta \, d\beta \\
&= -\rho R_d A_j a_{0'} \int_0^\theta \cos \beta \, d\beta \\
&= -\rho R_d A_j a_{0'} \sin \theta
\end{aligned}$$

3. The momentum flux relative to the control surface.

$$\begin{aligned}
\oint_S V_{x'} \rho \mathbf{V}_{x'y'} \cdot d\mathbf{S} &= \int_{A_{\beta=0}} V_{x'} \rho \mathbf{V}_{x'y'} \cdot d\mathbf{S} + \int_{A_{\beta=\theta}} V_{x'} \rho \mathbf{V}_{x'y'} \cdot d\mathbf{S} \\
&= -(V_j - V_c)^2 \rho A_j + (V_j - V_c)^2 \cos \theta \, \rho A_j \\
&= \rho A_j (V_j - V_c)^2 (\cos \theta - 1)
\end{aligned}$$

Substitution of the three preceding integrals into the momentum equation gives

$$-a_{0'}(m_c + R_d \theta A_j \rho) = -\rho R_d A_j a_{0'} \sin \theta + \rho A_j (V_j - V_c)^2 (\cos \theta - 1)$$

Solving for $a_{0'}$ yields

$$a_{0'} = \frac{\rho A_j (V_j - V_c)^2 (\cos \theta - 1)}{\rho R_d A_j \sin \theta - (m_c + R_d \theta A_j \rho)}$$ ◀

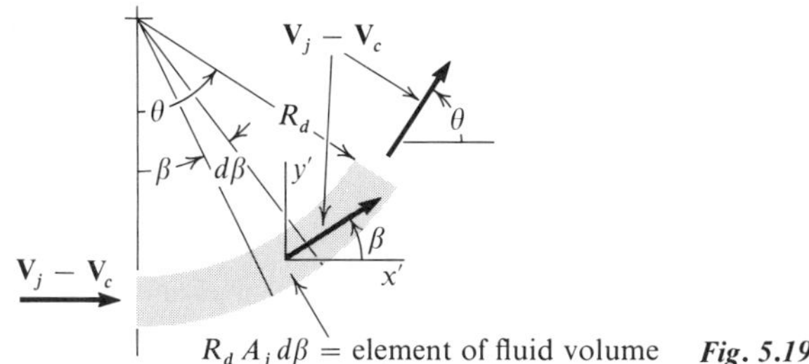

Fig. 5.19

which is the desired result. This expression can be used to obtain a velocity-time relation for the cart by writing $a_{0'}$ as dV_c/dt and specifying an initial condition for the velocity of the cart.

5.5 *Closing Remarks*

The linear momentum principle has been formulated for the inertial and noninertial control volumes. The principle of moment of momentum was formulated for the inertial control volume only. The extension of the latter principle to a noninertial control volume is analogous to the method applied to the linear momentum principle, with the use of inertial moments instead of inertial forces.

Self-Study Questions

5.1 State a physical meaning for

a. $\rho\mathbf{V}$ *b.* $\rho\mathbf{V}\, d\mathcal{V}$ *c.* $\dfrac{\partial(\rho\mathbf{V})}{\partial t} d\mathcal{V}$ *d.* $\displaystyle\int_{\mathcal{V}} \frac{\partial(\rho\mathbf{V})}{\partial t} d\mathcal{V}$

5.2 State a physical meaning for

a. $\rho\mathbf{V}\cdot d\mathbf{S}$ *b.* $\mathbf{V}\,\rho\mathbf{V}\cdot d\mathbf{S}$ *c.* $\displaystyle\oint_S \mathbf{V}\,\rho\mathbf{V}\cdot d\mathbf{S}$

d. $\displaystyle\int_{\mathcal{V}} \frac{\partial(\rho\mathbf{V})}{\partial t} d\mathcal{V} + \oint_S \mathbf{V}\,\rho\mathbf{V}\cdot d\mathbf{S}$ *e.* $\sum \mathbf{F}_S + \sum \mathbf{F}_B$

5.3 *a.* Give an integral expression for a gravity force acting on a fluid within a finite control volume.
b. Give an integral expression for the surface forces acting on an ideal fluid within a finite control surface.
c. Same as (*b*) except that the fluid is newtonian.

5.4 When the net linear-momentum flux of a fluid passing through a control surface (entering at section 1 and leaving at section 2) reduces to the following expressions, discuss what may be inferred about the flow field.
a. $\rho \oint_S \mathbf{V}\,\mathbf{V}\cdot d\mathbf{S}$ *b.* $\int_{S_2} \rho V^2\, d\mathbf{S} + \int_{S_1} \rho V^2\, d\mathbf{S}$ *c.* $\rho_2 V_2^{\,2}\mathbf{S}_2 - \rho_1 V_1^{\,2}\mathbf{S}_1$
d. $\rho_1 V_1 A_1(\mathbf{V}_2 - \mathbf{V}_1)$ *e.* $-\rho_1(V_x^{\,2})_1 A_1 - \rho_2(V_y^{\,2})_2 A_2$

5.5 Do the following conditions yield a zero value for the instantaneous time rate of change of linear momentum within the control volume?

a. Steady flow.

b. Steady flow and div $\rho\mathbf{V} = 0$.

c. Steady flow and a fluid with a uniform and steady density.

5.6 State mathematically and in words the transport equation applied to linear momentum.

5.7 What is the algebraic sign of the x component of the linear-momentum flux surface integral if

a. The mass flow is into that part of the control surface and in the assigned positive x direction?

b. The mass flow is out of that part of the control surface and in the assigned negative x direction?

c. The mass flow is into that part of the control surface and in the assigned negative x direction?

5.8 *a.* For horizontal flow with changing area in the x direction, obtain an algebraic expression for the force on the walls between the entering and exiting sections where the fluid properties are assumed to be uniform in value. Assume no change of momentum with time within the control volume.

b. Same as (*a*) except that the flow changes direction by entering in the positive x direction and exiting in the negative x direction.

5.9 What simplification in the use of the integral momentum equation for a control volume is achieved by choosing a control surface

a. Along a wall containing the fluid?

b. Normal to the direction of the flow?

c. Along a streamline?

d. Where some properties may be assumed to be uniform?

e. Fixed in an inertial frame of reference?

5.10 A device, restricted to motion in the x direction only, has an internal mechanism that causes the flow of a fluid in the directions shown in Fig. 5.20. If it is assumed that the pressure in the fluid at the entrance and the exit is essentially atmospheric, state, for each of the cases shown, whether the device will move to the right or to the left when released.

5.11 State a physical meaning for

a. $\mathbf{V}\rho\, d\mathcal{V}$ *b.* $\mathbf{r} \times \mathbf{V}\rho\, d\mathcal{V}$ *c.* $\dfrac{\partial}{\partial t}(\mathbf{r} \times \mathbf{V})\rho\, d\mathcal{V}$ *d.* $\displaystyle\int_{\mathcal{V}} \frac{\partial}{\partial t}(\mathbf{r} \times \mathbf{V})\rho\, d\mathcal{V}$

Fig. 5.20

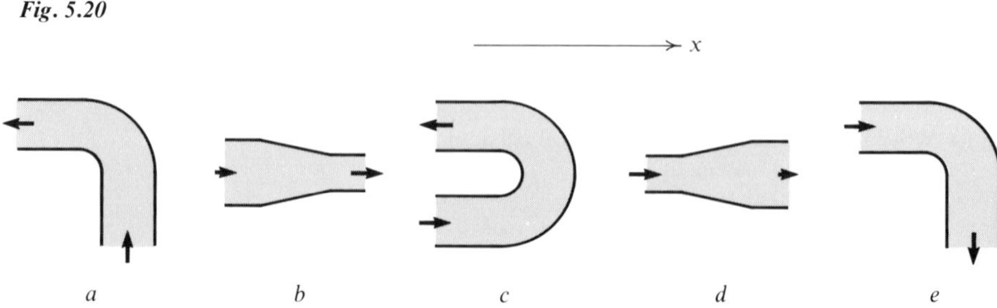

5.12 State a physical meaning for
a. $\mathbf{V}\,\rho\mathbf{V}\cdot d\mathbf{S}$
b. $(\mathbf{r}\times\mathbf{V})\,\rho\mathbf{V}\cdot d\mathbf{S}$
c. $\oint_S (\mathbf{r}\times\mathbf{V})\,\rho\mathbf{V}\cdot d\mathbf{S}$
d. The sum of Probs. 5.11*d* and 5.12*c*.

5.13 What are several convenient choices for an origin about which to take moments of momentum?

5.14 State mathematically and in words the transport equation applied to the moment of momentum.

5.15 If the origin of the coordinate system is taken at the center of mass of the fluid within a control volume,
a. What is the value of the moment of the gravity body force acting on a fluid within a finite control volume if ρ = a constant?
b. State an integral expression for the moment of the surface forces acting on an ideal fluid within a finite control surface.
c. State an integral expression for the moment of the surface forces due to a Newtonian fluid flowing normal to the control surface.

5.16 State three different conditions that might exist in flow situations through a finite control volume that would yield a zero time rate of change of moment of momentum within the control volume.

5.17 *a.* State the integral relation of the moment of forces and momentum about the z axis that is fixed in inertial space.
b. If a cylindrical control surface of radius R and height h is selected with its centerline along the z axis,
i. Express the moment of momentum flux through the control surface (about the z axis).
ii. Express the moment of momentum flux through the control surface (about the z axis) if the entering flow is in the axial direction and the exiting flow has a velocity of $V_0(\boldsymbol{\varepsilon}_r + \boldsymbol{\varepsilon}_\theta)$ for $0 \le z \le h/2$. Assume the density to have one value throughout the control volume, and that V_0 is a dimensional constant.

5.18 What is the algebraic sign of the moment of momentum surface integral for that surface where
a. The mass flow is into the control volume and the $\mathbf{r}\times\mathbf{V}$ contribution is positive (following the right-hand rule)?
b. The mass flow is out of the control volume and the $\mathbf{r}\times\mathbf{V}$ contribution is negative?
c. The mass flow is into the control volume and the $\mathbf{r}\times\mathbf{V}$ contribution is negative?

5.19 Obtain a reduced form of Eq. (5.11) for the situation where
a. The selected control volume is fixed in an accelerating frame of reference translating only in the x direction.
b. The selected control volume is fixed to a frame of reference whose origin is fixed to a stationary point but is rotating about the z axis at a constant angular velocity.
c. The selected control volume is fixed relative to a wing surface and it includes the surrounding flow. The wing is moving through still air at a constant velocity.

Problems

5.1 A gas passes through a stationary pipe that has an area and direction change. The fluid properties are assumed steady within the control volume and uniform at the selected cross sections. Sketch the control volume to be used and calculate the components of force on the internal surface of the pipe if

$\mathbf{A}_1 = -16\mathbf{i}$ in.^{2} $\quad$ $\mathbf{A}_2 = 4\mathbf{j}$ in.^{2}
$\rho_1 = 0.072$ lb$_m$/ft^3 $\quad$ $\rho_2 = 0.060$ lb$_m$/ft^3
$p_1 = 30$ lb$_f$/in.^{2} $\quad$ $p_2 = 20$ lb$_f$/in.^{2}
$\mathcal{V}_{\text{cont vol}} = 4$ ft^3 $\quad$ $\mathbf{f}_B = -32.2\mathbf{j}$ ft/sec^2
$\dot{m} = 0.8$ lb$_m$/sec

5.2 A pipeline has an expanding section that changes diameter from 6 to 12 in. All of the ideal fluid flow of 2 ft^3/sec is in the x direction. If the pressure is 45 lb$_f$/in.^{2} at its upstream section (i.e., the smaller diameter), estimate the resultant force on the expanding section due to the fluid inside (which has a density of 2.0 slug/ft^3) and the fluid outside (which is static at 1 atm).

5.3 *a.* Calculate the force (in the x direction) of the fluid on the walls between sections 1 and 2 in Example 3.6.
b. Calculate the force (in the x direction) of the interior fluid and the outside atmosphere (14.7 lb$_f$/in.^{2}) on the walls between sections 1 and 2 in Example 3.6.

5.4 Calculate the force (in the x direction) of the interior fluid on the wall between sections 1 and 2 in Example 3.7.

5.5 Use a control-volume analysis to obtain the force per length on the pipe wall ($R = 6$ in.) by the fluid flow in Prob. 4.14.

5.6 Water with a density of 62 lb$_m$/ft^3 flows steadily through a 180° bend while the cross-sectional area is halved as shown in Fig. 5.21. Assuming ideal fluid behavior, calculate the force in the x direction of the moving fluid on the containing wall between the cross sections.

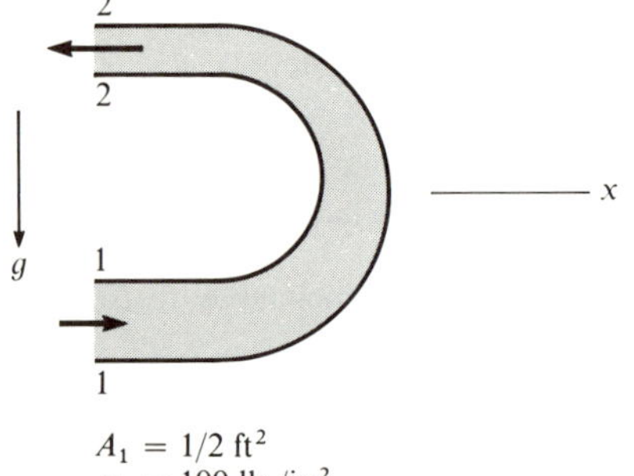

$A_1 = 1/2$ ft^2
$p_1 = 100$ lb$_f$/in.^{2}
$V_1 = 30$ ft/sec
$g = 32.2$ ft/sec^2

Fig. 5.21

5.7 Water flows horizontally in a straight 1-in.-ID pipe and out to the atmosphere through a well-rounded nozzle that has a reduction of area such that the ratio of the exit area to the pipe area is 0.8. At a point upstream, the pressure of the liquid is 200 lb$_f$/in.^{2} and the longitudinal tensile force in the pipe cross section is measured as 129 lb$_f$. If the fluid density is 62.4 lb$_m$/ft^3, calculate the mass flow rate.

5.8 Thrust reversers essentially turn the jet exhaust 180°. If the exit velocity of the exhaust is

3000 ft/sec and the mass flow rate is 100 lb_m/sec, estimate the force exerted on the aircraft for these conditions.

5.9 Gas empties from a tank held in place by hinged vertical and horizontal rods as shown in Fig. 5.22. The bars are instrumented to obtain the tensile force and its variation with time in the rods at A and B. The fluid exits to the atmosphere horizontally. Use a control-volume analysis to show how the exit velocity of the gas could be obtained by knowing the measurements of the tensile force at A and the time-varying tensile force at B.

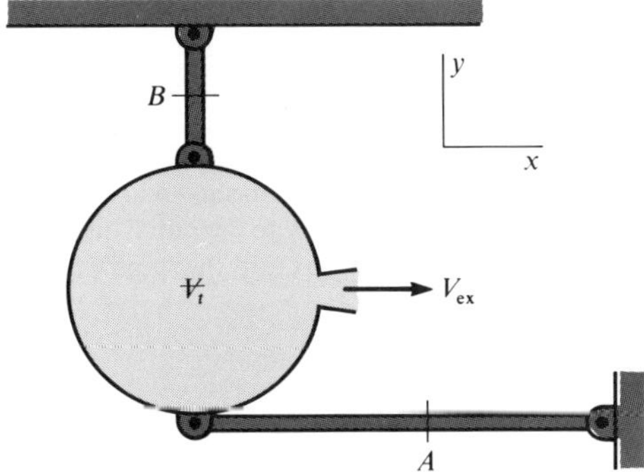

Fig. 5.22

5.10 Obtain the moment in the pipe at section C shown in Fig. 5.23 due to fluid forces on both of the surfaces of the stationary pipe. The surroundings are at 1 atm. Steady flow of 100°F liquid water passes through the area at C of 0.01 ft^2 and exits at B through an area of 0.005 ft^2. The volume flow rate is 0.5 ft^3/sec. Note that the fluid exits in the yz plane and at an angle of 45° to the y axis.

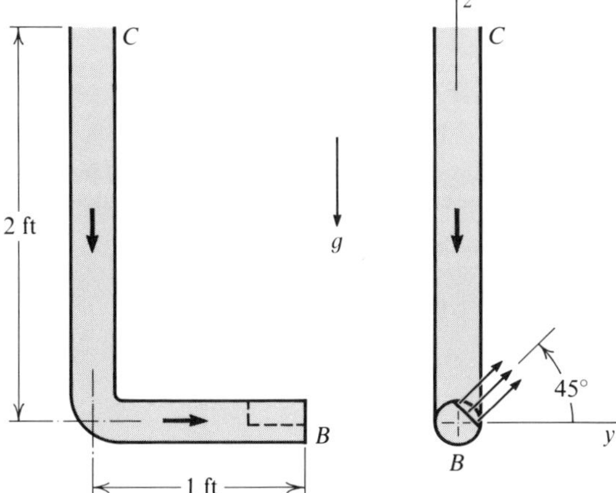

Fig. 5.23

5.11 Water flows steadily through a rotor at 200 ft^3/sec. The density is constant at 62.4 lb_m/ft^3 and the depth of the rotor is 1 ft. The absolute velocity of the fluid at the rotor exit is in a radial direction as shown in Fig. 5.24. Calculate the horsepower imparted by the fluid to the rotor.

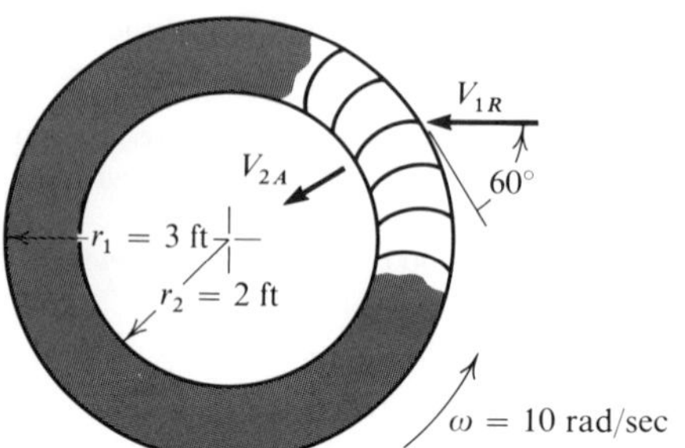

Fig. 5.24

5.12 A centrifugal pump impeller is rotating at 1200 rev/min in the direction shown in Fig. 5.25. The flow enters along an axis of rotation and leaves at an angle of 30° to the radial direction. The exit velocity V_{2A} is 90 ft/sec. Estimate the torque necessary to turn the impeller if the fluid density is 2.0 slug/ft^3.

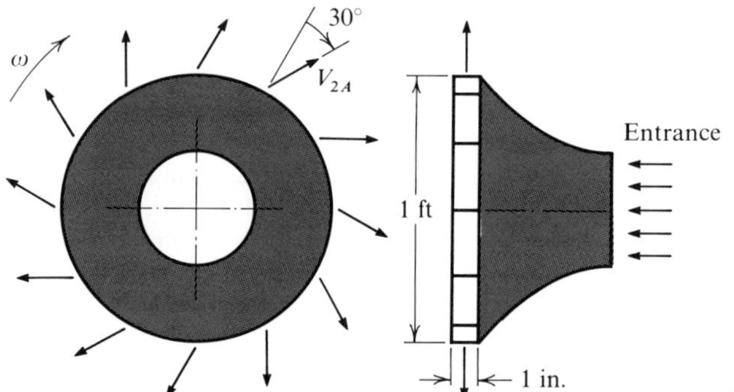

Fig. 5.25

5.13 The turbine shown in Fig. 5.26 has axial-entering and radial-exiting flow. A steady torque of 50 lb$_f$-ft is exerted by the fluid on the rotor, which turns at the rate of 600 rpm. The fluid density is 62.4 lb$_m$/ft^3. Neglecting the cross-sectional area of the shaft, determine the absolute exit velocity V_{2A}.

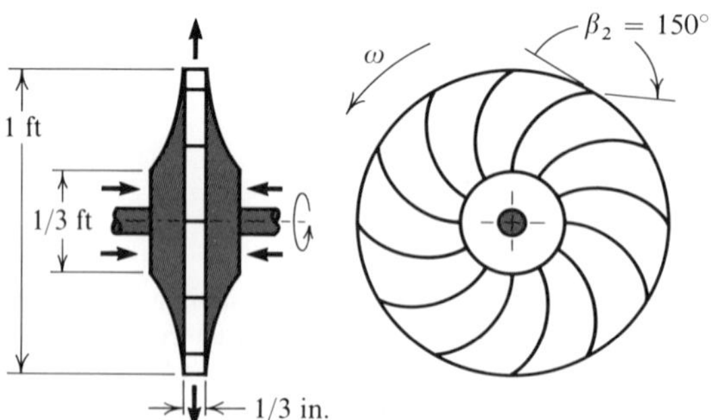

Fig. 5.26

6

ENERGY CONCEPTS AND LAWS

6.1 Introduction

The first law of thermodynamics relates the heat and work transfers at the boundaries of a system to the change of energy of the system. The expression of this principle in control-volume form and as a differential equation provides support for analyzing a large class of problems in fluid mechanics and heat transfer. The second law of thermodynamics is expressed and utilized in control-volume form, and Bernoulli's equation is compared to a form of the energy equation in the latter portion of the chapter.

6.2 First Law of Thermodynamics

For a system undergoing a process, the discipline of thermodynamics provides the following relation

$$Q - W = \Delta E \tag{6.1}$$

where Q = heat transferred to the system
W = work done by the system
ΔE = energy increase of the system

Work done by the surroundings on the system is considered negative, as is heat transferred from the system to the surroundings. These conventions are necessary for the *first law of thermodynamics* written in the form of Eq. (6.1). This expression may be put on a rate basis as follows

$$\frac{đQ‡}{dt} - \frac{đW}{dt} = \left(\frac{dE}{dt}\right)_{\text{syst}} = \frac{DE}{Dt} \tag{6.2}$$

The derivative of the energy is written as the material derivative, for it is the change of the energy of the system which is required in Eq. (6.1) and this energy is a property of the system.

Now consider a system that instantaneously occupies a control volume. According to Eq. (1.37), which applies to the change of an extensive scalar property such as E,

$$\frac{DE}{Dt} = \int_{V\!\!\!/} \frac{\partial(e\rho)}{\partial t}\, dV\!\!\!/ + \oint_S e\rho \mathbf{V}\cdot d\mathbf{S}$$

in which e is the specific energy, i.e., the energy per unit mass. Substitution of the last expression for DE/Dt into Eq. (6.2) gives

$$\frac{đQ}{dt} - \frac{đW}{dt} = \int_{V\!\!\!/} \frac{\partial(e\rho)}{\partial t}\, dV\!\!\!/ + \oint_S e\rho \mathbf{V}\cdot d\mathbf{S} \tag{6.3}$$

The term $đQ/dt$ is the rate at which heat is transferred to the system at its boundaries. Since the boundaries of the system are instantaneously coincident with the control surface, this derivative is the rate at which heat is transferred through the control surface to the material in the control volume at the given instant of time. A similar statement can be made concerning the term $-đW/dt$. The volume integral on the right-hand side of Eq. (6.3) represents the time change of energy of the material in the control volume, and the surface integral represents the net energy efflux over the control surface.

The last form of the energy equation is somewhat general in its implication. As a matter of fact, the form is too general for immediate application to engineering problems requiring elementary skills for their solution. It is therefore necessary to identify the physical entities to be considered and included in the terms heat, energy, and work. The next section discusses these terms.

6.3 Heat, Energy, and Work

Heat. The mechanisms of heat transmission considered in engineering heat-transfer problems are conduction, convection, and radiation.

‡The convention of $đ$ is used to denote inexact differentials for heat and work.

Heat transfer by conduction is usually formulated by Fourier's law, which may be expressed as

$$\dot{\mathbf{q}} = -\kappa\, \mathbf{grad}\, T \tag{6.4}$$

where $\dot{\mathbf{q}}$ = rate of heat transfer per unit area
T = temperature
κ = thermal conductivity of the material

Equation (6.4) expresses the conduction of heat in a three-dimensional temperature field, which is indicated by the gradient of the scalar temperature field. The minus sign is placed in front of the right-hand side so that $\dot{\mathbf{q}}$ will be in the direction of maximum decrease in temperature, which is in accord with the concept that heat is conducted from a higher to a lower temperature —a manifestation of the second law of thermodynamics. The basic dimensions for $\dot{\mathbf{q}}$ are (force)(length)/(length)^2(time) and those of thermal conductivity, κ, are the dimensions of $\dot{\mathbf{q}}$ divided by dimensions of **grad** T, with the result of force/time-temperature. The thermal conductivity varies with temperature for a given material, and for gases it is a function of pressure also. The study of heat transfer[1] contains a more complete discussion of thermal conductivity and its values as a function of temperature for various materials of engineering interest. Since $\dot{\mathbf{q}} = -\kappa\, \mathbf{grad}\, T$ represents the heat-transfer rate per unit area at any point in the temperature field (and hence any point on the control surface), the net rate of *heat transfer to the material* in the control volume may be written as

$$\left(\frac{đQ}{dt}\right)_{\text{cond}} = -\oint_S (-\kappa\, \mathbf{grad}\, T) \cdot d\mathbf{S} = \oint_S \kappa\, \mathbf{grad}\, T \cdot d\mathbf{S} \tag{6.5}$$

This is a form suitable for substitution in the energy equation.

The *transfer of "heat" by convection* at the control surface consists of the transfer of energy identified with material passing through the control surface. Since this flow of energy is already included in the surface integral of Eq. (6.3), it is excluded from the term $đQ/dt$. The discipline of heat transfer has good reason to formulate its problems in a manner so that convective "heat" transfer must be considered as a separate term in energy balances, but it is a subject that is better left to specialized studies beyond the scope of this text.

Radiant heat transfer is the transmission of energy by electromagnetic waves. Even though it is possible to express this transport of energy in a form suitable for placing it in the energy equation, the mechanism of radiation will not be considered here. Radiative effects of this nature are important in high-speed aerodynamics and in flow fields that contain chemically reacting

[1]Frank Kreith, "Principles of Heat Transfer," 2d ed., pp. 9–11, International Textbook Company, Scranton, Pa., 1965.

materials, but the nature of these phenomena prohibits their treatment in an elementary text.

The term $đQ/dt$ in Eq. (6.3) will be taken as *the rate of heat conducted at the control surface.*

Energy. The specific energy e, by convention, is composed of molecular internal (i.e., as observed relative to the material), kinetic, and gravitational potential energies, and these are respectively noted by u, $V^2/2$, and gz (z is the appropriate position coordinate in the gravitational field so that gz is the desired specific potential energy). The height z is usually measured above some arbitrary datum. By the convention of including $V^2/2$ and gz as terms for the energy of the material,[1] the net work in the energy equation does not include a work term associated with the conservative body force of a gravity field or the reversible velocity change of the fluid.

For specific energy $e = u + V^2/2 + gz$ to be applicable to flow fields of a defined fluid where internal energy u is considered as a function of two independent thermodynamic variables [for example, $u = u(\rho,T)$], there must be an absence of capillarity, electric, and magnetic field effects. This is in accordance with the state principle,[2] which specifies the number of independent variables in a thermodynamic equation of state for equilibrium conditions. Correspondingly, specific internal energy and enthalpy changes may be calculated using

$$u_2 - u_1 = \int_{T_1}^{T_2} c_v \, dT - \int_{\rho_1}^{\rho_2} \frac{1}{\rho^2}\left[T\left(\frac{\partial p}{\partial T}\right)_\rho - p\right] d\rho \tag{6.6}$$

and

$$h_2 - h_1 = \int_{T_1}^{T_2} c_p \, dT + \int_{p_1}^{p_2} \frac{1}{\rho}\left[1 + \frac{T}{\rho}\left(\frac{\partial \rho}{\partial T}\right)_p\right] dp \tag{6.7}$$

with enthalpy defined as

$$h = u + \frac{p}{\rho}$$

In two cases of particular interest where the fluid is conceived to behave as (*a*) a perfect gas (that is, where $p = \rho RT$) or (*b*) an incompressible fluid ($D\rho/Dt = 0$), it follows that

$$u_2 - u_1 = \int_{T_1}^{T_2} c_v \, dT \tag{6.8}$$

[1] William C. Reynolds, "Thermodynamics," 2d ed., p. 36, McGraw-Hill Book Company, New York, 1968.

[2] Kenneth Wark, "Thermodynamics," p. 36, McGraw-Hill Book Company, New York, 1966.

A constant density fluid is the limiting case where there is only one independent thermodynamic variable. A pressure change in an incompressible fluid does not cause volume variation and an associated internal energy change, so pressure is not a pertinent thermodynamic-state variable; however, it remains an important dynamic variable.

Work. The rate of work $đW/dt$ transmitted at the control surface is classified in two ways:

1. The rate of shaft work $đW_s/dt$ is the power transmitted from the fluid to its environment outside of the control surface at locations on this surface through which there is no fluid flow. For example, this can occur by the rotation of a shaft or the motion of a boundary. Consider the control volume shown in Fig. 6.1 in which there is a rotor turned by the fluid passing through the control volume. This rotor is rigidly attached to a shaft extending through the control surface. As the shaft rotates, the external suspended mass is raised against the force of gravity, and the material inside the control volume does work. This work is transmitted by the shaft from the fluid inside the control volume to the surroundings and results from the action of the normal and shear stresses on the moving blades. The rate of mechanical work at the control surface is determined by the product of the torque exerted on and the angular velocity of the shaft.

As another example, consider the flow of fluid between the parallel

Fig. 6.1 Control volume with shaft work transmitted across the control surface.

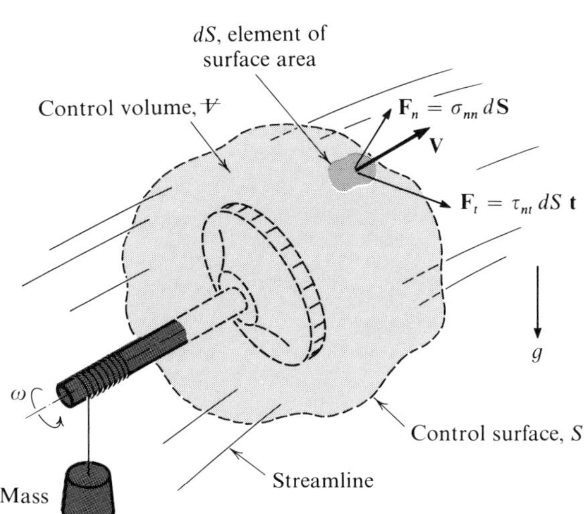

plates in Fig. 6.2. The force F must be applied to the moving plate to balance the fluid resistance encountered at the plate-fluid interface. There is shaft work done by the moving plate on the fluid inside the control volume. The rate of work done would be equal to the shear force acting over the surface AB, multiplied by the plate velocity. This power must be supplied by some external means; that is, some agent external to the material in the control volume must apply the force F in such a way that a boundary point on the line of action of this force moves to the right at a velocity equal to V_p. It should be noted that the force due to shear stresses acting over the surface CD does no work since the surface CD is not moving. Stated another way, there would be some external force necessary to hold the "stationary plate" fixed in space, but this force is not producing any displacement of the plate.

2. Work is performed on or by the material in the control volume, and this energy transfer can be related to the stresses in the fluid at points on the control surface through which fluid is passing. Consider the small element of area dS on the control surface shown in Fig. 6.1. The resultant force on this surface element has been represented by two components: one normal to dS, and the other tangent to dS. These are $\sigma_{nn}\, d\mathbf{S}$ and $\tau_{nt}\, dS\, \mathbf{t}$, respectively, where $\mathbf{t}$ in the latter expression is a unit vector tangent to the surface and in the direction of the force associated with τ_{nt}. The rate of work done by the material in the control volume associated with these forces is

$$-(\sigma_{nn}\, d\mathbf{S} + \tau_{nt}\, dS\, \mathbf{t}) \cdot \mathbf{V}$$

The minus sign in front of the parentheses indicates that the terms following this sign correspond to work done *on* the material within the control volume; hence these terms are prefixed by a negative sign before being inserted in the energy equation. If this last expression is integrated over the control surface, the following accrues:

$$-\oint_S \sigma_{nn}\, d\mathbf{S} \cdot \mathbf{V} - \oint_S \tau_{nt} \mathbf{t} \cdot \mathbf{V}\, dS$$

Fig. 6.2 *Control surface with shaft work transmitted to the fluid in the control volume.*

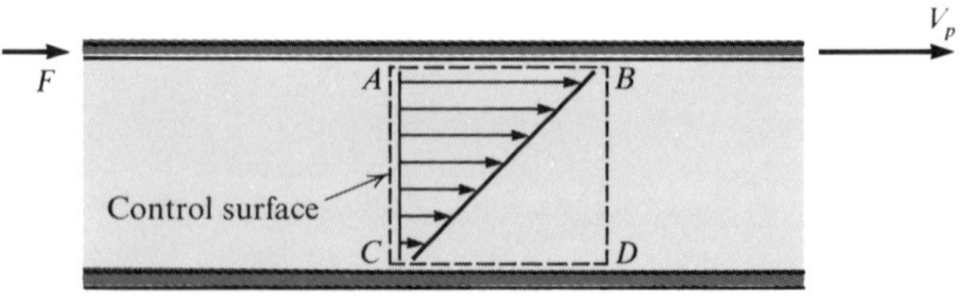

Objection might be raised to integrating over the entire control surface in the example associated with Fig. 6.2 since there is obviously no fluid passing through the portion of the control surface adjacent to the solid surface. However, at these points on the surface, the velocity of the fluid relative to the control surface is zero (no-slip condition) and these contributions to the integrals would be zero. The rate of work $đW/dt$ is then given by

$$\frac{đW}{dt} = \frac{đW_s}{dt}‡ - \oint_S \sigma_{nn} \mathbf{V} \cdot d\mathbf{S} - \oint_S \tau_{nt} \mathbf{t} \cdot \mathbf{V} \, dS \tag{6.9}$$

With the identification of the various work-transfer terms, it remains to insert these in the energy equation.

6.4 Energy Equation for a Control Volume

A useful form of the energy equation for control-volume analysis is obtained by substitution of Eq. (6.9) into (6.3)

$$\frac{đQ}{dt} - \frac{đW_s}{dt} + \oint_S \sigma_{nn} \mathbf{V} \cdot d\mathbf{S} + \oint_S \tau_{nt} \mathbf{t} \cdot \mathbf{V} \, dS = \int_{\mathcal{V}} \frac{\partial(e\rho)}{\partial t} \, d\mathcal{V} + \oint_S e\rho \mathbf{V} \cdot d\mathbf{S} \tag{6.10}$$

Two special cases of Eq. (6.10) provide a relatively simple form of the energy equation:

1. If the fluid is ideal, all of the shear stresses are zero. From previous discussion in Chapters 3 and 4, the absence of shear assures isotropy of the normal stresses at any point in the fluid, and $\sigma_{nn} = -p$. Then $\oint_S \sigma_{nn} \mathbf{V} \cdot d\mathbf{S}$ becomes $-\oint_S p\mathbf{V} \cdot d\mathbf{S}$, and $\oint_S \tau_{nt} \mathbf{t} \cdot \mathbf{V} \, dS = 0$ so that Eq. (6.10) may be written as

$$\frac{đQ}{dt} - \frac{đW_s}{dt} = \int_{\mathcal{V}} \frac{\partial(e\rho)}{\partial t} \, d\mathcal{V} + \oint_S \left(e + \frac{p}{\rho}\right) \rho \mathbf{V} \cdot d\mathbf{S} \tag{6.11}$$

which is valid if all τ's $= 0$.

2. If a control volume can be selected so that the velocity vector is everywhere normal to the control surface at points where fluid is passing through this surface, then

$$\oint_S \tau_{nt} \mathbf{t} \cdot \mathbf{V} \, dS = 0$$

‡Another interpretation for $đW_s/dt$ is the rate of all work at the control surface other than that specified by the two surface integrals shown in Eq. (6.9).

since **t** and **V** are perpendicular to each other. This selection of a control surface permits writing Eq. (6.10) in the form of

$$\frac{đQ}{dt} - \frac{đW_s}{dt} = \int_{V\!\!\!\!/} \frac{\partial(e\rho)}{\partial t}\, dV\!\!\!\!/ + \oint_S \left(e - \frac{\sigma_{nn}}{\rho}\right) \rho \mathbf{V} \cdot d\mathbf{S} \tag{6.12}$$

The following examples emphasize the use of the energy equation for these two cases.

Example 6.1

A perfect gas with a constant specific-heat ratio of $\gamma = c_p/c_v$ passes through a nozzle as shown in Fig. 6.3. The velocity, energy, and density are steady, that is, $\partial \mathbf{V}/\partial t = \partial e/\partial t = \partial \rho/\partial t = 0$. It is assumed that there is no heat transfer within or at the control surface and that the flow is reversible. The pressure at cross sections 1 and 2 are p_1 and p_2, respectively; and the flow and thermodynamic variables may be assumed uniform over each of the sections. The temperature at section 1 is T_1. Use the energy equation to obtain an expression for the velocity at the nozzle exit in terms of the variables T_1, V_1, and p_2/p_1.

Solution. Since the flow is given as reversible, there can be no internal irreversibilities such as fluid shear present in the fluid. Under this idealization, the energy equation (6.11) may be employed

$$\frac{đQ}{dt} - \frac{đW_s}{dt} = \int_{V\!\!\!\!/} \frac{\partial(e\rho)}{\partial t}\, dV\!\!\!\!/ + \oint_S \left(e + \frac{p}{\rho}\right) \rho \mathbf{V} \cdot d\mathbf{S} \tag{6.11}$$

The flow is adiabatic, so $đQ/dt = 0$; also $đW_s/dt = 0$. The volume integral on the right-hand side of the equation is zero from the given steady conditions on e and ρ, and the surface integral may be written as the sum of two integrals, so that

Fig. 6.3

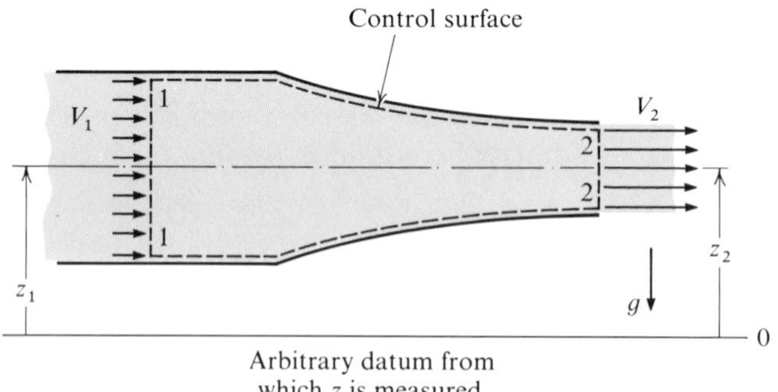

$$\int_{A_1}\left(e + \frac{p}{\rho}\right)\rho\mathbf{V}\cdot d\mathbf{S} + \int_{A_2}\left(e + \frac{p}{\rho}\right)\rho\mathbf{V}\cdot d\mathbf{S} = 0$$

Substitution for the specific energy ($e = u + V^2/2 + gz$) in the last equation gives

$$\int_{A_1}\left(u + \frac{V^2}{2} + gz + \frac{p}{\rho}\right)\rho\mathbf{V}\cdot d\mathbf{S} + \int_{A_2}\left(u + \frac{V^2}{2} + gz + \frac{p}{\rho}\right)\rho\mathbf{V}\cdot d\mathbf{S} = 0$$

Since the flow and state variables are specified as being uniform over each of sections 1 and 2,

$$-\left(h_1 + \frac{V_1^2}{2}\right)\rho_1 V_1 A_1 - \rho_1 V_1 g\int_{A_1} z\,dS + \left(h_2 + \frac{V_2^2}{2}\right)\rho_2 V_2 A_2 + \rho_2 V_2 g\int_{A_2} z\,dS = 0$$

in which specific enthalpy has been substituted for $u + p/\rho$. From mechanics,

$$\int_{A_1} z\,dS = \bar{z}_1 A_1 \qquad \text{and} \qquad \int_{A_2} z\,dS = \bar{z}_2 A_2$$

in which $\bar{z}_1$ and $\bar{z}_2$ are the z coordinates of the centroids of the areas A_1 and A_2, respectively. The last equation may then be written as

$$-\left(h_1 + \frac{V_1^2}{2} + g\bar{z}_1\right)\rho_1 V_1 A_1 + \left(h_2 + \frac{V_2^2}{2} + g\bar{z}_2\right)\rho_2 V_2 A_2 = 0$$

Since conservation of matter requires that $\rho_1 V_1 A_1 = \rho_2 V_2 A_2$, and since $g\bar{z}_1 = g\bar{z}_2$ for this horizontal flow, division of the last equation by $\rho_1 V_1 A_1$, and rearrangement for V_2 gives

$$V_2 = [2(h_1 - h_2) + V_1^2]^{1/2}$$

By using Eq. (6.7) this result may be expressed in terms of the given pressures and temperature by substituting $h_1 - h_2 = c_p(T_1 - T_2)$ for a perfect gas with constant specific heat c_p. Furthermore, the reversible and adiabatic conditions insure that the entropy of the fluid is the same at all points in the flow field. For a perfect gas undergoing an isentropic change of state,

$$\frac{T_2}{T_1} = \left(\frac{p_2}{p_1}\right)^{(\gamma-1)/\gamma}$$

where γ = a constant. Using these observations, and $c_p = \gamma R/(\gamma - 1)$, in which R is the specific-gas constant, yields

$$V_2 = \left\{\frac{2\gamma RT_1}{\gamma - 1}\left[1 - \left(\frac{p_2}{p_1}\right)^{(\gamma-1)/\gamma}\right] + V_1^2\right\}^{1/2} \qquad \blacktriangleleft$$

which is a form of the desired answer in terms of the given information. The term V_1^2 in the braces may be dropped from this result on the basis of negligible kinetic energy at the nozzle entrance; or V_1 may be eliminated from the energy equation by using $\rho_1 A_1 V_1 = \rho_2 A_2 V_2$. The result of Example 3.7, which was obtained by application of Bernoulli's equation and the appropriate barotropic relation, is the same as the above answer.

Example 6.2

The pump shown in Fig. 6.4 maintains steady values of the flow variables at sections 1 and 2 where the gage pressures are 2 and 50 lb_f/in^2, respectively. The fluid properties are assumed uniform over each of these sections. The heat transfer from the fluid through the pipe walls and pump casing may be neglected. The fluid with a specific heat at constant volume equal to 0.44 Btu/lb_m-°F is newtonian and has a density of 56.8 lb_m/ft^3. The temperature in the fluid at sections 1 and 2 is 80.0 and 81.3°F, respectively. Calculate the power imparted to the fluid passing through the pump at a flow rate of 1 ft^3/sec.

Solution. The control surface, indicated by the dashed line, encloses the pump and piping but "cuts" the shaft. Fluid only enters and leaves the control volume at sections 1 and 2, respectively, and the fluid velocity is assumed normal to the control surface at all points on these sections; therefore Eq. (6.12) may be employed in the analysis of this problem.

$$\frac{đQ}{dt} - \frac{đW_s}{dt} = \int_{\mathcal{V}} \frac{\partial(e\rho)}{\partial t} d\mathcal{V} + \oint_S \left(e - \frac{\sigma_{nn}}{\rho}\right)\rho \mathbf{V} \cdot d\mathbf{S} \tag{6.12}$$

The rate of heat transfer is neglected on the basis of given data. This is equivalent to stating that the temperature gradients are very small at the control surface and/or the thermal conductivity of the fluid is not very large.

The volume integral on the right side of the equation is considered as zero on the following basis. Consider the fixed point *A* indicated in Fig. 6.4. As one of the blades, *BC*, approaches point *A*, it is reasonable to assume that the flow variables at this point vary. This indicates that there are changes of these variables with respect to time at some

Fig. 6.4

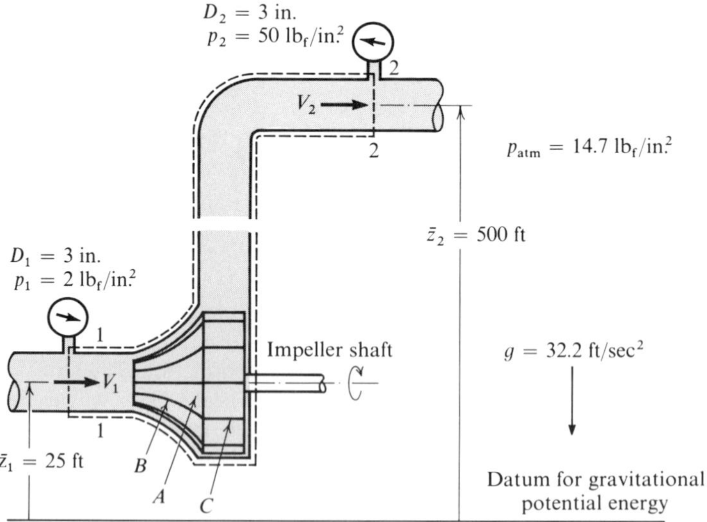

fixed points in the control volume; hence the variables are unsteady at such points. However, if the blade *BC* makes one revolution (or an integral number of revolutions) and the impeller moves at constant angular velocity, it is reasonable to assume that the flow variables at point *A* are restored to the values that prevailed prior to *BC* making the revolution. Under these considerations, all of the variables within the control volume may be considered "steady." While the previous reasoning may appear to lack the rigor of sound mathematical logic, note should be made that any unsteadiness within the control volume could be reflected by fluctuations in the readings of the pressure gages at sections 1 and 2. In this situation the observer taking test data would have to average these readings over a given cycle of pressure fluctuations or else employ an instrument that would provide a pressure-time record during the test. For a pump with a rotating impeller, these fluctuations are generally small compared to those produced by a reciprocating pump that utilizes the movement of a piston, so the flow in this problem will be considered "steady on the average" with the previous considerations understood.

One more term needs consideration before the power $đ W_s/dt$ is calculated from Eq. (6.12). This is the surface integral that includes the normal stress σ_{nn}. This term can be reduced to the pressure at each of the respective sections by referring to the constitutive equations. Consider the axis of the inlet pipe to be the x axis as shown in Fig. 6.5. From Eq. (4.18*a*),

$$\sigma_{xx} = -p - \frac{2}{3}\mu \operatorname{div} \mathbf{V} + 2\mu \frac{\partial V_x}{\partial x}$$

The continuity equation requires that

$$\operatorname{div} \mathbf{V} = \frac{\partial V_x}{\partial x} + \frac{\partial V_y}{\partial y} + \frac{\partial V_z}{\partial z} = 0$$

Furthermore, at section 1, $(V_y)_1 = (V_z)_1 = 0$; and $(\partial V_y/\partial y)_1 = (\partial V_z/\partial z)_1 = 0$ so that $(\partial V_x/\partial x)_1 = 0$. It is then apparent from Eq. (4.18*a*) that

$$(\sigma_{nn})_1 = (\sigma_{xx})_1 = -p_1$$

Similar reasoning may be applied at section 2, to obtain

$$(\sigma_{nn})_2 = -p_2$$

Fig. 6.5

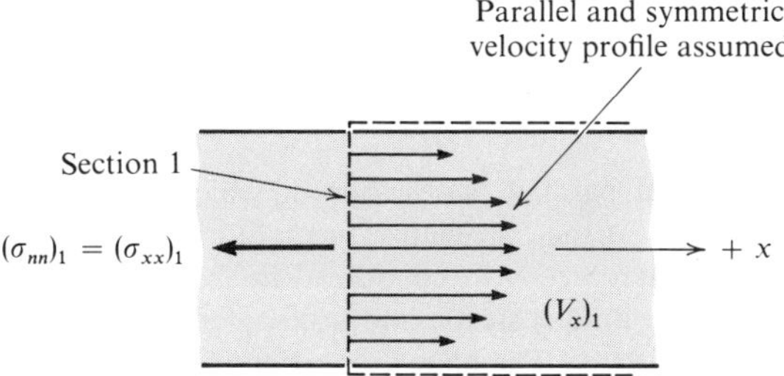

The previous observations allow Eq. (6.12) to be written as

$$-\frac{đW_s}{dt} = \oint_S \left(u + \frac{V^2}{2} + gz + \frac{p}{\rho}\right)\rho\mathbf{V}\cdot d\mathbf{S}$$

Following the procedures used in Example 6.1 to simplify the surface integral, the last equation becomes

$$-\frac{đW_s}{dt} = \left(u_2 + \frac{V_2^{\,2}}{2} + g\bar{z}_2 + \frac{p_2}{\rho_2}\right)\int_{A_2}\rho V\,dS - \left(u_1 + \frac{V_1^{\,2}}{2} + g\bar{z}_1 + \frac{p_1}{\rho_1}\right)\int_{A_1}\rho V\,dS$$

The subscripts for the densities are superfluous since the density field is steady and uniform. Conservation of matter requires

$$\rho\int_{A_2} V\,dS = \rho\int_{A_1} V\,dS = \dot{m}$$

Additionally, since $A_2 = A_1$, then $V_2 = V_1$ and there is zero net kinetic energy efflux. The specific internal energy change may be evaluated by using specific heat and temperature data since the fluid density is constant, so from Eq. (6.8),

$$(u_2 - u_1)\dot{m} = c_v(T_2 - T_1)\dot{m}$$

Using this information, the energy equation reduces to

$$\frac{đW_s}{dt} = \left[c_v(T_1 - T_2) + g(\bar{z}_1 - \bar{z}_2) + \frac{p_1 - p_2}{\rho}\right]\dot{m}$$

$$= \left\{\left(0.44\frac{\text{Btu}}{\text{lb}_m{}^\circ\text{F}}\right)(80.0^\circ\text{F} - 81.3^\circ\text{F})\left(778\frac{\text{ft-lb}_f}{\text{Btu}}\right)\right.$$

$$+\left(32.2\frac{\text{ft}}{\text{sec}^2}\right)(25\,\text{ft} - 500\,\text{ft})\frac{1\,\text{lb}_f\text{-sec}^2}{32.2\,\text{lb}_m\text{-ft}} + \frac{1}{56.8\,\text{lb}_m/\text{ft}^3}\left[\left(2\frac{\text{lb}_f}{\text{in}^2} + 14.7\frac{\text{lb}_f}{\text{in}^2}\right)\right.$$

$$\left.\left.-\left(50\frac{\text{lb}_f}{\text{in}^2} + 14.7\frac{\text{lb}_f}{\text{in}^2}\right)\right]144\frac{\text{in}^2}{\text{ft}^2}\right\}\left(56.8\frac{\text{lb}_m}{\text{ft}^3}\right)\left(1\frac{\text{ft}^3}{\text{sec}}\right)$$

$$= -59{,}100\,\text{ft-lb}_f/\text{sec} = -107\,\text{hp}$$ ◀

The minus sign indicates that shaft work is being performed at a constant rate by the pump impeller on the fluid in the control volume. The relative magnitude of the net internal energy flux emphasizes the need for accuracy in temperature measurements in this situation.

Example 6.3

Consider the flow between two horizontal parallel plates, one of which is moving. (This flow was analyzed in Example 4.4.) Further, specialize the solution to the zero pressure gradient case ($dp/dx = 0$). It is assumed heat conducted in the direction of flow may be neglected. The viscosity of the fluid is steady and uniform at the value of $10^{-4}\,\text{lb}_f\text{-sec/ft}^2$ and the plate velocity is 10 ft/sec. The plate separation is 0.02 ft. Select a control volume that has a 3-ft length in the z direction. If both plates are assumed to be adiabatic walls (almost perfect thermal insulators), what is the net internal energy efflux for a control volume that extends 1 ft in the direction of flow?

Solution. The control volume is shown in Fig. 6.6. From Example 4.4, the solution for the velocity field is

$$V_x = \frac{V_p}{2}\left(1 + \frac{y}{a}\right)$$

using the coordinate axes indicated. According to the discussion of shaft work in Section 6.3, the plate at $y = a$ is performing this type of work on the fluid in the control volume; and it may be obtained from

$$\frac{dW_s}{dt} = -\int_{A_{ABCD}} V_p \tau_{yx}\bigg|_{y=a} dS$$

The minus sign is prefixed to allow for the consideration that work is done on the fluid. Integration over the area is greatly simplified since τ_{yx} and V_p are each independent of position on the surface $ABCD$.

From the constitutive equation (4.18d), one may obtain

$$\frac{dW_s}{dt} = -V_p\mu\left(\frac{\partial V_x}{\partial y} + \frac{\partial V_y}{\partial x}\right)_{y=a} A_{ABCD} = -V_p\mu\left(\frac{V_p}{2a}\right)A_{ABCD}$$

$$= -\frac{(10\text{ ft/sec})^2(10^{-4}\text{ lb}_f\text{-sec/ft}^2)(3\text{ ft}^2)}{0.02\text{ ft}} = -1.5\text{ ft-lb}_f/\text{sec}$$

The energy equation for this simple flow reduces to

$$-\frac{dW_s}{dt} = \oint_S u\rho\mathbf{V}\cdot d\mathbf{S}$$

Fig. 6.6

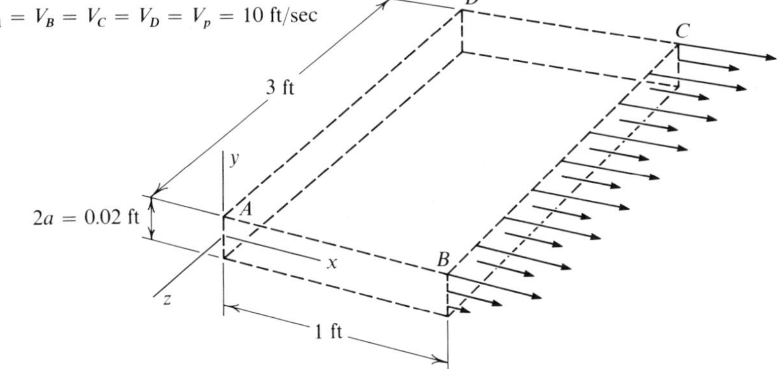

Since the net efflux of internal energy is the surface integral in the last equation

$$\oint_S u\rho\mathbf{V}\cdot d\mathbf{S} = -(-1.5 \text{ ft-lb}_f/\text{sec})$$

The net internal energy efflux = 1.5 ft-lb$_f$/sec ◀

The fact that the answer is a positive number indicates that the work done on the fluid has produced a net outflow of energy from the control volume.

The preceding examples introduced the following important features:

1. Selection of a control surface conveniently oriented with respect to the flow passing through it serves to eliminate the work term $\oint \tau_{nt}\mathbf{t}\cdot\mathbf{V}\,dS$ from the energy equation. This results in considerable simplification of analysis.
2. Use of constitutive equations to (*a*) express normal stresses in terms of the given pressures, and (*b*) ascertain shear stresses when these are required to formulate shaft work.
3. Treatment of an unsteady (but cyclic) case as *steady on the average*, which has great utilization in analyzing a large class of engineering problems associated with turbomachinery.

The assigned problems at the end of this chapter offer an assortment of applications of the first law of thermodynamics. Analyses of some of these will require consideration of the salient features of the last three examples.

6.5 Use of Differential Energy Equation

Constant-density viscous fluid. The first law of thermodynamics for a system may be formulated as

$$\frac{đQ}{dt} - \frac{đW}{dt} = \frac{DE}{Dt} \tag{6.2}$$

This may be applied to heat-transfer problems to find the temperature distribution in a fluid in motion subject to certain boundary conditions on the temperature function.

Consider the flow of a Newtonian fluid between two parallel plates with zero pressure gradient (the same flow configuration given in Example 6.3). The element of fluid, $\Delta x\,\Delta y\,\Delta z$, considered is indicated in Fig. 6.7, and the terms in the energy equation may be easily formulated. The temperature is assumed to be steady and to vary in the y direction only, and the temperature of the plates at $y = a$ and $y = -a$ is maintained at the steady and uniform values T_p and T_w, respectively. The conducted heat flux is

$$-\kappa\,\mathbf{grad}\,T = -\kappa\frac{\partial T}{\partial y}\mathbf{j}$$

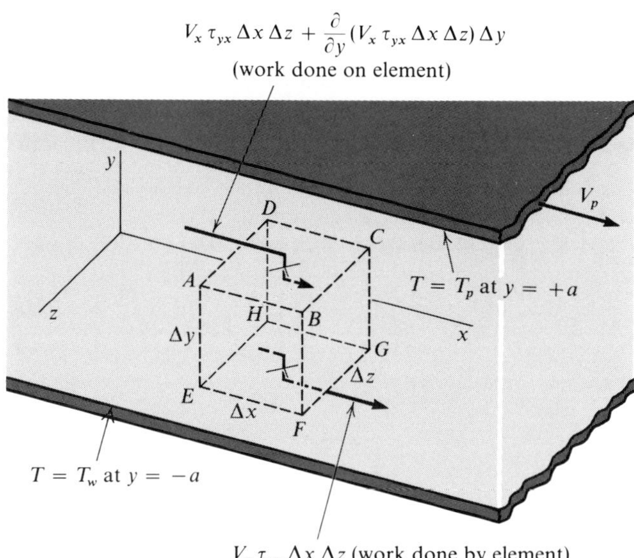

Fig. 6.7 *Element of fluid with shear work.*

The rate of heat transfer at the surface $-(\Delta x\, \Delta z\, \mathbf{j})$ is

$$(-\kappa\, \mathbf{grad}\, T) \cdot (-\Delta x\, \Delta z\, \mathbf{j}) = \kappa \frac{\partial T}{\partial y} \Delta x\, \Delta z$$

while that at the surface $\Delta x\, \Delta z\, \mathbf{j}$ is given by

$$\left[-\kappa\, \mathbf{grad}\, T + \frac{\partial}{\partial y}(-\kappa\, \mathbf{grad}\, T)\, \Delta y\right] \cdot \Delta x\, \Delta z\, \mathbf{j} = -\kappa \frac{\partial T}{\partial y} \Delta x\, \Delta z - \frac{\partial}{\partial y}\left(\kappa \frac{\partial T}{\partial y}\right) \Delta x\, \Delta y\, \Delta z$$

Whether or not each of these terms represents an influx or efflux depends on the sign of the temperature gradient $\partial T/\partial y$ on each face. However, each term was formulated by using the dot product with the outward-drawn normal to the surface and, as such, were cast as effluxes, so that each must be prefixed by a negative sign before substitution is made for $đQ/dt$ in Eq. (6.2). The velocity field for this flow (determined in Example 4.4) is given by

$$V_x = \frac{V_p}{2}\left(1 + \frac{y}{a}\right)$$

From this and the constitutive equation (4.18*d*) the shear stress is $\tau_{yx} =$

$(\mu V_p/2a)$. With the velocity and shear stress distribution known, the shear work on the face *ABCD* is done on the element, and that associated with the face *EFGH* is work done by the element, as indicated on Fig. 6.7. These are prefixed by a negative and positive sign, respectively, when placed in the energy equation. Substitution of the heat fluxes and the work terms into Eq. (6.2) gives

$$\left[\kappa\frac{\partial T}{\partial y}\Delta x\,\Delta z+\frac{\partial}{\partial y}\left(\kappa\frac{\partial T}{\partial y}\right)\Delta x\,\Delta y\,\Delta z-\kappa\frac{\partial T}{\partial y}\Delta x\,\Delta z\right]$$
$$-\left\{-\left[V_x\tau_{yx}\Delta x\,\Delta z+\frac{\partial(V_x\tau_{yx})}{\partial y}\Delta y\,\Delta x\,\Delta z\right]+V_x\tau_{yx}\Delta x\,\Delta z\right\}=\frac{DE}{Dt}$$

which simplifies to

$$\frac{\partial}{\partial y}\left(\kappa\frac{\partial T}{\partial y}\right)+\frac{\partial}{\partial y}(V_x\tau_{yx})=\rho\frac{De}{Dt} \tag{6.13}$$

The reduction of the energy equation to such a simple form is only indicative of the simple problem it represents. If the faces normal to the x direction are considered, the product $V_x\sigma_{xx}\,\Delta y\,\Delta z=-V_xp\,\Delta y\,\Delta z$ on face *AEHD* is equal but opposite in sign to $V_x\sigma_{xx}\,\Delta y\,\Delta z=-V_xp\,\Delta y\,\Delta z$ on the face *BFGC*, since $\partial V_x/\partial x=0$ and $\partial p/\partial x=0$. Then the net work associated with the normal stresses is zero. The remaining four faces of the element would have no work associated with the normal stresses acting over them since $V_y=V_z=0$. A similar line of reasoning may be employed to show that Eq. (6.13) accounts for all of the shear work.

The material derivative De/Dt can be shown to be zero for a fluid with c_v = a constant.

$$\frac{De}{Dt}=\underset{(1)}{V_x\frac{\partial e}{\partial x}}+\underset{(2)}{V_y\frac{\partial e}{\partial y}}+\underset{(3)}{V_z\frac{\partial e}{\partial z}}+\underset{(4)}{\frac{\partial e}{\partial t}}$$

Term (1) in expanded form is

$$V_x\frac{\partial e}{\partial x}=V_x\left[\frac{\partial(u+V_x^2/2)}{\partial x}\right]=V_x\left(\frac{\partial u}{\partial x}+V_x\frac{\partial V_x}{\partial x}\right)=0$$

since

$$\frac{\partial V_x}{\partial x}=0 \qquad \text{for } V_x=V_x(y)$$

and

$$\frac{\partial u}{\partial x}=\frac{\partial(c_vT)}{\partial x}=0 \qquad \text{for } T=T(y)$$

Terms (2) and (3) are zero since $V_y = V_z = 0$. Term (4) is zero because the velocity and temperature field are steady.

If the thermal conductivity can be assumed essentially constant with respect to a variation in y, Eq. (6.13) can be written as

$$\kappa \frac{d^2 T}{dy^2} = -\frac{\partial(V_x \tau_{yx})}{\partial y} = -\frac{\mu V_p{}^2}{4a^2}$$

by using the velocity profile given previously. The boundary-value problem for the temperature is, for $-a < y < a$, and all values of x and z:

$$\frac{d^2 T}{dy^2} = -\frac{\mu V_p{}^2}{4\kappa a^2} \tag{6.14}$$

For all values of x and z,

$$T(x,a,z) = T_p$$

$$T(x,-a,z) = T_w$$

Two integrations of Eq. (6.14) give

$$T = -\frac{\mu V_p{}^2}{8\kappa a^2} y^2 + C_1 y + C_2 \tag{6.15}$$

The boundary conditions require:

$$T_p = -\frac{\mu V_p{}^2}{8\kappa} + aC_1 + C_2$$

$$T_w = -\frac{\mu V_p{}^2}{8\kappa} - aC_1 + C_2$$

from which

$$C_1 = \frac{T_p - T_w}{2a}$$

and

$$C_2 = \frac{T_p + T_w}{2} + \frac{\mu V_p{}^2}{8\kappa}$$

Substitution for the constants C_1 and C_2 in Eq. (6.15) and rearrangement of terms gives

$$T = T(y) = \frac{\mu V_p{}^2}{8\kappa}\left(1 - \frac{y^2}{a^2}\right) + \left(\frac{T_p + T_w}{2}\right)\left(1 + \frac{T_p - T_w}{T_p + T_w}\frac{y}{a}\right) \tag{6.16}$$

which is the temperature distribution in the fluid.

The heat flux at the moving plate and at the stationary wall may be found from Eq. (6.4) as follows:

$$\begin{aligned}\text{Heat flux at moving plate} &= -\kappa \frac{\partial T}{\partial y}\bigg|_{y=a} = -\kappa\left(-\frac{\mu V_p^2}{4\kappa a^2}y + \frac{T_p - T_w}{2a}\right)_{y=a} \\ &= -\kappa\left(\frac{-\mu V_p^2}{4\kappa a} + \frac{T_p - T_w}{2a}\right) \end{aligned} \tag{6.17}$$

$$\text{Heat flux at stationary wall} = -\kappa\left(\frac{\mu V_p^2}{4\kappa a} + \frac{T_p - T_w}{2a}\right) \tag{6.18}$$

It is instructive to examine the temperature profile for various relative magnitudes of T_p and T_w and compare their differences to $\mu V_p^2/2\kappa$. Figures 6.8 and 6.9 present these conditions. For plate temperature greater than wall temperature, $\partial T/\partial y$ at the plate is positive if $T_p - T_w > \mu V_p^2/2\kappa$. The rate of heat flux at the plate (through an area dS_p) is negative, a fact substantiated by Eq. (6.17) or by

$$-\kappa\,\mathbf{grad}\,T \cdot d\mathbf{S}_p = -\kappa\frac{\partial T}{\partial y}\mathbf{j}\cdot dS_p\,\mathbf{j} = -\kappa\frac{\partial T}{\partial y}dS_p$$

The negative heat flux means that the conducted heat is "flowing" in a negative direction, and therefore represents a heat transfer from the plate to the fluid. This is restated here since a negative heat flux might mislead one into thinking the heat was rejected by the fluid on the basis of the sign

Fig. 6.8 Temperature profiles for $T_p > T_w$.

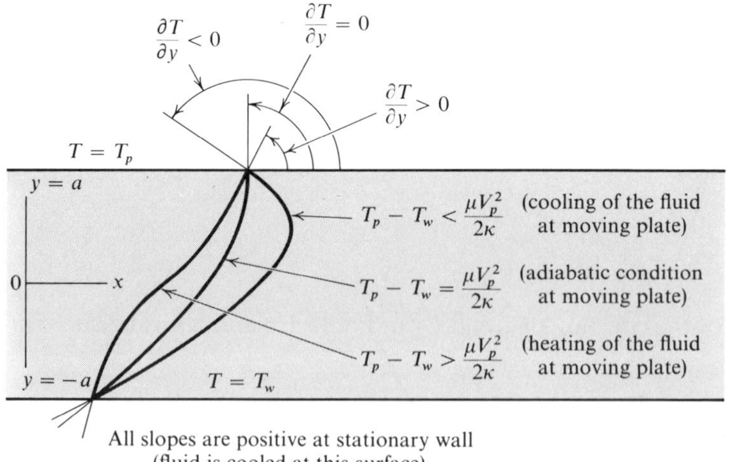

convention used for heat in the formulation of the first law of thermodynamics.

The slope $\partial T/\partial y < 0$ at the plate in Fig. 6.8 requires that there is a heat flow from the fluid to the plate if $T_p - T_w < \mu V_p^2/2\kappa$. This might appear strange since heat is also conducted from the fluid at the wall; that is, the fluid is cooled at both the plate and the wall. The physical explanation is that the rate of shear work performed by the moving plate on the fluid is large enough to require that the heat be conducted from the fluid at the plate if the internal energy of the fluid does not change with respect to time at any point in the field.

For the case where $T_p - T_w > \mu V_p^2/2\kappa$, with $T_p > T_w$, the internal energy of the fluid is maintained at a steady value by a transfer of heat to the fluid at the moving plate. If $T_p = T_w$, then the fluid is cooled at the plate and the wall. This result is readily discerned by reference to Eqs. (6.17) and (6.18).

The preceding analysis illustrates the use of the differential form of the energy equation and the simplicity that accrues to the determination of the temperature field when the density, viscosity, specific heat, and thermal conductivity are each assumed to be uniform and steady. Reference to Example 4.4 reveals that the reduced continuity equation and the Navier-Stokes equations are uncoupled since the density does not appear in the continuity equation, and this permits solving for the velocity field without concern for the density. This velocity distribution may then be substituted in an applicable form of the energy equation to assist in the determination

Fig. 6.9 *Temperature profiles for* $T_p < T_w$.

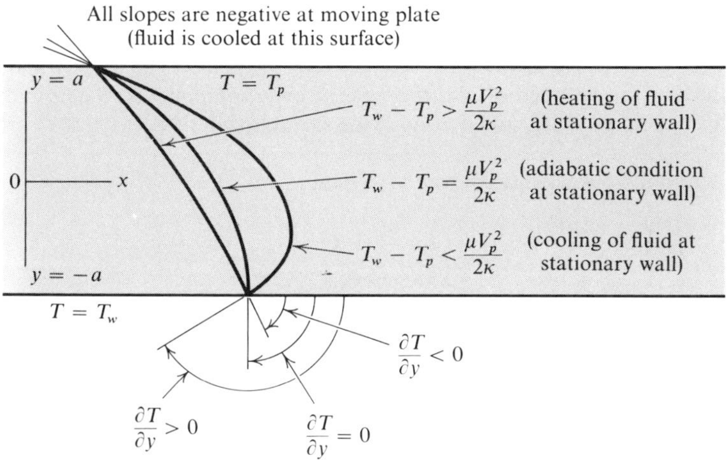

of the temperature field. In addition, the state principle in the form of $u = u(T,\rho)$ was employed in this analysis to handle the term $\partial u/\partial x$.

As might be expected, the analysis of more complicated velocity and temperature fields would require skills beyond those requisite for a first course in fluid mechanics. However, the reader should refer again to the statements following Eq. (4.22) in Section 4.8 to gain more appreciation of the general problem. These statements and the following example will provide some insight with regard to the difficulties involved when the fluid density is variable.

Compressible and viscous fluid. There are a few exact solutions of the Navier-Stokes equations for compressible fluids, in which the temperature gradient and shear work cause a variation in fluid viscosity. Illingworth[1] has found several simple solutions, one of which is given in the following example.

Example 6.4

The Newtonian fluid is a perfect gas with a thermal conductivity κ and viscosity μ, each functions of temperature only. The specific heat at constant pressure is c_p and is assumed constant (steady and uniform), which is reasonable if the range of temperature in the gas is low enough so that rotational or vibrational energies of the molecules are not appreciably changed. The flow is confined between two parallel plates; this motion is maintained steady by the upper plate, which moves at a constant velocity V_p as shown in Fig. 6.10. All of the dependent variables are considered to be functions of the y coordinate only. This would be true of μ and κ also since both are functions of T, and T is a function of y. The stationary wall is thermally insulated and the temperature of the moving plate is uniform and constant. Body forces are neglected since all of the fluid particles would have almost the same gravitational potential energies for $b > y > 0$, even if the gravity field were oriented in the y direction. Also, there would be negligible variation of pressure in the y direction for a small distance between the plates. The density and temperature of the fluid are assumed steady, as are the velocity components. Determine the velocity, temperature, pressure, and density fields as a function of y.

Solution. The continuity equation (2.3) may be written as

$$\frac{\partial(\rho V_x)}{\partial x} + \frac{\partial(\rho V_y)}{\partial y} + \frac{\partial(\rho V_z)}{\partial z} = 0 \qquad \text{for } \frac{\partial \rho}{\partial t} = 0$$

Since all dependent variables are functions of y only, Eq. (2.3) reduces to

$$\frac{\partial(\rho V_y)}{\partial y} = 0 = \frac{d(\rho V_y)}{dy}$$

[1]C. R. Illingworth, Some Solutions of the Equations of Flow of a Viscous Compressible Fluid, *Proc. Cambridge Phil. Soc.*, vol. 46, no. 3, pp. 469–478, 1950.

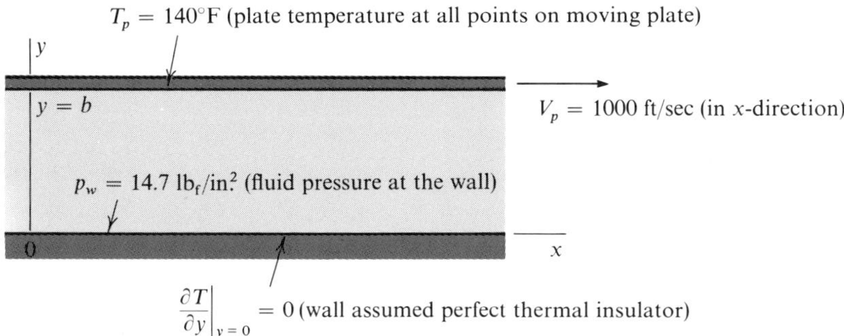

Fig. 6.10

and integration of this gives

$$\rho V_y = \text{a constant} = (\rho V_y)_{y=0} = 0$$

as the no-slip condition at the stationary plate requires V_y to vanish at $y = 0$. The last equation indicates that $V_y = 0$ since $\rho \neq 0$, an important result accruing from the conservation of matter for this problem.

The fluid is newtonian, so the Navier-Stokes equations apply; however, those used are Eqs. (4.19a, b, and c) since the viscosity is variable in this example. Equation (4.19a) is given by the following with the "zero-valued terms" indicated:

$$\rho\left(\cancelto{0}{V_x\frac{\partial V_x}{\partial x}} + \cancelto{0}{V_y}\frac{\partial V_x}{\partial y} + \cancelto{0}{V_z\frac{\partial V_x}{\partial z}} + \cancelto{0}{\frac{\partial V_x}{\partial t}}\right) = -\cancelto{0}{\frac{\partial p}{\partial x}} - \frac{\partial}{\partial x}\left(\frac{2}{3}\mu\,\cancelto{0}{\operatorname{div}\mathbf{V}} - 2\mu\cancelto{0}{\frac{\partial V_x}{\partial x}}\right)$$
$$+ \frac{\partial}{\partial y}\left[\mu\left(\cancelto{0}{\frac{\partial V_y}{\partial x}} + \frac{\partial V_x}{\partial y}\right)\right] + \frac{\partial}{\partial z}\left[\mu\left(\cancelto{0}{\frac{\partial V_x}{\partial z}} + \cancelto{0}{\frac{\partial V_z}{\partial x}}\right)\right] + \rho\,\cancelto{0}{(\mathbf{f}_B)_x} \qquad (4.19a)$$

The zero terms result from either or both of the following fact: the dependent variables are each functions of y only, and $V_y = 0$. The Navier-Stokes equations, when reduced in this manner, yield

$$\frac{d}{dy}\left(\mu\frac{dV_x}{dy}\right) = 0 \qquad (4.19a)_{\text{reduced}}$$

$$\frac{dp}{dy} = 0 \qquad (4.19b)_{\text{reduced}}$$

$$\frac{d}{dy}\left(\mu\frac{dV_z}{dy}\right) = 0 \qquad (4.19c)_{\text{reduced}}$$

which are written as ordinary differential equations because of the dependence on the single variable y. These integrate, respectively, as

$$\mu \frac{dV_x}{dy} = C_1 = \text{a constant} \tag{6.19}$$

$$p = \text{a constant} = p_w = 14.7 \text{ lb}_f/\text{in}^2 \tag{6.20}$$

and

$$\mu \frac{dV_z}{dy} = \text{a constant} = \tau_{yz}$$

However, τ_{yz} at $y = b$ is zero since there is no shear stress applied by the moving plate in the direction of the z axis. This requires $dV_z/dy = 0$ for $\mu \neq 0$. Since V_z is zero at $y = 0$, then

$$V_z = 0 \tag{6.21}$$

The energy equation may be applied in the form of Eq. (6.13) (with $DE/Dt = 0$ for the same reasons given earlier in this section):

$$\frac{d}{dy}\left(\kappa \frac{dT}{dy}\right) + \frac{d(V_x \tau_{yx})}{dy} = 0 \qquad (6.13)_{\text{reduced}}$$

From the constitutive equation (4.18d),

$$\tau_{yx} = \mu\left(\frac{\partial V_y}{\partial x} + \frac{\partial V_x}{\partial y}\right) = \mu \frac{dV_x}{dy}$$

since

$$V_y = 0 \qquad \text{and} \qquad V_x = V_x(y)$$

Substitution for τ_{yx} in the preceding form of Eq. (6.13) gives

$$\frac{d}{dy}\left(\kappa \frac{dT}{dy}\right) + \frac{d}{dy}\left(V_x \mu \frac{dV_x}{dy}\right) = 0$$

or

$$\frac{d}{dy}\left(\kappa \frac{dT}{dy}\right) + \mu\left(\frac{dV_x}{dy}\right)^2 + V_x \frac{d}{dy}\left(\mu \frac{dV_x}{dy}\right) = 0$$

The third term on the left side of the last equation is zero from Eq. $(4.19a)_{\text{reduced}}$, and $\mu(dV_x/dy)^2 = C_1(dV_x/dy)$ from Eq. (6.19) so that

$$\frac{d}{dy}\left(\kappa \frac{dT}{dy}\right) + C_1 \frac{dV_x}{dy} = 0 \tag{6.22}$$

Integrating the preceding expression yields

$$\kappa \frac{dT}{dy} + C_1 V_x = C_2 \tag{6.23}$$

and $C_2 = 0$ using the boundary conditions $dT/dy = 0$ and $V_x = 0$ at $y = 0$. Elimination of C_1 in Eq. (6.23) by use of Eq. (6.19) gives

$$\frac{\kappa}{\mu}\frac{dT}{dV_x} + V_x = 0 \tag{6.24}$$

which is a relation between temperature and velocity. To integrate this expression requires some knowledge of the variations of viscosity and thermal conductivity of the gas with temperature. For gases it has been found experimentally that these are linear variations over a reasonable temperature range,[1] if the gases are not highly compressed. If these variations are expressed as

$$\kappa = \alpha T \qquad \text{and} \qquad \mu = \beta T$$

in which α and β are suitable constants, then the ratio κ/μ is independent of temperature. Equation (6.24) may be integrated to give

$$\frac{\kappa}{\mu}T + \frac{V_x^2}{2} = C_3 \tag{6.25}$$

The constant C_3 may be determined from the conditions $V_x = V_p$ and $T = T_p$ for $y = b$, and

$$C_3 = \frac{\kappa}{\mu}T_p + \frac{V_p^2}{2}$$

so that Eq. (6.25) may be written as

$$T = T_p + \frac{\mu}{2\kappa}V_p^2 - \frac{\mu}{2\kappa}V_x^2 \tag{6.26}$$

which is a relation between temperature and velocity. Another relation between T and V_x may be obtained by expressing Eq. (6.19) as

$$\int_0^{V_x} \mu\, dV_x = C_1 \int_0^y dy$$

and substituting βT for μ. This gives

$$\beta \int_0^{V_x} T\, dV_x = C_1 y$$

Substituting for T in the last integral yields

$$\beta \int_0^{V_x} \left(T_p + \frac{\mu}{2\kappa}V_p^2 - \frac{\mu V_x^2}{2\kappa}\right) dV_x = C_1 y$$

which reduces to

$$T_p V_x + \frac{\mu}{2\kappa}V_p^2 V_x - \frac{\mu}{6\kappa}V_x^3 = \frac{C_1 y}{\beta} \tag{6.27}$$

[1]G. N. Patterson, "Molecular Flow of Gases," pp. 102–103, John Wiley & Sons, Inc., New York, 1956.

after performing the integration between the indicated limits. The constant C_1/β may be found from the condition $V_x = V_p$ for $y = b$, and it is given by

$$\frac{C_1}{\beta} = \frac{1}{b}\left(T_p V_p + \frac{\mu}{3\kappa} V_p^3\right)$$

The velocity profile is then given by

$$T_p V_x + \frac{\mu}{2\kappa} V_p^2 V_x - \frac{\mu}{6\kappa} V_x^3 = \frac{y}{b}\left(T_p V_p + \frac{\mu}{3\kappa} V_p^3\right) \qquad (6.28)$$

which is a cubic equation in V_x. In principle Eq. (6.28) could be solved, ascertaining V_x as a function of y. However, all three roots, i.e., the three functions of V_x, would have to be examined for their physical admissibility. The simpler and more instructive approach is to place Eq. (6.28) in a nondimensional form, select a value for some physical properties, and plot the three velocity profiles that lie between the two plates. The choice of the "correct" profile may be made on the basis of physical considerations.

Equation (6.28) may be put into the normal form of a cubic by rearranging its terms and dividing the result by the coefficient of V_x^3. This gives

$$V_x^3 - 3V_p^2 V_x - 6\frac{\kappa}{\mu} T_p V_x + \frac{y}{b}\left(6\frac{\kappa}{\mu} T_p V_p + 2V_p^3\right) = 0$$

Dividing through by V_p^3, one obtains the following nondimensional form:

$$\left(\frac{V_x}{V_p}\right)^3 - 3\frac{V_x}{V_p} - 6\frac{\kappa}{\mu}\frac{T_p}{V_p^2}\frac{V_x}{V_p} + \frac{y}{b}\left(\frac{6\kappa T_p}{\mu V_p^2} + 2\right) = 0$$

The coefficients in the last equation may be evaluated from the given data and reference to physical information pertaining to μ and κ,[1] and the cubic becomes

$$\left(\frac{V_x}{V_p}\right)^3 - 31.8\frac{V_x}{V_p} + 30.8\frac{y}{b} = 0$$

This equation has three real roots for $0 \leq y/b \leq 1$. These are shown in Fig. 6.11. The two velocity profiles indicated by dashed lines do not meet the boundary conditions of the problem, which requires $V_x/V_p = 0$ for $y/b = 0$, and $V_x/V_p = 1$ for $y/b = 1$. However, the slopes of these two profiles at $y = b$ open up another argument against their validity. Each of these slopes is negative, and this would require τ_{yx} to be negative at the moving plate, which contradicts the given fact that the plate is applying a positive τ_{yx} to the fluid. The profile *EF*, which satisfies the boundary conditions, is the only valid velocity distribution and is the graphical answer to the variation of velocity in the y direction.

The temperature distribution may be found from Eq. (6.26), and in nondimensional form this is given by

$$\frac{T}{T_p} = 1 + \frac{\mu V_p^2}{2\kappa T_p}\left[1 - \left(\frac{V_x}{V_p}\right)^2\right] \qquad (6.29)$$

[1] The value of κ/μ used in this problem was taken as 8000 ft^2/sec^2-°R. This value corresponds to air at 100°F.

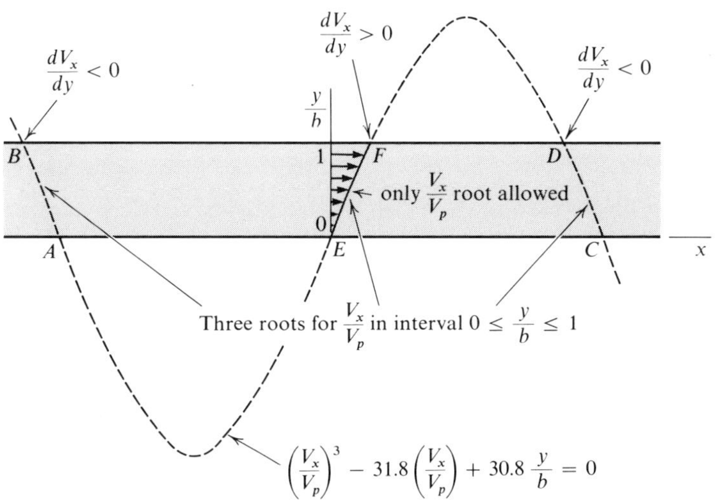

Fig. 6.11

The profile T/T_p as a function of y/b could be constructed from the velocity profile V_x/V_p as a function of y/b, and this would be the answer for $T(y)$.

To find the density profile requires the use of the equation of state,

$$\rho = \frac{p}{RT} \tag{6.30}$$

in which R is the specific gas constant. This is also valid for the fluid at the moving plate so that

$$\rho_p = \frac{p_p}{RT_p} \tag{6.31}$$

in which the subscript p refers to the state variables of the fluid at $y = b$. Dividing Eq. (6.30) by Eq. (6.31) gives

$$\frac{\rho}{\rho_p} = \frac{p}{p_p}\frac{T_p}{T}$$

and recalling from Eq. (6.20) that p is a constant,

$$\frac{\rho}{\rho_p} = \frac{T_p}{T} \qquad \blacktriangleleft \tag{6.32}$$

This is an answer for ρ/ρ_p as a function of y/b since Eq. (6.29) gives the distribution of T/T_p. The value of the density at the plate, ρ_p, may be calculated from Eq. (6.31) for a particular perfect gas.

The preceding example serves to illustrate the need for all of the basic equations and some physical data in solving even a simple problem when such variables as density, viscosity, and thermal conductivity are not restricted to constant values. The basic laws and equations employed are summarized as follows:

The conservation of matter	Eq. (2.3)
Three Navier-Stokes equations	Eqs. (4.19*a*, *b*, *c*)$_{\text{reduced}}$
The first law of thermodynamics	Eq. (6.13)$_{\text{reduced}}$
The second law of thermodynamics	Implied in the use of Fourier's law of heat conduction with minus sign in front of $\kappa\,\mathbf{grad}\,T$

In addition to the basic laws, the following subsidiary information was used:

Equation of state ($p = \rho RT$)	Eq. (6.30)
Fourier's law of heat conduction	Eq. (6.4)
$\kappa = \alpha T$ and $\mu = \beta T$	
State postulate $[u = u(\rho, T)]$	
Constitutive equations	Eqs. (4.18*a*, *b*, *c*, *d*, *e*, *f*)

In conclusion, solutions to physical problems generally require the analyst to use all of the governing basic laws and constitutive equations.

6.6 Entropy Considerations

The second law of thermodynamics for a system[1] may be used to obtain

$$d\$ \geq \frac{đQ}{T} \tag{6.33}$$

where $d\$$ = change of entropy of the system
$đQ$ = heat transferred to the system
T = absolute temperature at boundary of the system where heat is transferred

The equality holds for reversible processes in the system, while the inequality applies to irreversible processes. The specific entropy change between equilibrium states for a simple compressible substance is related to other fluid properties by:[2]

$$d¢ = \frac{du}{T} + \frac{p}{T}\,d\left(\frac{1}{\rho}\right) \tag{6.34}$$

[1]George N. Hatsopoulos and Joseph H. Keenan, "Principles of General Thermodynamics," p. 155, John Wiley & Sons, Inc., New York, 1965.
[2]*Ibid.*, p. 158.

which yields entropy expressions for two substances of particular interest:

A perfect gas.

$$\not{s}_2 - \not{s}_1 = \int_{T_1}^{T_2} c_v \frac{dT}{T} + R \ln \frac{\rho_1}{\rho_2} \tag{6.35a}$$

$$\not{s}_2 - \not{s}_1 = \int_{T_1}^{T_2} c_p \frac{dT}{T} - R \ln \frac{p_2}{p_1} \tag{6.35b}$$

An incompressible liquid.

$$\not{s}_2 - \not{s}_1 = \int_1^2 \frac{du}{T} = \int_{T_1}^{T_2} c_v \frac{dT}{T} \tag{6.36}$$

Equation (6.33) may be placed on a rate basis by following the system as it instantaneously occupies a control volume, thus

$$\frac{D\$}{Dt} \geq \oint_S \frac{1}{T} \frac{\text{đ}q}{dt} dS \tag{6.37}$$

in which $\text{đ}q/dt$ is the rate of heat transferred per unit area and the integration is carried out over the entire boundary of the system at a given instant of time.

A specialization of Eq. (6.37) accrues if the system's boundaries are impervious to heat (perfectly insulated boundaries), and such a system is described as *thermally isolated*. For a system of this nature, the appropriate specialization is

$$\left.\frac{D\$}{Dt}\right|_{\substack{\text{thermally}\\ \text{isolated}}} \geq 0 \tag{6.38}$$

Entropy is an extensive property of matter, which allows substitution for the material derivative of $\$$ in Eq. (6.37) through use of the transport equation (1.37). Denoting specific entropy by $\not{s}$, the transport relation becomes

$$\frac{D\$}{Dt} = \int_{\not{V}} \frac{\partial(\not{s}\rho)}{\partial t} d\not{V} + \oint_S \not{s}\, \rho \mathbf{V} \cdot d\mathbf{S}$$

Using the preceding expression for $D\$/Dt$ in Eq. (6.37) gives

$$\int_{\not{V}} \frac{\partial(\not{s}\rho)}{\partial t} d\not{V} + \oint_S \not{s}\, \rho \mathbf{V} \cdot d\mathbf{S} - \oint_S \frac{1}{T} \frac{\text{đ}q}{dt} dS \geq 0$$

with the surface integrals taken over the control surface since the boundaries of the system are coincident with the control surface at the given instant of time. If the only mechanism of heat transfer across the boundaries is conduction, the preceding equation may be written as

$$\int_{\mathcal{V}} \frac{\partial(\$\rho)}{\partial t}\, d\mathcal{V} + \oint_S \$\, \rho \mathbf{V} \cdot d\mathbf{S} - \oint_S \frac{\kappa\, \mathbf{grad}\, T}{T} \cdot d\mathbf{S} \geq 0 \tag{6.39}$$

through the use of Fourier's law; and, in the case of a thermally insulated control surface,

$$\int_{\mathcal{V}} \frac{\partial(\$\rho)}{\partial t}\, d\mathcal{V} + \oint_S \$\, \rho \mathbf{V} \cdot d\mathbf{S} \geq 0 \tag{6.40}$$

The second law of thermodynamics as applied to the analysis of fluid mechanics problems may be used to discriminate between several families of solutions. This aspect will be employed in Chapter 8 to show that a normal shock wave can only occur in supersonic flow.

The following examples will serve to familiarize the reader with applications of the principle introduced in this section.

Example 6.5

Show that the solution for velocity and temperature in Example 6.4 is compatible with the second law of thermodynamics.

Solution. Recall that the velocity profile has a positive slope at $y = b$. Equation (6.24) provides

$$\frac{dT}{dV_x} = -\frac{\mu}{\kappa} V_x$$

which is a negative quantity; thus if V_x is increasing with respect to y at $y = b$, then T must be decreasing with respect to y at this location. This indicates that $\mathbf{grad}\, T$ is negative at $y = b$ and for all values of x.

Consider the control volume shown in Fig. 6.12, which is a rectangular parallelepiped of arbitrary dimension in the x and z directions. The only portion of the control surface where heat is transferred is that which is in contact with the moving plate, and the rate of heat transfer there is given by $-\kappa\, \mathbf{grad}\, T \cdot \Delta x\, \Delta z\, \mathbf{j}$ which is in the positive y direction. Equation (6.39) requires that

$$\int_{\mathcal{V}} \frac{\partial(\$\rho)}{\partial t}\, d\mathcal{V} + \oint_S \$\, \rho \mathbf{V} \cdot d\mathbf{S} \geq \oint_S \frac{\kappa\, \mathbf{grad}\, T}{T} \cdot d\mathbf{S}$$

The first integral is zero since $\$$ and ρ are not functions of time. The second integral is zero since there is only flow through faces of the control surface that are normal to the

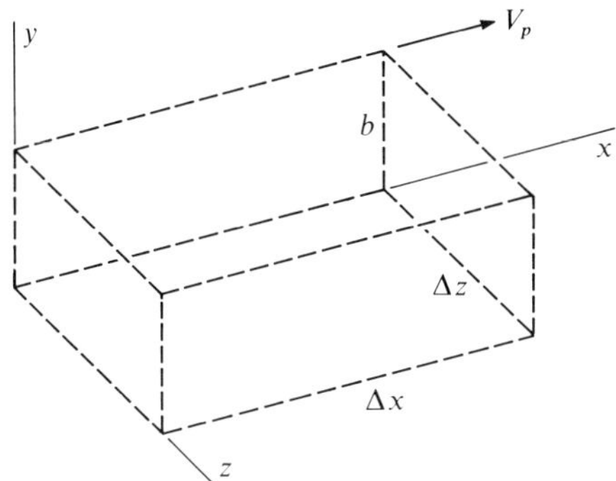

Fig. 6.12

x direction, and none of the dependent variables are functions of x. The second law then requires

$$0 \geq \oint_S \frac{\kappa \,\mathbf{grad}\, T}{T} \cdot d\mathbf{S} = \frac{\kappa}{T_p} \left. \frac{dT}{dy} \right|_{y=b} \mathbf{j} \cdot \Delta x \,\Delta z \,\mathbf{j}$$

and

$$0 > \frac{\kappa \,\Delta x \,\Delta z}{T_p} \left. \frac{dT}{dy} \right|_{y=b}$$

since the temperature derivative is negative on the basis of the opening observations concerning this problem, and the second law is reconciled with the inequality prevailing.

An alternative, and much simpler, method would be the recognition that heat is rejected by the material in the control volume if $dT/dy|_{y=b}$ is negative. For $T > 0$,

$$\oint_S \frac{1}{T} \frac{đq}{dt} \, dS < 0$$

because heat rejected by the fluid has been taken as negative. The fact that the inequality prevails should come as no surprise since there is an interchange of shear work and heat within the fluid—and these aspects contribute to irreversibilities.

Example 6.6

A perfect gas flows steadily in a horizontal and circular duct. The duct is thermally insulated and has a changing cross-sectional area. The velocity V_x varies over the cross section because of the shear stresses in a boundary layer on the inside wall of the duct, shown in Fig. 6.13. Outside of the boundary layer, a one-dimensional assumption is reasonable for the thermodynamic variables at sections 1 and 2. If it is known that there is a net efflux of kinetic energy from the control surface, and this efflux increases continuously between sections 1 and 2, show that $p_2 < p_1$ by using the second law of thermodynamics.

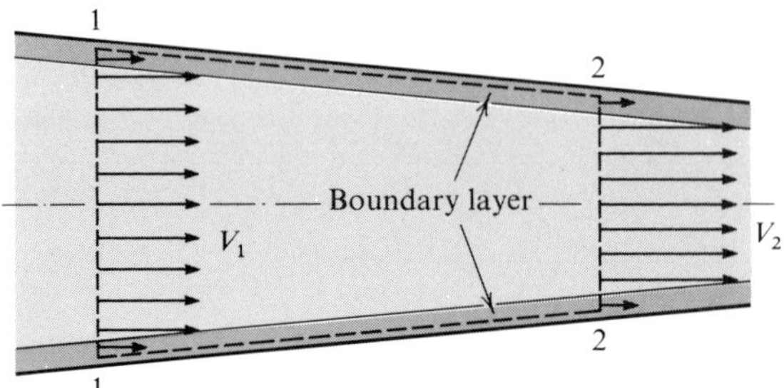

Fig. 6.13

Solution. The control surface is indicated by dashed lines on Fig. 6.13. Since the flow variables are steady and the enthalpy is uniform over each of the cross sections (selected normal to the flow), the energy equation (6.12) reduces to

$$\oint_S \left(h + \frac{V^2}{2}\right)\rho\mathbf{V}\cdot d\mathbf{S} = (h_2 - h_1)\,\dot{m} + \oint_S \frac{V^2}{2}\rho\mathbf{V}\cdot d\mathbf{S} = 0$$

in the absence of heat transfer and shaft work at the walls. The fluid is a perfect gas and so enthalpy is a function of temperature only. Since the net efflux of kinetic energy is positive, the enthalpy of the fluid (and hence the temperature) decreases on moving from section 1 to section 2.

From Eq. (6.40), the second law requires

$$\oint_S \not{s}\,\rho\mathbf{V}\cdot d\mathbf{S} > 0$$

and

$$\not{s}_2 - \not{s}_1 > 0$$

since $\not{s} = \not{s}(p, T)$ is uniform over each cross section.

From Eq. (6.35*b*), $\int c_p\,dT/T$ is negative since $T_2 < T_1$. If $\not{s}_2 - \not{s}_1 > 0$ is to be satisfied, then $-R\ln p_2/p_1$ must be positive. Hence p_2 must be less than p_1.

The result, $p_2 < p_1$, can be obtained by using the conservation of matter and the p-ρ-T relation with the energy equation, provided the cross-sectional area of the duct is the same at sections 1 and 2.

6.7 Comparison of the Energy Equation with Bernoulli's Equation

The energy equation for the steady, frictionless, and incompressible flow of a fluid reduces to Bernoulli's equation provided there is no transfer of heat and shaft work. This is readily seen by applying the appropriate form of the energy equation (6.11) to the flow along a streamline. This equation may be reduced to

$$\left(e_2 + \frac{p_2}{\rho_2}\right) - \left(e_1 + \frac{p_1}{\rho_1}\right) = 0$$

by the same considerations employed in Example 6.1; in expanded form it yields

$$\frac{V_2^2 - V_1^2}{2} + \frac{p_2 - p_1}{\rho} + g(z_2 - z_1) = u_1 - u_2$$

The subscript associated with ρ has been dropped since the density does not vary. Since there is no change of internal energy [see Eqs. (6.8) and (6.36) for s = const], the preceding expression can be written as

$$\frac{V_2^2 - V_1^2}{2} + \frac{p_2 - p_1}{\rho} + gz_2 - gz_1 = 0$$

which is the same as Eq. (3.17), the Bernoulli equation for steady incompressible flow of an ideal fluid.

6.8 Closing Remarks

At this point, the reader has been exposed to the basic equations of fluid mechanics. Some of the examples required only several of these relations in finding solutions; however, the Example 6.4 used every basic equation at the disposal of the analyst. The remainder of the text will draw on this material to analyze some special flow fields and introduce the concepts of turbulent flow.

Self-Study Questions

6.1 When applying the first law of thermodynamics, where do Q and W occur locally for
a. A system? *b.* A control volume?

6.2 *a.* What would be the algebraic sign convention for Q and W in the first law of thermodynamics if it were stated as $Q + W = \Delta E$?
b. Does the answer in (*a*) depend on what has been defined as a meaning for ΔE?

6.3 State a physical meaning for

a. $e\rho\mathbf{V} \cdot d\mathbf{S}$

b. $\oint_S e\rho\mathbf{V} \cdot d\mathbf{S}$

c. $e\rho\, d\mathcal{V}$

d. $\int_{\mathcal{V}} \frac{\partial(e\rho)}{\partial t} d\mathcal{V}$

e. $\int_{\mathcal{V}} \frac{\partial(e\rho)}{\partial t} d\mathcal{V} + \oint_S e\rho\mathbf{V} \cdot d\mathbf{S}$

6.4 Discuss the implications with regard to the description of the flow field if the applicable energy equation reduces to

a. $\dfrac{đQ}{dt} = \oint_S e\rho\mathbf{V}\cdot d\mathbf{S}$ *d.* $\dfrac{đQ}{dt} - \dfrac{đW}{dt} = e_2\rho_2 V_2 A_2 - e_1\rho_1 V_1 A_1$

b. $-\dfrac{đW}{dt} = \int_{\mathcal{V}} \dfrac{\partial(e\rho)}{\partial t}\, d\mathcal{V}$ *e.* $0 = \dfrac{\partial(e\rho)}{\partial t}\mathcal{V} + \int_{S_{exit}} e\rho\mathbf{V}\cdot d\mathbf{S}$

c. $\dfrac{đQ}{dt} = \rho \oint_S e\mathbf{V}\cdot d\mathbf{S}$

6.5 Discuss the physical meaning of **grad** T.

6.6 Express Fourier's law of heat conduction in
a. x, y, z coordinates. *b.* r, θ, z coordinates.

6.7 A temperature field is given as $T = 2x + T_0$ where x is a spatial coordinate and its coefficient has dimensions of deg/length. What is the algebraic sign and direction of heat transfer
a. From the Fourier heat-conduction equation?
b. From Eq. (6.5) for a differential area $dS\,\mathbf{i}$ at $x = 1$ on a control surface of a unit cube as shown in Fig. 1.18?
c. Briefly explain the difference in algebraic sign.

6.8 Show that mechanical work performed on a unit mass
a. By a force opposing gravity is $g(z_2 - z_1)$.
b. By a force causing a given change of velocity is $(V_2{}^2 - V_1{}^2)/2$.

6.9 What use is made of the state principle once a fluid flow problem is defined?

6.10 *a.* What categories are used for rate of work?
b. What type of forces are considered to do work?
c. What is the meaning of $đW_s/dt$?

6.11 State the physical meaning of

a. $-\sigma_{nn}\, d\mathbf{S}\cdot\mathbf{V}$ *c.* $-\oint_S (\sigma_{nn}\, d\mathbf{S} + \tau_{nt}\, dS\,\mathbf{t})\cdot\mathbf{V}$

b. $-\tau_{nt}\, dS\,\mathbf{t}\cdot\mathbf{V}$

6.12 Values of pertinent variables in a flow field are given in the following units:

u, Btu/lb$_m$ $V^2/2$, ft^2/sec^2 $\bar{z}$, ft

Show what proportionality and/or unit conversion factor is needed for an answer of energy change involving these variables in

a. ft-lb$_f$/lb$_m$ *b.* Btu/slug *c.* ft^2/sec^2

6.13 Does the energy equation apply to a control volume fixed in noninertial space?

6.14 Write out a reduced form of the energy equation (6.10) for the following:
a. An inviscid fluid in horizontal flow with all fluid variables steady. There is no shaft work.
b. All fluid variables are steady. The Newtonian fluid motion is normal to the control surface. The temperature is uniform.
c. An adiabatic and reversible fluid flow without shaft work. The fluid properties are uniform and constant over the one entrance and the one exit section of the control surface.
d. The fluid properties in the control volume are uniform and unsteady. There is no heat

transfer or shaft work. There is only an efflux of fluid and its properties are uniform at the control surface.

6.15 When applying Eq. (6.12), it may be desirable to substitute $-p$ for σ_{nn}. Comment on this possibility for a flow

a. In which the viscous effects are negligible.

b. In the x direction only. The fluid flow variables are steady and the density has the same value everywhere.

c. That has predominant viscous effects only in a thin layer near a stationary surface.

6.16 If flow cross sections were vertically separated by a large distance relative to the vertical dimension of each cross section, what would be a reasonable, approximate, algebraic relation for the potential energy flux at each of these sections?

6.17 Consider a variable-density fluid moving horizontally through a circular duct with the density and the velocity symmetric about the centerline. Show that the potential energy surface integral reduces to $\dot{m}g\bar{z}$.

6.18 For a fluid that behaves like a perfect gas (that is, $p = \rho RT$), show that (refer to a thermodynamics text if necessary!)

a. $u = u(T)$ *c.* $c_p = c_v + R$

b. $h = h(T)$ *d.* $c_p = \dfrac{\gamma R}{\gamma - 1}$

6.19 For a fluid of constant density and constant specific heat, show that

a. $\Delta u = c_v \Delta T$ *c.* $\Delta u = 0$ for $\mathcal{S}$ = constant

b. $\Delta h = c_v \Delta T + \dfrac{\Delta p}{\rho}$

6.20 What is meant by: "The space-averaged velocity based on continuity can be used in the expressions for the transport of energy and momentum with adequate accuracy"?

6.21 Demonstrate the homogeneity of the basic dimensions in the terms given in Eq. (6.13) and state in words a physical meaning for each.

6.22 Formulate an expression in cylindrical coordinates similar to Eq. (6.13) where $T = T(r)$ and $\mathbf{V} = V_z \boldsymbol{\varepsilon}_z$. Assume that each of these functions increases in the positive coordinate direction.

6.23 Briefly explain the lack of normal or shear stress work on the other four faces described in the flow of Section 6.5.

6.24 Obtain the Gibbs equation (6.34) from the second law of thermodynamics as formulated in Eq. (6.33). Why doesn't a kinetic energy and shear work term appear?

6.25 Consider a system within a closed and constant-volume container that is thermally insulated. The initial state is not in equilibrium, but the final state is in equilibrium. Since E and $\mathcal{V}$ are constant, is $\Delta\mathcal{S}$ for the process equal to or greater than zero?

6.26 Derive the following expressions (refer to a thermodynamics text if necessary)

a. $s = \displaystyle\int c_v \frac{dT}{T} + C$ for an incompressible fluid.

b. $s = c_v \ln T + R \ln\left(\dfrac{1}{\rho}\right) + C$ for a perfect gas with c_v = constant.

c. $s = c_p \ln T - R \ln p + C$ for a perfect gas with c_v = constant.
d. $pT^{\gamma/(\gamma-1)}$ = constant for a perfect gas with c_v = constant and undergoing an isentropic process.

6.27 State the flow conditions under which the energy and Bernoulli equations are identical.

6.28 Give a physical meaning for the following terms in the Bernoulli (or reduced energy) equation:

$$\frac{V_2^2 - V_1^2}{2} \qquad \frac{p_2 - p_1}{\rho} \qquad g(z_2 - z_1)$$

a. As energy terms. *b.* As dynamic terms.

Problems

6.1 A control volume is chosen where all flow variables are steady. The rate of heat transfer conducted is 10 Btu/sec, while the net energy transported out of the control surface is -7780 ft-lb$_f$/sec. What is the rate and direction of work done at the control surface?

6.2 A Newtonian fluid has a parallel motion along a stationary wall as shown in Fig. 6.14. The fluid is incompressible and all flow variables are steady. Given

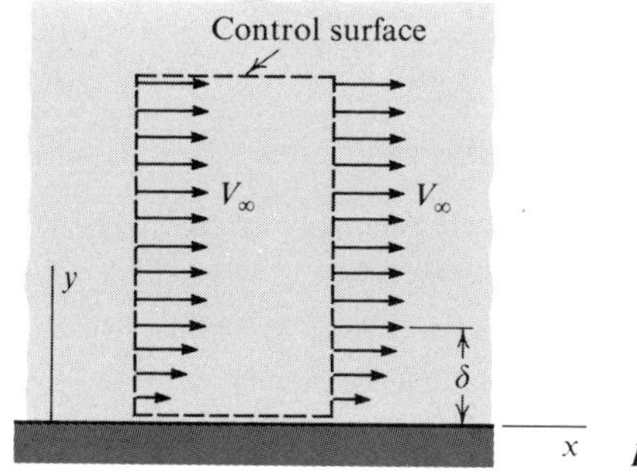

Fig. 6.14

$$\frac{dp}{dx} = 0$$

$$\mathbf{V} = \left[C_1 \frac{y}{\delta} + C_2 \left(\frac{y}{\delta}\right)^2\right]\mathbf{i} \qquad 0 \le \frac{y}{\delta} \le 1$$

$$\mathbf{V} = V_\infty \mathbf{i} \qquad \frac{y}{\delta} > 1$$

If the net internal energy transported by the fluid through the control surface is zero,
a. Use the control volume shown (of unit depth) to determine the possibilities of heat transfer at the control surface.
b. Would one expect the temperature to be uniform in value in the flow field?

6.3 Consider the flow of a Newtonian fluid in a long pipe of constant diameter. If the pressure is 15 lb$_f$/in.^{2} and the viscosity is 10^{-6} lb$_f$-sec/ft^2, refer to Eq. (4.18*c*) and estimate the value of $\partial V_z/\partial z$ to cause a ± 1 percent variation of σ_{zz} from $\bar{\sigma}$ (use z as the direction of flow). Comment on the reasonability of assuming that $\sigma_{nn} = -p$.

6.4 A long horizontal pipe of constant area has liquid water ($\rho = 62.4\ \text{lb}_m/\text{ft}^3$) passing through it. All flow variables are steady. The pressure is essentially uniform at the entrance and exit cross sections, with $p_1 = 30\ \text{lb}_f/\text{in.}^2$, and $p_2 = 20\ \text{lb}_f/\text{in.}^2$, respectively. The space-average velocity at section 1 is 10.0 ft/sec where the velocity is essentially uniform. The rate of kinetic energy transported from the exit section is 930 ft-lb_f/sec. If the mass flow rate is 300 lb_m/sec, evaluate the sum of the rate of heat and the net internal energy influx at the control surface.

6.5 The pump shown in Fig. 6.15 imparts 5 hp to the ideal fluid. If the density of the fluid is 60.0 lb_m/ft^3, estimate the flow rate.

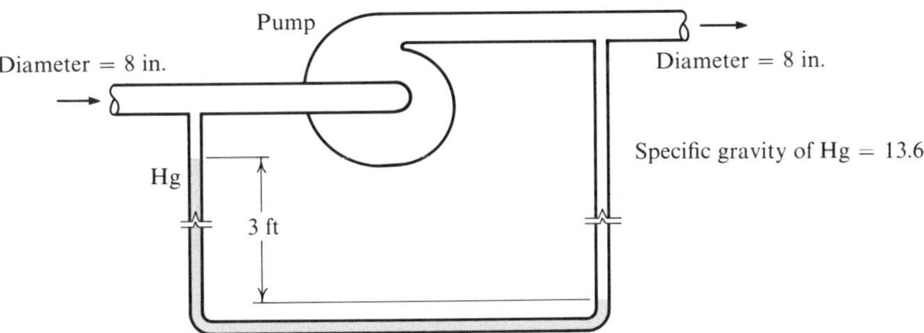

Fig. 6.15

6.6 A fluid enters a device at two locations and leaves at section 3 while the heat transfer into the device is 30,000 ft-lb_f/sec. Given that the fluid is a perfect gas and that

$\rho_1 = 0.002\ \text{slug/ft}^3$		
$T_1 = 100°\text{F}$	$T_2 = 100°\text{F}$	$T_3 = 100°\text{F}$
$A_1 = 0.1\ \text{ft}^2$	$A_2 = 0.1\ \text{ft}^2$	$A_3 = 0.2\ \text{ft}^2$
$p_1 = 15\ \text{lb}_f/\text{in.}^2$	$p_2 = 20\ \text{lb}_f/\text{in.}^2$	$đW_s/dt = 0$
$V_1 = 70\ \text{ft/sec}$	$V_2 = 40\ \text{ft/sec}$	

what is the velocity at section 3?

6.7 Nitrogen in gaseous form flows steadily and isothermally through a long constant-diameter pipe. Given the following information, calculate the heat transfer (Btu/lb) between the fluid and the pipe:

$p_1 = 100\ \text{lb}_f/\text{in.}^2$	$p_2 = 80\ \text{lb}_f/\text{in.}^2$
$z_1 = 0$	$z_2 = -1000\ \text{ft}$
$V_1 = 20\ \text{ft/sec}$	

6.8 A perfect gas flows steadily through an insulated constant-area pipe. A porous plug is in the line and a decrease in temperature of 5°F occurs in the fluid as it passes through the obstruction. If $R = 50.0$ ft-lb_f/lb_m-°R and $\gamma = 1.4$, calculate the corresponding kinetic energy change in ft-lb_f/lb_m.

6.9 Air expands through a turbine from 60 $\text{lb}_f/\text{in.}^2$ and 600°F to 15 $\text{lb}_f/\text{in.}^2$ and 300°F. During this steady process, 5 Btu/lb_m of heat are conducted to the surroundings at 14.7 lb_f/in^2 and 70°F. Calculate the shaft work by the fluid (on a per lb_m basis).

6.10 Steam enters a turbine steadily at 200 lb_f/in^2 and 600°F and has a specific enthalpy of 1322 Btu/lb_m. It exits at 15 lb_f/in^2 and with a specific enthalpy of 1192 Btu/lb_m. If the flow is essentially adiabatic with negligible kinetic and potential energy changes, evaluate the mass flow rate necessary for the turbine to produce 200 hp.

6.11 A nozzle is attached to a structural member at section A-A as shown in Fig. 6.16. The entering area is large compared to the exit area of 4 in^2. If the fluid flow variables are steady, estimate the largest force in the x direction that could be exerted on the structure at A-A by

a. A fluid with constant and uniform density of 60 lb_m/ft^3.

b. A perfect gas of $R = 53.0$ lb_f-ft/lb_m-°R, $\gamma = 1.4$, and an entering temperature of 300°F.

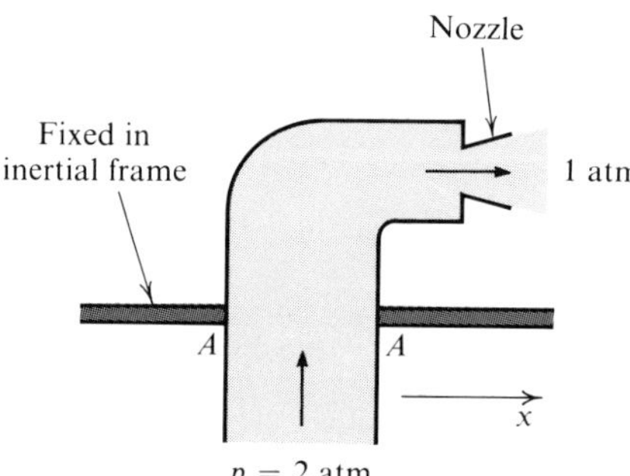

Fig. 6.16

6.12 Solve Prob. 5.6 for the given conditions except that the water is replaced by a perfect gas with $R = 53.3$ lb_f-ft/lb_m-°R, $\gamma = 1.4$, $T_1 = 100$°F, and the walls are adiabatic.

6.13 Reconsider Prob. 5.9 in which the tank contains a perfect gas with constant specific heats. Data as a function of time are available for the pressure p of the gas in the large tank, the temperature T of the tank, and the tensile forces F_A and F_B in the rods. The product of mass and specific heat for the metal tank is given as $(mc)_t$. Assume that the outside surface of the tank is insulated and that the temperature of the tank and the gas inside (but not in the nozzle) is essentially uniform during the emptying process, and show how to obtain

a. $\dot{m}_{\text{exit}} = -\dfrac{1}{g}\dfrac{dF_B}{dt}$

b. $V_{\text{exit}} = \dfrac{F_A}{\dot{m}_{\text{exit}}}$

c. $0 = \dfrac{c_v}{R}\dfrac{\partial p}{\partial t}\mathcal{V}_t + (mc)_t\dfrac{\partial T}{\partial t} + \left(c_p T_{\text{exit}} + \dfrac{V_{\text{exit}}^2}{2}\right)\dot{m}_{\text{exit}}$

d. $T_{\text{exit}} = \dfrac{-g^2 F_A{}^2}{2c_p(dF_B/dt)^2} + \dfrac{\mathcal{V}_t g c_v(dp/dt)}{Rc_p(dF_B/dt)} + \dfrac{g(mc)_t\,(dT/dt)}{c_p(dF_B/dt)}$

6.14 Estimate the enthalpy change of the fluid passing through the rotor in Prob. 5.11.

6.15 A perfect gas is contained in an insulated tank. The gas has a specific heat at constant pressure of 0.24 Btu/lb_m-°R and a specific heat ratio (c_p/c_v) of 1.4. The tank has an outlet valve to the surroundings at 1 atm and a device to do work on the gas within the tank. If shaft work is transferred at the rate of 0.1 hp to the gas during an emptying process as the valve opening is varied to maintain the temperature within the tank at 80°F, estimate the mass rate of flow from the tank at the instant the pressure is 40 lb_f/in^2. Assume that the pressure and temperature of the gas remain uniform within the tank.

6.16 A straight and constant-diameter pipe of length L, open at both ends, is moved in a quiescent body of water. The pipe as illustrated in Fig. 6.17 moves horizontally at a steady speed V_0 such that a laminar flow exists in the pipe and a parabolic velocity profile relative to the pipe is formed at the exit. Assume that the pipe has adiabatic surfaces, and obtain an expression for estimating the rate of enthalpy change in the fluid between the entrance and the exit in terms of the density of the fluid, the pipe speed, and the inner-pipe radius R.

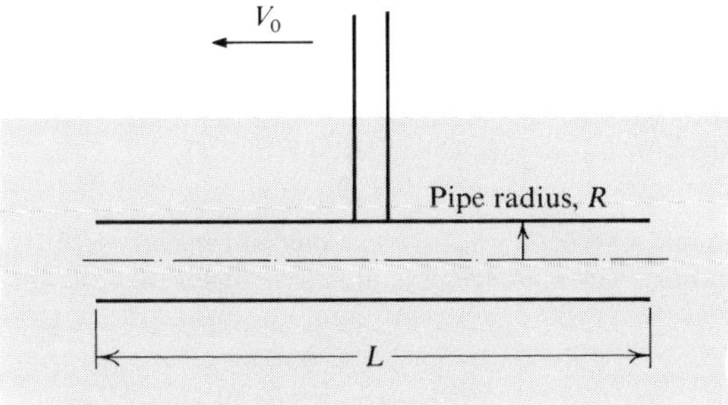

Fig. 6.17

6.17 A well-insulated straight pipe with a 6-in. diameter is fixed on a test stand. This is shown in Fig. 6.18. Air passes steadily through the pipe while fuel enters the pipe to a burner. Some data at the inlet and the exit are given in the figure. Calculate the x component of the force exerted on the test stand.

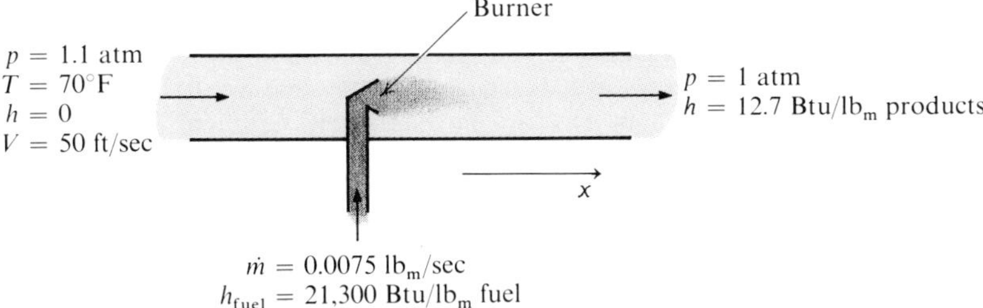

Fig. 6.18

6.18 Derive the following differential energy equation for a constant-density Newtonian fluid moving steadily in the x direction ($V_y = 0$ and $V_z = 0$) between fixed parallel plates (in the x, z plane); the pressure gradient is in the x direction and $T = T(x, y)$. The specific heats and coefficients of viscosity and thermal conductivity are constant.

$$\kappa\left(\frac{\partial^2 T}{\partial x^2} + \frac{\partial^2 T}{\partial y^2}\right) + \mu\left(\frac{\partial V_x}{\partial y}\right)^2 = V_x \rho c_v \frac{\partial T}{\partial x}$$

What term(s) might be negligible at low speeds?

6.19 Consider the same flow as presented in Section 6.5 (constant density fluid), except that there is a pressure gradient in the x direction, $V_p = 0$, and T is a function of y only.

a. Use a differential momentum equation and a solution for V_x to show that the applicable energy equation is

$$\kappa\frac{d^2 T}{dy^2} = -\mu\left(\frac{dV_x}{dy}\right)^2 = -\frac{y^2}{\mu}\left(\frac{dp}{dx}\right)^2$$

b. State the boundary conditions for temperature.
c. Obtain an expression for $T(y)$ in the fluid.
d. Obtain an expression for the heat flux at either wall.
e. Determine a condition when the heat flux at the plate is zero for any x.
f. What is the temperature profile for the case when $T_p = T_w$?

6.20 Consider the flow situation in Example 4.5 with $T = T(y, t)$ in a fluid of constant viscosity, thermal conductivity, and specific heat. The plate temperature T_p is greater than the surrounding fluid temperature.

a. Show that the applicable energy equation is

$$\kappa\frac{\partial^2 T}{\partial y^2} + \mu\left(\frac{\partial V_x}{\partial y}\right)^2 = c_v \rho \frac{\partial T}{\partial t}$$

b. If the viscous work were negligible, state the boundary conditions for the energy equation.

c. Using the assumption of (*b*), show that a solution for the temperature could be

$$T - T_\infty = (T_p - T_\infty)(1 - \operatorname{erf}\xi)$$

where $\xi = \dfrac{y}{2}\left(\dfrac{c_v \rho}{\kappa t}\right)^{1/2}$

Will the solutions for V_x and T be similar if a fluid is considered where $c_v \mu/\kappa = 1$?

6.21 A steady and laminar flow in a long horizontal tube of constant diameter is considered. The fluid is an incompressible Newtonian liquid and has a constant specific heat, viscosity, and thermal conductivity. The pressure varies only in the axial direction. The applicable differential energy equation is

$$\kappa\left[\frac{\partial^2 T}{\partial z^2} + \frac{1}{r}\frac{\partial}{\partial r}\left(r\frac{\partial T}{\partial r}\right)\right] + \frac{\mu}{r}\frac{d}{dr}\left(V_z r\frac{dV_z}{dr}\right) - V_z\frac{dp}{dz} = \rho V_z \frac{\partial e}{\partial z}$$

a. Show how to obtain this energy equation.
b. Show that the equation in (*a*) may be written as

$$\kappa\left[\frac{\partial^2 T}{\partial z^2} + \frac{1}{r}\frac{\partial}{\partial r}\left(r\frac{\partial T}{\partial r}\right)\right] + \mu\left(\frac{dV_z}{dr}\right)^2 = \rho c_v V_z \frac{\partial T}{\partial z}$$

c. If the wall temperature is uniform and constant, refer to Example 4.6 and obtain the temperature profile when

i. The term $\partial T/\partial z$ is negligible.
ii. The term $\partial T/\partial z$ is negligible and the viscous term is not appreciable.

6.22 Consider the flow situation in Prob. 6.21 for the case where the viscous effects are negligible in the energy equation and there is a linear temperature change in the z direction. The results of Example 4.6 may be used.

a. State applicable boundary conditions for $T(r,z)$.

b. Show that the heat flux $\left(\text{or } \left.\frac{\partial T}{\partial r}\right|_{r=R}\right)$ at the wall is constant if dp/dz is constant.

c. Show that the temperature profile is $T - T_{\text{wall}} = \frac{\rho c_v V_z}{16\kappa}\frac{\partial T}{\partial z}(r^2 - 3R^2)$

6.23 The following energy equation applies to the annular flow given in Prob. 4.17:

$$\mu\left(\frac{dV_z}{dr}\right)^2 + \frac{\kappa}{r}\frac{d}{dr}\left(r\frac{dT}{dr}\right) = 0$$

a. State all the applicable assumptions.
b. Determine the radial temperature variation in the fluid in terms of inner and outer wall temperatures given that

$$V_z = \frac{1}{4\mu}\frac{dp}{dz}(R_2{}^2 - R_1{}^2)\left[\frac{r^2 - R_1{}^2}{R_2{}^2 - R_1{}^2} + \frac{\ln(r/R_1)}{\ln(R_1/R_2)}\right]$$

6.24 In free convection, the motion of a fluid is the result of density variation because of heat transfer. For a two-dimensional flow along a vertical flat plate, the following equations are applicable:

Momentum.

$$\rho\left(V_x\frac{\partial V_x}{\partial x} + V_y\frac{\partial V_x}{\partial y}\right) = -\frac{\partial p}{\partial x} - \rho g + \mu\frac{\partial^2 V_x}{\partial y^2}$$

Energy.

$$\rho c_p\left(V_x\frac{\partial T}{\partial x} + V_y\frac{\partial T}{\partial y}\right) = \kappa\frac{\partial^2 T}{\partial y^2}$$

The coordinate y is in the direction of the normal to the plate.
a. Derive these equations and state all assumptions made.
b. Are the equations applicable to a compressible fluid, nonsteady flow, and variable properties such as ρ, c_p, κ, and μ?

6.25 Consider the flow of a constant density fluid described in Section 6.5 and apply the second law of thermodynamics in the form of Eq. (6.39) to establish that in a fluid of constant thermal conductivity

$$\left.\frac{d(\ln T)}{dy}\right|_{\text{wall}} > \left.\frac{d(\ln T)}{dy}\right|_{\text{plate}}$$

Select one of the profiles in Fig. 6.8 or 6.9, and show that the above relation is satisfied.

6.26 The value of the three terms on the left side of the inequality of Eq. (6.39) is called the *rate of entropy production.* Make an evaluation of this rate based on a unit plate area for the flow of a constant density fluid discussed in Section 6.5, if

$T_p = 200°\text{F}$ $\quad a = 3$ in.
$T_w = 100°\text{F}$ $\quad \kappa = 0.38$ Btu/hr-ft-°R
$V_p = 80$ ft/sec $\quad \mu = 2.94\,(10^{-4})$ lb_m/ft-sec

6.27 Show that the inequality (rate of entropy production per unit of plate area) in Eq. (6.39) reduces to the following for the flow in Example 6.4:

$$\frac{\mu V_p}{bT_p^2}\left(T_p V_p + \frac{\mu V_p^3}{3\kappa}\right) \geq 0$$

6.28 If a fluid moves steadily through a fixed control surface such that $\sigma_{nn} = -p$ with all of the heat transferred at one isothermal surface of T_0, show that the shaft power $đW_s/dt$ is

$$\frac{đW_s}{dt} \leq -\oint_S (e + p/\rho - T_0 \mathscr{s})\, \rho \mathbf{V} \cdot d\mathbf{S}$$

6.29 A liquid with a density of 60.0 lb_m/ft^3 moves through a turbine and all fluid properties are steady, at least cyclically steady. The inlet area is one-fourth that of the exit area. The inlet velocity is 60 ft/sec while the inlet pressure is 20 lb_f/in^2 and the exit pressure is 10 lb_f/in^2. The inlet and exit temperatures are the same. Calculate the maximum shaft work (per lb_m) that the fluid can accomplish under the given conditions by making use of the second law for a control-volume analysis (see Prob. 6.28).

6.30 Apply Eq. (6.39) to the flow in Prob. 6.9 to establish if the inequality is satisfied.

6.31 Use the second law of thermodynamics to determine the direction of flow in Prob. 6.7 if the heat transfer is from the nitrogen to the surroundings at 70°F.

6.32 Apply the derived result in Prob. 6.28 to the flow in Prob. 6.9 and demonstrate that the actual shaft work per pound is less than the maximum possible shaft work per pound. Use $T_0 = 530°\text{R}$.

7

IRROTATIONAL FLUID MOTION

7.1 Introduction

Some difficulties that arise in solving problems associated with real fluid behavior should now be apparent. Very often, if one can obtain an idea of the main features of a flow field with viscous effects neglected, some physical insight can be gained with respect to the more complicated viscous flow. The flow fields idealized in the above manner are referred to as *irrotational flow fields*. Figure 7.1 compares the motions of an inviscid and a viscous fluid, each with uniform and steady velocity at large distances from a cylinder, as the fluids pass over the cylinders. The difference between these two flows exists near the surface of the cylinder, where a boundary layer forms on the surface of the body, and in the wake, which is downstream of the point of separation. Outside of the wake and the boundary layer, the two fields are quite similar. The physical reason for this is associated with the small velocity gradients normal to the cylinder in the region outside of the boundary layer and wake. Since small velocity gradients in this direction imply small rate of strain γ_{xy}, the shear stress τ_{xy} would be small if the viscosity were not too large. At least such an argument would apply to a Newtonian fluid by referring to its constitutive equations. Furthermore, in regions where

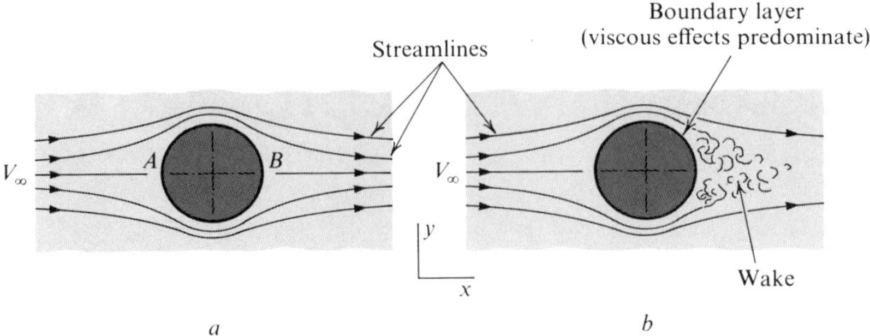

Fig. 7.1 Flow over a cylinder. Uniform velocity at large distances from the cylinder. (a) Idealized, inviscid flow. (b) Viscous flow.

$\partial V_x/\partial y$ and $\partial V_y/\partial x$ are very small, one would expect the curl of the velocity vector to be small; and if the **curl V** $= 0$ at all points in the flow, the velocity field would be irrotational (see Section 1.7).

It would appear that the study of irrotational motion could provide some intuition regarding fluid motion in general. The following sections introduce the subject and illustrate a few ideas through examples.

7.2 Velocity Potential from an Irrotational Velocity Field

An irrotational flow field is considered, so **curl V** $= 0$, and from Stokes' theorem, Eq. (1.30),

$$\oint_C \mathbf{V} \cdot d\mathbf{l} = \int_S \mathbf{curl\, V} \cdot d\mathbf{S} = 0$$

or

$$\oint_C \mathbf{V} \cdot d\mathbf{l} = 0 \qquad \text{for any closed path} \tag{7.1}$$

Two closed curves are arbitrarily chosen in this flow field as shown in Fig. 7.2. Applying Eq. (7.1) around each of these loops gives

$$\int_{1A}^{2} \mathbf{V} \cdot d\mathbf{l} + \int_{2C}^{1} \mathbf{V} \cdot d\mathbf{l} = 0$$

$$\int_{1B}^{2} \mathbf{V} \cdot d\mathbf{l} + \int_{2C}^{1} \mathbf{V} \cdot d\mathbf{l} = 0$$

and subtracting the second expression from the first yields

$$\int_{1A}^{2} \mathbf{V} \cdot d\mathbf{l} = \int_{1B}^{2} \mathbf{V} \cdot d\mathbf{l}$$

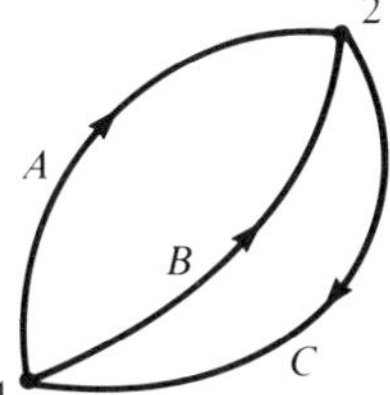

Fig. 7.2 *Closed curves* 1-A-2-C-1 *and* 1-B-2-C-1.

The preceding indicates that the integrals are independent of the path between points 1 and 2 since A and B are any two paths connecting these points, and $\mathbf{V} \cdot d\mathbf{l}$ is said to be an exact differential. This differential will be noted by $-d\phi$, and so

$$d\phi = -\mathbf{V} \cdot d\mathbf{l} \tag{7.2}$$

which may then be written as

$$\mathbf{grad}\ \phi \cdot d\mathbf{l} = -\mathbf{V} \cdot d\mathbf{l}$$

and[1]

$$\mathbf{V} = -\mathbf{grad}\ \phi \tag{7.3}$$

The function ϕ is called the *velocity potential*, and flows derived from ϕ are referred to as *potential flows*. An important observation pertaining to the last equation is that a vector function $\mathbf{V}$ may be exchanged for a single scalar function ϕ, if the motion is irrotational. In general, a vector function contains three scalar functions, the components of the vector, so substitution of $-\mathbf{grad}\ \phi$ for $\mathbf{V}$ should simplify the equations of motion. This is considered in Section 7.4.

Example 7.1

An irrotational velocity field is specified by the following components:

$$V_x = e^{-x} \sin y \qquad V_y = -e^{-x} \cos y \qquad V_z = 0$$

Find the velocity potential.

Solution. The exact differential expression for ϕ is

[1] The reader who has had a formal course in vector analysis will recall the theorem: "A necessary and sufficient condition that the curl of a vector vanish identically is that the vector be the gradient of some scalar field." It should be noted that the curl of the velocity $\mathbf{V}$ given by Eq. (7.3) does vanish identically; also Eq. (7.3), which exhibits the existence of a velocity potential, derives from $\mathbf{curl\ V} = 0$.

$$d\phi = \frac{\partial \phi}{\partial x} dx + \frac{\partial \phi}{\partial y} dy + \frac{\partial \phi}{\partial z} dz$$

Since $\mathbf{V} = -\mathbf{grad}\ \phi$,

$$V_x \mathbf{i} + V_y \mathbf{j} + V_z \mathbf{k} = -\left(\frac{\partial \phi}{\partial x} \mathbf{i} + \frac{\partial \phi}{\partial y} \mathbf{j} + \frac{\partial \phi}{\partial z} \mathbf{k} \right)$$

and

$$V_x = -\frac{\partial \phi}{\partial x} \qquad V_y = -\frac{\partial \phi}{\partial y} \qquad V_z = -\frac{\partial \phi}{\partial z}$$

The expression for $d\phi$ becomes

$$d\phi = -V_x\, dx - V_y\, dy - V_z\, dz$$

Substituting the given information concerning the velocity components gives

$$d\phi = -e^{-x} \sin y\, dx + e^{-x} \cos y\, dy + 0$$

This may be integrated between any two points along any path between these points since $d\phi$ is an exact differential. Letting these be the points x, y and x_0, y_0,

$$\phi - \phi_0 = -\int_{x_0, y_0}^{x, y} e^{-x} \sin y\, dx + \int_{x_0, y_0}^{x, y} e^{-x} \cos y\, dy$$

Using the path of integration indicated in Fig. 7.3, and setting $\phi_0 = 0$,

$$\begin{aligned} \phi &= -\sin y_0 \int_{x_0}^{x} e^{-x}\, dx + e^{-x} \int_{y_0}^{y} \cos y\, dy \\ &= -\sin y_0 \Big[-e^{-x} \Big]_{x_0}^{x} + e^{-x} \Big[\sin y \Big]_{y_0}^{y} \\ &= e^{-x} \sin y_0 - e^{-x_0} \sin y_0 + e^{-x} \sin y - e^{-x} \sin y_0 \\ &= e^{-x} \sin y - e^{-x_0} \sin y_0 \end{aligned}$$ ◀

The second term in the answer is a constant.

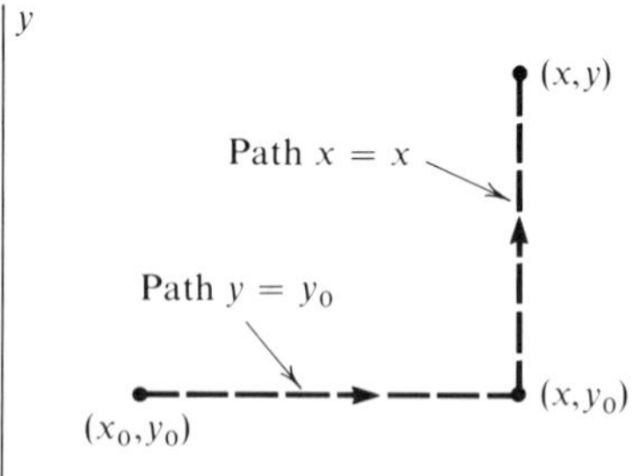

Fig. 7.3

7.3 Differential Equation Satisfied by a Velocity Potential

The continuity equation is the differential equation expressing the conservation of matter. If the velocity field in the continuity equation (2.3) is replaced by $-\mathbf{grad}\ \phi$ using Eq. (7.3), then

$$\operatorname{div}\left(-\rho\ \mathbf{grad}\ \phi\right) + \frac{\partial \rho}{\partial t} = 0 \tag{7.4}$$

For a steady density field this equation simplifies to

$$\operatorname{div}\left(\rho\ \mathbf{grad}\ \phi\right) = 0 \tag{7.5}$$

Finally, for the *incompressible fluid in irrotational flow*, Eq. (2.6) indicates

$$\operatorname{div}\left(\mathbf{grad}\ \phi\right) = 0 \tag{7.6}$$

which is *Laplace's equation.* This may be solved to ascertain the velocity potential when boundary conditions are specified and these conditions differ from the no-slip condition since the ideal fluid is not required to adhere to a solid boundary. The most frequent boundary condition imposed on ϕ at a point is the requirement that the normal component of the velocity at and relative to a solid boundary must vanish. This may be formalized by reference to Fig. 7.4, in which $\mathbf{V}_B$ is the velocity of point P on the solid boundary and $\mathbf{n}$ is the unit normal to the boundary surface at the point. This condition may be written as

$$\left[\left(-\mathbf{grad}\ \phi - \mathbf{V}_B\right)\cdot \mathbf{n}\right]_{\substack{\text{solid}\\ \text{boundary}}} = 0 \tag{7.7}$$

which requires the component of relative velocity of the fluid at the boundary to be tangent to the boundary, and this is equivalent to requiring no flow through the solid boundary. The magnitude of the tangential component must come from the solution for ϕ.

Fig. 7.4 *Boundary condition for velocity in potential flow.*

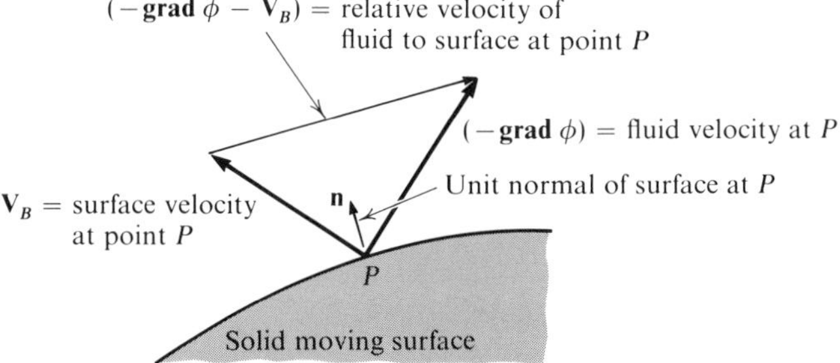

There are other conditions that are imposed on ϕ and reflect specific velocity field configurations. As an example, consider the cylinder in Fig. 7.1*a*. If the velocity at large distances from the cylinder is uniform and has a value of V_∞ in the positive x direction, then the boundary conditions ϕ must meet are given by

$$\lim_{x \to \pm\infty} (-\mathbf{grad}\ \phi) = V_\infty \mathbf{i}$$

and

$$\lim_{y \to \pm\infty} (-\mathbf{grad}\ \phi) = V_\infty \mathbf{i}$$

The velocity potential does not satisfy Laplace's equation if the fluid is compressible, which is obvious from the form of the continuity equation (2.3). *Irrotational compressible flow will not be considered in the text.*

Note that Eq. (7.6), which applies to the incompressible fluid, is a linear differential equation. In the rectangular cartesian coordinate system, this equation is

$$\frac{\partial^2 \phi}{\partial x^2} + \frac{\partial^2 \phi}{\partial y^2} + \frac{\partial^2 \phi}{\partial z^2} = 0 \tag{7.8}$$

As a consequence of the linearity, one may superpose solutions that satisfy Eq. (7.6) or (7.8), and any linear combination of solutions of these equations will also satisfy these equations. Solutions of Laplace's equation are called *harmonic functions.*

Example 7.2

If ϕ_1 and ϕ_2 are each harmonic functions, show that the linear combinations $K_1\phi_1 \pm K_2\phi_2$, in which K_1 and K_2 are constants, also satisfy Laplace's equation.

Solution. Consider ϕ_1 and ϕ_2 to be functions of x, y and z. Then it must be shown that

$$\frac{\partial^2 (K_1\phi_1 \pm K_2\phi_2)}{\partial x^2} + \frac{\partial^2 (K_1\phi_1 \pm K_2\phi_2)}{\partial y^2} + \frac{\partial^2 (K_1\phi_1 \pm K_2\phi_2)}{\partial z^2} = 0$$

The left-hand side of this equation may be written as

$$K_1 \frac{\partial^2 \phi_1}{\partial x^2} \pm K_2 \frac{\partial^2 \phi_2}{\partial x^2} + K_1 \frac{\partial^2 \phi_1}{\partial y^2} \pm K_2 \frac{\partial^2 \phi_2}{\partial y^2} + K_1 \frac{\partial^2 \phi_1}{\partial z^2} \pm K_2 \frac{\partial^2 \phi_2}{\partial z^2}$$

which can be rearranged as

$$K_1 \left(\frac{\partial^2 \phi_1}{\partial x^2} + \frac{\partial^2 \phi_1}{\partial y^2} + \frac{\partial^2 \phi_1}{\partial z^2} \right) \pm K_2 \left(\frac{\partial^2 \phi_2}{\partial x^2} + \frac{\partial^2 \phi_2}{\partial y^2} + \frac{\partial^2 \phi_2}{\partial z^2} \right)$$

The sums in each of the parentheses are zero since ϕ_1 and ϕ_2 each satisfy Laplace's equation. Therefore

$$\frac{\partial^2(K_1\phi_1 \pm K_1\phi_2)}{\partial x^2} + \frac{\partial^2(K_1\phi_1 \pm K_2\phi_2)}{\partial y^2} + \frac{\partial^2(K_1\phi_1 \pm K_2\phi_2)}{\partial z^2} = 0$$

which was to be demonstrated.

The velocity potential may be obtained by solving Laplace's equation for a flow of an ideal and incompressible fluid subject to certain boundary conditions. Even though this requires some knowledge of solving partial and ordinary differential equations, the procedure is not always difficult. The next example will demonstrate a formal method for acquiring such a velocity potential.

Example 7.3

A right-circular cylinder of radius a is shown in Fig. 7.5 translating with a speed V_c in the positive x direction. It is immersed in an "infinite" fluid which is otherwise at rest. If the fluid is incompressible, find the velocity potential when the axis of the cylinder is at 0.

Solution. The coordinate system shown is fixed, and the axis of the cylinder is at the origin 0 of the coordinate system at the given instant of time. For irrotational, two-dimensional motion, Laplace's equation in cylindrical coordinates[1] is

$$\frac{1}{r}\frac{\partial}{\partial r}\left(r\frac{\partial\phi}{\partial r}\right) + \frac{1}{r^2}\frac{\partial^2\phi}{\partial\theta^2} = 0 \tag{7.9}$$

noting that $\partial^2\phi/\partial z^2$ is zero for this problem.

[1] See Appendix B.

Fig. 7.5

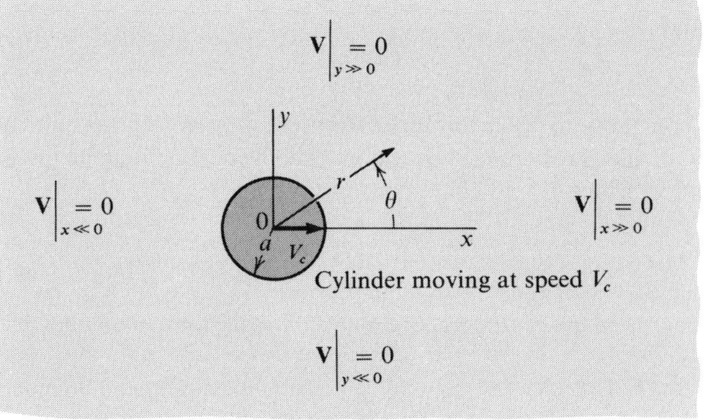

A solution of this equation may be found by separation of variables.[1] This consists of assuming a trial solution for $\phi = \phi(r,\theta)$ in the form of $\phi = R(r)\,\Theta(\theta)$ in which R is a function of r only and Θ is a function of θ only. By forming the derivatives in Laplace's equation, the following results

$$\frac{1}{r}\frac{\partial}{\partial r}\left(r\frac{\partial\phi}{\partial r}\right) = \Theta\frac{d^2R}{dr^2} + \frac{\Theta}{r}\frac{dR}{dr}$$

and

$$\frac{1}{r^2}\frac{\partial^2\phi}{\partial\theta^2} = \frac{R}{r^2}\frac{d^2\Theta}{d\theta^2}$$

in which the ordinary derivatives are appropriate since both R and Θ are each a function of a single variable only. Substitution of these in Eq. (7.9) gives

$$\frac{d^2R}{dr^2}\Theta + \frac{\Theta}{r}\frac{dR}{dr} + \frac{R}{r^2}\frac{d^2\Theta}{d\theta^2} = 0$$

and separation of variables yields

$$\frac{1}{\Theta}\frac{d^2\Theta}{d\theta^2} = -\frac{r^2}{R}\frac{d^2R}{dr^2} - \frac{r}{R}\frac{dR}{dr} \tag{7.10}$$

Since the left-hand side of Eq. (7.10) is a function of θ only, and the right-hand side is a function of r only, the variables r and θ must occur in this equation in some redundant manner, and both sides must be equal to a constant. This constant is noted by $-\beta^2$, in which β is a real, positive[2] number. The choice of a negative sign and the square of β is taken as a matter of convenience that will become evident as the solution progresses. Equation (7.10) may now be written as two ordinary differential equations, which are

$$\frac{d^2\Theta}{d\theta^2} + \beta^2\Theta = 0 \tag{7.11}$$

and

$$r^2\frac{d^2R}{dr^2} + r\frac{dR}{dr} - \beta^2R = 0 \tag{7.12}$$

A solution of Eq. (7.11) is well known from the study of simple harmonic motion in engineering mechanics, and

$$\Theta = C_1\cos\beta\theta + C_2\sin\beta\theta \tag{7.13}$$

in which C_1 and C_2 are constants to be determined from conditions imposed on the problem. Equation (7.12) is Euler's differential equation,[3] and a solution may be found by assuming that $R = Cr^K$ in which C and K are constants. Then

[1] Paul W. Berg and James L. McGregor, "Elementary Partial Differential Equations," pp. 14–16, Holden-Day, Inc., Publishers, San Francisco, 1966.

[2] The real number β could have been taken as a negative number and the same solution would have resulted.

[3] William E. Boyce and Richard C. DiPrima, "Elementary Differential Equations and Boundary Value Problems," pp. 165–169, John Wiley & Sons, Inc., New York, 1965.

$$\frac{dR}{dr} = CKr^{K-1}$$

and

$$\frac{d^2R}{dr^2} = CK(K-1)\,r^{K-2}$$

Substitution of these into Eq. (7.12) gives

$$r^2CK(K-1)\,r^{K-2} + rCKr^{K-1} - \beta^2R = 0$$

which reduces to

$$K^2 - \beta^2 = 0 \qquad \text{for } r \neq 0$$

Thus $K = \pm\beta$, and the general solution of Eq. (7.12) is given by

$$R = C_3r^{\beta} + C_4r^{-\beta} \tag{7.14}$$

in which C_3 and C_4 are constants to be found from the conditions imposed on the problem. The trial solution is then

$$\phi = R\Theta = (C_3r^{\beta} + C_4r^{-\beta})(C_1\cos\beta\theta + C_2\sin\beta\theta)$$

The boundary conditions which apply are:

Boundary condition 1. Equation (7.7) requires the following for $r = a$ and $0 \leq \theta < 2\pi$ at the given instant of time:

$$\left[\left(-\frac{\partial\phi}{\partial r}\,\boldsymbol{\varepsilon}_r - \frac{1}{r}\,\frac{\partial\phi}{\partial\theta}\,\boldsymbol{\varepsilon}_\theta\right) - (V_c\cos\theta\boldsymbol{\varepsilon}_r - V_c\sin\theta\boldsymbol{\varepsilon}_\theta)\right]\cdot\boldsymbol{\varepsilon}_r = 0$$

This reduces to

$$\frac{\partial\phi}{\partial r} = -V_c\cos\theta \qquad \text{at } r = a \qquad \text{and} \qquad 0 \leq \theta < 2\pi$$

Boundary condition 2. At large distances from the cylinder, the V_r component of fluid velocity vanishes for $0 \leq \theta < 2\pi$:

$$\lim_{r\to\infty} -\frac{\partial\phi}{\partial r} = \lim_{r\to\infty} V_r = 0$$

Boundary condition 3. At large distances from the cylinder, the V_θ component of fluid velocity vanishes for $0 \leq \theta < 2\pi$:

$$\lim_{r\to\infty} \frac{1}{r}\left(-\frac{\partial\phi}{\partial\theta}\right) = \lim_{r\to\infty} V_\theta = 0$$

Boundary condition 4. The velocity potential ϕ is a bounded function for

$$0 \leq \theta < 2\pi \qquad \text{and} \qquad a \leq r < \infty$$

The fourth boundary condition requires that $C_3 = 0$ since $C_3 r^{\beta}$ is undefined as $r \to \infty$ with β a positive real number; then

$$\phi = C_4 r^{-\beta} (C_1 \cos \beta\theta + C_2 \sin \beta\theta)$$

Differentiating gives

$$\frac{\partial \phi}{\partial r} = -C_4 \beta r^{-\beta-1} (C_1 \cos \beta\theta + C_2 \sin \beta\theta)$$

and boundary condition 1 imposes, at $r = a$,

$$-V_c \cos \theta = -C_4 \beta a^{-\beta-1} (C_1 \cos \beta\theta + C_2 \sin \beta\theta)$$

which is satisfied by setting $C_2 = 0$ and $\beta = 1$. This yields

$$-V_c \cos \theta = -\frac{C_4 C_1}{a^2} \cos \theta$$

from which

$$C_4 C_1 = a^2 V_c$$

and the solution becomes

$$\phi = \frac{a^2 V_c}{r} \cos \theta$$ ◀

This also satisfies boundary conditions 2 and 3. The fact that the cosine function appears in the solution comes as no surprise, if one considers the symmetry of flow over the cylinder. Stated briefly, this symmetry would require

$$\phi(r,\theta) = \phi(r, -\theta)$$

and the cosine function meets this requirement, while the sine function does not. The lines of *constant potentials*, called *equipotentials*, are shown as solid lines in Fig. 7.6 for the given instant of time, and the streamlines are indicated by dashed lines.

Example 7.4

Find the velocity potential for a flow field characterized by

$$V_x = -V_\infty \qquad V_y = 0 \qquad V_z = 0$$

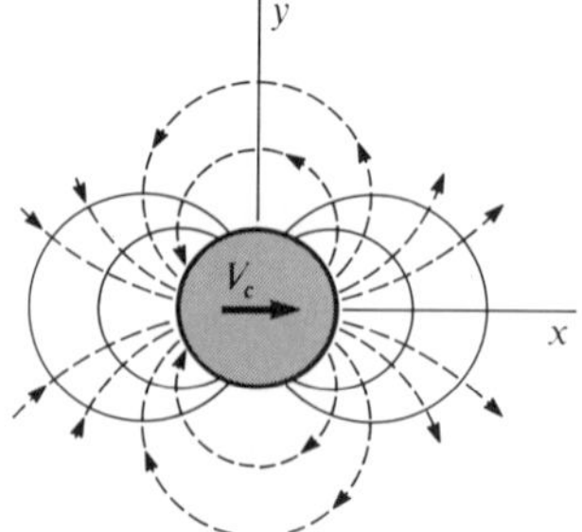

Fig. 7.6

if a velocity potential exists. V_∞ is a constant.

Solution. By inspection, **curl V** $= 0$; therefore a velocity potential exists. Since

$$V_y = -\frac{\partial \phi}{\partial y} = 0 \qquad \text{and} \qquad V_z = -\frac{\partial \phi}{\partial z} = 0$$

ϕ is a function of x only. Then

$$V_x = -\frac{d\phi}{dx} \qquad \text{or} \qquad \frac{d\phi}{dx} = -V_x = V_\infty$$

Integrating this,

$$\phi = V_\infty x + \text{a constant}$$ ◀

The constant depends on the arbitrary specification of ϕ. If ϕ is taken as 0, when $x = 0$, then the constant is zero; and $\phi = V_\infty x$.

Since Laplace's equation was not used in acquiring the potential for this simple flow, the potential function obtained is valid for a compressible as well as an incompressible fluid. However, the solution $\phi = V_\infty x$ satisfies Laplace's equation, which is obvious by inspection.

Example 7.5

If the uniform motion of the fluid in Example 7.4 is superposed on the motion imparted by the cylinder to the fluid in Example 7.3, describe the resulting motion and check the boundary condition at the surface of the cylinder for $V_c = V_\infty$.

Solution. The resultant motion is described by a velocity potential given by

$$\phi = V_\infty x + \frac{a^2 V_\infty}{r} \cos\theta$$

which is obtained by superposition of the two given solutions.

This potential may be expressed in plane polar coordinates by noting that $x = r\cos\theta$; and

$$\phi = V_\infty r\cos\theta + \frac{a^2 V_\infty}{r}\cos\theta = V_\infty \cos\theta \left(r + \frac{a^2}{r}\right)$$

which describes the resulting motion. Since the cylinder was translating at a velocity $V_c\mathbf{i}$, the effect of superposing a velocity field of $-V_\infty\mathbf{i}$ produces another problem in which the cylinder is stationary; and the fluid, which was at rest at large distances from the cylinder in Example 7.3, is now moving with a velocity $-V_\infty\mathbf{i}$. A check on the velocities will reconcile the preceding statements.

$$V_r = -\frac{\partial \phi}{\partial r} = -\frac{\partial}{\partial r}\left[V_\infty \cos\theta\left(r + \frac{a^2}{r}\right)\right]$$

$$V_r = -V_\infty \cos\theta\left(1 - \frac{a^2}{r^2}\right)$$

For $r = a$ (the radius of the cylinder) and all values of θ, $V_r = 0$. This, of course, satisfies the boundary condition specified by Eq. (7.7) for the stationary cylinder.

To check the boundary conditions for velocity as $r \to \infty$, the velocity components are

$$V_r = -\frac{\partial \phi}{\partial r} = -V_\infty \cos\theta \left(1 - \frac{a^2}{r^2}\right)$$

$$V_\theta = -\frac{1}{r}\frac{\partial \phi}{\partial \theta} = V_\infty \sin\theta \left(1 + \frac{a^2}{r^2}\right)$$

The velocity components V_x and V_y may be expressed as functions of V_r and V_θ by decomposition of the former into components in the direction of the latter. This yields

$$V_x = V_r \cos\theta - V_\theta \sin\theta$$

and

$$V_y = V_r \sin\theta + V_\theta \cos\theta$$

so that

$$V_x = -V_\infty \cos^2\theta \left(1 - \frac{a^2}{r^2}\right) - V_\infty \sin^2\theta \left(1 + \frac{a^2}{r^2}\right)$$

and

$$V_y = -V_\infty \sin\theta\cos\theta \left(1 - \frac{a^2}{r^2}\right) + V_\infty \sin\theta\cos\theta \left(1 + \frac{a^2}{r^2}\right)$$

Taking the limits of these components as $r \to \infty$ gives

$$V_x = -V_\infty(\cos^2\theta + \sin^2\theta) = -V_\infty$$

$$V_y = -V_\infty \sin\theta\cos\theta + V_\infty \sin\theta\cos\theta = 0$$

which indicates that the fluid is moving parallel to the negative x axis at large distances from the cylinder. The streamlines and equipotentials are shown as dashed and solid lines, respectively, in Fig. 7.7.

7.4 Bernoulli's Equation

It will be recalled that the equations of motion (3.10), (3.11), and (3.12),

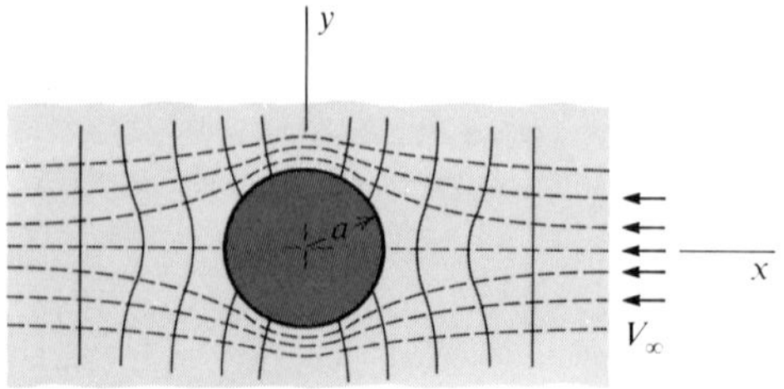

Fig. 7.7

with the shear stresses neglected, were written as a single vector relation and identified as Euler's equation

$$-\mathbf{grad}\, p + \rho \mathbf{f}_B = \rho \frac{D\mathbf{V}}{Dt} \tag{3.13}$$

By using a body-force potential Φ_B, and by assuming a conservative body-force field, Euler's equation was integrated along a streamline for a steady velocity field; and the result Eq. (3.15), was identified as a form of Bernoulli's equation.

The idealization of irrotational flow and its attendant velocity potential enables the analyst to remove the restriction of integrating along a streamline, as well as providing a convenient means of handling an unsteady velocity field. Equation (3.13) may be rewritten as

$$-\frac{\mathbf{grad}\, p}{\rho} - \mathbf{grad}\, \Phi_B = \mathbf{grad} \frac{V^2}{2} - \mathbf{V} \times \mathbf{curl}\, \mathbf{V} + \frac{\partial \mathbf{V}}{\partial t} \tag{7.15}$$

by using the vector identity

$$\frac{D\mathbf{V}}{Dt} = \mathbf{grad} \frac{V^2}{2} - \mathbf{V} \times \mathbf{curl}\, \mathbf{V} + \frac{\partial \mathbf{V}}{\partial t}$$

and continuing the assumption of a conservative body-force field. For irrotational motion, $\mathbf{curl}\, \mathbf{V} = 0$; and

$$\frac{\partial \mathbf{V}}{\partial t} = \frac{\partial}{\partial t}(-\mathbf{grad}\, \phi) = -\mathbf{grad} \frac{\partial \phi}{\partial t}$$

granting continuity and existence of the partial derivatives of ϕ with respect to space coordinates and time. Equation (7.15) can then be rearranged as

$$\frac{\mathbf{grad}\, p}{\rho} + \mathbf{grad}\, \Phi_B + \mathbf{grad} \frac{V^2}{2} - \mathbf{grad} \frac{\partial \phi}{\partial t} = 0 \tag{7.16}$$

If the density is constant or uniform, Eq. (7.16) can be integrated by noting that the taking of the gradient is a distributive operation with respect to addition, and

$$\mathbf{grad}\left(\frac{p}{\rho} + \Phi_B + \frac{V^2}{2} - \frac{\partial \phi}{\partial t}\right) = 0$$

so that

$$\frac{p}{\rho} + \Phi_B + \frac{V^2}{2} - \frac{\partial \phi}{\partial t} = g(t) \tag{7.17}$$

in which $g(t)$ is an arbitrary function of time. This is another form of Bernoulli's equation.

To obtain a similar integrated result for the case of variable density, the scalar product of Eq. (7.16) may be taken with $d\mathbf{r}$, a small displacement in any direction (this could be, but doesn't have to be, in the direction of a streamline). Forming the scalar product of Eq. (7.16) with $d\mathbf{r}$ gives

$$\frac{\mathbf{grad}\, p \cdot d\mathbf{r}}{\rho} + \mathbf{grad}\, \Phi_B \cdot d\mathbf{r} + \mathbf{grad}\, \frac{V^2}{2} \cdot d\mathbf{r} - \mathbf{grad}\, \frac{\partial \phi}{\partial t} \cdot d\mathbf{r} = 0$$

Since the gradient of a scalar quantity is the maximum change in the scalar with respect to distance, the gradient "dotted" into $d\mathbf{r}$ gives the change of the scalar in the direction of $d\mathbf{r}$, and

$$\frac{dp}{\rho} + d\Phi_B + d\frac{V^2}{2} - d\frac{\partial \phi}{\partial t} = 0$$

This may be integrated to yield

$$\int \frac{dp}{\rho} + \Phi_B + \frac{V^2}{2} - \frac{\partial \phi}{\partial t} = f(t) \tag{7.18}$$

where $f(t)$ is an arbitrary function of time. It was pointed out in Section 3.5 that a barotropic relation of the form $\rho = \rho(p)$ must be known to evaluate $\int (dp/\rho)$ in this equation. Equation (7.18) is another form of Bernoulli's equation and the only restrictions on it are a conservative body-force field and irrotational motion.

It is now apparent that if the velocity potential can be found, the velocity field follows from $-\mathbf{grad}\, \phi$; and from Bernoulli's equation and a barotropic relation, the pressure field may be ascertained. This is analogous to the use of the Navier-Stokes equations and the continuity equation in solving a viscous flow problem, but the procedure is simpler with the idealization of irrotational motion.

Example 7.6

The fluid shown in Fig. 7.8 has a uniform velocity of $-V_\infty \mathbf{i}$ at "infinity." Its density ρ is one value everywhere, and the velocity field is steady. Body forces may be neglected.

Fig. 7.8

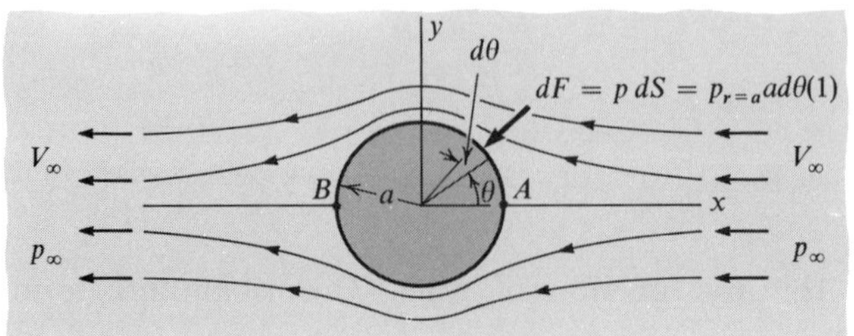

If the pressure at large distances from the stationary cylinder is known to be p_∞, obtain an expression for the pressure at any point in the field in terms of r, θ, p_∞, V_∞, and ρ. Also formulate the resultant surface force exerted by the fluid on the cylinder per unit length of z axis.

Solution. The velocity components for this flow were found in Example 7.5, and they are

$$V_r = -V_\infty \cos\theta\left(1 - \frac{a^2}{r^2}\right)$$

$$V_\theta = V_\infty \sin\theta\left(1 + \frac{a^2}{r^2}\right)$$

$$V_z = 0$$

Bernoulli's equation may be applied in the form of Eq. (7.17) by noting that $\partial\phi/\partial t = 0$ for a steady velocity field, and $g(t)$ becomes a constant

$$\frac{p}{\rho} + \frac{V^2}{2} = C$$

The constant C can be evaluated since the pressure and the velocity are known for $r \to \infty$:

$$C = \frac{p_\infty}{\rho} + \frac{(-V_\infty \cos\theta)^2 + (V_\infty \sin\theta)^2}{2}$$

$$= \frac{p_\infty}{\rho} + \frac{V_\infty{}^2}{2}$$

Also

$$V^2 = \left[-V_\infty \cos\theta\left(1 - \frac{a^2}{r^2}\right)\right]^2 + \left[V_\infty \sin\theta\left(1 + \frac{a^2}{r^2}\right)\right]^2$$

$$= V_\infty{}^2 + 2V_\infty{}^2\frac{a^2}{r^2}(\sin^2\theta - \cos^2\theta) + \frac{V_\infty{}^2 a^4}{r^4}$$

Substituting for C and V^2 in Bernoulli's equation and solving for p gives

$$p = p_\infty - \frac{\rho V_\infty{}^2}{2}\left[\frac{2a^2}{r^2}(\sin^2\theta - \cos^2\theta) + \frac{a^4}{r^4}\right] \quad \blacktriangleleft$$

This is an expression for the pressure at any point in the fluid flow field.

To find the resultant surface force exerted by the fluid on a unit length of the cylinder, the pressure on the surface must be known. This is found from the last equation by setting $r = a$, and

$$p_{r=a} = p_\infty + \frac{\rho V_\infty{}^2}{2}(1 - 4\sin^2\theta)$$

The x component of the incremental force acting on a small element of area dS is $dF_x = (-p_{r=a}\,dS)\cos\theta$, and

$$F_x = -\int_0^{2\pi}\left[p_\infty + \frac{\rho V_\infty^{\ 2}}{2}(1 - 4\sin^2\theta)\right]a\cos\theta\,d\theta = 0$$ ◀

Similarly

$$F_y = -\int_0^{2\pi} p_{r=a}\,a\sin\theta\,d\theta = 0$$ ◀

The total resultant surface force that the fluid exerts on the cylinder is zero. This result might be expected on an intuitive basis since there are no shear stresses in the ideal fluid and the pressure is symmetrically distributed over the surface of the cylinder.

Example 7.7

Are there any stagnation points in the flow field given in Example 7.6? If so, locate them. Where on the surface of the cylinder is the pressure a minimum?

Solution. According to the definition of a stagnation point (also called a critical point, see Section 1.5), the velocity must be zero at that point, and the components of a zero or null vector must be zero. It is then necessary to equate each velocity component to zero and find the value of r and θ from the resulting equations:

$$V_r = -V_\infty\cos\theta\left(1 - \frac{a^2}{r^2}\right) = 0$$

$$V_\theta = V_\infty\sin\theta\left(1 + \frac{a^2}{r^2}\right) = 0$$

The second equation indicates that either $\sin\theta = 0$ or $1 + a^2/r^2 = 0$. The latter of these alternatives would indicate that the value of r would be imaginary, so $\sin\theta = 0$ is selected; then $\theta = 0$ or π. With these values of θ, the equation for V_r yields

$$1 - \frac{a^2}{r^2} = 0 \qquad \text{or} \qquad r = a$$

Hence there are two stagnation points on the surface of the cylinder. One is located at $r = a$ and $\theta = 0$, and the other at $r = a$ and $\theta = \pi$. They are identified as points A and B in Fig. 7.8. From Bernoulli's equation one would expect the pressure to be a maximum at a stagnation point, and this is a correct observation. The maximum pressure in this flow field is

$$p_{\text{stag}} = p_{\text{max}} = p_\infty + \frac{\rho V_\infty^{\ 2}}{2}$$

To locate the minimum pressure on the surface of the cylinder, the derivative of $p_{r=a}$ (obtained in Example 7.6) with respect to θ must be set equal to zero. That is,

$$\frac{dp_{r=a}}{d\theta} = \frac{dp_\infty}{d\theta} + \rho\frac{V_\infty^2}{2}(-8\sin\theta\cos\theta)$$

$$= 0 - 4\rho V_\infty^2 \sin\theta\cos\theta = 0$$

which requires $\cos\theta = 0$ or $\sin\theta = 0$. For $\sin\theta = 0$, $\theta = 0$ or π, and these values of θ correspond to p_{max} or p_{stag}, so that the minimum pressure on the surface of the cylinder is located where $\cos\theta = 0$ or $\theta = \pi/2$ or $3\pi/2$. This minimum pressure is given by

$$p_{min} = p_\infty - \frac{3}{2}\rho V_\infty^2$$ ◀

7.5 Stream Function for Two-Dimensional Flow

The advantage of a function such as the velocity potential should now be apparent. There exists another function that may be used to obtain velocity components and discharges through surfaces. This is called a *stream function* and noted by ψ; it is a consequence of the conservation of matter. The equation for a streamline in x, y coordinates is

$$\frac{dx}{V_x} = \frac{dy}{V_y} \tag{1.8}$$

and this can be rearranged as

$$V_y\,dx - V_x\,dy = 0 \tag{7.19}$$

or

$$\rho V_y\,dx - \rho V_x\,dy = 0 \tag{7.20}$$

The left side of either Eq. (7.19) or (7.20) is an exact differential[1] if

$$\frac{\partial V_y}{\partial y} = -\frac{\partial V_x}{\partial x} \tag{7.21}$$

or

$$\frac{\partial(\rho V_y)}{\partial y} = -\frac{\partial(\rho V_x)}{\partial x} \tag{7.22}$$

The continuity equation (2.3) requires condition (7.21) for a two-dimensional, incompressible flow, as well as requiring Eq. (7.22) for a two-dimensional flow with a steady density field ($\partial\rho/\partial t = 0$). It is evident that Eqs. (7.21)

[1] Wilfred Kaplan, "Advanced Calculus," p. 248, Addison-Wesley Publishing Company, Inc., Reading, Mass., 1952.

and (7.22) define an exact differential, and this is noted by $d\psi$. Then $d\psi = 0$ or ψ = a constant is the *equation of a streamline*. Using

$$d\psi = V_y\,dx - V_x\,dy$$

by previous definition for a two-dimensional flow of an incompressible fluid yields

$$d\psi = \frac{\partial\psi}{\partial x}\,dx + \frac{\partial\psi}{\partial y}\,dy$$

and

$$V_y\,dx - V_x\,dy = \frac{\partial\psi}{\partial x}\,dx + \frac{\partial\psi}{\partial y}\,dy \tag{7.23}$$

Equation (7.23) must hold for all variations in x and/or y, so that

$$V_x = -\frac{\partial\psi}{\partial y} \tag{7.24a}$$

$$V_y = \frac{\partial\psi}{\partial x} \tag{7.24b}$$

Defining

$$d\psi = \rho V_y\,dx - \rho V_x\,dy$$

for a two-dimensional and steady density flow, one can obtain

$$V_x = -\frac{1}{\rho}\frac{\partial\psi}{\partial y}$$

$$V_y = \frac{1}{\rho}\frac{\partial\psi}{\partial x}$$

However, the stream functions for a compressible flow, as well as three-dimensional flows, require more complex considerations that are not generally included in a beginning study of fluid mechanics. The interested reader may pursue this further by reference to articles listed at the bottom of the page.[1] *Hereafter in this text, the symbol ψ will refer to a two-dimensional, incompressible flow.*

The stream function may be related to discharges across surfaces. Figure 7.9 shows a family of streamlines, each of which is given by ψ = a

[1]Chia-Shun Yih, Stream Functions in Three-Dimensional Flow, *Houille Blanche*, vol. 12, no. 3, pp. 445–450, 1957. Victor L. Streeter (ed.-in-chief), "Handbook of Fluid Dynamics," sec. 4, pp. 6–8, McGraw-Hill Book Company, New York, 1961.

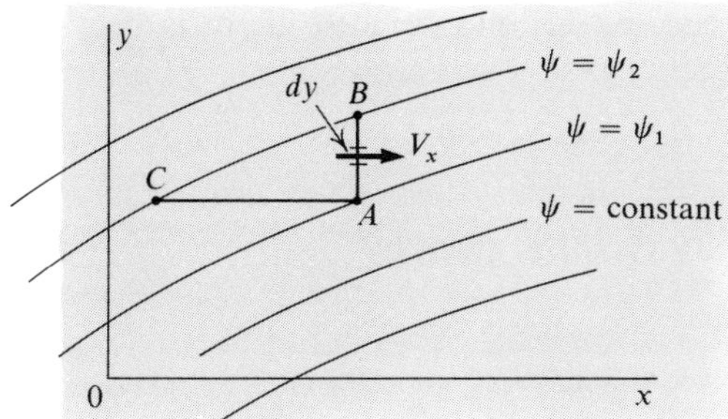

Fig. 7.9 A family of streamlines, each of which represents a constant value of the stream function.

constant. The flow between the two streamlines, ψ_1 and ψ_2, can be found by integrating V_x over the area AB, and this volume rate of flow is given by

$$\int_{S_{AB}} V_x(1)\,dy = \int_A^B -\frac{\partial \psi}{\partial y}\,dy$$

where Eq. (7.24*a*) has been used to substitute for V_x. Carrying out the integration,

$$\int_A^B -\frac{\partial \psi}{\partial y}\,dy = -\int_A^B d\psi = \psi_A - \psi_B = \psi_1 - \psi_2$$

The same discharge, $\psi_1 - \psi_2$, could have been obtained by integrating V_y over the area CA. The difference between the two stream functions is the volume flow per unit depth between the two streamlines. The dimensions of the two-dimensional stream function for incompressible flow are $(\text{length})^2/\text{time}$.

It should be noted that the stream function does not derive from the concept of irrotational motion. ψ is a logical consequence of the velocity field and conservation of matter. If, however, the velocity field is irrotational, then

$$\frac{\partial V_y}{\partial x} - \frac{\partial V_x}{\partial y} = 0$$

and substituting the derivatives of ψ in Eqs. (7.24*a* and *b*) for V_x and V_y yields

$$\frac{\partial^2 \psi}{\partial x^2} + \frac{\partial^2 \psi}{\partial y^2} = 0$$

so that ψ also satisfies Laplace's equation for the two-dimensional, irrotational, incompressible stream function. The obvious boundary condition for ψ is that a streamline may be replaced by a solid stationary boundary in a steady velocity field. The preceding follows because there is no flow across a streamline.

7.6 Relation between ϕ and ψ

For irrotational motion, the velocity vector is given by

$$\mathbf{V} = -\mathbf{grad}\ \phi$$

and from Eqs. (7.24*a* and *b*)

$$\mathbf{V} = -\frac{\partial \psi}{\partial y}\mathbf{i} + \frac{\partial \psi}{\partial x}\mathbf{j}$$

Equating these two expressions, component for component, gives

$$\frac{\partial \phi}{\partial x} = \frac{\partial \psi}{\partial y} \tag{7.25}$$

$$\frac{\partial \phi}{\partial y} = -\frac{\partial \psi}{\partial x} \tag{7.26}$$

These are called the *Cauchy-Riemann conditions* after the mathematicians who discovered their significance with regard to analysis. The two equations are also called *conjugate relations* since they join together the functions ϕ and ψ.

A geometric relation between the equipotential lines and the streamlines is readily discerned by noting that the velocity vector is given by $-\mathbf{grad}\ \phi$, which requires the velocity to be perpendicular to lines of constant ϕ. Since the velocity vectors are tangent to streamlines, the streamlines and equipotentials must be orthogonal at their points of intersection. The only exception to this can be a critical or stagnation point, where the velocity is zero and hence has no direction.

The conjugate relations (7.25) and (7.26) in plane polar coordinates are

$$\frac{\partial \phi}{\partial r} = \frac{1}{r}\frac{\partial \psi}{\partial \theta} \tag{7.27}$$

$$\frac{1}{r}\frac{\partial \phi}{\partial \theta} = -\frac{\partial \psi}{\partial r} \tag{7.28}$$

and the velocity components are

$$V_r = -\frac{1}{r}\frac{\partial \psi}{\partial \theta} \tag{7.29}$$

$$V_\theta = \frac{\partial \psi}{\partial r} \tag{7.30}$$

These may be obtained by direct transformation of the corresponding expressions in the x, y coordinate system.

Example 7.8

Find the stream function for the flow around the cylinder of Example 7.5; also, check the boundary condition, ψ = a constant, over the surface of the cylinder.

Solution. From the example,

$$\phi = V_\infty \cos\theta\left(r + \frac{a^2}{r}\right)$$

From Eq. (7.27),

$$\frac{\partial \psi}{\partial \theta} = r\frac{\partial \phi}{\partial r} = r\left[V_\infty \cos\theta\left(1 - \frac{a^2}{r^2}\right)\right]$$

$$= V_\infty \cos\theta\left(r - \frac{a^2}{r}\right)$$

Integrating with respect to θ gives

$$\psi = V_\infty \sin\theta\left(r - \frac{a^2}{r}\right) + f(r)$$

where $f(r)$ is an arbitrary function of r. However, Eq. (7.28) requires

$$\frac{1}{r}\frac{\partial \phi}{\partial \theta} = -\frac{\partial \psi}{\partial r}$$

$$-\frac{\partial \psi}{\partial r} = -V_\infty \sin\theta\left(1 + \frac{a^2}{r^2}\right) - \frac{df}{dr}$$

and from $\phi = V_\infty \cos\theta\ (r + a^2/r)$,

$$\frac{1}{r}\frac{\partial \phi}{\partial \theta} = \frac{1}{r}\left[-V_\infty \sin\theta\left(r + \frac{a^2}{r}\right)\right] = -V_\infty \sin\theta\left(1 + \frac{a^2}{r^2}\right)$$

Hence Eq. (7.28) can be expressed as

$$-V_\infty \sin\theta\left(1 + \frac{a^2}{r^2}\right) = -V_\infty \sin\theta\left(1 + \frac{a^2}{r^2}\right) - \frac{df}{dr}$$

which indicates that $df/dr = 0$, or $f(r)$ = a constant. To the extent of this additive constant,

$$\psi = V_\infty \sin\theta\left(r - \frac{a^2}{r}\right)$$ ◀

The constant would contribute nothing to the velocity components obtained by partial differentiation of ψ, nor would the constant matter if the difference between two stream functions were used to ascertain the flow between two corresponding streamlines.

To check the boundary condition, ψ = a constant on the surface of the cylinder, note that this surface is given by $r = a$ for all values of θ.

$$\psi\bigg|_{r=a} = V_\infty \sin\theta\left(a - \frac{a^2}{a}\right) = 0 \qquad ◀$$

and ψ = a constant over the surface $r = a$. Note also that the portions of the x axis given by $x \leq a$ and $x \geq a$ correspond to $\psi = 0$. This streamline intersects itself at two points, which are stagnation points. Figure 7.7 shows the dashed streamlines plotted from $\psi = V_\infty \sin\theta\,(r - a^2/r)$.

Example 7.9

Figure 7.10 contains some streamlines and equipotentials for a flow defined by

$$\phi = x^2 - y^2 \qquad \text{and} \qquad \psi = 2xy$$

The direction of flow is indicated on the streamlines. It is observed that the lines $\phi = 0$ and $\psi = 0$ intersect at an angle of 45°. Explain this. What is the flow per unit of z axis between the streamlines passing through the points (8, 1) and (4, 4)?

Solution. Since streamlines and equipotentials must be orthogonal with the possible exception of a stagnation point, it is natural to find the velocity at the point (the origin in this case) where these lines are not orthogonal. The velocity is

$$-\mathbf{grad}\,\phi = -\mathbf{grad}\,(x^2 - y^2) = -2x\mathbf{i} + 2y\mathbf{j}$$

which immediately reveals that the origin is a stagnation or critical point.

To find the value of ψ along two different streamlines, from the given information,

$$\psi(x, y) = 2xy$$

and

$$\psi(4, 4) = (2)(4)(4) = 32$$

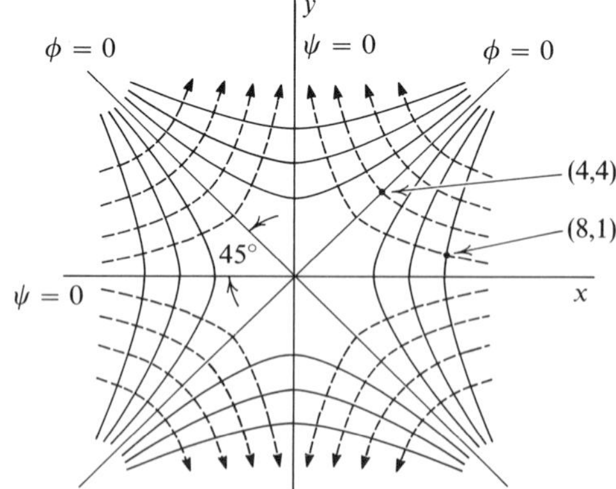

Fig. 7.10

while

$$\psi(8,1) = (2)(8)(1) = 16$$

The flow per unit of z axis between these two streamlines is

$$\psi(4,4) - \psi(8,1) = 32 - 16 = 16 \text{ units of flow}$$ ◀

for a unit dimension in the z direction.

7.7 *Two-Dimensional Vortex*

By way of introduction to a vortex, attention is directed to the stationary cylinder of Example 7.6 in which it was found that the fluid exerted a net force of zero on this surface. It is natural to inquire what flow configuration would produce a "lifting force" on the surface, that is, a force at right angles to the uniform velocity field at infinity. The force F_y in this example was zero because the pressure field was symmetric with respect to the x axis. If the pressure distribution were asymmetric with respect to the x axis, then it would be reasonable to expect a nonzero net resultant force in the y direction. A convenient way to accomplish this consists of altering the velocity field around the cylinder without changing the boundary conditions of the problem. If a circulation Γ, in the counterclockwise direction, were superposed on this problem, the speed of the fluid would be increased over the "top-half" of the cylinder, while a reduction in speed would result over the "bottom half"; and Bernoulli's equation indicates that such velocity changes would result in a decrease of pressure on the "top half" with an increase of pressure on the "bottom half." This would lead to a net positive F_y on integration over the surface of the cylinder. The problem then reduces to one of finding what flow may be superposed to produce a circulation. Such a flow must be irrotational, and hence have its velocity vector given by a potential; also, the boundary conditions imposed by the circulation must leave the velocity unchanged at infinity and the normal component of velocity on the surface of the cylinder unaltered. A logical and simple choice is an irrotational flow field with a velocity potential that is a function of θ only. To test this selection, consider $\phi = \phi(\theta)$. The simplest harmonic function of this form (where K is a constant) is

$$\phi = \pm K\theta \tag{7.31}$$

It can be easily verified that ϕ so defined satisfies Laplace's equation written as

$$\frac{1}{r}\frac{\partial}{\partial r}\left(r\frac{\partial \phi}{\partial r}\right) + \frac{1}{r^2}\frac{\partial^2 \phi}{\partial \theta^2} = 0 \tag{7.9}$$

Also $V_r = -\partial\phi/\partial r = 0$ at all points in the field, so that $V_r = 0$ for $r = a$ and $0 \leq \theta < 2\pi$. Therefore the boundary condition on the surface of the cylinder is unchanged. Since $V_\theta = -\dfrac{1}{r}\dfrac{\partial\phi}{\partial\theta}$, the velocity component in the θ direction would be altered by the superposition of the potential in Eq. (7.31) on the potential of Example 7.6. However, this makes no difference since the tangential component of the fluid velocity relative to a solid boundary at the surface is not a boundary condition; indeed, this component must come from the solution for ϕ.

The contribution to the velocity at infinity is zero from $\phi = \pm K\theta$ since

$$\lim_{r\to\infty}\left[-\frac{\partial(\pm K\theta)}{\partial r}\right] = 0$$

and

$$\lim_{r\to\infty}\left[-\frac{1}{r}\frac{\partial(\pm K\theta)}{\partial\theta}\right] = 0$$

The function $\phi = \pm K\theta$ is the desired velocity potential that will produce the circulation and give a lifting force F_y. If the plus sign is chosen, F_y will be negative; the opposite choice yields a positive F_y. Equation (7.31) gives the velocity potential for a vortex distributed along the z axis. In the two-dimensional velocity field, this vortex is referred to as being "located at the origin"; however, this may be misleading if one does not recall the definition of a two-dimensional velocity field. The streamlines and equipotentials for $\phi = -K\theta$ are shown in Fig. 7.11. The stream function may be found by

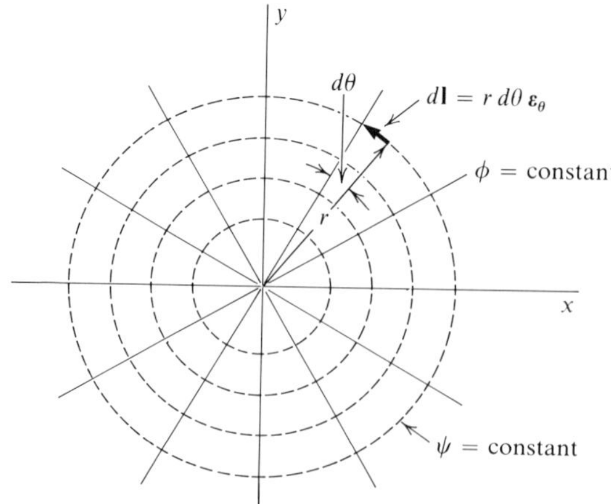

Fig. 7.11 Streamlines and equipotentials for $\phi = -K\theta$, a vortex distributed along the z axis.

using the conjugate relations of Eqs. (7.27) and (7.28). For this potential, the corresponding stream function is $\psi = K \ln r$.

Example 7.10

Obtain an expression for the circulation around any streamline for the vortex $\phi = -K\theta$. Are there any singular or stagnation points in the velocity field?

Solution. The circulation is defined by

$$\Gamma = \oint_C \mathbf{V} \cdot d\mathbf{l} \tag{1.31}$$

For a typical streamline,

$$d\mathbf{l} = r\, d\theta\, \boldsymbol{\varepsilon}_\theta$$

(see Fig. 7.11), and

$$\mathbf{V} = -\mathbf{grad}\, \phi = \frac{K}{r} \boldsymbol{\varepsilon}_\theta$$

Substituting these values in Eq. (1.31) gives

$$\Gamma = \int_0^{2\pi} \frac{K}{r} \boldsymbol{\varepsilon}_\theta \cdot r\, d\theta\, \boldsymbol{\varepsilon}_\theta = K \int_0^{2\pi} d\theta$$

and

$$\Gamma = 2\pi K$$ ◀

There is a singular point in this flow at the origin, and each point on the z axis is also a singular point. This is readily discerned by noting that as r approaches zero, V_θ is unbounded for all values of z. There are no stagnation or critical points at finite distances from the z axis, which follows by noting that

$$\lim_{r \to \infty} V_\theta = 0$$

An important feature of this vortex is revealed if θ is allowed to range over an interval equal to or greater than 2π. Consider the value of ϕ that would accrue by allowing θ to range from 0 to 2π.

$$\begin{aligned}\phi(r,2\pi) &= \phi(r,0) + \oint_C d\phi \\ &= \phi(r,0) + \oint_C -\mathbf{grad}\, \phi \cdot d\mathbf{l}\end{aligned}$$

in which C is any path enclosing the origin. From the previous example,

$$\oint_C -\mathbf{grad}\,\phi \cdot d\mathbf{l} = \Gamma = 2\pi K$$

so that

$$\phi(r,2\pi) = \phi(r,0) + 2\pi K$$

Since the points $r, 0$ and $r, 2\pi$ are identical for the same values of r, ϕ is a multivalued function of θ unless θ is restricted according to $\beta < \theta < 2\pi + \beta$, where β is any constant angle. To insure that $\phi = -K\theta$ was single-valued, θ was allowed to range from 0 up to any value less than 2π.

7.8 Flow Around a Cylinder with Circulation

The superposition of a vortex on the flow around a cylinder given in Example 7.6 adds a circulation that produces a lifting force on the cylinder. If this circulation is counterclockwise when viewed from the positive z axis, the potential and stream functions are given by

$$\phi = V_\infty \cos\theta \left(r + \frac{a^2}{r}\right) - K\theta$$

and

$$\psi = V_\infty \sin\theta \left(r - \frac{a^2}{r}\right) + K \ln r$$

with velocity components given by

$$V_r = -\frac{\partial\phi}{\partial r} = -V_\infty \cos\theta \left(1 - \frac{a^2}{r^2}\right) \tag{7.32}$$

$$V_\theta = -\frac{1}{r}\frac{\partial\phi}{\partial\theta} = V_\infty \sin\theta \left(1 + \frac{a^2}{r^2}\right) + \frac{K}{r} \tag{7.33}$$

$$V_z = 0 \tag{7.34}$$

The stagnation points of this flow may be found by setting Eqs. (7.32) and (7.33) equal to zero, which yields

$$V_\infty \cos\theta \left(1 - \frac{a^2}{r^2}\right) = 0$$

and

$$V_\infty \sin\theta \left(1 + \frac{a^2}{r^2}\right) + \frac{K}{r} = 0$$

The first of these indicates that either $\cos\theta = 0$ or $1 - a^2/r^2 = 0$. For $\cos\theta \neq 0$, $r = a$; and the second equation gives

$$\sin\theta = -\frac{K}{2V_\infty a}$$

which is valid if $K \leq 2V_\infty a$. The streamline patterns for each of these relative magnitudes of K are shown in Fig. 7.12. The stagnation points are labeled A and B, and both of these are on the surface of the cylinder.

For $r \neq a$ and $\cos\theta = 0$, $\sin\theta = \pm 1$; and $V_\theta = 0$, so that

$$\pm V_\infty\left(1 + \frac{a^2}{r^2}\right) + \frac{K}{r} = 0$$

This leads to two quadratic equations for r:

$$r^2 + \frac{K}{V_\infty}r + a^2 = 0$$

and

$$r^2 - \frac{K}{V_\infty}r + a^2 = 0$$

The solution of the first of these gives two negative values of r, which are rejected. The second equation has the roots

$$r = \frac{K}{2V_\infty}\left[1 \pm \left(1 - \frac{4a^2 V_\infty{}^2}{K^2}\right)^{1/2}\right]$$

Fig. 7.12 Flow around a cylinder with circulation $\Gamma = 2\pi K$. (a) $K < 2V_\infty a$, two distinct stagnation points. (b) $K = 2V_\infty a$, two stagnation points coincide.

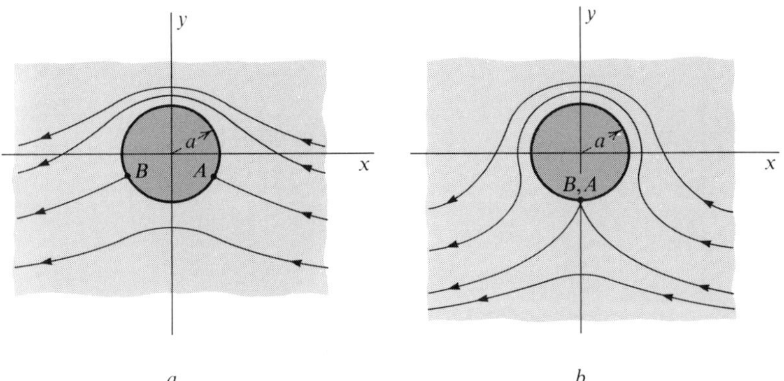

and for $K > 2V_\infty a$, there are two real values of r (one greater, and one less than a, the radius of the cylinder). The point inside the cylinder is of no interest in this example. The stagnation point given by

$$r = \frac{K}{2V_\infty}\left[1 + \left(1 - \frac{4a^2V_\infty{}^2}{K^2}\right)^{1/2}\right]$$

and $\theta = 3\pi/2$ is shown in Fig. 7.13.

By use of Bernoulli's equation, the pressure distribution in the fluid may be found; then the pressure on the surface of the cylinder may be integrated over this surface to obtain

$$F_x = 0$$

$$F_y = \Gamma\rho V_\infty$$

which clearly indicates that the addition of circulation to the flow of Example 7.6 produces a lifting force on the cylinder. The details of finding F_x and F_y are left as an assigned problem at the end of the chapter.

7.9 Sources and Sinks

The velocity potential for the vortex in Section 7.7 was found to be a function of θ only, and it gave rise to a circulating flow. At this point it appears natural to examine the flow given by a potential that is a function of r only. Under this condition, Laplace's equation reduces to $\dfrac{1}{r}\dfrac{\partial}{\partial r}\left(r\dfrac{\partial \phi}{\partial r}\right) = 0$,

Fig. 7.13 *Flow around a cylinder with circulation $\Gamma = 2\pi K$ for $K > 2V_\infty a$.*

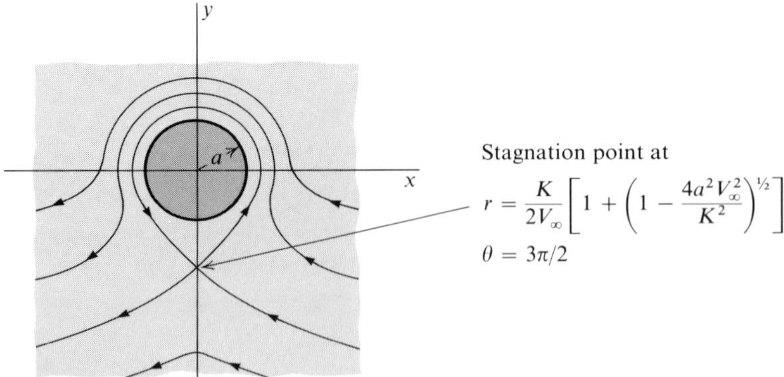

which may be written as the following ordinary differential equation:

$$\frac{d}{dr}\left(r\frac{d\phi}{dr}\right) = 0$$

Two integrations of the preceding equation yield

$$\phi = C \ln r + C_1$$

where C and C_1 are constants. Since the latter is an additive constant, it will be dropped; and the flows given by $\phi = C \ln r$ with $C < 0$ and $C > 0$ will be examined.

For $C < 0$, the negative constant is replaced by $-K$ and

$$\phi = -K \ln r \tag{7.35}$$

Reference to Fig. 7.14 will assist in the interpretation of this flow. The volume rate of flow, $\dot{V}$, through a cylindrical surface of length L and general radius $r > 0$, is given by

$$\dot{V} = \int_S -\textbf{grad}\ \phi \cdot d\mathbf{S} = \int_0^{2\pi} \frac{K}{r} \boldsymbol{\varepsilon}_r \cdot Lr\ d\theta\ \boldsymbol{\varepsilon}_r = K2\pi L$$

and $K = \dot{V}/2\pi L$. For steady incompressible flow, $\dot{V}$ would be the same through any cylindrical surface of length L with its axis coincident with the z axis. Furthermore, any control volume that contained any portion of the z axis would appear to have matter created within it. For this reason the z axis is referred to as a *line source*. The continuity equation is not satisfied

Fig. 7.14 *Two-dimensional source flow* ($V_\theta = V_z = 0$), $\phi = -K \ln r$.

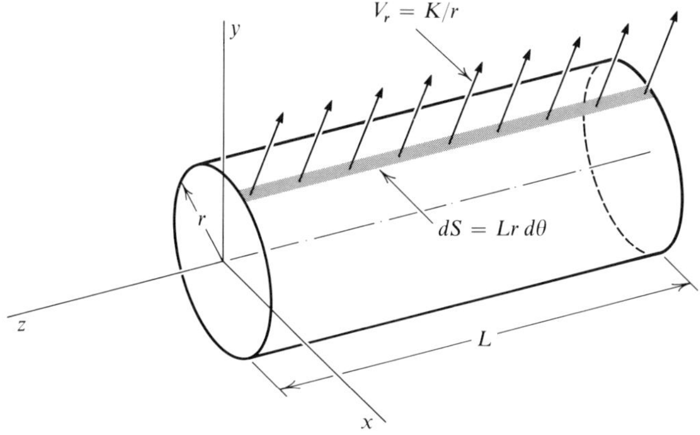

at any point on the z axis; all of the points on this axis are singular points, which is readily discerned from $V_r = K/r$. The equipotentials for this flow are given by $r =$ a constant, and the streamlines are radial lines. The constant K is referred to as the *strength of the source*. The stream function for this flow is given by

$$\psi = -K\theta \tag{7.36}$$

which can be obtained from the conjugate relations (7.27) and (7.28).

If the constant C is greater than zero, say K, then

$$\phi = K \ln r \tag{7.37}$$

and

$$\psi = K\theta \tag{7.38}$$

and this corresponds to a *line sink*. For this case the flow is radially inward. Again, the z axis contains only singular points.

Since both sources and sinks are characterized by the contradiction of continuity at certain singular points, the reader might be dubious of the value of such mathematical solutions of Laplace's equation. The use of these flows rests in their superposition on other flows. This is amplified in the following example.

Example 7.11

Consider the two-dimensional irrotational steady flow of an incompressible fluid, resulting from the superposition of a line source along the z axis on a uniform flow given by

$$V_x = -V_\infty \qquad V_y = 0 \qquad V_z = 0$$

Locate any stagnation points; and find the equation of the streamline passing through each of these.

Solution. The potential for the source and uniform flow are, respectively, $\phi = -K \ln r$ and $\phi = V_\infty x = V_\infty r \cos\theta$. Superposing these gives

$$\phi = V_\infty r \cos\theta - K \ln r$$

from which

$$V_r = -\frac{\partial \phi}{\partial r} = -V_\infty \cos\theta + \frac{K}{r}$$

and

$$V_\theta = -\frac{1}{r}\frac{\partial \phi}{\partial \theta} = V_\infty \sin\theta$$

For $V_\theta = 0$, $\sin\theta$ must vanish; and θ must equal 0 or π. From $V_r = 0$,

$$r = \frac{K}{V_\infty \cos\theta}$$

For $\theta = 0$, $\cos\theta = 1$ and $r = K/V_\infty$. The only stagnation point is given by the coordinates

$$r = \frac{K}{V_\infty} \qquad \text{and} \qquad \theta = 0$$ ◀

To find the value of ψ for the streamline that passes through the stagnation point, the stream function must be ascertained by either using the conjugate relations (7.27) and (7.28) or superposing the stream functions for the line source and the uniform flow. Either method gives the same result of course, and

$$\psi = V_\infty r \sin\theta - K\theta$$

This is the equation for all streamlines in the field, and each may be obtained by assigning a constant to ψ. The one passing through the point $(K/V_\infty, 0)$ is

$$\psi = V_\infty\left(\frac{K}{V_\infty}\right)\sin(0) - K(0) = 0$$

and the streamline passing through this point is given by

$$V_\infty r \sin\theta - K\theta = 0$$ ◀

The streamlines and equipotentials for this flow are shown in Fig. 7.15. The line $\psi = 0$ is the entire positive x axis (origin excluded) and the locus BAC. The negative x axis (origin excluded) is given by $\psi = -K\pi$. The portion labeled BAC extends infinitely far from the origin in the direction of the negative x axis; however, points on this locus are a finite distance from the x axis. This is made clear by noting that BAC is given by $V_\infty r \sin\theta - K\theta = 0$, and then substituting y for $r\sin\theta$ to obtain $y = K\theta/V_\infty$. As $x \to -\infty$, $\theta \to \pi$, and $y = K\pi/V_\infty$, which is finite.

Fig. 7.15

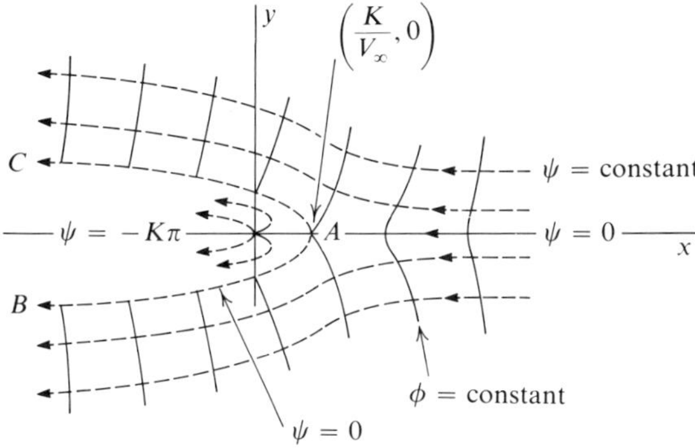

Since BAC is given by ψ = a constant, it may be replaced by a solid body; and since there would be no concern for the source, the flow would simply be that of a uniform stream at infinity passing over a body of the shape of BAC. Such a body is referred to as a *two-dimensional half-body*.

7.10 Doublet

The *doublet* is the fluid mechanics counterpart of the electric dipole. The model consists of a source and a sink of equal strength approaching coincidence in such a manner that the product of the absolute value of the strength of either one and the distance between them remains constant. Figure 7.16 shows the source and sink with a distance $2b$ between them. The potentials at point P due to the presence of the source and sink are, respectively,

$$\phi_{\text{source}} = -K \ln r_1 \qquad \text{and} \qquad \phi_{\text{sink}} = K \ln r_2$$

Using superposition, the potential at P is given by

$$\phi = \phi_{\text{source}} + \phi_{\text{sink}} = -K \ln r_1 + K \ln r_2 = K \ln \frac{r_2}{r_1}$$

$$= K \ln\left(1 + \frac{\Delta r}{r_1}\right)$$

in which $\Delta r = r_2 - r_1$. For b small compared to r (and it is in the limit), Δr is approximately equal to $2b \cos \theta_2$, so that

$$\phi \simeq K \ln\left(1 + \frac{2b \cos \theta_2}{r_1}\right)$$

The logarithm may be expanded[1] in a series provided $-1 < (2b \cos \theta_2)/r_1 \leq 1$, and

$$\phi \simeq K\left\{\frac{2b \cos \theta_2}{r_1} - \frac{[(2b \cos \theta_2)/r_1]^2}{2} + \frac{[(2b \cos \theta_2)/r_1]^3}{3} - \cdots\right\}$$

$$\simeq 2bK\left(\frac{\cos \theta_2}{r_1} - \frac{2b}{r_1^{\,2}} \frac{\cos^2 \theta_2}{2} + \frac{4b^2}{r_1^{\,3}} \frac{\cos^3 \theta_2}{3} - \cdots\right)$$

The limit of the last expression is taken holding $2bK = A$, a constant, as $b \to 0$. This yields

$$\phi = \frac{A \cos \theta}{r} \tag{7.39}$$

by noting that $\theta_2 \to \theta$ and $r_1 \to r$ as $b \to 0$.

[1] Milton Abramowitz and Irene A. Stegun (eds.), "Handbook of Mathematical Functions," p. 68 (entry 4.1.24), Dover Publications, Inc., New York, 1965.

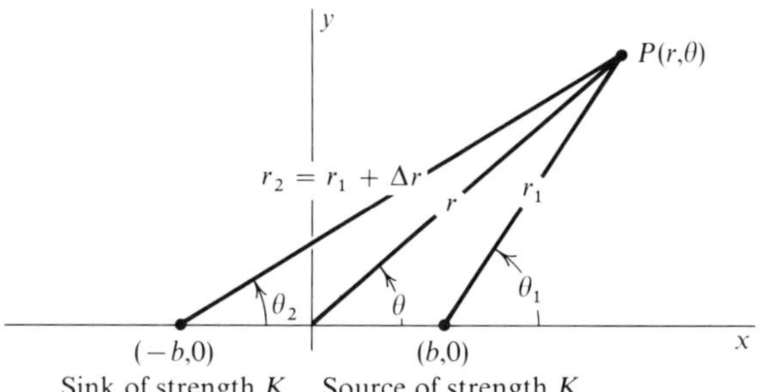

Fig. 7.16 *Model for a doublet, as $b \to 0$.*

The doublet has a directional characteristic specified by its axis. The doublet whose potential is given by Eq. (7.39) has an axis in the positive x direction, the direction from sink to source. The constant $A = 2bK$ is referred to as the *strength* of the doublet. The stream function corresponding to Eq. (7.39) is

$$\psi = -\frac{A \sin \theta}{r} \tag{7.40}$$

which is found from the conjugate relations of Eqs. (7.27) and (7.28). The equipotentials and streamlines are shown in Fig. 7.17. The origin is a singular point.

The doublet when superposed on a uniform flow field given by $\phi = V_\infty x$ can be used to analyze flow around a cylinder. This exercise is left for the assigned problems at the end of the chapter.

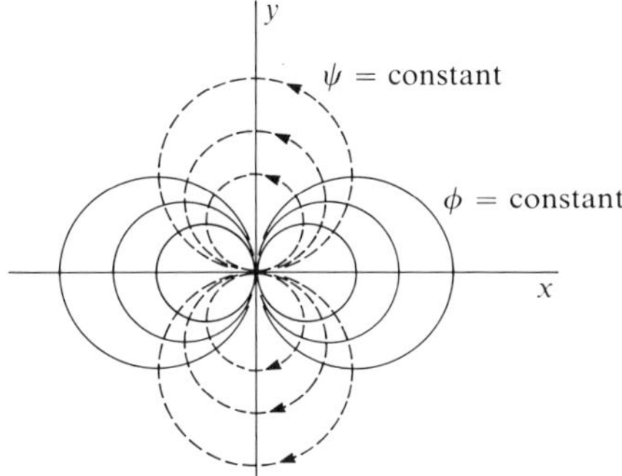

Fig. 7.17 *Equipotentials and streamlines for a doublet with axis in the positive x direction.*

7.11 A Numerical Approach to Potential Flow Problems

Very often the exact solution of Laplace's equation by formal and mathematical means is difficult. The troubles arise from the configuration of the boundaries over which conditions are specified. Hence it is plausible to explore a numerical method that enables the solution for ψ or ϕ to be approximated. One simple technique that lends itself to hand calculation or use of a computer is presented here. Consider the problem of finding the streamlines for the flow in a right-angled bend. The motion of an incompressible fluid will be assumed two dimensional, irrotational, and steady. Figure 7.18 shows the geometry of the boundaries. The dimensions of the lines on the figure (with units unspecified) are given as follows:

$$AB = BC = CD = 2 \qquad DE = 3 \qquad FE = 4$$

The flow coming into the bend has a given uniform velocity, $-V_\infty \mathbf{j}$, over AB, and the velocity is given as uniform over DE. From the conservation of matter, the velocity over DE is $\mathbf{V} = (2/3)\, V_\infty \mathbf{i}$. To find the streamlines within the domain bounded by A-B-C-D-E-F-A, loci characterized by $\psi =$ a constant must be found, which is information obtained by solving

$$\frac{\partial^2 \psi}{\partial x^2} + \frac{\partial^2 \psi}{\partial y^2} = 0$$

subject to values of ψ specified on the boundaries. AF and FE are obviously streamlines since the velocity is tangent to these lines. The value $\psi = 0$ is assigned to each of these segments since it is only necessary to know the value of the stream function to the extent of an additive constant. The values of ψ along AB can be found from

$$V_y = \frac{\partial \psi}{\partial x} = -V_\infty$$

and $\psi = -V_\infty x$ along AB. From this, $\psi_B = -2V_\infty$, and the value of ψ along BC and CD is also $-2V_\infty$. The value of ψ along ED can be obtained from

$$V_x = -\frac{\partial \psi}{\partial y} = \frac{2}{3} V_\infty$$

and $\psi = -(2/3)V_\infty y$ along ED. The value of the stream function is now known at all points on the boundary, and it remains to satisfy Laplace's equation within these boundaries. The partial derivatives in this equation may be approximated by Taylor's theorem[1] for a function of two variables.

[1]Harold K. Crowder and S. W. McCuskey, "Topics in Higher Analysis," pp. 223–225, The Macmillan Company, New York, 1964.

The theorem gives

$$\psi(x+h, y+k) = \psi(x,y) + \left(h\frac{\partial\psi}{\partial x} + k\frac{\partial\psi}{\partial y}\right) + \frac{1}{2!}\left(h^2\frac{\partial^2\psi}{\partial x^2} + 2hk\frac{\partial^2\psi}{\partial x\,\partial y} + k^2\frac{\partial^2\psi}{\partial y^2}\right) + \frac{1}{3!}\left(h^3\frac{\partial^3\psi}{\partial x^3} + 3h^2k\frac{\partial^3\psi}{\partial x^2\,\partial y} + 3hk^2\frac{\partial^3\psi}{\partial x\;\partial y^2} + k^3\frac{\partial^3\psi}{\partial y^3}\right) + \frac{1}{4!}\left(h^4\frac{\partial^4\psi}{\partial x^4} + 4h^3k\frac{\partial^4\psi}{\partial x^3\,\partial y} + 6h^2k^2\frac{\partial^4\psi}{\partial x^2\,\partial y^2} + 4hk^3\frac{\partial^4\psi}{\partial x\,\partial y^3} + k^4\frac{\partial^4\psi}{\partial y^4}\right) + \cdots \tag{7.41}$$

and it is assumed that the partial derivatives of ψ are continuous up to an order required by accuracy of representation of the function. Equation (7.41) may be used to approximate both $\partial^2\psi/\partial x^2$ and $\partial^2\psi/\partial y^2$; for example, with $k = 0$,

$$\psi(x+h,y) = \psi(x,y) + h\frac{\partial\psi}{\partial x} + \frac{h^2}{2!}\frac{\partial^2\psi}{\partial x^2} + \frac{h^3}{3!}\frac{\partial^3\psi}{\partial x^3} + \frac{h^4}{4!}\frac{\partial^4\psi}{\partial x^4} + \cdots \tag{7.42}$$

Also with $k = 0$

$$\psi(x-h,y) = \psi(x,y) - h\frac{\partial\psi}{\partial x} + \frac{h^2}{2!}\frac{\partial^2\psi}{\partial x^2} - \frac{h^3}{3!}\frac{\partial^3\psi}{\partial x^3} + \frac{h^4}{4!}\frac{\partial^4\psi}{\partial x^4} - \cdots \tag{7.43}$$

and addition of Eqs. (7.42) and (7.43) yields

$$\psi(x+h,y) + \psi(x-h,y) = 2\psi(x,y) + \frac{2h^2}{2!}\frac{\partial^2\psi}{\partial x^2} + \frac{2h^4}{4!}\frac{\partial^4\psi}{\partial x^4} + \cdots$$

If h^2 is large compared to h^4, a condition fulfilled by making h "suitably small," then

$$\frac{\partial^2\psi}{\partial x^2} \simeq \frac{\psi(x+h,y) + \psi(x-h,y) - 2\psi(x,y)}{h^2}$$

and similarly

$$\frac{\partial^2\psi}{\partial y^2} \simeq \frac{\psi(x,y+k) + \psi(x,y-k) - 2\psi(x,y)}{k^2}$$

for k^2 large compared to k^4.

Requiring $\partial^2\psi/\partial x^2 + \partial^2\psi/\partial y^2 = 0$, and setting $k = h$ gives

$$\psi(x + h, y) + \psi(x - h, y) + \psi(x, y + h) + \psi(x, y - h) - 4\psi(x, y) = 0 \quad (7.44)$$

This equation must be satisfied at points within the domain of solution of Laplace's equation. In this problem, 47 interior points have been selected, and these have been designated by $\psi_{11}, \psi_{12}, \ldots$, etc., on Fig. 7.18. Each interior point is a distance $h = 1/2$ from four neighboring points. The interior points are called *nodes*. The problem now reduces to writing Eq. (7.44) in the appropriate form for each of the nodes. For example, consider node 11; Eq. (7.44) would be written as

$$\psi_{12} + 0 + \psi_{21} + 0 - 4\psi_{11} = 0$$

for node 53,

$$\psi_{54} + \psi_{52} + \psi_{63} + \psi_{43} - 4\psi_{53} = 0$$

After all 47 of these equations have been written the collection must be solved to find the 47 ψ_{ij}'s. This would be a lengthy task if undertaken using hand calculation; however, computer centers have programs readily available for matrix inversion, which is a direct way of solving a system of simultaneous equations. The computer solution of the system of equations for the nodes in Fig. 7.18 has been posted below each node; for example, $\psi_{53} = -1.055V_\infty$. The heavy dashed lines $\psi = -0.500V_\infty$, $-1.000V_\infty$, and $-1.500V_\infty$ were drawn by interpolation. These three streamlines in addition to $\psi = 0$ and $\psi = -2.000V_\infty$ have equal quantities of flow among them in the amount of $\Delta\psi = 0.500V_\infty$. It is very apparent that $\psi = -2.000V_\infty$ and $\psi = -1.500V_\infty$ are a minimum distance apart in the vicinity of point C, so it is reasonable to expect that the space-average velocity would be the greatest in this vicinity. This line of reasoning continued and applied to the flow between a streamline $\psi = -1.990V_\infty$ and $\psi = -2.000V_\infty$ would indicate that the fluid velocity is very large close to C. At point C the velocity is infinite and this is a singular point. A similar line of reasoning may be employed near point F, to conclude that F is a stagnation point.

By decreasing the size of h, greater accuracy can be obtained in the approximation of ψ at each of the nodes; however, decreasing h would increase the number of nodes and hence increase the number of simultaneous equations.

If a computer is not available to solve the node equations, there is another method of approximating ψ. This consists of assuming a distribution of ψ, that is, assuming a value of ψ at each node. The initial values are then adjusted to satisfy Eq. (7.44); however, satisfaction of this equation by

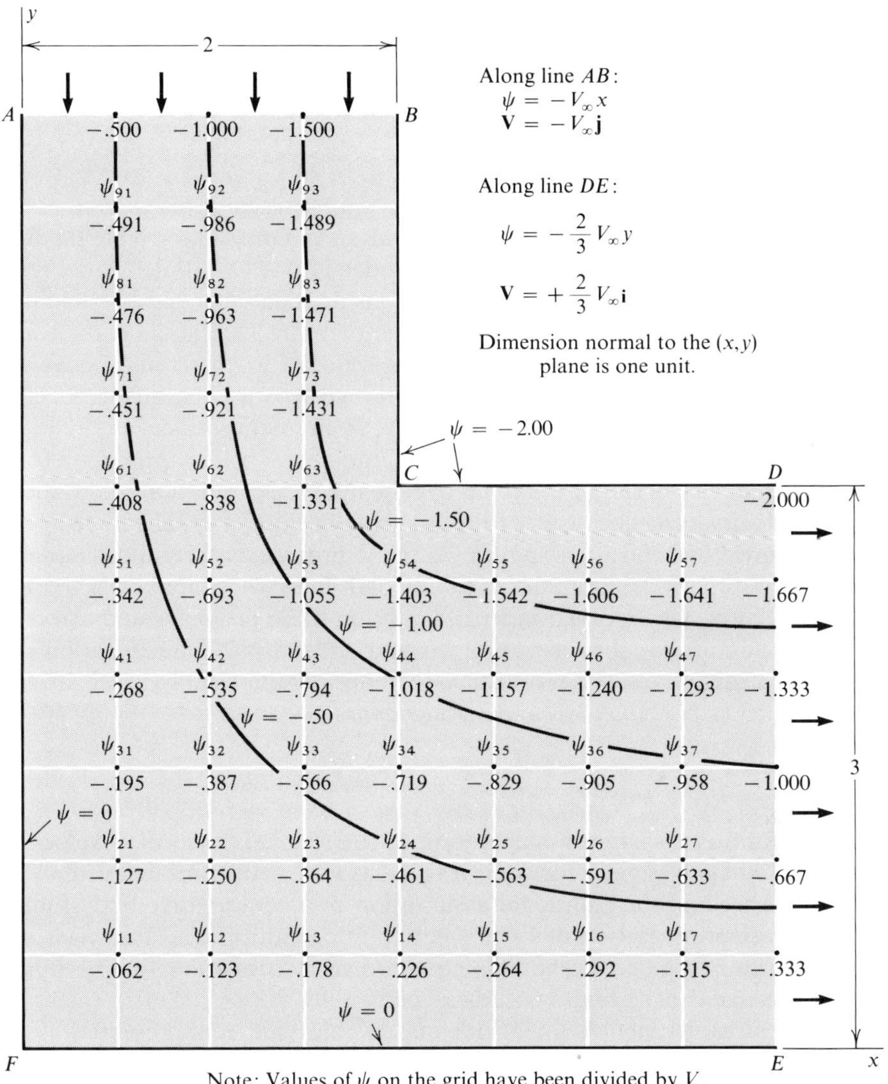

Fig. 7.18 *Flow grid to solve div* **grad** $\psi = 0$ *for flow in the right-angled bend.*

adjusting the value of ψ at any given node will be reflected throughout the entire domain. The object of this method is to minimize the absolute value of the left-hand side of Eq. (7.44) in order that it be as near zero as calculation time permits and accuracy requires. There are some expedient techniques associated with this method, which is referred to as the *relaxation method.* The interested reader can refer to the literature[1] on this subject.

7.12 Foils and Lift

In Section 7.8 it was found that the addition of circulation to the two-dimensional flow around a cylinder produced a lifting force distributed over the surface of the cylinder. Methods of analysis in complex variable theory provide a means of transforming the cylinder into a two-dimensional foil as shown in Fig. 7.19. These transformations are characterized by yielding flow patterns with infinite velocities at one or more locations on the foil. Figure 7.19*a* shows a foil with an infinite velocity at the trailing edge *B* of the foil. This theoretical result, even though it is a consequence of an idealized flow field, departs considerably from reality since an infinite velocity would require an infinite pressure gradient. A better approximation to flow around a foil can be obtained by adjusting the circulation to produce a finite velocity at *B*. The result of this is shown in Fig. 7.19*b*. The method of transforming flows and adjusting circulation to provide mathematical models of flow around foils is credited to Nicolai Egorovich Joukowsky and Wilhelm Kutta, who worked independently on these problems and arrived at an explanation for the relation of lift and circulation. The generation of lifting surfaces and foil shapes is treated in considerable detail in specialized courses such as *potential flow* and *aerodynamics*.

7.13 Closing Remarks

The assumption of irrotational motion led to the existence of a velocity potential, which was valid for compressible as well as incompressible flows. This potential and the additional assumption of a conservative body-force field permitted integration of Euler's equation of motion, and the result of this integration yielded Bernoulli's equation in a form valid for the fluid with a variable density and unsteady velocity field.

The stream function for an incompressible fluid was presented as a consequence of the conservation of matter, and the discussion of this was restricted to two-dimensional flows. For irrotational and two-dimensional flow of an incompressible fluid, the conjugate or Cauchy-Riemann relations were presented.

A numerical method for finding flow patterns was introduced. The reader who desires a knowledge of three dimensional and compressible

[1] R. V. Southwell, "Relaxation Methods in Engineering Science," Oxford University Press, London, 1940. D. N. de G. Allen, "Relaxation Methods," McGraw-Hill Book Company, New York, 1954.

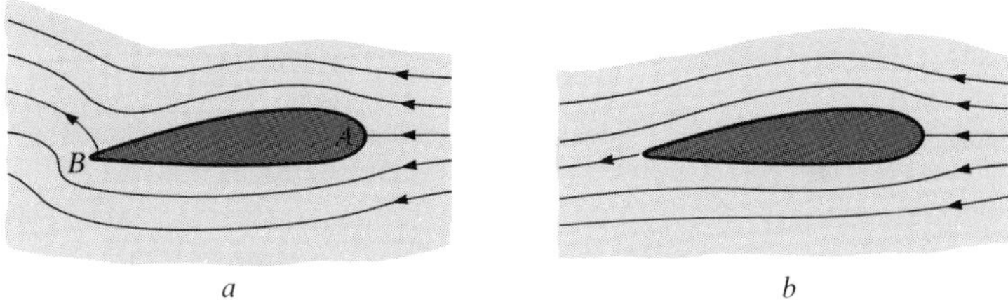

Fig. 7.19 (a) Two-dimensional foil resulting from the transformation of flow around a cylinder. (b) Same foil with circulation adjusted to produce a finite velocity at the trailing edge.

irrotational motion should consult more advanced and specialized treatises on the subject.

Self-Study Questions

7.1 *a.* What conditions must exist on the velocity of a fluid in irrotational flow?
b. What is the expected value of **curl V** for an ideal fluid?

7.2 Given a one-dimensional irrotational flow field in the x direction,

a. Determine a value for $\oint_C \mathbf{V} \cdot d\mathbf{l}$.

b. What can be said about the spatial variation of V_x in the field?

7.3 If $\oint_C V_x\,dx + V_y\,dy = 0$ for two-dimensional flow, show that

$$\frac{\partial V_y}{\partial x} = \frac{\partial V_x}{\partial y}$$

7.4 *a.* What is the mathematical definition of a velocity potential?
b. What type of flow is called a potential flow?

7.5 *a.* State in words the meaning of **grad** ϕ with respect to $d\phi/dl$.
b. Is potential flow in the direction of decreasing or increasing ϕ values?

7.6 If $\dfrac{\partial^2\phi}{\partial x^2} + \dfrac{\partial^2\phi}{\partial y^2} + \dfrac{\partial^2\phi}{\partial z^2} = 0$ for a given flow field,

a. Is the flow field necessarily steady?
b. Is the density field necessarily steady?

7.7 Show that the continuity equation for a compressible fluid in a steady irrotational flow may be expressed as

$$\text{div}\,\mathbf{grad}\,\phi = -\frac{\mathbf{grad}\,\phi \cdot \mathbf{grad}\,\rho}{\rho}$$

7.8 State the flow condition at a stationary solid boundary in terms of

a. ϕ *b.* $\mathbf{V}$

7.9 *a.* When could a velocity potential for a flow field exist and not be a harmonic function?
b. Is it possible for a velocity potential to exist for an incompressible flow field and not be a harmonic function?

7.10 Show that a function obtained by adding a constant to a harmonic function is a harmonic function.

7.11 What is Laplace's equation for a velocity potential in
a. x, y, z coordinates. *b.* r, θ, z coordinates.

7.12 Review the reason for choice of $-\beta^2$ in Example 7.3 in order that the trial solution would meet the boundary condition, and briefly note the trial solution for a choice of β^2?

7.13 *a.* In what way does the derivation of Bernoulli's equation for steady flow of an inviscid fluid differ from that for irrotational flow?
b. Which derivation allows using

$$\frac{p_1}{\rho} + \frac{V_1^{\,2}}{2} + \Phi_{B_1} = \frac{p_2}{\rho} + \frac{V_2^{\,2}}{2} + \Phi_{B_2}$$

for any two points (subscripted 1 and 2) in the flow field?

7.14 A pressure distribution in a two dimensional and irrotational flow field is desired at a finite-sized solid boundary in order to obtain a value for the total force by the fluid. What additional information about the flow field is needed for use with the Bernoulli equation?

7.15 Show that the ψ function obtained in Example 7.8 has an exact differential.

7.16 *a.* State the expressions for V_x and V_y in terms of a stream function and in terms of a velocity potential.
b. If both ψ and ϕ exist, what is the description of the flow field?

7.17 If ψ exists for a given flow field, is it a harmonic function?

7.18 *a.* Obtain the slope of a ϕ = constant line from $d\phi = 0$.
b. Obtain the slope of a ψ = constant line from $d\psi = 0$.
c. Show that the intersections of streamlines and equipotential lines are orthogonal by showing

$$\left(\frac{dy}{dx}\right)_{\psi=\text{const}} = -\left(\frac{dx}{dy}\right)_{\phi=\text{const}}$$

Is this true at a stagnation point?

7.19 *a.* Obtain expressions for V_x and V_y for given ϕ or ψ in Example 7.9 and establish that the flow is in directions shown in Fig. 7.10.
b. When streamlines lie closer together, are the velocities higher or lower?

7.20 In Section 7.7 a vortex flow field was described by $\psi = K \ln r$.
a. Is this flow irrotational everywhere?
b. What is the value of the circulation for a closed path excluding any singular points?

7.21 Obtain V_r and V_θ for the vortex flow field described in Section 7.7 and note the variation with r.

7.22 Briefly discuss how stagnation points are located in a potential flow field if ψ is given.

7.23 Establish that circulation is zero everywhere in the potential flow field of Section 7.8 if the cylinder is not enclosed by the path for the line integral.

7.24 Given the velocity potential of a line source, $-K \ln r$,
a. Obtain expressions for V_r and V_θ.
b. Sketch a two-dimensional view of the flow at two radii for a given z location.
c. Compare the volume flow rate per unit depth through the cylindrical surfaces at two different radii.

7.25 *a.* Describe a doublet in terms of a source and a sink.
b. Obtain expressions for velocity in a doublet flow field and substantiate that the direction of flow is correctly shown in Fig. 7.17.

7.26 *a.* Use the doublet location as an origin, and show that the volume flow rate per unit length of z axis of a doublet flow field through a cylindrical surface of any radius is zero.
b. Does this result allow consideration of the cylindrical surface as a solid with a potential flow around it?

7.27 Sketch several streamlines for a potential flow field due to a sink and source of equal strength
a. That are a distance b apart.
b. That are a distance b apart in the limit as $b \to 0$.

7.28 Show that the streamlines for a vortex flow are circles.

7.29 Compare the flow field of a doublet to that of a right circular cylinder steadily moving in an incompressible fluid as given in Example 7.3.

7.30 Briefly discuss how V_θ and V_r vary with r for a source, a vortex, and a doublet.

Problems

7.1 Find a velocity potential for an irrotational velocity field where

a. $V_x = -10x \qquad V_y = 10y \qquad V_z = 0$

b. $V_x = \dfrac{-5x}{x^2 + y^2} \qquad V_y = \dfrac{-5y}{x^2 + y^2} \qquad V_z = 0 \qquad \text{where } x \neq 0,\ y \neq 0$

7.2 *a.* Show that the resultant velocity field in Example 7.3 is

$$\mathbf{V} = \frac{a^2 V_c}{r^2} (\cos\theta\, \boldsymbol{\varepsilon}_r + \sin\theta\, \boldsymbol{\varepsilon}_\theta)$$

b. Is the velocity of the fluid at the solid boundary equal to $V_c\mathbf{i}$? Why?

7.3 *a.* Show mathematically, using

$$\left.\frac{\mathbf{r}\, d\theta}{dr}\right|_{\psi=C} = -\left.\frac{dr}{\mathbf{r}\, d\theta}\right|_{\phi=C}$$

that the intersection of a streamline with an equipotential line ($d\phi = 0$) is one of 90°.
b. For a spatial displacement $d\mathbf{l}$ along a streamline, establish that the distance between equipotential lines is smaller in regions of higher speeds.

7.4 A velocity potential exists for flow in a region where $x \geq 0$. The potential is $A(x^2 - y^2)$ where A is a dimensional constant.

a. Show that the potential is a harmonic function.
b. Show that ϕ satisfies the continuity equation for an incompressible fluid.
c. Show that the y axis could be considered a solid boundary.
d. Sketch streamlines near the solid boundary.

7.5 Refer to the velocity field in Prob. 1.18 and

a. Show that a velocity potential exists.
b. Obtain the velocity-potential function.
c. Show that the velocity potential is the sum of two velocity potentials.

7.6 Refer to the velocity field in Prob. 1.20 and

a. Show that a velocity potential does or does not exist.
b. Obtain the velocity-potential function if it does exist.
c. Change the sign of V_x and again answer (*a*) and (*b*).

7.7 For a given velocity potential, $\phi = A \ln r$, and a constant density fluid, show that the horizontal pressure variation in terms of the stagnation pressure p_∞ is

$$p = p_\infty - \rho \frac{A^2}{2r^2}$$

7.8 A wind of 60 mph blows transversely across a building that has a semicircular cross section with a radius of 20 ft. Atmospheric pressure is 14.2 $\mathrm{lb}_f/\mathrm{in.}^2$ and the temperature is $-30°$F.

a. Estimate the vertical- and horizontal-force components exerted on the outer surface by the fluid per ft-length of the building.
b. What is the minimum external pressure acting on the roof of the building?

7.9 Obtain a stream function (if it exists) for each of the velocity fields given in Prob. 7.1.

7.10 *a.* Does a velocity potential exist for the velocity field given in Prob. 1.25?
b. Obtain a stream function for the velocity field in (*a*) assuming an incompressible fluid.

7.11 A two-dimensional flow of an ideal incompressible fluid is represented between $\pi/4 > \theta > 0$ (that is, there is a solid wall for $\pi/4 \leq \theta \leq 2\pi$) by

$$\phi = C(x^4 - 6x^2y^2 + y^4)$$

where C is a dimensional constant.

a. Obtain the corresponding stream function.
b. Determine the velocity at (2, ½) and (2, 3/2), and show these on a sketch.

7.12 Obtain a stream function for an incompressible flow field given by

$$\phi = \frac{a^2 V_\infty}{r} \cos\theta$$

where a and V_∞ are positive constants.

7.13 *a.* If $V = f(y)\mathbf{i}$, does $\psi = g(x,y)$? Does $\psi = g(x)$? Does $\psi = g(y)$?
b. Show that $\mathbf{V} = f(x,y)\mathbf{i}$ is not physically possible for an incompressible fluid.

7.14 Given a velocity potential $\phi = -K\theta$, show that the corresponding stream function is $\psi = K \ln r$.

7.15 Show that $F_x = 0$ and $F_y = \Gamma \rho V_\infty$ in potential flow around a cylinder with circulation, where F_x and F_y are the resultant components of force exerted by the fluid on the cylinder.

7.16 *a.* For the potential flow described in Section 7.8, obtain an algebraic expression for the circulation in the fluid using a circular path that includes the cylinder.
b. Show that the circulation is zero for potential flow around the cylinder described in Example 7.6.

7.17 Given a velocity potential for a line source, $\phi = -K \ln r$, obtain a corresponding stream function and evaluate the circulation in the fluid around a closed circular path about the origin. K is a positive constant.

7.18 A steady potential flow is represented by combination of a sink and a vortex for which $\phi = K \ln r - K\theta$ with $K = 30\ \text{ft}^3/\text{sec}$. Calculate the pressure in lb_f/in^2 at $(2, \pi/2)$ if $p = 2000\ \text{lb}_f/\text{ft}^2$ at $(1, 0)$ for a liquid density of $60\ \text{lb}_m/\text{ft}^3$

7.19 Consider a half-body with a surface of $\psi = 0$ as discussed in Example 7.11.
a. Sketch the surface pressure on the half-body as a function of x for $-\infty < x < \infty$ with $p_\infty = 0$.
b. What is the value of circulation for the flow in the neighborhood of a half-body?

7.20 Obtain Eq. (7.40) beginning with Eq. (7.39).

7.21 Consider a potential flow for

$$\phi = -V_\infty r \sin\theta - K \ln r$$

where V_∞ and K are positive constants.
a. Locate any stagnation points.
b. Find the equation of the streamline through each of these.
c. Sketch several streamlines and equipotential lines.

7.22 Consider a potential flow

$$\phi = -V_\infty r \cos\theta + \frac{A\cos\theta}{r}$$

in which A is a positive constant.
a. Find a corresponding stream function for a two-dimensional flow of an incompressible fluid.
b. Locate any stagnation points and evaluate ψ at these points.
c. Sketch several streamlines for $\psi \le -2(V_\infty A)^{1/2}$.
d. What simple flow fields have been superposed?

7.23 Consider a potential flow given by

$$\phi = -V_\infty r \sin\theta + \frac{A\cos\theta}{r}$$

a. Find a corresponding stream function for a two-dimensional flow of an incompressible fluid.
b. Locate any stagnation points and evaluate ψ at these points.
c. Sketch several streamlines for

$$\psi \le -(2V_\infty A)^{1/2} \quad \text{and} \quad \psi \ge (2V_\infty A)^{1/2}$$

7.24 Tabulate the stream function ψ and the velocity potential ϕ, and make a brief sketch of the flow field for

a. Uniform flow in a positive x direction.
b. Uniform flow in a positive y direction.
c. A vortex with clockwise motion.
d. A line source.
e. A line sink.
f. A doublet.
g. A source and a sink of equal strength and a distance of $2a$ apart on the x axis.
h. A pair of vortices of equal strength and a distance of $2a$ apart on the x axis with opposite circulation.

7.25 Use superposition of several potential fields to establish a spiral flow that yields a logarithmic path of $C\theta = \ln r$ where C is a positive constant.

7.26 Potential flow fields are superposed to obtain

$$\phi = V_\infty r \cos\theta - K \ln r_1 + K \ln r_2$$

where r_1 and r_2 are described as in Fig. 7.16.

a. Obtain a comparable expression for the stream function.
b. Show that there are two stagnation points along the x axis.
c. Sketch the streamline that passes through the stagnation points.
d. Sketch several streamlines that represent the flow field around a body.

7.27 A stream function is given by

$$\psi = \sin\frac{y}{H}\sinh\frac{x}{H}$$

for the region $x \geq 0$ and $H\pi \geq y \geq 0$. H is a constant.

a. Does this function satisfy continuity and irrotational criteria?
b. Locate any stagnation points in the flow field.
c. Sketch several stream lines, including one through the stagnation point(s).

8

SIMPLIFIED APPROACH TO FLUID FLOW PROBLEMS

8.1 Introduction

At this point, the reader has realized the difficulty involved in the mathematical solution of viscous flow problems. A natural question that begs answering is "Does there exist some simple and rational way in which fluid mechanics problems can be resolved; and, if so, does this method lend itself to experimental interpretation?" In the subsequent sections, it will be found that affirmative answers may be given to both parts of the question.

First it will be assumed that the differential equations governing the flow are known. These equations will be put into a nondimensional form, and by an elementary argument some basic similarity laws will be uncovered to enable modeling the flow. The use of models to maintain similarity of flows is frequently employed in wind-tunnel work. In such experiments, data are taken which pertain to the model, and this information is used to predict corresponding parameters that apply to the full-scale design or prototype. The civil engineer will take flow measurements from a model of a dam before the structure is actually erected, since discovery of weak features in design or undesirable flow patterns are less expensively corrected in the model than in the full-scale structure. Models of ocean-going vessels are towed in a towing basin for the purpose of predicting performance of the prototype ship. These are just a few examples of the use of models and similarity laws in engineering fluid mechanics.

The second portion of this chapter will examine a rational approach to the formulation of dimensionless (and similarity) parameters without reference to the differential equations governing the flow. The final portion treats compressible and open-channel flows as well as simplified frictional flows by application of a one-dimensional assumption.

8.2 Similarity from a Differential Equation

As a point from which to argue, consider the flow field around a fixed obstacle placed within a fluid that is otherwise uniform at a speed V_∞ as shown in Fig. 8.1. Let L be a length that characterizes a dimension of the obstacle, and assume that the fluid is Newtonian with a steady uniform density and viscosity. The appropriate forms of the Navier-Stokes equations are (4.20*a*, *b*, and *c*). Considering only Eq. (4.20*a*) for the time being, the equation reduces to

$$V_x \frac{\partial V_x}{\partial x} + V_y \frac{\partial V_x}{\partial y} + V_z \frac{\partial V_x}{\partial z} + \frac{\partial V_x}{\partial t} = -\frac{1}{\rho}\frac{\partial p}{\partial x} + \nu\left(\frac{\partial^2 V_x}{\partial x^2} + \frac{\partial^2 V_x}{\partial y^2} + \frac{\partial^2 V_x}{\partial z^2}\right) + (\mathbf{f}_B)_x \qquad (8.1a)$$

This equation is next put into nondimensional form by introducing the following change of variables:

$$x = Lx' \qquad y = Ly' \qquad z = Lz' \qquad V_x = V_\infty V'_x \qquad V_y = V_\infty V'_y$$

$$V_z = V_\infty V'_z \qquad p = p_\infty p'$$

in which p_∞ is the pressure at large distances from the obstacle. The quantities with the primes are dimensionless variables. The time variable and the body force per unit mass may be expressed in terms of dimensionless quantities by

$$t = \tau t' \qquad \text{and} \qquad (\mathbf{f}_B)_x = g(\mathbf{f}'_B)_x$$

in which τ is some characteristic time; for example, it could correspond to a period of oscillation of the small obstacle. The use of a characteristic time becomes necessary only in the case of an unsteady velocity field. For the steady case, $\partial V_x/\partial t$ in Eq. (8.1*a*) is zero. The g in the body-force expression is simply the weight per unit mass which characterizes the usual body force considered in elementary fluid mechanics. Each term of Eq. (8.1*a*) is now

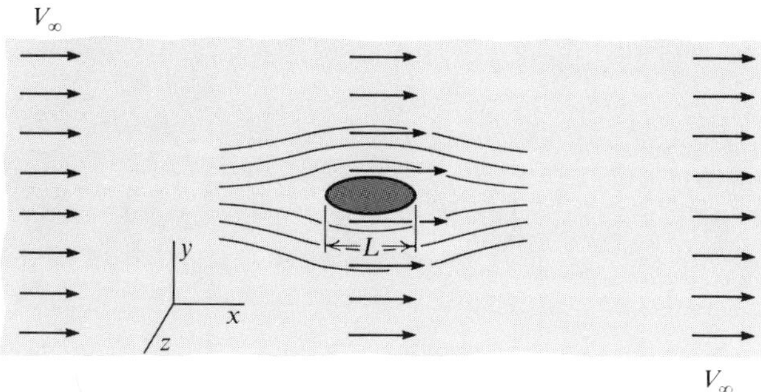

Fig. 8.1 *A small obstacle in a uniform flow field.*

transformed to the variables with primes. Considering the characterizing quantities L, V_∞, p_∞, τ, and g to be constant, the procedure is simple and straightforward, and it is typified by the following operations:

$$V_x \frac{\partial V_x}{\partial x} = V_\infty V'_x \frac{\partial (V_\infty V'_x)}{\partial x} = V_\infty{}^2 V'_x \frac{\partial V'_x}{\partial x}$$

$$= V_\infty{}^2 V'_x \frac{\partial V'_x}{\partial x'} \frac{\partial x'}{\partial x}$$

From $x = Lx'$,

$$\frac{\partial x'}{\partial x} = \frac{1}{L}$$

and

$$V_x \frac{\partial V_x}{\partial x} = \frac{V_\infty{}^2}{L} V'_x \frac{\partial V'_x}{\partial x'}$$

The unsteady term $\partial V_x/\partial t$ is transformed accordingly,

$$\frac{\partial V_x}{\partial t} = \frac{\partial (V_\infty V'_x)}{\partial t} = V_\infty \frac{\partial V'_x}{\partial t} = V_\infty \frac{\partial V'_x}{\partial t'} \frac{\partial t'}{\partial t}$$

From $t = \tau t'$,

$$\frac{\partial t'}{\partial t} = \frac{1}{\tau}$$

and

$$\frac{\partial V_x}{\partial t} = \frac{V_\infty}{\tau} \frac{\partial V'_x}{\partial t'}$$

Similarly,

$$-\frac{1}{\rho}\frac{\partial p}{\partial x} = -\frac{p_\infty}{\rho L}\frac{\partial p'}{\partial x'}$$

$$\nu\left(\frac{\partial^2 V_x}{\partial x^2} + \frac{\partial^2 V_x}{\partial y^2} + \frac{\partial^2 V_x}{\partial z^2}\right) = \frac{\nu V_\infty}{L^2}\left(\frac{\partial^2 V_x'}{\partial x'^2} + \frac{\partial^2 V_x'}{\partial y'^2} + \frac{\partial^2 V_x'}{\partial z'^2}\right)$$

and

$$(\mathbf{f}_B)_x = g(\mathbf{f}_B')_x$$

Substituting for each of the terms in Eq. (8.1*a*) and dividing both sides by $V_\infty{}^2/L$ yields

$$V_x'\frac{\partial V_x'}{\partial x'} + V_y'\frac{\partial V_x'}{\partial y'} + V_z'\frac{\partial V_x'}{\partial z'} + \frac{L}{\tau V_\infty}\frac{\partial V_x'}{\partial t'} = -\frac{p_\infty}{\rho V_\infty{}^2}\frac{\partial p'}{\partial x'}$$
$$+ \frac{\nu}{LV_\infty}\left(\frac{\partial^2 V_x'}{\partial x'^2} + \frac{\partial^2 V_x'}{\partial y'^2} + \frac{\partial^2 V_x'}{\partial z'^2}\right) + \frac{Lg}{V_\infty{}^2}(\mathbf{f}_B')_x \quad (8.1b)$$

Equation (8.1*b*) is the nondimensional form of Eq. (8.1*a*), and the physical significance of the coefficients of the various terms may be discerned by reference to each of the terms before they were transformed to their respective dimensionless forms. For example, $V_x\dfrac{\partial V_x}{\partial x}$, $\dfrac{\partial V_x}{\partial t}$, $\dfrac{1}{\rho}\dfrac{\partial p}{\partial x}$, etc., are each dimensionally equivalent to a force per unit mass. Since $p_\infty/\rho V_\infty{}^2$ is the coefficient of the dimensionless term $-\partial p'/\partial x'$, this coefficient must be the ratio of the pressure force to the inertia force. This ratio occurs quite often in the analysis of problems and is called the *Euler number*. The dimensionless coefficients (or their reciprocals) in Eq. (8.1*b*) and the name and interpretation of each are:

Euler number $N_E = \dfrac{p_\infty}{\rho V_\infty{}^2}$ = ratio of pressure force to inertia force

Reynolds number $N_R = \dfrac{LV_\infty}{\nu} = \dfrac{L\rho V_\infty}{\mu}$ = ratio of inertia force to viscosity force

Froude number $N_F = \dfrac{V_\infty{}^2}{Lg}$ = ratio of inertia force to gravity force

Strouhal number $N_S = \dfrac{\tau V_\infty}{L}$ = ratio of inertia force to force associated with period τ

It is apparent that Eqs. (4.20*b* and *c*) applied to an incompressible fluid may be placed in nondimensional forms similar to Eq. (8.1*b*). The continuity equation (2.3) can also be expressed as

$$\frac{\partial V'_x}{\partial x'} + \frac{\partial V'_y}{\partial y'} + \frac{\partial V'_z}{\partial z'} = 0 \tag{8.2}$$

If a family of flows with uniform and steady density and with geometrically similar boundaries is considered, while the Euler, Reynolds, Froude, and Strouhal numbers are each held constant for all flows of the family, the solution of the nondimensional differential equations for V'_x, V'_y, V'_z, and p' will be the same for each flow of the family. Since the dimensionless velocities are the same, these flows are said to possess *kinematic similarity*.

If the constitutive equations (4.18*a* to *f*) are transformed to a nondimensional form using $\sigma_{nn} = \rho V_\infty{}^2 \sigma'_{nn}$ for the normal stresses, and $\tau_{nt} = \rho V_\infty{}^2 \tau'_{nt}$ for the shear stresses, then for $\operatorname{div} \mathbf{V} = 0$:

$$\sigma'_{xx} = -N_E p' + \frac{2}{N_R} \frac{\partial V'_x}{\partial x'} \tag{8.3a}$$

$$\sigma'_{yy} = -N_E p' + \frac{2}{N_R} \frac{\partial V'_y}{\partial y'} \tag{8.3b}$$

$$\sigma'_{zz} = -N_E p' + \frac{2}{N_R} \frac{\partial V'_z}{\partial z'} \tag{8.3c}$$

$$\tau'_{xy} = \frac{1}{N_R}\left(\frac{\partial V'_y}{\partial x'} + \frac{\partial V'_x}{\partial y'}\right) \tag{8.3d}$$

$$\tau'_{xz} = \frac{1}{N_R}\left(\frac{\partial V'_z}{\partial x'} + \frac{\partial V'_x}{\partial z'}\right) \tag{8.3e}$$

$$\tau'_{yz} = \frac{1}{N_R}\left(\frac{\partial V'_z}{\partial y'} + \frac{\partial V'_y}{\partial z'}\right) \tag{8.3f}$$

The corresponding nondimensional stresses for each flow of the family are equal since the Reynolds number was held equal for each flow of the family, as was the Euler number, to preserve kinematic similarity. Since the stresses given by Eqs. (8.3) are the nondimensional surface forces per unit area, these are the same for all flows at the same point given by the dimensionless coordinates x', y', and z'. If the Froude number is held constant for all the flows, it follows that the nondimensional body force per unit mass is the same for all the flows of the family. This equality of nondimensional forces is referred to as *dynamic similarity*.

From the preceding discussion it is seen that equating the Reynolds, Froude, Strouhal, and Euler numbers for each flow of the family ensures kinematic and dynamic similarity for the incompressible flow of a Newtonian fluid. The similarity concepts provide a partial answer to the question posed in the introduction to this chapter in the sense that the nondimensional coefficients of the terms in the nondimensional differential equations may be measured in the laboratory by using a model geometrically similar to a full-scale prototype, and this information may be applied to predict corresponding variables in the full-scale flow. The answer is partial since the differential equation must be known before it can be placed in nondimensional form. Note that a form of the energy equation when placed in the x', y', z' variables would yield other dimensionless groups. Concerning this last comment, the reader is referred to books that treat dimensional analysis and similitude as specialized subjects.

An interesting observation concerning the Euler number for an incompressible flow is that any pressure, other than p_∞, may be employed in the transformation $p = p_\infty p'$ since the pressure level[1] does not affect the flow of an incompressible fluid. Indeed, the pressure difference is the important influence here [see Eq. (8.1a)] and the datum or reference pressure would "subtract out" in the difference of two pressures. A convenient selection of reference pressure could be ρV_∞^2, and the Euler number would become unity.

The following examples will illustrate some of the points discussed in this section.

Example 8.1

A model of an airplane wing is being tested in a wind tunnel. The model scale is 1:20. The plane flies at a speed of 100 mph in air at a pressure and temperature of 14.7 $\text{lb}_f/\text{in.}^2$ and 60°F, respectively. The tunnel air-supply temperature is 60°F. If it is desired to maintain a tunnel speed equal to flight speed, calculate the tunnel air pressure.

Solution. If the plane flies at a steady speed relative to the air, the flow may be modeled without consideration of the Strouhal number. Note that N_S is associated with the unsteady term in Eq. (8.1b). Body forces will be omitted on the basis that they contribute only a small amount to the variation of surface forces over the wing. This means that the Froude number does not have to be considered. To maintain kinematic and dynamic similarity under these circumstances, it is only necessary to equate tunnel and flight Reynolds numbers, that is,

$$\frac{L_M V_M \rho_M}{\mu_M} = \frac{L_P V_P \rho_P}{\mu_P}$$

[1]This does not imply that if the absolute pressure of a liquid is low enough, the liquid will not vaporize; for if the pressure is lowered below the saturation pressure corresponding to the temperature of the liquid, bubbles of vapor will form. This is referred to as *cavitation*.

in which the subscripts M and P denote the model and prototype, respectively. The densities may be expressed[1] as $\rho = p/RT$, using the equation of state for a perfect gas; and if μ is assumed to be a function of temperature only, then $\mu_M = \mu_P$. Note that the given temperature information assures that $(N_E)_M = (N_E)_P$, since p/ρ is proportional to T. The Reynolds number equation then yields

$$p_M = p_P \frac{L_P}{L_M} = (14.7 \text{ lb}_f/\text{in}^2)(20)$$

$$= 294 \text{ lb}_f/\text{in}^2 \qquad ◀$$

Example 8.2

What value of tunnel speed and pressure would be necessary to preserve dynamic similarity in Example 8.1 if a temperature of 100°F is maintained in the tunnel ahead of the fluid and Freon-12 vapor is used as a fluid in the model?

Solution. Equating the Euler number for model and prototype, and assuming perfect-gas behavior,

$$\left(\frac{p}{\rho V_\infty^2}\right)_M = \left(\frac{p}{\rho V_\infty^2}\right)_P \qquad \left(\frac{RT}{V_\infty^2}\right)_M = \left(\frac{RT}{V_\infty^2}\right)_P$$

$$V_{\infty M} = V_{\infty P}\left(\frac{R_M T_M}{R_P T_P}\right)^{1/2}$$

Using physical data from Appendix E,

$$V_{\infty M} = 100\left[\frac{(12.8)(560)}{(53.3)(520)}\right]^{1/2} = 50.9 \text{ mph} \qquad ◀$$

Next, matching Reynolds numbers,

$$\left(\frac{\rho V_\infty L}{\mu}\right)_M = \left(\frac{\rho V_\infty L}{\mu}\right)_P \qquad \left(\frac{p V_\infty L}{RT\mu}\right)_M = \left(\frac{p V_\infty L}{RT\mu}\right)_P$$

$$p_M = p_P \frac{R_M T_M \mu_M V_{\infty P} L_P}{R_P T_P \mu_P V_{\infty M} L_M}$$

and with physical data from Appendix E,

$$p_M = 14.7\frac{(12.8)(560)(2.57)(10^{-7})(100)(20)}{(53.3)(520)(3.75)(10^{-7})(50.9)(1)} = 102 \text{ lb}_f/\text{in}^2 \qquad ◀$$

Example 8.3

It is known that electric power transmission lines vibrate when exposed to the wind.

[1]Even though air is a compressible fluid, for low fluid speeds the flow of gases may be treated on an incompressible basis if pressure changes are not large in the direction of flow.

This action is due in part to the shedding of vortices in the wake side of the wire, and the flow is characterized by the Strouhal and Reynolds numbers. It is proposed to model a flow that has a speed of 30 ft/sec over a wire of 0.075-in. diameter using air with the same pressure and temperature flowing at 10 ft/sec. What should be the diameter of the wire used for the model? Predict the frequency of vibration of the actual transmission line if the measured frequency for the model is found to be 1120 cycles/sec.

Solution. Equating the Reynolds numbers of the model and prototype for dynamic similarity yields

$$D_M = \frac{D_P V_P}{\nu_P} \frac{\nu_M}{V_M} = D_P \frac{V_P}{V_M}$$

$$= (0.075 \text{ in.})\left(\frac{30 \text{ ft/sec}}{10 \text{ ft/sec}}\right) = 0.225 \text{ in.} \quad \blacktriangleleft$$

Note that the diameters of the wires have been taken as characterizing dimensions. Next, equating the Strouhal numbers gives

$$\frac{\tau_M V_M}{D_M} = \frac{\tau_P V_P}{D_P}$$

Since the period τ is the reciprocal of the frequency,

$$f_P = \left(\frac{V_P}{V_M}\right)\left(\frac{D_M}{D_P}\right) f_M$$

in which f denotes frequency; and

$$f_P = \left(\frac{30 \text{ ft/sec}}{10 \text{ ft/sec}}\right)\left(\frac{0.225 \text{ in.}}{0.075 \text{ in.}}\right) 1120 \text{ cycles/sec}$$

$$= 10{,}080 \text{ cycles/sec} \quad \blacktriangleleft$$

Another example of Strouhal modeling can be found in the design of suspension bridges, if there is a possibility of aerodynamic forces due to wind being in resonance with the elastic forces in the bridge. The reader is probably aware of the well-known "aerodynamic failure" of the Tacoma Narrows suspension bridge.[1] In the western plains of the United States, telephone wires are frequently weighted by a large mass hanging from the line between each pair of poles. This added mass prevents the elastic forces in the wire from resonating with the aerodynamic forces.

Example 8.4

The drag on an ocean vessel is of two recognizable forms: the viscous friction which is a consequence of the motion of the ship relative to the water, and the resistance offered by the wave pattern of the ocean surface through which the ship is moving. The restoring

[1] Failure of the Tacoma Narrows Bridge, *Am. Soc. Civil Engr. Proc.*, vol. 69, no. 10, pp. 1555–1586, 1943.

force of the wave pattern is gravity. What must be the model scale for the preservation of kinematic and dynamic similarity if the model is towed in a basin that contains ocean water?

Solution. Since viscous and gravity forces are both involved, the Reynolds and Froude numbers are used in the modeling. The Reynolds number equation is

$$\frac{L_M V_M}{\nu_M} = \frac{L_P V_P}{\nu_P}$$

from which

$$\frac{L_M}{L_P} = \frac{V_P}{V_M} \qquad (i)$$

since $\nu_M \simeq \nu_P$. Equating the Froude numbers gives

$$\frac{V_M^2}{L_M g_M} = \frac{V_P^2}{L_P g_P}$$

and

$$\frac{V_P}{V_M} = \left(\frac{L_P}{L_M}\right)^{1/2} \qquad (ii)$$

since g does not vary "appreciably" over the surface of the earth. Satisfaction of both Eqs. (*i*) and (*ii*) requires

$$L_M = L_P$$

and the model scale is 1:1, so that the model is the same size as the prototype. ◀

Since it would be quite expensive to build a "full" scale model, the thought of towing the model in a fluid with a different kinematic viscosity appears to be a natural and constructive suggestion. Trying this approach to the problem would require

$$\frac{L_M}{L_P} = \left(\frac{\nu_M}{\nu_P}\right)^{2/3}$$

If the model were towed in mercury at room temperature, the geometric-scale factor L_M/L_P would be approximately

$$\left(\frac{\nu_M}{\nu_P}\right)^{2/3} \simeq \left[\frac{3.25(10^{-5})/845}{1.67(10^{-5})/62.2}\right]^{2/3} = 0.275$$

An observation of the preceding would indicate that there is not much of a reduction in the size of the model; furthermore, mercury would hardly be safe or inexpensive to use in model studies. Since there are no liquids well suited for this purpose, the model studies are generally conducted by duplicating the Froude number and then applying a Reynolds number correction. This is referred to as *incomplete similarity*, and it is not the intention of this book to delve into this subject. The interested reader may consult some of the references listed in the bibliography for further details.

8.3 Dimensional Analysis

If the form of the differential equation is unknown, there is another method whereby the similarity parameters may be ascertained. This alternate approach to physical problems is known as the *method of dimensional analysis*, and it requires some experimental knowledge concerning the variables involved in the problem before the method may be employed.

Dimensional analysis is based on a very simple principle, *dimensional homogeneity*. This principle requires that each term in an equation representing a physical problem must have the same dimension as all the other terms of the equation. For example, consider the simple equation for the distance s a particle travels in time t, if the initial velocity is V_0, and the acceleration a is constant. Then

$$s = V_0 t + \frac{1}{2} a t^2 \tag{8.4}$$

and it is noted that the terms on both sides of the equation have the dimension of length, so Eq. (8.4) is dimensionally homogeneous.

Before illustrating the method of dimensional analysis, it will be convenient to recognize that problems in elementary fluid mechanics require only four basic dimensions for a description of the variables involved. These are the dimensions force F, length L, time t, and temperature T; or the dimension mass M could be used in place of force. There would be little need to use both force and mass dimensions in the same problem since Newton's law ($F = Ma$) defines a dimension of force in terms of mass and acceleration dimensions. The dimensional equation relating these is

$$F \overset{D}{=} M \frac{L}{t^2}$$

in which $\overset{D}{=}$ means dimensionally equivalent. Table 8.1 lists a number of physical variables employed in fluid mechanics problems and their corresponding dimensions in the F, L, t, T and M, L, t, T systems of dimensions. To convert from a system involving a force dimension to one with a mass dimension, Newton's law in the form of $Ma/F = 1$ is used as a dimensional conversion factor, for example, consider pressure,

$$\frac{F}{L^2} \overset{D}{=} \frac{F}{L^2} \frac{ML}{Ft^2} \overset{D}{=} \frac{M}{Lt^2}$$

With the preceding preliminaries it is possible to illustrate a method of dimensional analysis. Assume that an investigator knows from experimental data that the drag force F_D on a high-speed airfoil depends on the air density ρ, the air viscosity μ, the modulus of elasticity of the air E, the speed of the

Table 8.1 Dimensions

Variables	F, L, t, T dimensions	M, L, t, T dimensions
Time	t	t
Length	L	L
Area	L^2	L^2
Volume	L^3	L^3
Velocity	L/t	L/t
Acceleration	L/t^2	L/t^2
Force	F	ML/t^2
Mass	Ft^2/L	M
Pressure, stress, or modulus of elasticity	F/L^2	M/Lt^2
Density	Ft^2/L^4	M/L^3
Viscosity	Ft/L^2	M/Lt
Kinematic viscosity	L^2/t	L^2/t
Surface tension	F/L	M/t^2
Work	FL	ML^2/t^2
Power	FL/t	ML^2/t^3
Torque or moment	FL	ML^2/t^2
Temperature	T	T
Heat	FL	ML^2/t^2
Specific heat	L^2/t^2T	L^2/t^2T
Thermal conductivity	F/tT	ML/t^3T

foil relative to the undisturbed air ahead of the foil V, and the length of the foil l. It is desired to establish the similarity parameters (the dimensionless groups or numbers) that apply to the problem. The drag force is first formulated as a function of the variables involved, that is,

$$F_D = F_D(l,\rho,V,\mu,E) \tag{8.5}$$

This function is assumed to be a form of a power series in the independent variables and is written as

$$F_D = C_1 l^{\alpha_1}\rho^{\beta_1}V^{\gamma_1}\mu^{\delta_1}E^{\varepsilon_1} + C_2 l^{\alpha_2}\rho^{\beta_2}V^{\gamma_2}\mu^{\delta_2}E^{\varepsilon_2} + \cdots + C_n l^{\alpha_n}\rho^{\beta_n}V^{\gamma_n}\mu^{\delta_n}E^{\varepsilon_n} \tag{8.6}$$

in which the C's are dimensionless constants or, simply, pure numbers. Each of the terms in the series must have the dimensions of force using the principle of dimensional homogeneity and noting that F_D on the left-hand side of Eq. (8.6) is a force. Considering the nth term of Eq. (8.6),

$$F \overset{D}{=} L^{\alpha_n}\left(\frac{Ft^2}{L^4}\right)^{\beta_n}\left(\frac{L}{t}\right)^{\gamma_n}\left(\frac{Ft}{L^2}\right)^{\delta_n}\left(\frac{F}{L^2}\right)^{\varepsilon_n}$$

Equating the exponents of the dimensions F, L, and t on both sides of the equation to ensure dimensional homogeneity yields

$$\text{Force} \quad 1 = \beta_n + \delta_n + \varepsilon_n$$

$$\text{Length} \quad 0 = \alpha_n - 4\beta_n + \gamma_n - 2\delta_n - 2\varepsilon_n$$

$$\text{Time} \quad 0 = 2\beta_n - \gamma_n + \delta_n$$

Solving for any three of the exponents in terms of the remaining two (say, δ_n and ε_n) gives

$$\alpha_n = 2 - \delta_n$$

$$\beta_n = 1 - \delta_n - \varepsilon_n$$

$$\gamma_n = 2 - \delta_n - 2\varepsilon_n$$

Inserting these expressions for the exponents into $C_n l^{\alpha_n} \rho^{\beta_n} V^{\gamma_n} \mu^{\delta_n} E^{\varepsilon_n}$ provides the following for the nth term,

$$C_n l^{2-\delta_n} \rho^{1-\delta_n-\varepsilon_n} V^{2-\delta_n-2\varepsilon_n} \mu^{\delta_n} E^{\varepsilon_n}$$

and grouping the preceding according to exponents yields

$$C_n l^2 \rho V^2 \left(\frac{\mu}{l\rho V}\right)^{\delta_n} \left(\frac{E}{\rho V^2}\right)^{\varepsilon_n}$$

Similarly, the first term could be arranged as

$$C_1 l^2 \rho V^2 \left(\frac{\mu}{l\rho V}\right)^{\delta_1} \left(\frac{E}{\rho V^2}\right)^{\varepsilon_1}$$

By arranging all terms of the series it is possible to write Eq. (8.6) as

$$\frac{F_D}{l^2 \rho V^2} = f\left(\frac{\mu}{l\rho V}, \frac{E}{\rho V^2}\right) \tag{8.7}$$

which expresses the dimensionless group on the left-hand side[1] as a function of the two dimensionless groups $\mu/l\rho V$ and $E/\rho V^2$. The term $\mu/l\rho V$ is recognized as the reciprocal of the Reynolds number. The dimensionless group $\rho V^2/E$ is called the *Cauchy number*, N_C; it is the ratio of an inertia force to an elastic force. Equation (8.7) could be written as

$$C_D = g(N_R, N_C) \tag{8.8}$$

where g is another function.

[1]The group $F_D/l^2\rho V^2$ is sometimes written in the dimensionless form $F_D/(\frac{1}{2})(\rho V^2 S)$ in which S is the projected area of the foil as viewed from the approaching stream. This latter form is referred to as a *drag coefficient*, C_D.

The advantages of dimensional analysis become apparent at this point. Equation (8.5) is a relation among six variables; while Eq. (8.8) has three dimensionless groups: C_D, N_R, and N_C. It is obvious that varying two parameters in a given experiment is certainly simpler than varying five. However, the experimentation is only part of the difficulty involved with six variables. There is also the matter of presenting the experimental results in some useful and simple manner. Equation (8.8) could be represented as shown in Fig. (8.2). An important use of dimensional analysis is the organization of variables into dimensionless groups, reducing the amount of experimentation necessary with a corresponding simplification in the presentation of experimental results. Another advantage is that the corresponding dimensionless groups obtained may be equated for similarity considerations. Equation (8.8) indicates that equal Reynolds numbers and equal Cauchy numbers for model and prototype would ensure equal drag coefficients because of the functional dependence implied by the equation.

The formalization of the results of dimensional analysis is given by the *Buckingham Pi theorem*[1] which may be stated as follows: *If an equation in n variables is dimensionally homogeneous with respect to m fundamental dimensions, it can be expressed as a relation between a minimum of n − m independent dimensionless groups.* The proof of this theorem requires some knowledge of the theory of equations and matrices, and will not be presented here; however, the interested reader familiar with the theory of equations is referred to the excellent paper by Brand[2] which discusses the Pi theorem and its subsequent development.

Applying the theorem to the drag-force problem of this section gives $n = 6$ and $m = 3$, and $n - m = 3$, which is the number of independent dimensionless groups. Note that each of the three groups in Eq. (8.8) is

[1] E. Buckingham, On Physically Similar Systems; Illustrations of the Use of Dimensional Equations, *Phys. Rev.*, vol. 4, no. 4, pp. 345–378, 1914.

[2] Louis Brand, The Pi Theorem of Dimensional Analysis, *Arch. Ration. Mech. Anal.*, vol. 1, no. 1, pp. 35–45, 1957.

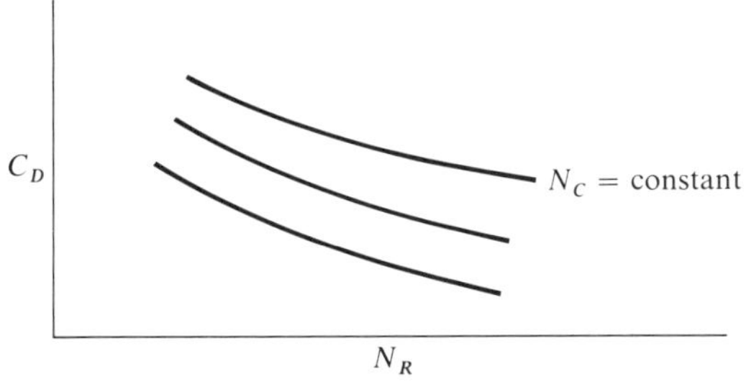

Fig. 8.2 *Variation of drag coefficient with Reynolds number using Cauchy number as a parameter.*

indeed independent since F_D, μ, and E occur only in C_D, N_R, and N_C, respectively; and the theorem predicted a minimum of three independent dimensionless groups.

A case will now be considered which will involve more groups than the minimum called for by the Pi theorem. To illustrate this situation, consider that the maximum torsional shear stress, τ_{nt}, in a circular shaft is known to be a function of the torque T applied to the shaft, the polar moment of inertia of the cross-sectional area of the shaft J, and the radius of the shaft c. A dimensional analysis of this problem would start with the functional formulation

$$\tau_{nt} = \tau_{nt}(T,J,c) = C_1 T^{\alpha_1} J^{\beta_1} c^{\gamma_1} + C_2 T^{\alpha_2} J^{\beta_2} c^{\gamma_2} + \cdots + C_n T^{\alpha_n} J^{\beta_n} c^{\gamma_n}$$

and the nth term would have the dimensions of stress. Substitution of dimensions yields

$$\frac{M}{Lt^2} \stackrel{D}{=} \left(\frac{ML^2}{t^2}\right)^{\alpha_n} (L^4)^{\beta_n} L^{\gamma_n}$$

using the M, L, t system. Setting up the equations involving the exponents gives

$$\begin{aligned} &\text{Mass} && 1 = \alpha_n \\ &\text{Length} && -1 = 2\alpha_n + 4\beta_n + \gamma_n \\ &\text{Time} && -2 = -2\alpha_n \end{aligned} \tag{8.9}$$

from which

$$\alpha_n = 1 \qquad \text{and} \qquad \gamma_n = -3 - 4\beta_n$$

Substituting for the exponents in the nth term, $C_n T^{\alpha_n} J^{\beta_n} c^{\gamma_n}$, and grouping according to exponents, the following accrues:

$$C_n \frac{T}{c^3}\left(\frac{J}{c^4}\right)^{\beta_n}$$

As in the previous problem, each term would be of the form of the nth term, so that the series could be written as

$$\frac{\tau_{nt} c^3}{T} = f\left(\frac{J}{c^4}\right)$$

in functional form, and there are two independent dimensionless groups. For this problem, the Pi theorem would predict a minimum of one group since $n = 4$ and $m = 3$. To discern why one more group was obtained than that predicted by the theorem, reference is made to the set of three equations identified as (8.9). Of these three, only two are independent, as the third may be obtained by multiplying the first by the constant, minus two. Since only two of the equations are independent, it was impossible to solve for the three quantities α_n, β_n, and γ_n. Instead, only two of these could be resolved.

If the F, L, t system of dimensions had been used instead of the M, L, t system, dimensional homogeneity would have required

$$\frac{F}{L^2} \stackrel{D}{=} (FL)^{\alpha_n} L^{4\beta_n} L^{\gamma_n} \qquad \text{for the } n\text{th term}$$

The Pi theorem with $n = 4$ and $m = 2$ would predict a minimum of two independent dimensionless groups, and carrying out the problem again, in the F, L, t system, would produce two groups, the minimum number. The reader is invited to check this as an exercise.

There is a systematic method of finding the number of independent equations among a system of equations. This consists of determining the rank of the matrix formed from the coefficients of the variables. The rank of a matrix is the highest order nonvanishing determinant that can be formed from the elements of the matrix. The rank of this matrix is the number of independent equations.

Example 8.5

Find the number of independent equations that result from equating the exponents of force, length, and time in the drag-force problem of Section 8.3, and do the same for the system of Eqs. (8.9).

Solution. The three equations are

$$\beta_n + \delta_n + \varepsilon_n = 1$$

$$\alpha_n - 4\beta_n + \gamma_n - 2\delta_n - 2\varepsilon_n = 0$$

$$2\beta_n - \gamma_n + \delta_n = 0$$

The matrix formed from the coefficients of α_n, β_n, γ_n, δ_n, and ε_n is

$$\begin{vmatrix} 0 & 1 & 0 & 1 & 1 \\ 1 & -4 & 1 & -2 & -2 \\ 0 & 2 & -1 & 1 & 0 \end{vmatrix}$$

There are several third-order determinants that can be formed from the elements. Consider one obtained by deleting the last two columns, and

$$\begin{vmatrix} 0 & 1 & 0 \\ 1 & -4 & 1 \\ 0 & 2 & -1 \end{vmatrix} = 1 \neq 0$$

The rank of the matrix is three, and the three equations from which the matrix was formed are independent.

If the set of Eqs. (8.9) are used to form a matrix from the coefficients of α_n, β_n, and γ_n, the following results:

$$\begin{vmatrix} 1 & 0 & 0 \\ 2 & 4 & 1 \\ -2 & 0 & 0 \end{vmatrix}$$

The only third-order determinant that can be formed from the elements of this matrix is the one using all nine elements. The value of this determinant is zero, so the rank of the matrix is certainly not three. A second-order determinant that does not vanish may be obtained by deleting the third row and the third column:

$$\begin{vmatrix} 1 & 0 \\ 2 & 4 \end{vmatrix} = 4 \neq 0$$

so the rank of the matrix is two; and only two of the three equations of the set (8.9) are independent—an obvious fact, which has been previously noted.

Buckingham's Pi theorem may now be stated in a more precise form by removal of "a minimum of" and replacing $n - m$ by $n - r$, where r is the rank of the matrix associated with the relations formed by equating exponents. The word *Pi* in the theorem simply refers to each of the dimensionless groups. Equation (8.8) could be written as $\pi_1 = g(\pi_2, \pi_3)$ instead of $C_D = g(N_R, N_C)$. The symbol π is frequently used to denote a product for purposes of mathematical symbolism.

8.4 Flow of Liquids in Pipes

Dimensional analysis may be applied to many problems in fluid mechanics. The flow of a fluid with uniform and steady density in a constant-area passage, such as a pipe, is quite common throughout a considerable range of engineering applications. The transport of liquid in pipelines and other fluid circuits is frequently encountered. The case of fully developed laminar or turbulent flow lends itself to the methods of analysis presented in this section (fully developed laminar flow was discussed in Example 5.4).

The pressure drop per unit length of pipe, $\dfrac{\Delta p_L}{L}$, for fully developed flow is known to be a function of the space-average velocity V_{SA} of the flowing fluid; the fluid density and viscosity, ρ and μ; the diameter D of the conduit; and the average roughness e of the interior of the conduit. The last variable, e, is the height of the average protuberance inside of the pipe, and it has the dimension of length. Note that Δp_L is defined as the magnitude of *pressure decrease* in the direction of flow. Functionally, this problem may be represented by

$$\frac{\Delta p_L}{L} = g_1(V_{SA}, \rho, \mu, D, e)$$

If methods of Section 8.3 are used,

$$\frac{\Delta p_L}{L} = C_1 V_{SA}{}^{\alpha_1} \rho^{\beta_1} \mu^{\gamma_1} D^{\delta_1} e^{\varepsilon_1} + C_2 V_{SA}{}^{\alpha_2} \rho^{\beta_2} \mu^{\gamma_2} D^{\delta_2} e^{\varepsilon_2} + \cdots + C_n V_{SA}{}^{\alpha_n} \rho^{\beta_n} \mu^{\gamma_n} D^{\delta_n} e^{\varepsilon_n}$$

Supplying dimensions and considering the nth term, which must have the dimensions of $\Delta p_L/L$, yields the following dimensional equation,

$$\frac{M}{L^2 t^2} \overset{D}{=} \left(\frac{L}{t}\right)^{\alpha_n} \left(\frac{M}{L^3}\right)^{\beta_n} \left(\frac{M}{Lt}\right)^{\gamma_n} L^{\delta_n} L^{\varepsilon_n}$$

Equating the exponents of mass, length, and time gives, respectively,

$$1 = \beta_n + \gamma_n$$
$$-2 = \alpha_n - 3\beta_n - \gamma_n + \delta_n + \varepsilon_n$$
$$-2 = -\alpha_n - \gamma_n$$

The three equations are obviously independent; however, determining the rank of the matrix of the coefficients would provide a formal check. The Pi theorem would then predict a functional relationship among three dimensionless groups. Carrying out the solution for the exponents, say in terms of γ_n and ε_n, and regrouping according to exponents, permits writing the relationship

$$\frac{\Delta p_L D}{\rho L V_{SA}{}^2} = g_2\left(\frac{\rho D V_{SA}}{\mu}, \frac{e}{D}\right) \tag{8.10}$$

The first group on the right-hand side of Eq. (8.10) is recognized as the Reynolds number; the second group, e/D, is called the *relative roughness* of the pipe or conduit. Twice the group on the left-hand side is defined as the *friction factor*, noted by the symbol f:

$$f = \frac{2D\, \Delta p_L}{\rho L V_{SA}{}^2} \tag{8.11}$$

Equation (8.10) can be written functionally as

$$f = h\left(N_R, \frac{e}{D}\right) \tag{8.12}$$

Measurements of the variables in this equation can be taken in the laboratory and presented in a generalized form. Figure 8.3 shows a simple test setup that could be used for this purpose. The pressure drop $\Delta p_L = p_1 - p_2$ may be calculated from the manometer reading d. The space-average velocity could be obtained by "weighing" the amount of fluid discharged to a weighing tank during a measured time interval. From the mass rate of flow, the density of the fluid, and the cross-sectional area through which the fluid flows, V_{SA} could be calculated using $V_{SA} = \dot{m}/\rho A$. The distance L between the manometer connections at 1 and 2 is a simple matter to ascertain. The viscosity and density of the fluid would have to be taken from tables of physical data for the fluid at its temperature, or the investigator would need a viscometer and a pycnometer bottle or gravity balance to determine these variables. The roughness or relative roughness can be found from information supplied by the manufacturer of the conduit. Some roughness data are given in Fig. 8.4. The variable e is somewhat of an uncertainty unless the conduit is relatively new, since some liquids will foul and corrode the interior of the walls. Again, uncertainties such as these are left to the judgment of the engineer using the best information he can obtain. The experiments suggested by the test setup in Fig. 8.3 would be carried out by making runs at different flow rates to vary V_{SA} and by using different fluids at the various flow rates to vary viscosity and density. In order to vary the relative roughness, different test sections could be used. Figure 8.4 presents the results of many experiments[1] similar to one just described. Some salient features of these results are:

1. The region of *laminar* flow prevails up to a Reynolds number of approximately 2100, and

$$f = \frac{64}{N_R} \qquad N_R < 2100 \tag{8.13}$$

[1]Lewis F. Moody, Friction Factors for Pipe Flow, *Trans. ASME*, vol. 66, Nov., pp. 671–684, 1944. By permission.

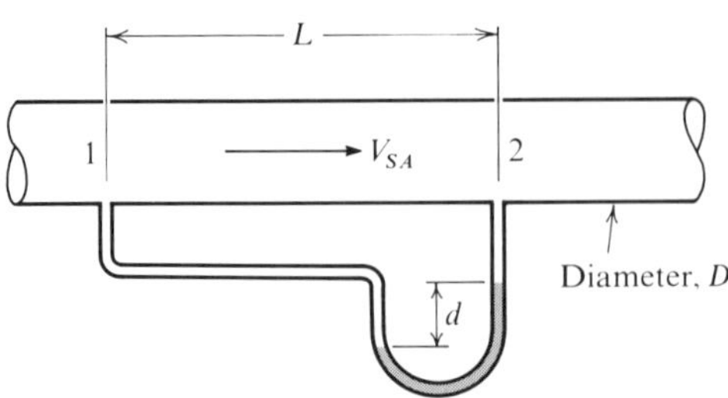

Fig. 8.3 *A test setup for determination of the variation of the friction factor with the Reynolds number and relative roughness.*

The friction factor is independent of the relative roughness for laminar flow.
2. There is a *transition* region for $N_R > 2100$, which extends to some upper limit of N_R and depends on the relative roughness.[1] In this zone, or region, the friction factor is dependent on both the Reynolds number and the relative roughness.
3. There is a region of interest labeled *complete turbulence, rough pipes*. In this zone the friction factor is practically independent of the Reynolds number and depends only on the relative roughness of the pipe.
4. Figure 8.4 applies to incompressible Newtonian fluids. Equation (8.13) can be derived by using the Navier-Stokes equations. From results in Example 4.6, the expression for the space-average velocity can be written as

$$V_{SA} = -\frac{R^2}{8\mu}\frac{dp}{dz} = \frac{D^2}{32\mu}\frac{\Delta p_L}{L} \tag{8.14}$$

[1] For $2100 < N_R < 4000$, there is a critical zone in which it is difficult to infer generality from experimental data since extraneous disturbances play a role in the stability of the flow. The study of hydrodynamic stability is one for advanced consideration.

Fig. 8.4 Variation of friction factor with Reynolds number.

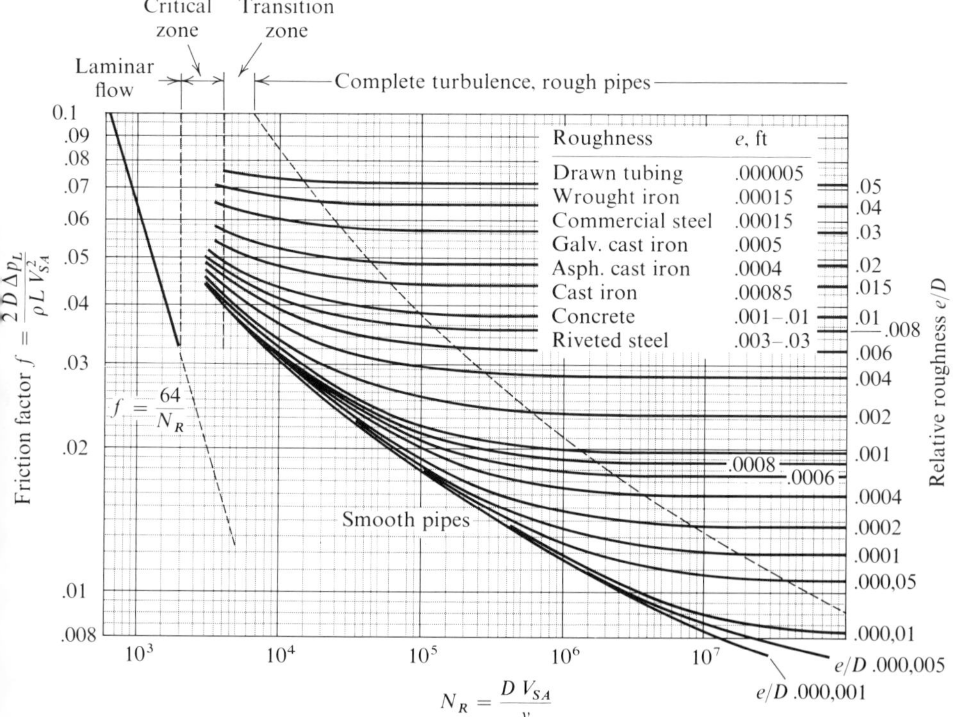

since the pressure drop is proportional to the length for fully developed flow, and a negative pressure gradient requires the pressure drop to be positive. Using the definition of the friction factor f, Eq. (8.11), solving for $\Delta p_L/L = f\rho V_{SA}^2/2D$, and substituting in Eq. (8.14) yields

$$f = \frac{64\mu}{DV_{SA}\rho} = \frac{64}{N_R}$$

This is identical to Eq. (8.13), which was obtained by writing the equation from the graph of experimental data. It will be recalled that the solution of Example 4.6 was predicated by the assumption of laminar flow of an incompressible Newtonian fluid. Comparison of experimental results with those obtained analytically should help establish some level of confidence in the methods presented in Chapter 4.

The pressure drop calculated by use of Fig. 8.4 is the pressure drop due to *fluid friction only*, and does not include the effects of pressure changes due to variation of cross-sectional areas in the conduit in the direction of flow, nor does it include pressure changes due to a change of position of the fluid in a gravitational field. The term *head loss* is frequently used in the literature treating pressure drop in pipes. The term is noted by h_L and is defined by *Darcy's equation*, which is

$$h_L = \frac{fLV_{SA}^2}{2D} \tag{8.15}$$

Upon reference to the definition of f [see Eq. (8.11)], it becomes apparent that $h_L = \Delta p_L/\rho$. The utility of such a term as head loss is realized by writing Bernoulli's equation in the form of (3.17), as

$$\frac{V_1^2}{2} + \frac{p_1}{\rho} + gz_1 = \frac{V_2^2}{2} + \frac{p_2}{\rho} + gz_2 + h_L \tag{8.16}$$

Actually, Eq. (3.17) is not applicable to flow in a pipe when shear stresses in the fluid are considered, hence the term h_L is appended in writing Eq. (8.16). The latter equation is sometimes referred to as the *modified Bernoulli equation*. The nature of h_L can be discerned by comparing Eq. (8.16) to the energy equation written as

$$-\frac{đW_s}{dm} + \frac{V_1^2}{2} + \frac{p_1}{\rho} + gz_1 + u_1 + \frac{đQ}{dm} = \frac{V_2^2}{2} + \frac{p_2}{\rho} + gz_2 + u_2 \tag{8.17}$$

in which $đQ/dm$ is heat transferred per unit mass of fluid flowing through the pipe. Comparison of the preceding equation ($đW_s/dm$ is not applicable) with Eq. (8.16) indicates that

$$h_L = u_2 - u_1 - \frac{đQ}{dm} \tag{8.18}$$

It can be concluded that the head loss due to viscous action appears as an increase of internal energy of the fluid and/or a transfer of heat from the fluid. If the walls of the pipe were insulated to minimize the heat transferred, then the head loss would appear almost entirely as an increase in internal energy. This interpretation is consistent with the thoughts of the second law of thermodynamics since viscosity precludes mechanical internal reversibility; and the head loss represents an energy converted into other forms of limited thermodynamic availability.

The practical advantage of using head loss is that it removes the necessity of measuring or estimating changes of internal energy of the fluid and the heat transferred. The reader may refer to Example 6.2 in which the very small temperature difference corresponded to a large change of internal energy.

Example 8.6

What is the pressure drop in 200 ft of 3-in.-diameter galvanized cast-iron pipe if it handles 0.4 ft^3/sec? The kinematic viscosity of the Newtonian fluid at a flow temperature of 50°F is 1.42 (10^{-5}) ft^2/sec. The density is constant at 62.4 lb_m/ft^3.

Solution. The Reynolds number is calculated in order to obtain the friction factor from Fig. 8.4.

$$V_{SA} = \frac{0.4 \text{ ft}^3/\text{sec}}{(\pi/4)(3 \text{ ft}/12)^2} = 8.15 \text{ ft/sec}$$

$$N_R = \frac{V_{SA}D\rho}{\mu} = \frac{V_{SA}D}{\nu} = \frac{(8.15 \text{ ft/sec})(3 \text{ ft}/12)}{1.42\,(10^{-5}) \text{ ft}^2/\text{sec}}$$

$$= 1.44\,(10^5)$$

and so the flow is turbulent. With $D = 3$ in., Fig. 8.4 is used to obtain $e/D = 0.002$ for galvanized-iron pipe; and a friction factor f of 0.025 is found for this N_R and e/D. From Eq. (8.11),

$$\Delta p_L = \frac{\rho f L V_{SA}^2}{2D} = \frac{(62.4 \text{ lb}_m/\text{ft}^3)(0.025)(200 \text{ ft})(8.15 \text{ ft/sec})^2}{(2)(3 \text{ ft}/12)(32.2 \text{ ft-lb}_m/\text{lb}_f\text{-sec}^2)}$$

$$= 1290 \text{ lb}_f/\text{ft}^2 = 8.95 \text{ lb}_f/\text{in.}^2$$ ◀

8.5 Minor Losses and Noncircular Conduits

Minor losses. Most pipe systems contain bends, valves, and other fittings that impose additional pressure drops or losses with respect to the fluid.

These losses may be estimated by the use of the following equation, in which K_L is referred to as a *loss coefficient*,

$$h_L = K_L \frac{V_{SA}^2}{2} \tag{8.19}$$

The space-average velocity ahead of the fitting is used in the preceding equation except for entrances.

Table 8.2 gives values[1] of K_L for several different fittings. Even though these losses are termed minor, they can become the major sources of pressure drops in systems that contain relatively small runs of straight pipe.

Table 8.2 Loss coefficients

Fitting	K_L
90° elbow (short radius)	0.90
45° elbow (short radius)	0.42
Return bend	2.2
Globe valve, wide open	10.00
Gate valve, wide open	0.19
Gate valve, ¾ open	1.15
Gate valve, ½ open	5.6
Gate valve, ¼ open	24.00
Inward-projecting pipe entrance	0.78
Sharp-edged entrance	0.50
Sharp-edged exit	1.0
Well-rounded entrance	0.04

Example 8.7

What horsepower must be transmitted to the liquid by the pump impeller to maintain a steady flow of water through the system shown in Fig. 8.5? Properties of the fluid are $\rho = 62.4\ \text{lb}_m/\text{ft}^3$ and $\nu = 1.42\ (10^{-5})\ \text{ft}^2/\text{sec}$ at the flow temperature of 50°F. The pumping rate is 0.4 ft³/sec. The vertical distance between the surface of the lake and the pump is negligible.

Solution. The power required to maintain the flow must be adequate to change the gravitational potential and kinetic energies of the liquid, pump the fluid through the required pressure difference, and overcome the resistance of the straight runs of pipe and the minor losses.

[1]The values given are satisfactory for purposes of this text, but a designer would need more specific information of flow losses since K_L is a function of size, shape, Reynolds number, design, etc. One such source of information is: Crane Company, Eng. Div., "Flow of Fluids through Valves, Fittings, and Pipe," Crane Company, Tech. Paper no. 410, 1969.

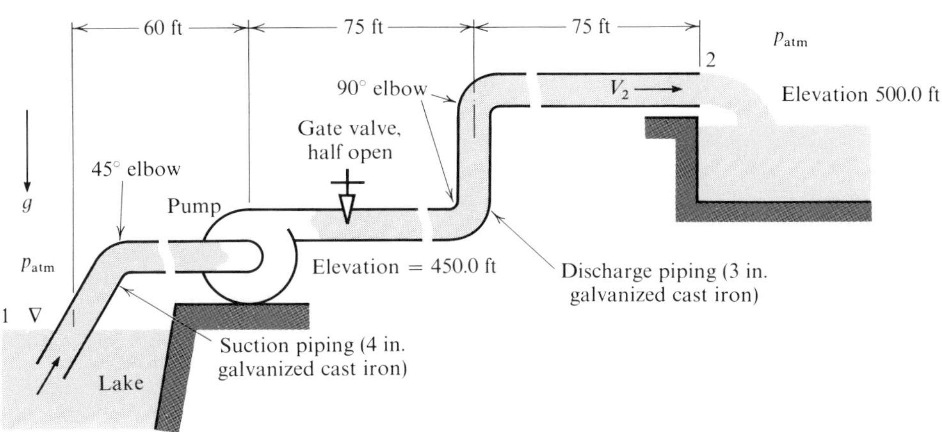

Fig. 8.5

For the 3-in. discharge line, there are 200 ft of straight pipe. The head loss for this may be calculated from the results of Example 8.6. For the 3-in. pipe,

$$h_L = \frac{\Delta p_L}{\rho} = \frac{1290 \text{ lb}_f/\text{ft}^2}{62.4 \text{ lb}_m/\text{ft}^3} = 20.7 \text{ ft-lb}_f/\text{lb}_m$$

The minor losses in the discharge pipe can be calculated from Eq. (8.19), Table 8.2, and $V_{SA} = 8.15$ ft/sec. For one 90° elbow,

$$h_L = \frac{(0.90)(8.15 \text{ ft/sec})^2}{(2)(32.2 \text{ ft-lb}_m/\text{lb}_f\text{-sec}^2)} = 0.929 \text{ ft-lb}_f/\text{lb}_m$$

For the half-open gate valve

$$h_L = \frac{(5.6)(8.15 \text{ ft/sec})^2}{(2)(32.2 \text{ ft-lb}_m/\text{lb}_f\text{-sec}^2)} = 5.77 \text{ ft-lb}_f/\text{lb}_m$$

In the 4-in. suction line there are 60 ft of straight pipe in which the space-average velocity is

$$V_{SA} = \frac{0.4 \text{ ft}^3/\text{sec}}{(\pi/4)(4 \text{ ft}/12)^2} = 4.59 \text{ ft/sec}$$

and the Reynolds number is

$$N_R = \frac{DV_{SA}}{\nu} = \frac{(4 \text{ ft}/12)(4.59 \text{ ft/sec})}{1.42\ (10^{-5}) \text{ ft}^2/\text{sec}} = 1.08\ (10^5)$$

From Fig. 8.4, $e/D = 0.0015$, and the friction factor is $f = 0.0235$. The head loss for the 4-in. suction line, by Eq. (8.15), is

$$h_L = \frac{(0.0235)(60 \text{ ft})(4.59 \text{ ft/sec})^2}{(2)(4 \text{ ft}/12)(32.2 \text{ ft-lb}_m/\text{lb}_f\text{-sec}^2)} = 1.38 \text{ ft-lb}_f/\text{lb}_m$$

and the minor losses in the suction pipe are, for the 45° elbow,

$$h_L = \frac{(0.42)(4.59 \text{ ft/sec})^2}{2(32.2 \text{ ft-lb}_m/\text{lb}_f\text{-sec}^2)} = 0.137 \text{ ft-lb}_f/\text{lb}_m$$

and for the entrance,

$$h_L = \frac{(0.78)(4.59 \text{ ft/sec})^2}{2(32.2 \text{ ft-lb}_m/\text{lb}_f\text{-sec}^2)} = 0.255 \text{ ft-lb}_f/\text{lb}_m$$

The work done by the impeller per unit mass of fluid flowing from the fluid surface 1 to discharge from the pipe 2 is

$$-\frac{đW_s}{dm} = gz_2 - gz_1 + \frac{V_2^2 - V_1^2}{2} + \frac{p_2 - p_1}{\rho} + \text{the total head loss}$$

$$= \frac{(32.2 \text{ ft/sec}^2)(50.0 \text{ ft})}{32.2 \text{ ft-lb}_m/\text{lb}_f\text{-sec}^2} + \frac{(8.15 \text{ ft/sec})^2}{(2)(32.2 \text{ ft-lb}_m/\text{lb}_f\text{-sec}^2)} + 0$$

$$+ [20.7 + 2(0.929) + 5.77 + 1.38 + 0.137 + 0.255] \text{ ft-lb}_f/\text{lb}_m$$

Note that $p_2 = p_1$, if it is assumed that the liquid leaves the discharge line as a free jet; hence the pressure in the jet is that of the surrounding atmosphere. Calculating the individual changes in energy for the unit mass of fluid yields

$$-\frac{đW_s}{dm} = (50.0 + 1.03 + 30.1) \text{ ft-lb}_f/\text{lb}_m$$

$$= 81.1 \text{ ft-lb}_f/\text{lb}_m$$

Multiplying the work per unit mass by the steady mass flow rate will give the power,

$$\frac{đW_s}{dt} = \frac{-(81.1 \text{ ft-lb}_f/\text{lb}_m)(0.4 \text{ ft}^3/\text{sec})(62.4 \text{ lb}_m/\text{ft}^3)}{550 \text{ ft-lb}_f/\text{hp-sec}}$$

$$= -3.68 \text{ hp}$$ ◀

The preceding example not only serves to illustrate computational procedure, but also presents a case where friction losses account for more than a third of the total power requirements.

Noncircular conduits. For flow in conduits with other than circular cross sections, the curves showing the variation of friction factor with Reynolds number in Fig. 8.4 may be used to give calculated results that compare roughly with experimental data, provided that the hydraulic diameter is used for the variable D. The *hydraulic diameter* is defined in terms of cross-sectional area A, and wetted perimeter P, as

$$D_H = \frac{4A}{P} \tag{8.20}$$

For a circular conduit,

$$D_H = \frac{(4)(\pi D^2/4)}{\pi D} = D$$

For flow in an annulus with inner and outer diameters D_1 and D_2, respectively,

$$D_H = \frac{4[(\pi D_2{}^2/4) - (\pi D_1{}^2/4)]}{\pi D_2 + \pi D_1} = D_2 - D_1 \tag{8.21}$$

Example 8.8

Oil flows in the annular passage of the two-pipe heat exchanger shown in Fig. 8.6. The space-average velocity of the oil is 50 ft/sec, and its kinematic viscosity and density are $7(10^{-5})$ ft^2/sec and 53.6 lb$_m$/ft^3, respectively. The walls of the annulus are similar to galvanized iron and have a roughness of 0.0005 ft. The heat-transfer analyst desires to know the pressure drop per foot in the direction of flow since this will have a bearing on the power required to pump the oil through the heat exchanger. Calculate the pressure drop per unit length.

Solution. The hydraulic diameter from Eq. (8.21) is

$$D_H = (3 - 2) \text{ in.} = 1 \text{ in.}$$

$$\frac{e}{D_H} = \frac{(0.0005 \text{ ft})(12 \text{ in./ft})}{1 \text{ in.}} = 0.006$$

$$N_R = \frac{D_H V_{SA}}{\nu} = \frac{(1 \text{ ft}/12)(50 \text{ ft/sec})}{7(10^{-5}) \text{ ft}^2/\text{sec}} = 5.95\,(10^4)$$

Fig. 8.6

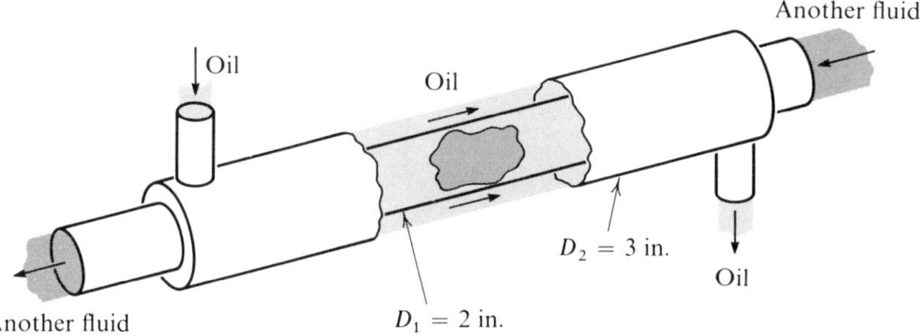

From Fig. 8.4,

$$f = 0.034$$

$$\Delta p_L = \frac{\rho f L V_{SA}^2}{2D_H} = \frac{(53.6\ \text{lb}_m/\text{ft}^3)(0.034)(1\ \text{ft})(50\ \text{ft/sec})^2}{(2)(1\ \text{ft}/12)(32.2\ \text{ft-lb}_m/\text{lb}_f\text{-sec}^2)}$$

$$\Delta p_L = 850\ \text{lb}_f/\text{ft}^2 = 5.90\ \text{lb}_f/\text{in}^2$$ ◀

The designer of heat exchangers of this nature will usually seek an optimum design, which makes the best compromise between two mutually antagonistic factors: pressure drop associated with high velocities and higher heat-transfer rates which are also associated with higher velocities. This example examines pressure drop per unit length assuming constant viscosity. Since the value of fluid temperature may change considerably, inclusion of the corresponding variation of viscosity and density may become important in determining an overall pressure drop.

Example 8.9

In Section 8.4, a friction factor and Reynolds number relationship, $f = 64/N_R$, was derived for fully developed laminar flow in a pipe by using the solution of the Navier-Stokes equations of Example 4.6. Obtain the solution for fully developed laminar flow of an incompressible Newtonian fluid in the annulus shown in Fig. 8.7. Then obtain a friction factor–Reynolds number relation by using $N_R = D_H V_{SA} \rho/\mu$, and compare this relation to $f = 64/N_R$.

Solution. By placing the appropriate assumptions on the Navier-Stokes equations, the same ordinary differential equation that was obtained in Example 4.6 would result,

$$\frac{d}{dr}\left(r\frac{dV_z}{dr}\right) = \frac{r}{\mu}\frac{dp}{dz}$$

The general solution is

$$V_z = \frac{1}{4\mu}\frac{dp}{dz}r^2 + K_1 \ln r + K_2$$

Fig. 8.7

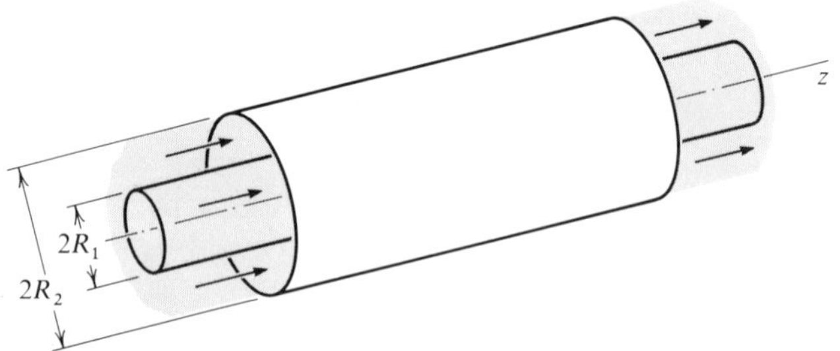

and the constants K_1 and K_2 may be determined from the no-slip conditions:

$$\left.\begin{aligned} V_z(R_1) &= 0 \\ V_z(R_2) &= 0 \end{aligned}\right\} \quad \text{for all values of } \theta \text{ and } z$$

The particular solution that satisfies the problem is

$$V_z = \frac{1}{4\mu}\frac{dp}{dz}\left[r^2 - R_1{}^2 + (R_1{}^2 - R_2{}^2)\frac{\ln(r/R_1)}{\ln(R_2/R_1)}\right]$$

The space-average velocity may be formulated from

$$V_{SA} = \int_{r=R_1}^{r=R_2} \frac{V_z\, 2\pi r\, dr}{\pi(R_2{}^2 - R_1{}^2)}$$

$$= \frac{1}{4\mu}\frac{dp}{dz}\left[\frac{R_2{}^2 - R_1{}^2}{\ln(R_2{}^2/R_1{}^2)} - \frac{R_2{}^2 + R_1{}^2}{2}\right] \qquad (i)$$

The definition of the friction factor gives

$$\frac{\Delta p_L}{L} = \frac{f\rho V_{SA}{}^2}{2D_H} \qquad (ii)$$

Substituting $-\Delta p_L/L$, given by expression (*ii*), for dp/dz in (*i*) yields after considerable manipulation,

$$f = \frac{64}{N_R}\left\{\frac{(1 - D_1/D_2)^2}{1 + (D_1/D_2)^2 + [1 - (D_1/D_2)^2]/\ln(D_1/D_2)}\right\} \quad ◀$$

in which

$$N_R = \frac{D_H V_{SA}\rho}{\mu} \qquad \text{and} \qquad D_H = D_2 - D_1$$

Comparison of this answer to the f–N_R relation for laminar flow in a pipe indicates that the use of D_H for an annulus yields an additional factor in the bracketed term—a function of D_1 and D_2. For the case of flow in a pipe with $D_1 = 0$, or the case where $D_2 \gg D_1$, the answer reduces to the familiar form of $f = 64/N_R$ with $D_H = D_2 - D_1 \simeq D_2$. The term in the brackets deviates from unity for $D_1/D_2 \neq 0$. With values of $D_1/D_2 = 0.2$ and 0.4, the relation between f and N_R is $f \simeq 92/N_R$ and $f \simeq 95/N_R$, which reflects the error incurred by employing D_H for the conversion of noncircular conduits to circular cross sections and using Fig. 8.4 for purposes of estimating head loss. The reason for this error can be seen qualitatively by reference to expression (*ii*): $\Delta p_L/L$ is inversely proportional to D_H, and D_H decreases with an increasing area of the solid surface to which the fluid is exposed. The head loss would then be proportional to this area.

For the case of laminar flow, the wetted area is only a portion of the resistance to flow, and considerable flow resistance is manifest in the part of the flow more remote from the wall. In the case of turbulent flow, the resistance of the wall is important (note that f is a function of e/D for turbulent flow in Fig. 8.4). This would indicate that the recommendation of using D_H for turbulent flow in noncircular ducts would produce more realistic head losses than when applied to laminar flow. In either case only a rough approximation can be obtained by conversion of the noncircular-conduit problem to that of a circular problem by using the hydraulic diameter. For a detailed discussion of the reliability of using the hydraulic diameter, see the papers by Owen[1] and Walker et al.[2] The use of D_H in estimating head loss in conduits of rectangular cross section is more reliable than the corresponding application to annular conduits.

8.6 Introduction to Free-Surface Flow

There are many applications of engineering interest encountered in fluid mechanics where a flowing liquid has at least one surface exposed to a gas such as the atmosphere. This surface is referred to as a *free surface*. Examples of free-surface flow are found in the transportation of liquids in open channels such as culverts, flumes, and irrigation canals, as well as in the runoff of rainfall in watersheds, and on highways and airports. Like flow in pipes, open-channel flow may be laminar or turbulent and steady or unsteady. In addition to these descriptions, the terms *uniform* and *nonuniform* (or *varied*) are employed in free-surface analysis. Unsteady flows will not be treated in this book, and the reader is referred to more advanced and specialized literature for discussion of this subject.

As a preliminary for gaining some insight concerning uniform and varied open-channel flow, consider a fluid in a channel of constant cross-sectional area. Figure 8.8 shows three different regions, two with varied flow and one with uniform flow. In the upstream portion of the channel, called the *upper reaches*, the force of gravity is causing the fluid to accelerate. The motion is slow here so that the flow resistance that depends on velocity is small. As the fluid gains speed in the upper reaches, the flow resistance due to shear on the bottom and sides of the channel increases until this resistance is balanced by the force of gravity. At this point the fluid moves with a constant space-average velocity. In region I where the flow is varied, the depth of the free surface is decreasing, which would be required from the conservation of matter for constant density and cross-sectional area with an increasing space-average velocity. In region II the depth of the free surface remains constant as well as the space-average velocity, with the shear forces and gravity force in balance. At the lower reaches of the channel, the flow becomes varied before it exits from the channel. If an ideal fluid were flowing through the channel of Fig. 8.8, there would be no shear stresses to

[1] W. M. Owen, Experimental Study of Water Flow in Annular Pipes, *Trans. ASCE*, vol. 117, pp. 485–496, 1952.

[2] J. E. Walker, G. A. Whan, and R. R. Rothfus, Fluid Friction in Noncircular Ducts, *J. AIChE.*, vol. 3, no. 4, pp. 484–489, 1957.

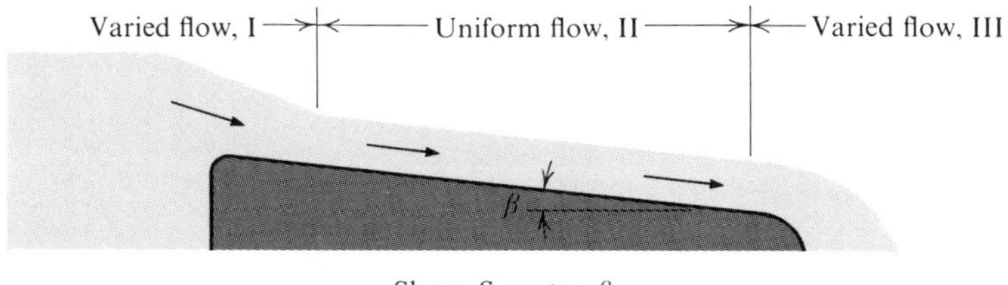

Fig. 8.8 The development of uniform flow in a channel of constant slope.

balance the force of gravity, and the flow would be varied throughout the entire channel. In this case the space-average velocity of the flow would increase with a decrease in depth from the upper to the lower reaches. In channels with short runs, the uniform flow may never be developed; however, the investigation of uniform-flow problems provides a basis for open-channel flow analysis.

Uniform and steady open-channel flow. To relate space-average velocity and hence flow to the geometric characteristics of the channel, the principle of linear momentum must be applied to a control volume that involves the shear stress at the bottom and walls of the channel and the force of gravity. Figure 8.9 shows a volume of fluid in a portion of the channel where the flow is uniform. The surface forces F_u and F_l at each end of the control volume are equal, and the average shear stress at the walls and bottom is indicated by τ_0. The space-average velocity is constant for uniform flow so that the net linear momentum efflux for the control surface is zero.

Fig. 8.9 Control volume for linear momentum in steady and uniform open-channel flow.

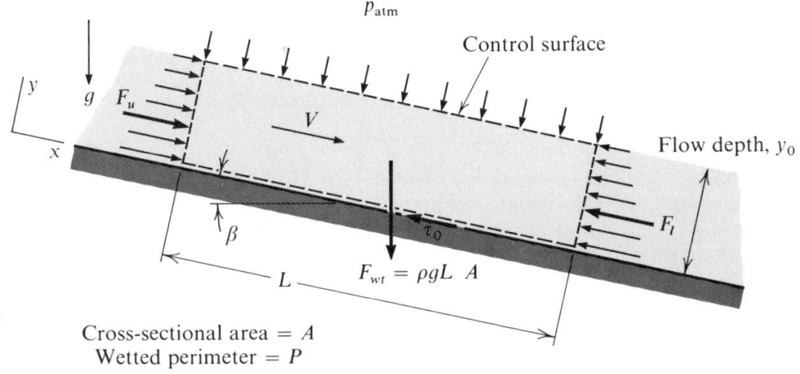

Application of Eq. (5.3) yields the following x component of the momentum equation:

$$\rho g L A \sin \beta - \tau_0 L P = 0$$

from which

$$\tau_0 = \frac{A}{P} \rho g \sin \beta$$

The term A/P is called the *hydraulic radius* and is noted by R_H. It is one-fourth of the hydraulic diameter D_H defined previously in Eq. (8.20). For small angles, $\beta < 10°$, $\sin \beta \simeq \tan \beta =$ slope S_0. The equation for the average shear stress may then be written as

$$\tau_0 = R_H \rho g S_0 \tag{8.22}$$

If the mechanism of viscous fluid motion is assumed to be similar in open-channel and pipe flow, and if it is further assumed that R_H accounts for differences in geometry, then the average shear stress may be eliminated from Eq. (8.22) in favor of the space-average velocity and the friction factor f used in pipe-friction problems.

Consider a section of pipe with fully developed turbulent flow shown in Fig. 8.10. Application of the momentum equation (5.3) to the control volume gives

$$p \frac{\pi}{4} D^2 - (p - \Delta p_L) \frac{\pi}{4} D^2 - \tau_0 \pi D L = 0$$

from which

$$\tau_0 = \frac{\Delta p_L D}{4L} \tag{8.23}$$

Fig. 8.10 *Fully developed turbulent flow in a circular pipe.*

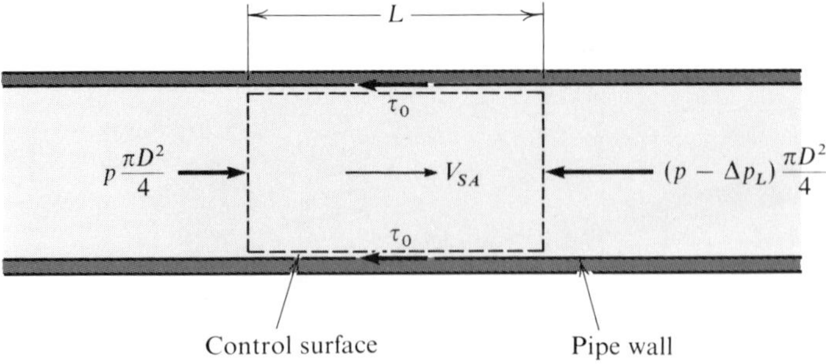

Solving for Δp_L from Eq. (8.11) and substituting in (8.23) yields

$$\tau_0 = \frac{f\rho V_{SA}^2}{8} \tag{8.24}$$

Equating the expressions for τ_0 given by Eqs. (8.22) and (8.24) provides the following useful result:

$$V_{SA} = \left(\frac{8g}{f} R_H S_0\right)^{1/2} \tag{8.25}$$

which is often written as

$$V_{SA} = C(R_H S_0)^{1/2} \tag{8.26}$$

The variable C, which depends on surface roughness and flow parameters, is called the *Chezy coefficient* after the French engineer Antoine Chezy (1718–1798). He developed the first and most enduring resistance formula for open-channel flow. For an interesting description of Chezy's work, see the account given by Rouse and Ince.[1] Note that the Chezy coefficient has dimensions of $L^{1/2}t^{-1}$. Robert Manning, an Irish engineer, evolved an expression for the space-average velocity of the form

$$V_{SA} = \frac{1.49}{n} R_H^{2/3} S_0^{1/2} \tag{8.27}$$

in which n is a roughness parameter. The constant 1.49 has the units $\text{ft}^{1/2}\text{-sec}^{-1}$, and n is given in Table 8.3 for various surfaces (n is a measure of the channel roughness and has units of $\text{ft}^{1/6}$). The space-average velocity for *turbulent flow* may be calculated from Eq. (8.27). This equation is based on the observation that the friction factor is relatively independent of N_R for fully developed turbulent flow and a strong function of surface roughness—a fact pointed out in the discussion of Fig. 8.4 concerning pipe flow. A recommended check for turbulent flow is to calculate the Reynolds number. If $N_R = D_H V_{SA}\rho/\mu < 2(10^3)$, the flow is considered laminar; if $N_R > 5(10^4)$, the flow is considered turbulent.[2] The value of n varies considerably for a given surface, which is similar to the uncertainty concerning the roughness of the interior of pipe surfaces. However, experience on the part of the designer or analyst usually serves as a guide. Table 8.3 gives some values commonly used in design.

[1] Hunter Rouse and Simon Ince, "History of Hydraulics," pp. 117–120, 142, Dover Publications, Inc., New York, 1963.

[2] Victor L. Streeter (ed.), "Handbook of Fluid Dynamics," sec. 24, p. 7, McGraw-Hill Book Company, New York, 1961.

Table 8.3 Values of roughness coefficient for Manning's equation n, $ft^{1/6}$

Surface	Range	Value commonly used in designing
Lumber, planed	0.010–0.014	0.012
Lumber, unplaned	0.011–0.015	0.013
Concrete pipe	0.012–0.016	0.015
Cement mortar surfaces	0.011–0.015	0.013
Concrete-lined channels	0.012–0.018	0.016
Vitrified sewer pipe	0.010–0.017	0.013
Glazed brickwork	0.011–0.015	0.013
Dry-rubble surface	0.025–0.035	—
Rivited and spiral steel pipe	0.013–0.017	0.015 and 0.017
Commercial wrought-iron pipe, black	0.012–0.015	—
Commercial wrought-iron pipe, galvanized	0.013–0.017	—

SOURCE: H. W. King and E. F. Brater, "Handbook of Hydraulics," 5th ed., sec. 7, p. 17, McGraw-Hill Book Company, New York, 1963. By permission of the publisher.

Example 8.10

Calculate the steady volume flow rate through a concrete pipe with a diameter of 10 ft and a slope of 0.001. The fluid has a kinematic viscosity $\nu = 1.4\,(10^{-5})$ ft²/sec. The pipe is half-full of the fluid as shown in Fig. 8.11.

Solution. Equation (8.27) may be used to calculate the space-average velocity

$$V_{SA} = \frac{1.49}{n} R_H^{2/3} S_0^{1/2}$$

$$R_H = \frac{(\pi/8)(10)^2}{(\pi/2)(10)} = 2.5 \text{ ft}$$

$$D_H = 4R_H = 10 \text{ ft}$$

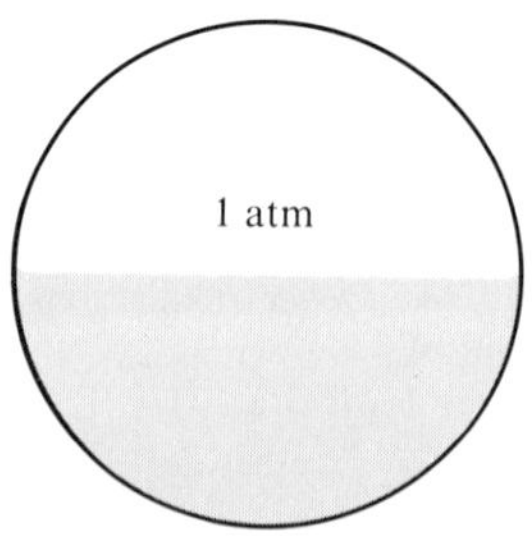

Fig. 8.11

From Table 8.3, an average value of n over the range is 0.014 ft$^{1/6}$.

$$V_{SA} = \frac{1.49}{0.014}(2.5)^{2/3}(0.001)^{1/2} = 6.16 \text{ ft/sec}$$

$$\dot{V} = AV_{SA} = \frac{\pi}{8}(10 \text{ ft})^2 (6.16 \text{ ft/sec}) = 241 \text{ ft}^3\text{/sec} \qquad \blacktriangleleft$$

The Reynolds number is checked for turbulent flow,

$$N_R = \frac{D_H V_{SA}}{\nu} = \frac{(10 \text{ ft})(6.16 \text{ ft/sec})}{1.4\,(10^{-5}) \text{ ft}^2\text{/sec}} = 4.41\,(10^6) > 5\,(10^4)$$

and the flow is turbulent.

Equation (8.25) may also be used to estimate the velocity since the flow is turbulent.

$$V_{SA} = \left(\frac{8g}{f} R_H S_0\right)^{1/2}$$

$$\frac{e}{D_H} = 0.0005 \qquad \text{an average value for the relative roughness from Fig. 8.4}$$

$$f = 0.0168 \qquad \text{from Fig. 8.4}$$

$$V_{SA} = \left[\frac{(8)(32.2 \text{ ft/sec}^2)(2.5 \text{ ft})(0.001)}{0.0168}\right]^{1/2}$$

$$= 6.20 \text{ ft/sec}$$

A close check would be expected since the point $[N_R = 4.41\,(10^6),\ e/D_H = 0.0005]$ is in the complete turbulence or rough pipe region, and f is not strongly dependent on the Reynolds number.

Hydraulic jump. The *hydraulic jump* is an example of nonuniform flow. It occurs in a rapidly flowing stream, causes a large amount of energy dissipation, and appears as a sudden rise in free-surface elevation of the flow. This is shown in Fig. 8.12. The flow is essentially uniform at sections 1 and 2,

Fig. 8.12 *Hydraulic jump.*

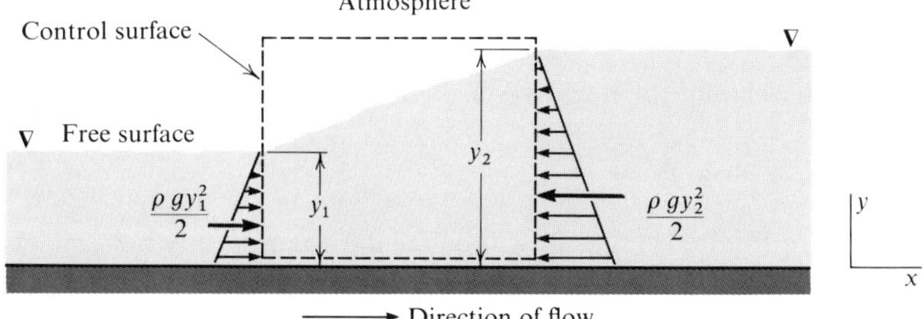

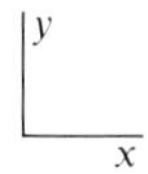

but there is severe turbulence in-between, causing some entrainment of air at the free surface.

A relation between the two depths y_1 and y_2 may be obtained by applying the conservation of matter and the principle of linear momentum to the control volume shown. The conservation of matter may be written as

$$\rho y_1 V_1 = \rho y_2 V_2 \tag{8.28}$$

for a unit width normal to the direction of flow. Neglecting shear forces at the solid boundaries, the momentum equation is

$$\frac{\rho g y_1^2}{2} - \frac{\rho g y_2^2}{2} = \rho y_1 V_1 (V_2 - V_1) \tag{8.29}$$

For a constant density fluid, division of Eq. (8.29) by $\rho g/2$ gives

$$y_1^2 - y_2^2 = \frac{2}{g} V_1^2 y_1 \left(\frac{V_2}{V_1} - 1\right)$$

Dividing both sides of the preceding equation by y_1^2 yields

$$\frac{y_1^2 - y_2^2}{y_1^2} = 2\frac{V_1^2}{g y_1}\left(\frac{V_2}{V_1} - 1\right)$$

By substituting y_1/y_2 for V_2/V_1 from Eq. (8.28) and noting the definition of the Froude number in Section 8.2,

$$\frac{y_1^2 - y_2^2}{y_1^2} = 2N_{F_1}\left(\frac{y_1 - y_2}{y_2}\right) \tag{8.30}$$

which is satisfied if $y_1 = y_2$; however, this would be the case of continuous uniform flow (without the jump), a situation of no interest at this point.

For the case where $y_1 \neq y_2$, Eq. (8.30) may be divided by the difference in depths. This results in the following biquadratic equation

$$y_2^2 + y_1 y_2 - 2N_{F_1} y_1^2 = 0$$

Solving for y_2 yields

$$y_2 = \frac{1}{2}[-y_1 \pm (y_1^2 + 8N_{F_1} y_1^2)^{1/2}]$$

and necessarily discarding the negative of the double sign gives

$$\frac{y_2}{y_1} = \frac{1}{2}[(1 + 8N_{F_1})^{1/2} - 1] \tag{8.31}$$

It is apparent from Eq. (8.31) that y_2 will be greater than y_1 only if $N_{F_1} > 1$. Uniform open-channel flow with a Froude number greater than unity is called *rapid* flow.

The Froude number of the flow after the jump can be related to that ahead by dividing the Froude number by its definition and equating at two sections:

$$\frac{N_{F_1} y_1 g}{V_1^2} = \frac{N_{F_2} y_2 g}{V_2^2}$$

Multiplying the left side by y_1^2/y_1^2 and the right side by y_2^2/y_2^2, and noting that $y_1^2 V_1^2 = V_2^2 y_2^2$ from Eq. (8.28), the following results:

$$N_{F_1} y_1^3 = N_{F_2} y_2^3 \tag{8.32}$$

Substituting y_2/y_1 using Eq. (8.31) yields

$$N_{F_2} = \frac{8N_{F_1}}{[(1 + 8N_{F_1})^{1/2} - 1]^3} \tag{8.33}$$

Since $N_{F_1} > 1$ if $y_2 > y_1$, Eq. (8.33) requires $N_{F_2} < 1$. A flow with a Froude number less than unity is called *tranquil* flow.

The use of hydraulic jumps to reduce the velocity of a stream leaving a spillway or chute is common practice in hydraulic engineering. The jump merely converts the rapid flow into tranquil flow by dissipation of kinetic energy.

Since the hydraulic jump is alleged to be dissipative in character, it is desirable to show that this premise is consistent with energy and entropy considerations. Returning to the control volume of Fig. 8.12, and applying a simplified form of the energy equation (6.11), the following is obtained:

$$\frac{gy_1}{2} + \frac{V_1^2}{2} + u_1 + \frac{gy_1}{2} = \frac{gy_2}{2} + \frac{V_2^2}{2} + u_2 + \frac{gy_2}{2}$$

assuming the heat transferred is negligible. The first term on each side of the equation is the gravitational potential energy and the last term on each side represents the work due to normal fluid forces at sections 1 and 2 of the control surface. The change of specific internal energy may then be written as

$$u_2 - u_1 = \frac{1}{2}(V_1^2 - V_2^2) + g(y_1 - y_2)$$

Entropy considerations using Eq. (6.36) may be simplified to

$$\$_2 - \$_1 = \frac{1}{T}\left[\frac{1}{2}(V_1^2 - V_2^2) + g(y_1 - y_2)\right] \tag{8.34}$$

with the absolute temperature T of the fluid in the control volume assumed to be slightly nonuniform.

Equation (8.34) may be altered by using the definition of the Froude numbers N_{F_1} and N_{F_2}, Eq. (8.31), and Eq. (8.32) to give

$$\$_2 - \$_1 = \frac{gy_1}{T}\left(\frac{N_{F_1}}{2}\left\{1 - \frac{4}{[(1 + 8N_{F_1})^{1/2} - 1]^2}\right\} + 1 - \frac{(1 + 8N_{F_1})^{1/2} - 1}{2}\right) \tag{8.35}$$

This indicates that N_{F_1} must be greater than unity to satisfy the requirement of the second law of thermodynamics, that $\$_2 - \$_1|_{Q=0} > 0$. Figure 8.13 presents a plot of $T(\$_2 - \$_1)/gy_1$ as a function of the Froude number ahead of the jump. Since $\$_2 - \$_1$ cannot be less than zero, hydraulic jumps cannot be produced in a free-surface flow with a Froude number less than unity. Indeed, these jumps have never been observed in tranquil flow.

Example 8.11

Fluid in a channel 10 ft wide has a depth of 2 ft and a velocity of 10 ft/sec. The fluid is water at 60°F and $g = 32.2$ ft/sec². Calculate (*a*) the depth after the hydraulic jump, (*b*) the change of specific internal energy across the jump, and (*c*) the horsepower dissipated by the jump.

Solution. *a.*

$$N_{F_1} = \frac{V_1^2}{y_1 g} = \frac{(10 \text{ ft/sec})^2}{(2 \text{ ft})(32.2 \text{ ft/sec}^2)} = 1.55$$

Fig. 8.13 *Variation of entropy change across a hydraulic jump as a function of the Froude number ahead of the jump.*

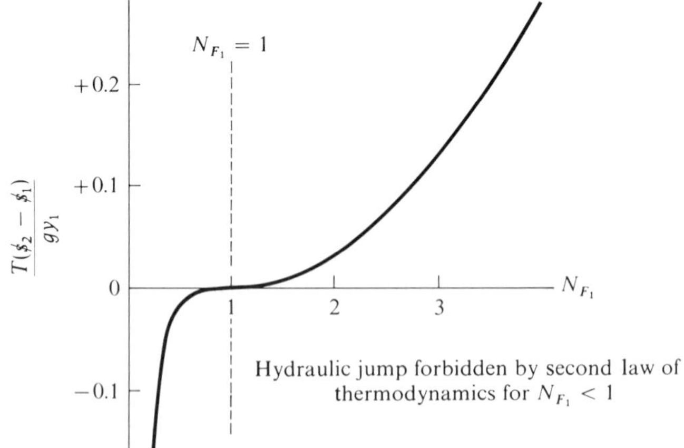

From Eq. (8.31),

$$y_2 = \frac{y_1}{2}[(1 + 8N_{F_1})^{1/2} - 1] = \frac{2\text{ ft}}{2}\{[1 + (8)(1.55)]^{1/2} - 1\}$$

$$= 2.66\text{ ft} \quad ◀$$

b. Assuming negligible transfer of heat and using Eqs. (6.36) and (8.35),

$$u_2 - u_1 = T(\$_2 - \$_1)$$

$$= gy_1\left(\frac{N_{F_1}}{2}\left\{1 - \frac{4}{[(1 + 8N_{F_1})^{1/2} - 1]^2}\right\} + 1 - \frac{(1 + 8N_{F_1})^{1/2} - 1}{2}\right)$$

With $N_{F_1} = 1.55$ and $y_1 = 2$ ft,

$$u_2 - u_1 = 0.45\text{ ft}^2/\text{sec}^2 = 0.45\text{ ft-lb}_f/\text{slug} \quad ◀$$

c. On a rate basis,

$$\dot{m}(u_2 - u_1) = \frac{(0.45\text{ ft-lb}_f/\text{slug})(10\text{ ft/sec})(2\text{ ft})(10\text{ ft})(1.94\text{ slug/ft}^3)}{550\text{ ft-lb}_f/\text{hp-sec}}$$

$$= 0.32\text{ hp} \quad ◀$$

The previous discussion did not include an explanation of how or where the hydraulic jump is formed. The positioning of a given jump in a rapid flow involves procedures and calculations that are presented in more specialized treatments of the subject. However, it is noteworthy that the entropy curve in Fig. 8.13 has a horizontal tangent at $N_F = 1.0$. This indicates that small oscillations of the free surface are permissible if the fluid velocity is $(gy_1)^{1/2}$. These motions of a free surface are referred to as *gravity waves*, and are observed in channels with small depths relative to the wavelengths.

8.7 One-Dimensional, Steady Compressible Flow

Before any basic equations for this simplified flow are derived, it is desirable to discuss the celerity at which pressure changes travel in compressible and incompressible fluids. In the incompressible fluid each fluid particle cannot be compressed, and a pressure perturbation at the location of one particle must be felt at every other particle in the field at the same instant of time. Stated another way, when one particle of a constant density fluid is disturbed, the signal travels with infinite celerity or speed. In the compressible fluid each fluid particle can be compressed so that the signal or pressure disturbance travels at a finite speed. Since this speed is important in the study of compressible fluid motion, it is appropriate to express it as a function of some variables that can be measured. To do this will require a formulation of the principles of linear momentum and conservation of matter for the model shown in Fig. 8.14. If the piston at the left end of the tube is struck an impulsive blow, the disturbance starts a weak pressure wave traveling through the fluid in the tube with a velocity c. The gas behind the wave is now compressed to a pressure $p + dp$ and a density $\rho + d\rho$, and it now has a small velocity dV; while the pressure and density of the undisturbed gas to the right of the wave are p and ρ, respectively, and the velocity of this gas is zero.

The control volume used for setting up the momentum and conservation of mass equations is indicated by the dashed lines in Fig. 8.14. This control volume and the control surface are moving to the right at a speed c. The fluid velocities relative to the control surface are shown in Fig. 8.15. If the speed c is constant, then the control volume is an inertial one,[1] and Eq. (5.3) may be written as

$$(p + dp)A - pA = \rho Ac[-(c - dV) - (-c)]$$

taking the direction of wave travel as positive. A is the cross-sectional area of the wave or tube. The shear stress acting on the wave between the inside tube wall and the wave is neglected since it acts over an exceedingly small area. The preceding momentum equation may be solved for dV. This gives

$$dV = \frac{dp}{\rho c} \tag{8.36}$$

Applying conservation of matter to the flow through the control volume results in

$$(\rho + d\rho)(c - dV)A = \rho cA$$

[1]If c is not constant, the noninertial terms in Eq. (5.3) may be neglected since the mass in the small control volume is negligibly small.

Fig. 8.14 *Weak pressure disturbance propagating through a tube of gas.*

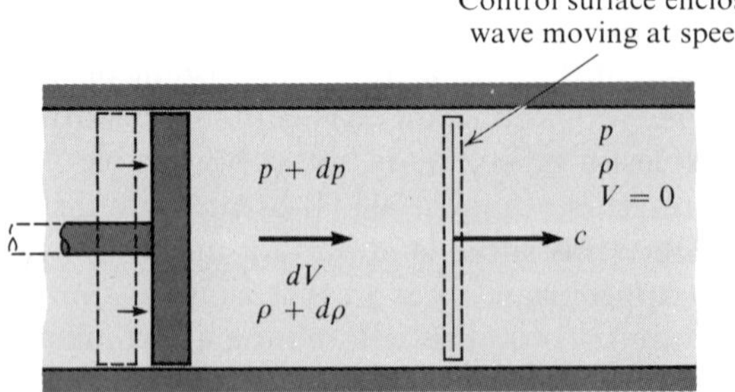

or

$$dV = c\frac{d\rho}{\rho} \tag{8.37}$$

neglecting the higher-order term $d\rho\, dV$. Eliminating dV from Eqs. (8.36) and (8.37) yields

$$c = \left(\frac{dp}{d\rho}\right)^{1/2}$$

Since the perturbation velocity dV is small, the internal dissipation in the fluid is minimal, and the wave disturbance travels too rapidly for any appreciable heat-transfer effects. The pressure and density changes are then assumed to be characterized by constant entropy. For these reasons, the last equation is written

$$c = \left(\frac{\partial p}{\partial \rho}\right)_s^{1/2} \tag{8.38}$$

the partial differentiation being carried out with the entropy held constant. Equation (8.38) gives the celerity at which the pressure perturbation travels through the medium in the tube. It is shown in physics that the speed of sound in an elastic medium is given by Eq. (8.38); hence c is often referred to as *sonic celerity*.

Example 8.12

At what speed would sound travel in air at a temperature of 60° F? Assume the air to be a perfect gas with constant specific heats.

Solution. With constant entropy, the p–ρ relation is $p\rho^{-\gamma}$ = constant for a perfect gas with constant specific heats. Then

$$\left(\frac{\partial p}{\partial \rho}\right)_s = (\text{const})\,\gamma\rho^{\gamma-1} = \frac{p}{\rho^\gamma}\gamma\rho^{\gamma-1} = \frac{\gamma p}{\rho}$$

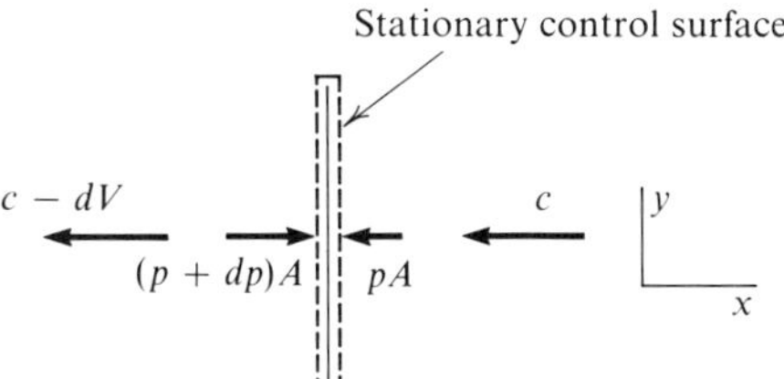

Fig. 8.15 *Control volume showing relative velocities and surface forces.*

However, the equation of state gives $p/\rho = RT$ so that

$$\left(\frac{\partial p}{\partial \rho}\right)_s = \gamma RT$$

Then,

$$c = \left(\frac{\partial p}{\partial \rho}\right)_s^{1/2} = (\gamma RT)^{1/2}$$

For a temperature of $60^\circ\text{F} = 520^\circ\text{R}$,

$$c = \left[(1.4)\left(53.3\,\frac{\text{ft-lb}_f}{\text{lb}_m\text{-}^\circ\text{R}}\right)(520^\circ\text{R})\left(32.2\,\frac{\text{ft-lb}_m}{\text{lb}_f\text{-sec}^2}\right)\right]^{1/2}$$

$$= 1120 \text{ ft/sec}$$ ◀

An important dimensionless ratio in compressible flow is the *ratio of the local fluid speed to local sonic celerity*. This is called the *Mach number* and noted by N_M. For a perfect gas with constant specific heats,

$$N_M = \frac{V}{c} = \frac{V}{(\gamma RT)^{1/2}} \tag{8.39}$$

Flows with $N_M > 1$ are called *supersonic*, while those with $N_M < 1$ are referred to as *subsonic*. A vehicle that travels through a fluid at supersonic speeds generates disturbances in the fluid, which cannot travel fast enough to signal the fluid ahead of it. For travel at subsonic speeds, the disturbance travels faster than the vehicle, and hence warns the fluid ahead that it is approaching. This is an important difference between supersonic and subsonic flow.

Isentropic flow*[1] *of a perfect gas with constant c_p. The basic equations pertaining to this steady flow may be readily obtained by reference to the control volume shown in Fig. 8.16. These are, for conservation of mass,

$$\rho_1 A_1 V_1 = \rho_2 A_2 V_2 \tag{8.40}$$

for the linear momentum,

$$p_1 A_1 - p_2 A_2 + F_w = \rho_1 A_1 V_1 (V_2 - V_1) \tag{8.41}$$

for the energy equation,[2]

$$h_1 + \frac{V_1^2}{2} = h_2 + \frac{V_2^2}{2} \tag{8.42}$$

[1]The term *isentropic* as used here means uniform and steady entropy.

[2]The change of gravitational potential energy is assumed negligible, and this equation is applicable to adiabatic and irreversible one-dimensional flows.

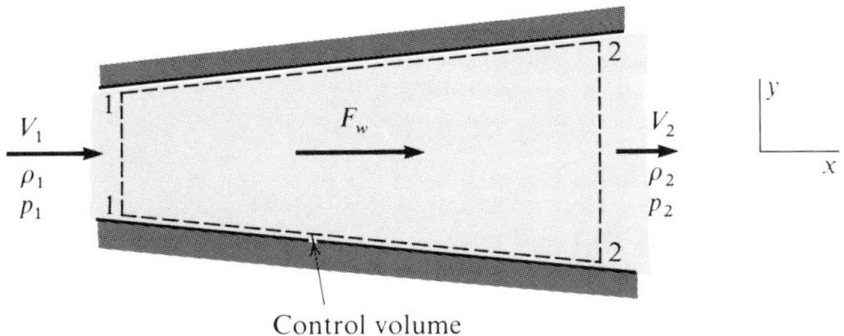

Fig. 8.16 *Control volume for one-dimensional flow.*

in which F_w in the momentum equation is the component of the surface force in the x direction, which the interior of the walls exerts on the fluid passing between the walls. In addition to these basic equations, there is an equation of state:

$$\frac{p_1}{\rho_1 T_1} = \frac{p_2}{\rho_2 T_2} \tag{8.43}$$

and an isentropic, pressure-density relation for a perfect gas:

$$\frac{p_1}{\rho_1{}^\gamma} = \frac{p_2}{\rho_2{}^\gamma} \tag{8.44}$$

These basic relations may be expressed in differential form as follows:

1. Conservation of mass.

$$\rho A V = \text{a constant}$$

and taking the logarithm of both sides of this equation yields

$$\ln \rho + \ln A + \ln V = \text{a constant}$$

Implicit differentiation of both sides of this equation results in

$$\frac{d\rho}{\rho} + \frac{dA}{A} + \frac{dV}{V} = 0 \tag{8.45}$$

2. Energy equation.

$$h + \frac{V^2}{2} = \text{a constant}$$

or

$$dh + VdV = 0 \tag{8.46}$$

3. Equation of state.

$$\frac{p}{\rho T} = \text{a constant}$$

or

$$\ln p - \ln \rho - \ln T = \text{a constant}$$

and

$$\frac{dp}{p} - \frac{d\rho}{\rho} - \frac{dT}{T} = 0 \tag{8.47}$$

4. Isentropic, pressure-density relation.

$$\frac{p}{\rho^{\gamma}} = \text{a constant}$$

or

$$\ln p - \gamma \ln \rho = \text{a constant}$$

and

$$\frac{dp}{p} - \gamma \frac{d\rho}{\rho} = 0 \tag{8.48}$$

5. Linear momentum equation.

$$pA - F_w + (\rho A V)V = p_1 A_1 + (\rho_1 A_1 V_1)V_1 = \text{a constant}$$

or

$$A\,dp + p\,dA - dF_w + (\rho A V)\,dV = 0$$

Since dA is the projected area of the wall in the direction of flow, and $p\,dA = dF_w$,

$$-dp = \rho V\,dV \tag{8.49}$$

Equation (8.49) is the same as Eq. (8.46) for isentropic flow. This can be seen by substituting $dh = d(u + p/\rho)$ in Eq. (8.46). Then

$$du + p\,d\frac{1}{\rho} + \frac{1}{\rho}dp = -V\,dV$$

However, a combination of the first and second laws of thermodynamics, Eq. (6.34), requires

$$T\,ds = du + p\,d\frac{1}{\rho}$$

which is equal to zero for constant entropy. This reduces Eq. (8.46) to $(1/\rho)\,dp = -V\,dV$. Hence the energy and momentum equations are not independent for one-dimensional, steady, isentropic flow.

One important difference between supersonic and subsonic flow was noted after Example 8.12. Other differences may be found by combining the basic governing equations, which have just been derived. The variation of Mach number, area, and velocity will be considered first. Equation (8.49) may be written as

$$\frac{dp}{d\rho}\frac{d\rho}{\rho} + V\,dV = 0$$

For constant entropy, $dp/d\rho = c^2$, and substitution for this derivative yields

$$c^2\frac{d\rho}{\rho} + V\,dV = 0$$

or

$$\frac{d\rho}{\rho} + \frac{V^2}{c^2}\frac{dV}{V} = 0$$

Noting that $V^2/c^2 = N_M{}^2$, and substituting for $d\rho/\rho$ by using Eq. (8.45), gives, after rearranging,

$$\frac{dA}{dV} = \frac{A}{V}(N_M{}^2 - 1) \qquad (8.50)$$

This would indicate that for a subsonic flow ($N_M < 1$), dA/dV is negative, and a decrease in area is required to produce an increase in velocity. For supersonic flow ($N_M > 1$), dA/dV is positive, and an increase in area is required to produce an increase in velocity. Note also that with $N_M = 1$, dA/dV is zero, which means that the fluid has a velocity that requires the area to be a maximum or minimum. The reader should observe that the relation (8.50) was obtained without using any information concerning a perfect gas; therefore this equation and the conclusions drawn from it are not restricted to the perfect gas.

Example 8.13

Show how a fluid may be accelerated continuously with constant entropy from a subsonic Mach number to supersonic speeds.

Solution. To accelerate a fluid with a subsonic Mach number would require a convergent passage such as shown in Fig. 8.17. If the fluid is accelerated to a Mach number of unity and is then received by a divergent passage, it is possible to accelerate the fluid

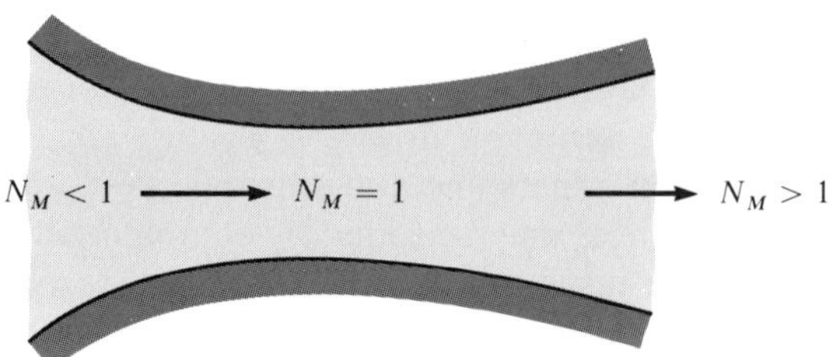

Fig. 8.17 A convergent passage followed by a divergent passage.

further and attain a supersonic Mach number. All the Mach numbers indicated on Fig. 8.17 are consistent with the conclusions drawn from Eq. (8.50).

Next, the variation of Mach number, area, and density may be obtained by eliminating dV/V from Eq. (8.50) by using the conservation of mass in the form of Eq. (8.45). This results in

$$\frac{dA}{d\rho} = \frac{A}{\rho}\left(\frac{1 - N_M{}^2}{N_M{}^2}\right) \tag{8.51}$$

which indicates that a convergent passage will produce a decrease of density in a subsonic stream. This is referred to as an *expansion*. To expand a supersonic stream, a divergent passage is required. For a sonic flow ($N_M = 1$), $dA/d\rho = 0$, which indicates that the density at this location corresponds to a maximum or minimum area. Note that the fluid is undergoing an expansion throughout the entire passage of Fig. 8.17. Equation (8.51), like (8.50), is *not restricted to a perfect gas*. An equation for the variation of Mach number, area, and pressure can be obtained by substituting $\dfrac{1}{\gamma}\dfrac{dp}{p}$ for $\dfrac{d\rho}{\rho}$ in Eq. (8.51). This requires the use of Eq. (8.48), and hence the result is *restricted to a perfect gas*. The relation is

$$\frac{dA}{dp} = \frac{1}{\gamma}\frac{A}{p}\left(\frac{1 - N_M{}^2}{N_M{}^2}\right) \tag{8.52}$$

Because of the similarity of the last two equations, what has been said about the variation of density with area for subsonic and supersonic flows applies to the variation of pressure with area. Observe that the fluid is undergoing a decrease in pressure throughout the entire passage in Fig. 8.17. The preceding discussion serves to describe the isentropic flow in a qualitative sense in that Eqs. (8.50), (8.51), and (8.52) were left in differential forms. These equations can be integrated by expressing the variables involved in terms of other variables and constants of the flow. However, the integration of these will be left for problems at the end of the chapter; while the constants of the flow will be considered immediately, for these constants greatly simplify calculations pertaining to compressible flows.

The first constant considered is the *stagnation temperature* T_0, which is the temperature of the fluid at a point in the stream where the fluid has zero velocity. This point may exist in fact or it can be imagined to exist. It is referred to as a *stagnation point*. For example, one might insert a small thermocouple in a moving stream. The moving fluid would come to rest relative to the thermocouple, and the temperature registered by the thermocouple would be higher than that measured by a thermometer moving at the speed of the fluid. Point 0, the junction in Fig. 8.18, senses the temperature of the fluid after the fluid has been brought to rest. Application of the energy equation (8.42) between a point upstream and the stagnation point gives

$$h + \frac{V^2}{2} = h_0$$

For a perfect gas with constant specific heats, the enthalpies may be replaced by $c_p T$ and $c_p T_0$, respectively, and

$$c_p T + \frac{V^2}{2} = c_p T_0$$

However,

$$c_p = \frac{\gamma R}{\gamma - 1}$$

so that

$$\frac{T_0}{T} = 1 + \frac{\gamma - 1}{2} \frac{V^2}{\gamma R T}$$

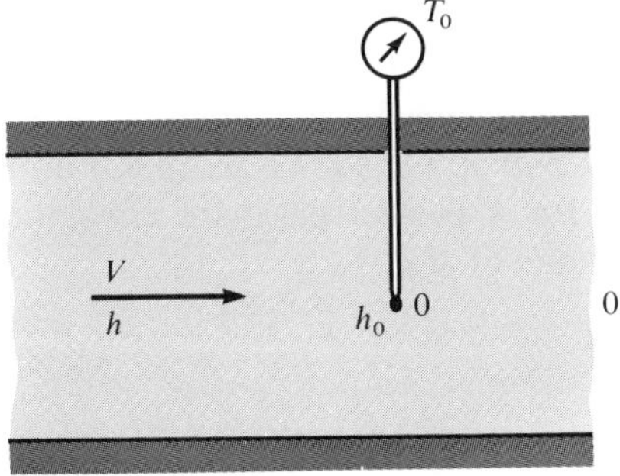

Fig. 8.18

Utilizing Eq. (8.39),

$$\frac{T_0}{T} = 1 + \frac{\gamma - 1}{2} N_M{}^2 \tag{8.53}$$

Table F.1 gives values of the temperature ratio as a function of the Mach number for $\gamma = 1.4$. To show that T_0 is a constant of the flow, the energy equation is written between two stagnation points, say 0 and 0′. Then $h_0 = h_{0'}$. Obviously the stagnation enthalpy is constant provided there is no shaft work, heat, or change in gravitational potential energy between points 0 and 0′. With the additional consideration of a perfect gas with constant specific heats, $h_0 = h_{0'}$ reduces to $T_0 = T_{0'}$, and the stagnation temperature is a constant of the flow.

Example 8.14

A missile moves in the earth's atmosphere at a Mach number of 3.00 relative to the undisturbed air. Consider the air a perfect gas with $\gamma = 1.4$. If the air temperature at flight altitude is $-60°$F, estimate the temperature at the nose of the missile.

Solution. The air at the nose of the missile is stagnant relative to the missile, therefore the air temperature should be close to the stagnation temperature.

$$T_0 = T\left(1 + \frac{\gamma - 1}{2} N_M{}^2\right) \tag{8.53}$$

$$= (-60 + 460)°\text{R}\left[1 + \frac{1.4 - 1}{2}(3)^2\right]$$

$$= (400°\text{R})(2.8) = 1120°\text{R}$$

$$= 660°\text{F}$$ ◀

This appears to be a high temperature; however, recall that meteorites "burn up" frequently on entering the earth's atmosphere.

The *isentropic stagnation pressure* p_0, and the *isentropic stagnation density* ρ_0, are the respective values of pressure and density when the fluid is decelerated to zero velocity at constant entropy. From Eqs. (8.43) and (8.44), the following relation is obtained for a perfect gas with constant specific heats undergoing an isentropic change of state:

$$\frac{p_2}{p_1} = \left(\frac{T_2}{T_1}\right)^{\gamma/(\gamma - 1)} \tag{8.54}$$

Substitution of Eq. (8.53) into (8.54) yields

$$\frac{p_0}{p} = \left(1 + \frac{\gamma - 1}{2} N_M{}^2\right)^{\gamma/(\gamma - 1)} \tag{8.55}$$

and use of Eq. (8.44) yields

$$\frac{\rho_0}{\rho} = \left(1 + \frac{\gamma - 1}{2} N_M{}^2\right)^{1/(\gamma - 1)} \tag{8.56}$$

Table F.1 gives values of p/p_0 and ρ/ρ_0 as functions of the Mach number for $\gamma = 1.4$. Since

$$\frac{p_{02}}{p_{01}} = \left(\frac{T_{02}}{T_{01}}\right)^{\gamma/(\gamma - 1)} = \left(\frac{\rho_{02}}{\rho_{01}}\right)^{\gamma}$$

for a perfect gas with $\$$ = a constant, it can be concluded that p_0 and ρ_0 are constants of the flow recalling that T_0 is constant.

Equations (8.53), (8.55), and (8.56) provide a convenient aid in the solving of compressible flow problems in the realm to which these equations apply, especially if T_0, p_0, and ρ_0 are constant.

Example 8.15

A perfect gas with $\gamma = 1.4$ flows steadily through the nozzle shown in Fig. 8.19. The entropy of the gas remains constant, and $R = 53.3$ ft-lb$_f$/lb$_m$-°R. The inlet temperature and pressure are 200°F and 100 lb$_f$/in.^{2}, respectively, and the exit pressure is 20 lb$_f$/in.^{2} The inlet is relatively large in comparison with the exit area which is 12 in.^{2} Obtain the exit Mach number, temperature, density, the mass flow rate through the nozzle, and the velocity at the smallest cross-sectional area.

Solution. In a large inlet, the velocity is low and the inlet pressure and temperature may be taken as stagnation values. Then, $p_0 = 100$ lb$_f$/in.^{2} and $T_0 = 200°\text{F} = 660°\text{R}$.

Fig. 8.19

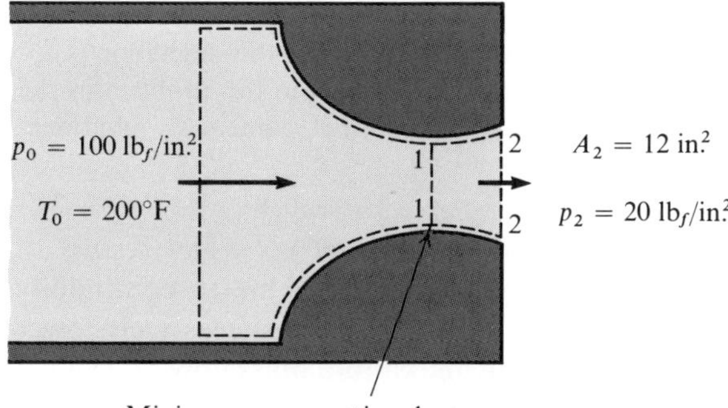

These are constants for the flow through the nozzle since the flow is isentropic and without shaft work.

$$\frac{p_2}{p_{02}} = \frac{20 \text{ lb}_f/\text{in}^2}{100 \text{ lb}_f/\text{in}^2} = 0.20$$

From Table F.1,

$$N_{M_2} = 1.71 \quad ◀$$

and

$$\frac{T_2}{T_{02}} = 0.631 \qquad \text{and} \qquad \frac{\rho_2}{\rho_{02}} = 0.316$$

$$T_2 = T_0(0.631) = (660)(0.631) = 416°\text{R} \quad ◀$$

$$V_2 = N_{M_2} c_2 = N_{M_2} (\gamma R T_2)^{1/2}$$
$$= (1.71)[(1.4)(53.3)(416)(32.2)]^{1/2} = 1718 \text{ ft/sec}$$

$$\rho_0 = \frac{p_0}{RT_0} = \frac{(100)(144)}{(53.3)(660)} = 0.410 \text{ lb}_m/\text{ft}^3$$

$$\rho_2 = \rho_{02}(0.316) = (0.410)(0.316) = 0.130 \text{ lb}_m/\text{ft}^3 \quad ◀$$

$$\dot{m} = \rho_2 A_2 V_2$$
$$= (0.130)\left(\frac{12}{144}\right)(1718) = 18.6 \text{ lb}_m/\text{sec} \quad ◀$$

Since the flow passes from the subsonic to the supersonic condition through the nozzle, the Mach number at the minimum cross-sectional area must be 1. With $N_{M_1} = 1.00$,

$$\frac{T_1}{T_{01}} = 0.833 \qquad \text{from Table F.1}$$

$$T_1 = T_0(0.833) = (660)(0.833) = 550°\text{R}$$

$$V_1 = N_{M_1}(\gamma R T_1)^{1/2} = [(1.4)(53.3)(550)(32.2)]^{1/2}$$
$$= 1151 \text{ ft/sec} \quad ◀$$

The student should rework this problem without the aid of the stagnation properties being constant to convince himself of the brevity accrued to this problem by the use of them. Of course, reference to the values in Table F.1 confers an additional advantage.

Isentropic choked flow of a perfect gas with constant c_p. The results of Example 8.15 might lead one to believe that it is possible to continually increase the mass flow by reducing the exit pressure p_2 to successively lower values. This is not the case. There is a definite maximum mass flow rate which the nozzle (with a fixed geometry) can pass at constant entropy and a given stagnation state (T_0, p_0). To examine this, consider the flow through a convergent nozzle. Figure 8.20 shows the nozzle with exit pressure p_{ex}. The mass rate of flow is given by

$$\dot{m} = \rho_{ex} A_{ex} V_{ex}$$

Since

$$\rho_{ex} = \rho_0 \left(\frac{p_{ex}}{p_0}\right)^{1/\gamma} = \frac{p_0}{RT_0}\left(\frac{p_{ex}}{p_0}\right)^{1/\gamma}$$

and

$$V_{ex} = [2(h_0 - h_{ex})]^{1/2} = \left[2c_p T_0 \left(1 - \frac{T_{ex}}{T_0}\right)\right]^{1/2}$$

$$= \left\{\frac{2\gamma R T_0}{\gamma - 1}\left[1 - \left(\frac{p_{ex}}{p_0}\right)^{(\gamma - 1)/\gamma}\right]\right\}^{1/2} \tag{8.57}$$

these may be substituted into the expression for the mass rate of flow, yielding

$$\dot{m} = \frac{p_0 A_{ex}}{RT_0}\left\{\frac{2\gamma R T_0}{\gamma - 1}\left[\left(\frac{p_{ex}}{p_0}\right)^{2/\gamma} - \left(\frac{p_{ex}}{p_0}\right)^{(\gamma + 1)/\gamma}\right]\right\}^{1/2} \tag{8.58}$$

Taking the derivative of $\dot{m}$ with respect to p_{ex}/p_0, and setting this derivative equal to zero for a maximum $\dot{m}$, yields

$$\frac{p_{ex}}{p_0} = \left(\frac{2}{\gamma + 1}\right)^{\gamma/(\gamma - 1)} \tag{8.59}$$

This gives the pressure at exit for a given stagnation state that will produce maximum mass flow. The velocity at exit may be obtained by substitution of Eq. (8.59) into (8.57), which on simplification, results in $V_{ex} = (\gamma R T)^{1/2}$; so that the flow at the exit has a Mach number of unity. The exit pressure that produces maximum mass flow is referred to as the *critical pressure*. If

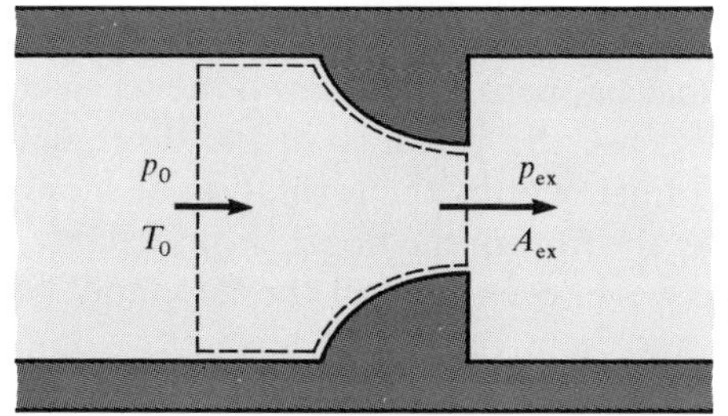

Fig. 8.20 Convergent nozzle.

the exit pressure in the medium receiving the emerging jet is reduced below this value, the pressure in the fluid at the exit area will remain at the critical value. Obviously a variation in pressure is transmitted at the speed of sound, which happens to be the speed at which the fluid jet emerges from the nozzle; thus a disturbance downstream cannot propagate upstream into the nozzle. After the fluid leaves the nozzle passage, the pressure is then reduced to the receiver pressure by expanding irreversibly down to this lower pressure.

If the receiver pressure is raised above the critical value, the fluid velocity at exit is reduced and the nozzle no longer passes the maximum mass flow. A nozzle that is passing maximum mass flow is said to be operating in a *choked condition*. If a divergent passage is added to the nozzle of Fig. 8.20 and the convergent nozzle is operating in a choked condition, it is possible to further expand the flow at constant entropy to a lower receiver pressure (see Example 8.15). As a matter of fact the convergent-divergent passage in Fig. 8.19 is operating in a choked condition, and the pressure at the minimum cross-sectional area (where $N_M = 1.00$) is the critical pressure, which is

$$p_{\text{crit}} = p_0 \left(\frac{2}{\gamma + 1} \right)^{\gamma/(\gamma - 1)} = (100)(0.528) = 52.8 \text{ lb}_f/in^2$$

Figure 8.21 shows the variation of pressure through a range of receiver pressures for flow through a convergent-divergent nozzle. The nozzle shape is indicated above the figure to note the axial position in the passage. The operating locus *A-B-C* is characterized by subsonic flow in the convergent and divergent portions of the nozzle with sonic flow at *B*. That depicted by *A-B-D* is subsonic in the convergent portion of the nozzle, sonic at *B*, and supersonic in the divergent portion of the nozzle. *A-B-D* is an expansion throughout the entire nozzle, while *A-B-C* is an expansion from *A* to *B* and a compression from *B* to *C*. Both loci represent isentropic flow with the nozzle choked. *A-E-F* represents a subsonic flow in the convergent and divergent portions, and at *E*. This flow is isentropic, but the flow is *not choked*. If the receiver pressure is lowered below the value p_D, the flow in the nozzle is still represented by *A-B-D* and the nozzle is again choked. The fluid exits at the pressure p_D and expands to the lower pressure outside the nozzle.

The question should arise as to what happens to the flow if the receiver pressure is between p_C and p_D. In this case the pressure in the divergent portion of the nozzle undergoes some adjustment by the formation of shock waves. Since these waves are dissipative, the flow in the nozzle is not characterized by constant entropy across the wave, even though the flow may be considered isentropic up to the wave and isentropic (at a higher entropy) after the wave. Flow through normal shock waves will be discussed subsequently in this section.

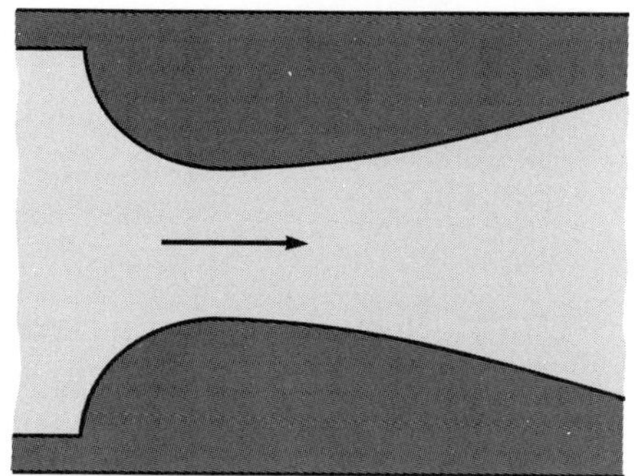

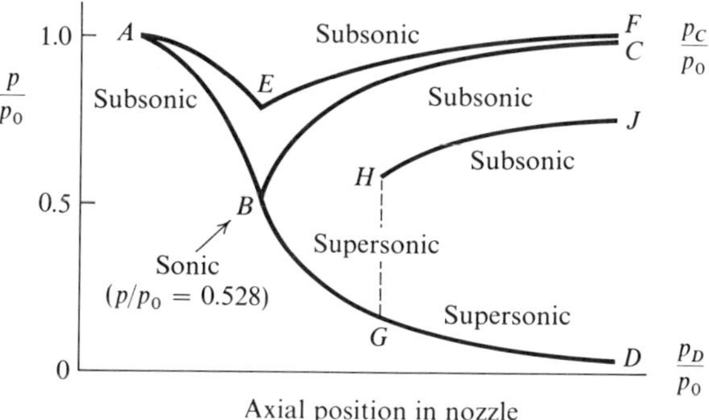

Fig. 8.21 *Pressure variation through a convergent-divergent nozzle. Perfect gas:* $\gamma = 1.4$.

Prior to making calculations concerning the operation of a nozzle, it will be useful to derive an area–Mach number relation for isentropic flow of a perfect gas. Conservation of mass for the steady one-dimensional flow is expressed by

$$\rho_1 A_1 V_1 = \rho_2 A_2 V_2$$

The densities may be expressed as a function of ρ_0 and N_M by use of Eq. (8.56), and the velocities may be written as

$$V_1 = N_{M1}(\gamma R T_1)^{1/2} \qquad \text{and} \qquad V_2 = N_{M2}(\gamma R T_2)^{1/2}$$

in which the temperature may be exchanged for functions of T_0 and N_M. Thus,

$$\frac{\rho_0 A_1 N_{M1}(\gamma R T_0)^{1/2}}{\left(1 + \dfrac{\gamma - 1}{2} N_{M1}{}^2\right)^{[1/(\gamma-1)]+1/2}} = \frac{\rho_0 A_2 N_{M2}(\gamma R T_0)^{1/2}}{\left(1 + \dfrac{\gamma - 1}{2} N_{M2}{}^2\right)^{[1/(\gamma-1)]+1/2}}$$

Noting that ρ_0 and T_0 are constants of the flow and solving for A_2/A_1 yields

$$\frac{A_2}{A_1} = \frac{N_{M1}}{N_{M2}} \left(\frac{1 + \dfrac{\gamma - 1}{2} N_{M2}{}^2}{1 + \dfrac{\gamma - 1}{2} N_{M1}{}^2} \right)^{(\gamma+1)/2(\gamma-1)} \tag{8.60}$$

which expresses the variation of area with Mach number during an isentropic process. Dropping the subscript 2, and setting $N_{M1} = 1.00$ with A_1 designated by A_1^* gives a form of Eq. (8.60) that lends itself to a tabulation of A/A^* as a function of the Mach number. This form is

$$\frac{A}{A^*} = \frac{1}{N_M} \left(\frac{1 + \dfrac{\gamma - 1}{2} N_M{}^2}{1 + \dfrac{\gamma - 1}{2}} \right)^{(\gamma+1)/2(\gamma-1)} \tag{8.61}$$

Table F.1 gives values of A/A^* as a function for N_M for $\gamma = 1.4$.

Example 8.16

Reconsider the nozzle flow given in Example 8.15 and calculate the minimum cross-sectional area A_1. What exit pressure p_2 would have to be maintained if isentropic flow is desired throughout the nozzle with subsonic streams in the convergent and divergent portions and sonic flow at A_1?

Solution. Since the flow is sonic at the throat,

$$\frac{A_2}{A_1} = \frac{A_2}{A^*} \qquad \text{and} \qquad \frac{A_2}{A^*} = 1.35$$

from Table F.1 with $N_{M_2} = 1.71$,

$$A^* = A_{\min} = \frac{12}{1.35} = 8.9 \text{ in}^2 \qquad ◀$$

Since the pressure p_2 that will produce a subsonic flow at the exit is desired, an exit Mach number must be found that is subsonic. Reference to Fig. 8.21 reveals that there are two exit pressures, p_C and p_D, which will produce sonic flow at the minimum area of the nozzle. The exit pressure corresponding to p_C is the required pressure. A_2/A^* is still equal to 1.35 and this corresponds to a subsonic Mach number $N_{M_2} = 0.495$ (Table F.1) and $p_2/p_0 = 0.846$. From this ratio,

$$\begin{aligned} p_2 &= 0.846\, p_0 = (0.846)(100) \\ &= 84.6 \text{ lb}_f/\text{in}^2 \qquad ◀ \end{aligned}$$

This example points out that there are two values of N_M that satisfy Eq. (8.61) for a given value of A/A^*, a supersonic and a subsonic Mach number. Whether $N_M > 1$ or < 1 depends on the exit pressure.

Normal shock wave. Very little has been stated concerning the line *A-B-G-H-J* in Fig. 8.21. The line *G-H* represents a pressure rise across a normal shock wave, which is a consequence of imposing some pressure p_J on the flow at the nozzle exit. The entropy of the flow is greater after the wave than it is ahead of the wave, as a result of irreversible effects in the wave. The *normal shock wave* in compressible flow is analogous to the hydraulic jump in free-surface flow. The ratio of various properties across the wave can be formulated as functions of the Mach number ahead of the wave and the constant specific-heat ratio γ. The normal shock wave is not very thick in the direction of the one-dimensional flow through it, so that the area normal to the flow ahead of the wave may be taken as equal to that after the wave. There is negligible heat transfer compared to other energy changes across the wave.

A control volume for analyzing the flow is shown in Fig. 8.22, and the basic equations are, for conservation of mass:

$$\rho_1 V_1 = \rho_2 V_2 \tag{8.62}$$

for linear momentum:

$$p_1 - p_2 = \rho_2 V_2^2 - \rho_1 V_1^2 \tag{8.63}$$

for energy:

$$\frac{V_1^2}{2} + h_1 = \frac{V_2^2}{2} + h_2 \tag{8.64}$$

for the equation of state:

$$\frac{p_1}{\rho_1 T_1} = \frac{p_2}{\rho_2 T_2} \tag{8.65}$$

The energy equation again indicates that the stagnation temperature T_0 is a constant of the flow; however, p_0 and ρ_0 are not constants since the

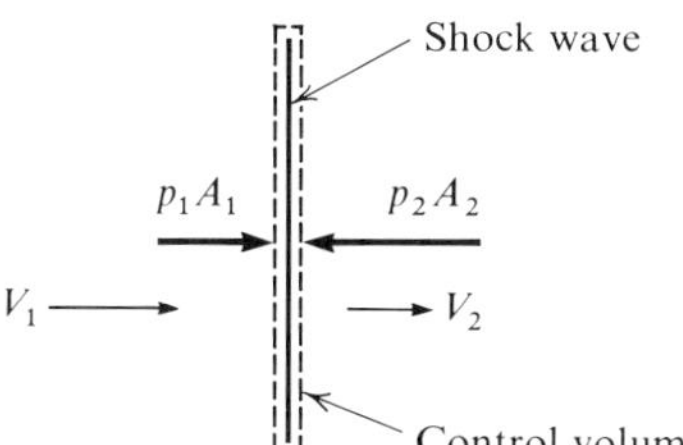

Fig. 8.22 *Control volume for normal shock wave.*

entropy increases across the wave. Also note that the momentum and energy equations are independent since the entropy is not constant.

The expression for p_2/p_1 may be obtained as a function of the Mach number ahead of the wave by using Eqs. (8.62) through (8.65). A consequence of the equation of state, the conservation of mass, and $h = c_p T$, is

$$\frac{T_2}{T_1} = \frac{h_2}{h_1} = \frac{p_2}{p_1}\frac{\rho_1}{\rho_2} = \frac{p_2}{p_1}\frac{V_2}{V_1} \tag{8.66}$$

Rearranging Eq. (8.63) to

$$1 - \frac{p_2}{p_1} = \frac{\rho_2 V_2^2 - \rho_1 V_1^2}{p_1} = \frac{\rho_1 V_1^2}{p_1}\left(\frac{\rho_2 V_2^2}{\rho_1 V_1^2} - 1\right)$$

and using Eq. (8.62) yields

$$1 - \frac{p_2}{p_1} = \gamma N_{M1}^2\left(\frac{V_2}{V_1} - 1\right) \tag{8.67}$$

by recognizing that

$$\frac{\rho_1 V_1^2}{p_1} = \gamma N_{M1}^2$$

The energy equation may be expressed in the following convenient form:

$$1 - \frac{h_2}{h_1} = \frac{\gamma - 1}{2} N_{M1}^2\left(\frac{V_2^2}{V_1^2} - 1\right) \tag{8.68}$$

Eliminating the velocity ratio and the enthalpy ratio from Eqs. (8.66), (8.67), and (8.68) yields, after algebraic manipulation,

$$\left(1 - \frac{p_2}{p_1}\right)\left(1 - \frac{1}{\gamma N_{M1}^2}\frac{p_2}{p_1}\right) = \left(1 - \frac{p_2}{p_1}\right)\left[\frac{\gamma - 1}{\gamma} + \frac{\gamma - 1}{2\gamma^2 N_{M1}^2}\left(1 - \frac{p_2}{p_1}\right)\right] \tag{8.69}$$

An obvious solution of Eq. (8.69) is $p_1 = p_2$, which is the case of continuous flow with no shock wave, a flow of no interest here. For $p_1 \neq p_2$, Eq. (8.69) may be divided by $1 - (p_2/p_1)$ and

$$1 - \frac{1}{\gamma N_{M1}^2}\frac{p_2}{p_1} = \frac{\gamma - 1}{\gamma} + \frac{\gamma - 1}{2\gamma^2 N_{M1}^2}\left(1 - \frac{p_2}{p_1}\right)$$

Solving for the pressure ratio,

$$\frac{p_2}{p_1} = \frac{2\gamma}{\gamma + 1} N_{M1}^2 - \frac{\gamma - 1}{\gamma + 1} \tag{8.70}$$

The velocity ratio as a function of N_{M1} may be obtained by substituting Eq. (8.70) into (8.67), yielding

$$\frac{V_2}{V_1} = \frac{\gamma - 1}{\gamma + 1} + \frac{2}{(\gamma + 1)N_{M1}^2} \tag{8.71}$$

and this is also equal to ρ_1/ρ_2 from Eq. (8.62).

Further algebraic manipulations of the preceding equations give

$$N_{M2}^2 = \frac{1 + \dfrac{\gamma - 1}{2} N_{M1}^2}{\gamma N_{M1}^2 - \dfrac{\gamma - 1}{2}} \tag{8.72}$$

and since

$$\frac{p_{02}}{p_{01}} = \frac{p_2}{p_1} \left(\frac{1 + \dfrac{\gamma - 1}{2} N_{M2}^2}{1 + \dfrac{\gamma - 1}{2} N_{M1}^2} \right)^{\gamma/(\gamma - 1)}$$

by definition of the isentropic stagnation pressure,

$$\frac{p_{02}}{p_{01}} = \left(\frac{2\gamma}{\gamma + 1} N_{M1}^2 - \frac{\gamma - 1}{\gamma + 1} \right)^{1/(1 - \gamma)} \left(\frac{2}{(\gamma + 1)N_{M1}^2} + \frac{\gamma - 1}{\gamma + 1} \right)^{\gamma/(1 - \gamma)} \tag{8.73}$$

Table F.2 in the appendix presents the temperature, density, pressure, and isentropic stagnation pressure ratios across a normal shock wave for $\gamma = 1.4$ as a function of N_{M1}. The Mach number after the wave, N_{M2}, is also tabulated.

An examination of the entropy change across the wave provides a basis for drawing further conclusions pertaining to this flow. Equation (6.35*b*) is applicable in the form

$$\$_2 - \$_1 = c_p \ln \frac{T_2}{T_1} - R \ln \frac{p_2}{p_1}$$

If one imagines isentropic stagnation states ($_{01}$) ahead of the wave and ($_{02}$) after the wave, then

$$\$_{02} - \$_{01} = \$_2 - \$_1 = c_p \ln \frac{T_{02}}{T_{01}} - R \ln \frac{p_{02}}{p_{01}}$$

Since T_0 is a constant of this flow, $\ln (T_{02}/T_{01}) = \ln 1 = 0$, and

$$\$_2 - \$_1 = -R \ln \frac{p_{02}}{p_{01}} \tag{8.74}$$

which can be expressed as a function of the initial Mach number by substitution for p_{02}/p_{01} using Eq. (8.73).

The variation of $(\$_2 - \$_1)/R$ with N_{M1} for a given value of γ is presented in Fig. 8.23. The similarity between this figure and Fig. 8.13 for the hydraulic jump is quite apparent. It appears that $\$_2 - \$_1 < 0$ for $N_{M1} < 1$, which is physically forbidden by the second law of thermodynamics for an adiabatic change of state. Indeed, a shock wave has never been observed in a subsonic flow. Since any dissipative action associated with the wave would require $\$_2 > \$_1$, Eq. (8.74) indicates that $p_{02} < p_{01}$, and the irreversibility of this flow through the wave produces a decrease of isentropic stagnation pressure. The entropy curve in Fig. 8.23 has a horizontal tangent at $N_M = 1$ indicating that small reversible pressure perturbations are permissible when the fluid velocity is equal to sonic speed. This is physically evidenced by alternate compressions and rarefactions at a point in the fluid through which sound waves are passing. The reader may readily observe from the equations presented previously and by reference to Table F.2, that velocity, Mach number, and stagnation pressure decrease across the wave, while temperature, pressure, density, and entropy increase. Since $\$_2 - \$_1$ is a rather strong function of N_{M1}, the aerodynamicist, when faced with accepting a normal shock wave, will try to design in such a way that the wave is accepted at the lowest practical supersonic Mach number, since he is aware that the wave with its attendant loss of stagnation pressure will impose a thrust penalty on the design of his vehicle.

Example 8.17

For the nozzle in Example 8.15, what pressure must be imposed on the nozzle exit to produce a normal shock wave at the exit?

Solution. From the calculations in Example 8.15, $N_{M_2} = 1.71$, and the pressure p_2

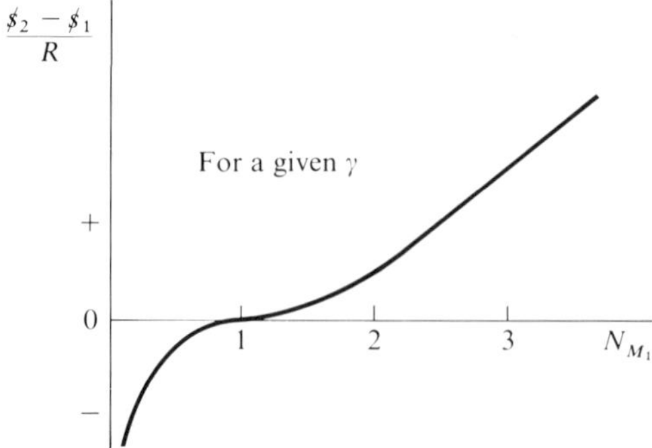

Fig. 8.23 Entropy change across a normal shock wave.

is given as 20 lb_f/in^2. Since isentropic flow prevails in the nozzle, these will be the Mach number and pressure ahead of the normal shock wave. Designating the receiver pressure (which is the pressure after the wave) as p_3, and using Table F.2,

$$\frac{p_3}{p_2} = 3.245 \qquad \text{at } N_{M_2}$$

$$p_3 = 3.245\,(20) = 64.9 \text{ lb}_f/\text{in}^2$$ ◀

The operating line for this nozzle would be similar to the locus *A-B-G-H-J* in Fig. 8.21 except that the wave at *G-H* would be at nozzle exit with $p_G = 20$ lb_f/in^2 and $p_H = p_J = 64.9$ lb_f/in^2.

Example 8.18

What force is necessary to hold the rocket in Fig. 8.24 on a static test stand? Assume isentropic expansion through the exit nozzle. Take $R = 53.3$ ft-lb_f/lb_m-°R and $\gamma = 1.4$ for the exit gas. The exit area is 12 in^2.

Solution. Select a control surface around the entire rocket. This is indicated by the dashed lines in Fig. 8.24. F_x is the force the static test stand exerts on the rocket to keep it stationary. Steady flow of the gas from the exit will be assumed since it is possible to supply both fuel and oxidizer at a steady rate from supply tanks on the test stand. The given data from Example 8.15 is the same as for this problem so that $V_2 = 1718$ ft/sec and $\dot{m} = 18.6$ lb_m/sec. The ambient pressure of 20 lb_f/in^2 acts over the entire control surface so that the net effect is zero. Applying the momentum equation (5.3) to the control volume of Fig. 8.24,

$$F_x = \dot{m}V_2 = \frac{(18.6)(1718)}{32.2} = 993 \text{ lb}_f$$ ◀

It is interesting to note that if the ambient pressure were 64.9 lb_f/in^2 in this example, a normal shock wave would exist at the nozzle exit and the force F_x would be considerably reduced, which would mean a reduction in thrust of the rocket motor (see

Fig. 8.24

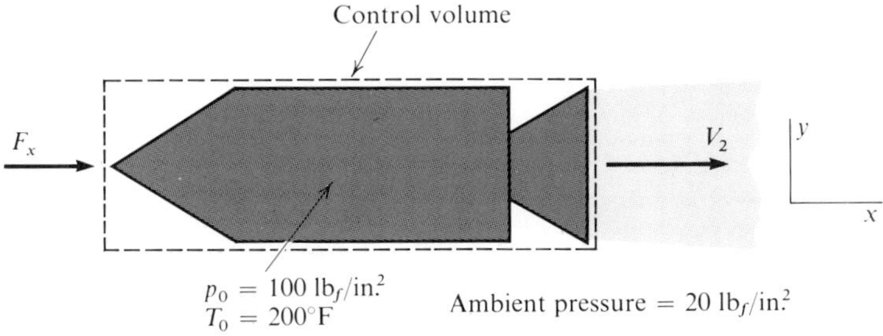

Example 8.17). The Mach number after the shock wave would be $N_{M_3} = 0.638$ (from Table F.2). Table F.1 yields

$$\frac{T_{03}}{T_3} = 0.924 \qquad \text{at } N_{M_3}$$

$$T_3 = (0.924)(660) = 611^\circ\text{R}$$

$$V_3 = N_{M_3}(\gamma R T_3)^{1/2} = 0.638\,[(1.4)(53.3)(611)(32.2)]^{1/2}$$

$$= 779 \text{ ft/sec}$$

$$F_x = \dot{m} V_3 = \frac{18.6}{32.2}(779) = 450 \text{ lb}_f$$

which indicates a considerable reduction in thrust. Note that the same mass flow of 18.6 lb_m/sec was used since the nozzle is choked and passing the maximum mass flow. From these results it is evident that in aircraft with jets exhausting at supersonic speeds, a normal shock wave may develop in the divergent portion of the tail-cone nozzle when the aircraft descends to a lower altitude where ambient pressure is higher. For this reason, more sophisticated aircraft engines are designed with variable areas at exit.

Adiabatic frictional flow of a perfect gas with constant c_p. The adiabatic flow of a perfect gas with constant c_p in a constant-area passage may be analyzed by a simplified one-dimensional treatment involving the basic equations; frictional effects may be included by using a relation between the friction factor and the wall shear stress. This is normally identified as *Fanno flow*.[1]

Figure 8.25 represents a constant-area passage through which a perfect gas flows steadily. Again the gas is assumed to have constant specific heats. The shear stress τ is indicated as being equivalent to $f\rho V^2/8$ since the assumed one-dimensional velocity is the space-average velocity. This shear stress–friction factor relation, derived as Eq. (8.24), assumed that the pressure forces were in balance with the shear forces, which is a consequence of assuming that the fluid is not accelerating as it flows through the passage.

[1]Named after Gino Fanno, an Italian engineer (1882–1962), whose thesis was concerned with frictional effects in compressible flows.

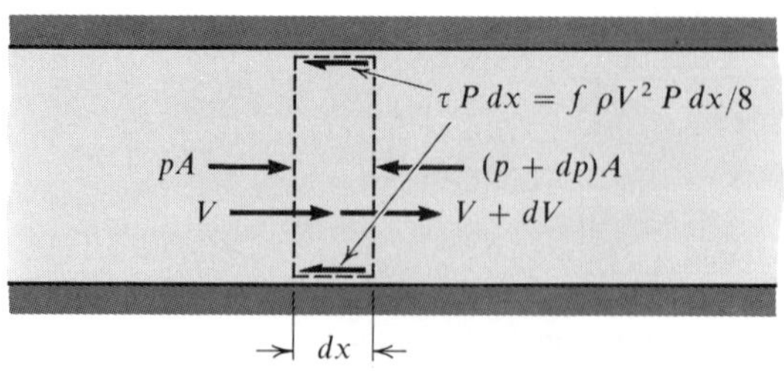

Fig. 8.25 Flow in a constant-area passage. P is the perimeter of the passage.

For the compressible fluid, the density changes in the direction of flow, so that the fluid velocity may change with time as the fluid passes through the conduit; hence Eq. (8.24) is only an approximation for the shear stress. However, this has been found to be reasonable for the subsonic Mach number range below $N_M = 0.5$. The basic equations of this flow are, for conservation of mass:

$$\rho V = \text{a constant} \tag{8.75}$$

for linear momentum:

$$-dp - \frac{\tau P}{A}\,dx = \rho V\,dV \tag{8.76}$$

for energy:

$$h + \frac{V^2}{2} = \text{a constant} \tag{8.77}$$

for the equation of state:

$$p = \rho RT \tag{8.78}$$

From Eq. (8.77) it is clear that the stagnation temperature is a constant of the flow.

Substituting $f\rho V^2/8$ for τ in Eq. (8.76) and noting that the hydraulic diameter $D_H = 4A/P$, changes the momentum equation to

$$dp = \rho\left(-\frac{fV^2}{2D_H}\,dx - V\,dV\right)$$

which can be written as

$$\frac{dp}{p} = \frac{1}{RT}\left(-\frac{fV^2}{2D_H}\,dx - V\,dV\right) \tag{8.79}$$

Next, dp/p can be eliminated from Eq. (8.79) by differentiation of Eq. (8.75) after substituting for ρ with Eq. (8.78). This procedure is

$$\frac{pV}{RT} = \text{a constant}$$

and $\ln p + \ln V = \ln T +$ another constant; thus, implicit differentiation yields

$$\frac{dp}{p} = \frac{dT}{T} - \frac{dV}{V}$$

Combining the preceding equation with (8.79) and rearranging terms gives

$$\frac{dT}{T} = \frac{V^2}{RT}\left(-\frac{f}{2D_H}\,dx - \frac{dV}{V}\right) + \frac{dV}{V} \tag{8.80}$$

which relates temperature and velocity. Noting that T_0 is a constant of this flow, the energy equation in the form of (8.53) may be differentiated to obtain

$$\frac{dT}{T} + \frac{(\gamma - 1)\,N_M\,dN_M}{1 + [(\gamma - 1)/2]\,N_M^{\,2}} = 0 \tag{8.81}$$

and differentiation of Eq. (8.39) yields

$$\frac{dN_M}{N_M} = \frac{dV}{V} - \frac{1}{2}\,\frac{dT}{T} \tag{8.82}$$

Finally, combining Eqs. (8.80), (8.81), and (8.82) gives

$$\frac{\gamma f}{2D_H}\,dx = \frac{(1 - N_M^{\,2})\,dN_M}{N_M^{\,3}\left(1 + \dfrac{\gamma - 1}{2}N_M^{\,2}\right)} \tag{8.83}$$

From this equation it becomes apparent that $dN_M/dx > 0$ for $N_M < 1$, which means that the role of the wall shear stress is to increase the Mach number in the direction of flow for a subsonic stream. For supersonic flow the opposite is true, that is, $dN_M/dx < 0$ when $N_M > 1$. For sonic flow ($N_M = 1$), $dx/dN_M = 0$, which indicates that the local Mach number of unity exists at the exit of the passage for the same constants of the flow, $\dot{m}$ and T_0. The second law of thermodynamics will bring to light some interesting facts concerning this flow.

The entropy function Eq. (6.34) may be expressed in differential form by noting that $dh = d(u + p/\rho)$, and

$$T\,d\$ = dh - \frac{dp}{\rho} \tag{8.84}$$

Substituting for dh and dp/ρ in Eq. (8.84) by using Eq. (8.77) and Eq. (8.79), respectively, gives

$$d\$ = \frac{fV^2}{2TD_H}\,dx = RN_M^{\,2}\left(\frac{\gamma f\,dx}{2D_H}\right)$$

The term in the parentheses may be eliminated by using Eq. (8.83) to give

$$\frac{d\$}{dN_M} = \frac{R(1 - N_M^{\,2})}{N_M\left(1 + \dfrac{\gamma - 1}{2}N_M^{\,2}\right)} \tag{8.85}$$

Considering $N_M < 1$, so that $dN_M > 0$, and noting that $d\$/dN_M > 0$ from Eq. (8.85), one concludes that $d\$ > 0$, a fact required by the second law of thermodynamics for irreversible adiabatic flow. For $N_M > 1$, with $dN_M < 0$, Eq. (8.85) requires that $d\$/dN_M < 0$, and again $d\$ > 0$. If the local Mach number is unity, then $d\$/dN_M = 0$; and the entropy is a maximum value when this flow becomes sonic. For a duct or passage of such a length that the exit Mach number is unity, any additional length placed at the exit will cause conditions at the entrance to change if the flow is subsonic. In the case of supersonic flow, any additional length of passage placed at the exit of the duct with an exit Mach number of unity will cause a normal shock wave to be formed in this passage. This wave may or may not be disgorged from the passage entrance, and the flow variables are altered by the presence of the shock wave as indicated by the previously derived equations which gave the ratios of these variables. A flow passage with an exit Mach number of unity is referred to as being *choked.*

Equation (8.83) can be integrated between limits that range from some position in the passage where the Mach number is known to a *critical length* where N_M is unity. This may be indicated as

$$\frac{\gamma f}{2D_H}\int_0^{L_{\text{crit}}} dx = \int_{N_M}^{1} \frac{(1 - N_M{}^2)\, dN_M}{N_M{}^3\left(1 - \dfrac{\gamma - 1}{2} N_M{}^2\right)} \tag{8.86}$$

provided the friction factor is assumed to be independent of x. By carrying out the integration, Eq. (8.86) becomes

$$L_{\text{crit}} = \frac{D_H}{f}\left[\frac{1 - N_M{}^2}{\gamma N_M{}^2} + \frac{\gamma + 1}{2\gamma}\ln\frac{(\gamma + 1)\, N_M{}^2}{2 + (\gamma - 1)\, N_M{}^2}\right] \tag{8.87}$$

The friction factor is taken from Fig. 8.4 by using a Reynolds number corresponding to conditions in the fluid at a location where the Mach number is N_M. This is obviously not correct in a strict sense since the curves of Fig. 8.4 do not consider the influence of the Mach number on the friction factor; also, the friction factor does vary some from the upstream location to the position in the passage where $N_M = 1$. However, Eq. (8.87) predicts a choking length L_{crit} to a reasonable degree of accuracy for subsonic flow with an upstream Mach number of 0.5 or less. Table F.3 gives a tabulation of critical or choking lengths as a function of the Mach number for a perfect gas with $\gamma = 1.4$. Also tabulated are the ratios p/p^*, $p_0/p_0{}^*$, T/T^*, and V/V^* as functions of N_M. The starred variables are the values of the variables at the location in this Fanno flow where the Mach number is unity, that is, where $N_{M^*} = 1$. These ratios are easily derived from the basic equations and the definition of the Mach number for a perfect gas. For example, to derive the p/p^* relation, conservation of matter may be expressed for the constant-area passage by

$$\rho V = \rho^* V^*$$

and using the equation of state to substitute for ρ,

$$\frac{p}{RT}(\gamma RT)^{1/2} N_M = \frac{p^*}{RT^*}(\gamma RT^*)^{1/2} N_{M^*}$$

so that

$$\frac{p}{p^*} = \frac{1}{N_M}\left[\frac{T}{T^*}\right]^{1/2}$$

Recall that T_0 is a constant of the flow, so

$$\frac{p}{p^*} = \frac{1}{N_M}\left[\frac{T_0\left(1 + \frac{\gamma - 1}{2}N_{M^*}^2\right)}{T_0\left(1 + \frac{\gamma - 1}{2}N_M^2\right)}\right]^{1/2}$$

and

$$\frac{p}{p^*} = \frac{1}{N_M}\left(\frac{\frac{\gamma + 1}{2}}{1 + \frac{\gamma - 1}{2}N_M^2}\right)^{1/2} \tag{8.88}$$

The ratio of stagnation pressures can be formulated as a function of the upstream Mach number by use of Eqs. (8.55) and (8.88). This yields

$$\frac{p_0}{p_0^*} = \frac{1}{N_M}\left(\frac{1 + \frac{\gamma - 1}{2}N_M^2}{\frac{\gamma + 1}{2}}\right)^{(\gamma + 1)/2(\gamma - 1)} \tag{8.89}$$

From Eq. (8.53), one may evolve

$$\frac{T}{T^*} = \frac{\gamma + 1}{2 + (\gamma - 1)N_M^2} \tag{8.90}$$

Finally, from conservation of matter,

$$\frac{V}{V^*} = \frac{\rho^*}{\rho} = \frac{p^*/RT^*}{p/RT} = \frac{p^*}{p}\frac{T}{T^*}$$

Equations (8.88) and (8.90) may be substituted directly into the preceding expression to give

$$\frac{V}{V^*} = N_M \left[\frac{\gamma + 1}{2 + (\gamma - 1) N_M{}^2}\right]^{1/2}$$

Example 8.19

A circular duct with an inside diameter of 1 ft handles air at subsonic speeds with $T_0 =$ 150°F. The pressure at the exit of the duct is 20 lb_f/in^2. If the duct is operating with an exit Mach number of unity, and is made from galvanized iron, (*a*) At what length upstream from the exit is the pressure 60 lb_f/in^2? (*b*) What is the velocity at the upstream location? (*c*) Calculate the temperature at exit. (*d*) What is the steady mass flow through the duct?

Solution. Since the exit Mach number is unity, $p^* = 20\ \text{lb}_f/\text{in}^2$. At the location where $p = 60\ \text{lb}_f/\text{in}^2$,

$$\frac{p}{p^*} = \frac{60}{20} = 3.00$$

$$N_M = 0.36 \qquad \text{from Table F.3}$$

$$\frac{T}{T_0} = 0.975 \qquad \text{from Table F.1}$$

$$T = (0.975)(610) = 595°\text{R}$$

The corresponding velocity is

$$V = N_M\,(\gamma RT)^{1/2} = (0.36)\,[(1.4)(53.3)(595)(32.2)]^{1/2}$$

$$= 431\ \text{ft/sec}$$ ◀

Calculating the Reynolds number with the value of μ at 135°F from Fig. E.1,

$$N_R = \frac{DV\rho}{\mu} = \frac{(1)(431)[(60)(144)/(53.3)(595)]}{4.15\,(10^{-7})\,(32.2)}$$

$$= 8.81\,(10^6)$$

At $e/D = 0.0005$, $f = 0.017$ from Fig. 8.4. Either Table F.3 or Eq. (8.87) may be used to obtain

$$L_{\text{crit}} = \frac{D_H}{f}\left(\frac{fL_{\text{crit}}}{D_H}\right) = \frac{1}{0.017}\,(3.18)$$

$$= 187\ \text{ft}$$ ◀

The temperature at exit is obtained by using T/T^* from Table F.3,

$$T^* = \left(\frac{T^*}{T}\right) T = \frac{595}{1.17} = 508^\circ \mathrm{R} \quad ◀$$

and the mass flow rate is obtained by

$$\dot{m} = \rho V A = (0.273)(431)\frac{\pi}{4}(1)^2 = 92.4 \ \mathrm{lb}_m/\mathrm{sec} \quad ◀$$

8.8 Closing Remarks

Two simplified approaches to the solution of fluid mechanics problems were presented in this chapter. The first method, dimensional analysis, provides a basic procedure whereby the pertinent variables may be organized into dimensionless groups that provide for similarity considerations and experiment design. The second approach consists of a one-dimensional idealization that can be applied to problems in which frictional effects are minimal, so that the fluid domain under consideration is essentially characterized by variables that change only in the direction of flow.

Self-Study Questions

8.1 What is the definition and the interpretation of each of the following dimensionless numbers:

a. Euler.
b. Reynolds.
c. Froude.
d. Strouhal.
e. Cauchy.

8.2 Identify the types of forces in the numerator and denominator of the following dimensionless numbers in which A is an area:

a. $\dfrac{p_\infty A}{\rho V_\infty{}^2 A}$ *b.* $\dfrac{\rho V_\infty{}^2 A}{A\mu V_\infty/L}$ *c.* $\dfrac{\rho A V_\infty{}^2}{\rho A L g}$

8.3 What characteristic is necessary if flow fields are

a. Kinematically similar?
b. Dynamically similar?
c. Geometrically similar?

8.4 When a term in a basic differential equation is expressed in a nondimensional form, why is a constant parameter chosen, for example, p_∞?

8.5 Name the dimensionless coefficients that should be important for modeling the following:

a. Watershed runoff.
b. Wind effects on smokestack effluent.
c. Drag on a high-speed submarine.
d. Acoustical effects on the transition flow region.

8.6 *a.* What is meant by dimensional homogeneity?
b. What is a dimensional equivalent of force F?

8.7 Briefly discuss an important use of dimensional analysis in experimentation.

8.8 State the Buckingham Pi theorem and briefly discuss the characteristics of the Pi terms.

8.9 Considering steady flow of a liquid in a long rough pipe,
a. Name six variables that influence the pressure drop in the fluid.
b. Identify the two Pi terms in

$$\frac{\Delta p_L}{\rho V_{SA}^2} = f(\pi_1, \pi_2)$$

8.10 State the definition of
a. Friction factor f.
b. Minor loss coefficient K_L.
c. Head loss h_L.
d. Modified Bernoulli equation.
e. Hydraulic diameter.
f. Relative roughness.

8.11 What is the region(s) of flow through rough pipes of constant diameter when the friction factor f is
a. Inversely proportional to the Reynolds number N_R?
b. Essentially independent of N_R and dependent on relative roughness?
c. Independent of relative roughness?

8.12 Consider a liquid flowing through a pipe whose diameter and elevation may change. What piping arrangement might yield

a. $\dfrac{\Delta p_L}{\rho} + h_L = 0$ *b.* $g\,\Delta z + h_L = 0$ *c.* $g\,\Delta z + \dfrac{\Delta V_{SA}^2}{2} = h_L$

8.13 Show that Eq. (8.18) is dimensionally homogeneous.

8.14 Obtain a relation between h_L and $\bar{d}Q/dm$ for steady isothermal flow of a liquid through a horizontal constant-diameter rough pipe.

8.15 Show that a minor loss may be expressed in an equivalent length of pipe that would cause the same head loss as $L_{eq} = K_L D/f$.

8.16 The steady flow of a fluid through a globe valve is an example of a throttling process where the enthalpy has the same value at entrance and exit. What is the relation of head loss to the internal energy change in the case of a liquid?

8.17 When is the pump work on an incompressible fluid equal to $\Delta p/\rho$?

8.18 How is uniform flow achieved in a fluid with a free surface?

8.19 What given information for open-channel and uniform flow is needed in order to obtain the space-average velocity using
a. Equation (8.27) by Manning? Is the solution direct?
b. Equation (8.25)? Is the solution direct?

8.20 Describe the flow to which Eq. (8.24) is applicable.

8.21 In the phenomena of hydraulic jump,
a. What does "hydraulic" imply?
b. What assumptions are allowed because of the rapid discontinuity in the flow?
c. Was discarding the negative of the double sign to obtain Eq. (8.31) a consequence of the second law of thermodynamics?

8.22 What criteria distinguishes tranquil from rapid flow in an open channel?

8.23 With respect to sonic celerity,
a. Describe the fluid for which this speed equals $(\gamma p/\rho)^{1/2}$.
b. Is it measured relative to any observer or to the fluid?
c. Establish that it is a property of the fluid.
d. Show that it depends on the molecular weight for a perfect gas.

8.24 Which of the Eqs. (8.40) through (8.44) are restricted to
a. Adiabatic flow?
b. Adiabatic and reversible flow?

8.25 In forming Eq. (8.49), could dF_w be written alternatively as $(p + dp/2)\,dA$?

8.26 Does the corresponding velocity of fluid particles increase or decrease as area increases for
a. Isentropic and subsonic flow of a perfect gas?
b. Isentropic and supersonic flow of water vapor?
c. Flow of an incompressible ideal fluid?

8.27 A perfect gas is stored in a high-pressure tank at temperature T. As the tank is emptied, the fluid within the tank remains essentially at constant temperature.
a. What is the maximum-possible exit Mach number?
b. What is the maximum-possible exit speed?

8.28 Derive Eq. (8.53) and clearly state all definitions and assumptions used.

8.29 *a.* State an important difference in the definition of stagnation temperature and stagnation pressure.
b. State the flow requirements for constant stagnation state, that is, p_0 = constant and T_0 = constant.

8.30 Consider the steady isentropic flow of a compressible fluid through a convergent-divergent passage. Sketch the expected variation of speed, density, and mass rate of flow per unit area with axial location.

8.31 Show how to obtain Eq. (8.54).

8.32 *a.* How is choked flow in a nozzle identified?
b. Is N_M always equal to unity at the minimum cross-sectional area?

8.33 What is the critical pressure of a monatomic gas flowing in a nozzle in terms of the stagnation pressure?

8.34 Obtain Eq. (8.59) by using

$$\frac{d\dot{m}}{d\,(T_{ex}/T_0)} = 0$$

8.35 Review the definition of the starred state used in Eq. (8.61).

8.36 What simplifying assumptions were used to derive flow and property relations for a normal shock wave?

8.37 *a.* For fluid particles passing through a normal shock wave, determine which of the following variables increase: T_0, T, p_0, p, ρ, V, N_M, A^*, $\$$.
b. Is it possible for a shock wave to form in a convergent passage?

8.38 *a.* Show that the "strength" of a shock, defined as the fractional increase in pressure, is equal to $\frac{2\gamma}{\gamma + 1}(N_M{}^2 - 1)$.
b. Show that the strength of a shock at $N_M = 5$ is eight times greater than at $N_M = 2$.

8.39 As a jet aircraft descends to higher ambient pressures, should the variable area at exit be increased or decreased?

8.40 What connection is there between critical length and a starred reference state in Fanno-type flow?

8.41 Consider the following variables: T_0, T, p_0, p, V, N_M, A^*, p^*; determine if their values for fluid particles in Fanno-type flow
a. Increase in subsonic conditions.
b. Decrease in supersonic conditions.
c. Remain constant.

8.42 Sketch the following processes on a *p-h* diagram like Fig. 8.26 for a gas at low pressure and identify a p_0 and T_0 for the end states:
a. $1 \rightarrow 2_\$$ by isentropic supersonic expansion.
b. $1 \rightarrow 2_f$ by constant area, adiabatic, and subsonic flow with friction.

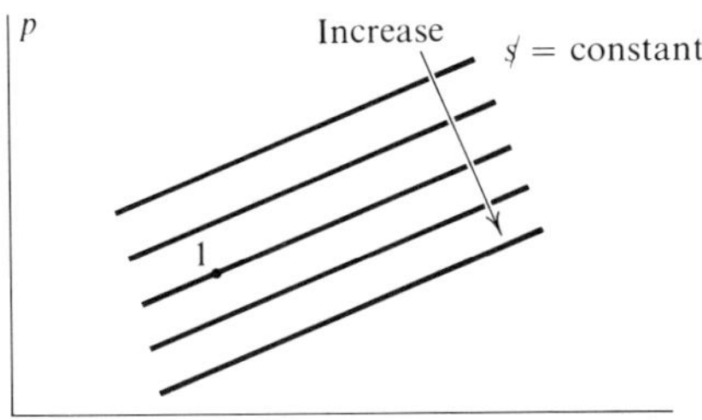

Fig. 8.26

8.43 What is choked flow in a constant-area insulated duct?

8.44 Establish the difference in definition of starred states (at $N_M{}^* = 1$) used in Tables F.1 and F.3.

Problems

8.1 Obtain these dimensionless coefficients

$$\frac{\kappa T_\infty}{\mu V_\infty{}^2} \quad \text{and} \quad \frac{c_v T_\infty L}{\nu V_\infty}$$

by expressing the following energy equation in nondimensional form:

$$\kappa\left(\frac{\partial^2 T}{\partial x^2} + \frac{\partial^2 T}{\partial y^2}\right) + \mu\left(\frac{\partial V_x}{\partial y}\right)^2 = V_x \rho c_v \frac{\partial T}{\partial x}$$

using

$$T = T_\infty T' \qquad V = V_\infty V' \qquad x = Lx' \qquad y = Ly'$$

8.2 Arrange the Bernoulli equation (3.17) in a dimensionless form and obtain nondimensional coefficients. Does a similarly applicable reduced form of nondimensional equation (8.1) yield the same coefficients?

8.3 The applicable momentum equation for a given fluid motion is

$$\rho\frac{\partial V_x}{\partial t} = -\frac{\partial p}{\partial x} + \mu\frac{\partial^2 V_x}{\partial y^2}$$

Use

$$t = \frac{L}{V_\infty}t' \qquad p = \rho V_\infty{}^2 p' \qquad V = V_\infty V' \qquad x = Lx' \qquad y = Ly'$$

to obtain a corresponding nondimensional form and identify the dimensionless coefficient(s).

8.4 A model study of water flow is to be conducted at a geometric scale of 1:25. Knowing that $N_F = f(N_E)$, what is the ratio of p_M/p_P if mercury with a specific gravity of 13.6 is used in the model study under the same gravity conditions?

8.5 A specialized study is to be made under different gravity conditions to predict pressure distributions in an air flow. Water at 90°F is used in a smaller model with a scale of 1:10. The Euler, Reynolds, and Froude numbers are considered important. Determine the required g_M/g_P and evaluate the ratio of p_M/p_P given $\nu_{air} = 0.4\ (10^{-3})$ ft²/sec, and $\rho_{air} = 0.046$ lb$_m$/ft³.

8.6 *a.* With reference to Prob. 6.18, obtain applicable differential equations for momentum and energy in nondimensional form by using

$$p = \rho V_{SA}{}^2 p' \qquad V = V_{SA} V' \qquad T = T_{SA} T' \qquad x = Lx' \qquad y = Ly'$$

b. Assume negligible viscous work in the energy equation, and show that two dimensionless coefficients, namely, Reynolds number N_R, and Peclet number $N_{Pe}\left(=\frac{c_v \rho V_{SA} L}{\kappa}\right)$, would be important for similarity.

c. Use air near room temperature and 1 atm in a model with $L_M/L_P = 1/5$ to represent a given flow of glycerin near room temperature, and estimate the test velocity needed for a prototype velocity of 5.0 ft/sec using Peclet modeling only.

8.7 An experiment is to determine if cavitation might occur in flow through variable-area cross sections. In the design of the experiment, the following are considered important:

p = fluid pressure $\qquad V$ = velocity
p_v = vapor pressure $\qquad D$ = diameter
L = length $\qquad \nu$ = kinematic viscosity
ρ = density

Determine an acceptable set of Pi terms for this set of variables.

8.8 Spherical objects are observed falling at terminal velocity in a body of quiescent and viscous fluid contained in a vertical cylinder. Obtain the Pi terms for pertinent variables of viscosity of the fluid, density of the fluid and sphere, terminal velocity, and diameters of sphere and cylinder.

8.9 The unsteady velocity of a fluid in the region near a large flat plate is considered to be dependent on plate velocity, normal distance from the plate, viscosity, density, and time. Show that there are three Pi terms and define their makeup.

8.10 When a valve is closed as fluid is passing through a pipe, inertial effects in the fluid may establish a phenomenon called *water hammer*. Establish the Pi terms for the following assumed pertinent variables: time for valve closure, diameter of pipe, length of pipe, fluid density, fluid velocity, pipe elasticity, pressure change at valve, pipe thickness, and sonic celerity in the fluid.

8.11 The pressure drop measured across a half-opened gate valve is 30 lb_f/in^2. Water passes steadily through the valve, which has piping diameters upstream and downstream of 1 in. and ½ in., respectively. Assume that the loss coefficient is based on entrance speed, and calculate the volume flow rate in ft^3/sec.

8.12 Calculate the horsepower imparted to the fluid to maintain the flow described in Example 8.6.

8.13 Oil is to be pumped from one vented tank to an elevated storage tank through 400 ft of pipe at a rate of 300 gal/min. The oil level in the vented storage tank varies from 70 to 90 ft higher than that in the lower tank. The pump is located near the outlet of the lower tank and can develop a pressure change of 60 lb_f/in^2 for this flow. The density and kinematic viscosity of the oil are 53.0 lb_m/ft^3 and 5 (10^{-5}) ft^2/sec, respectively. Use a friction factor of 0.02 to estimate the pipe diameter needed.

8.14 A liquid flows steadily through a horizontal 1-in. ID smooth pipe. The fluid exits as a free jet with a velocity of 50 ft/sec to the atmosphere through a nozzle which reduces the cross-sectional area to 0.4 of the pipe area. The density and kinematic viscosity are 2.0 $slug/ft^3$ and 4 (10^{-5}) ft^2/sec. What is the gage pressure of the fluid 20 ft upstream of the nozzle?

8.15 Water at 60° F flows steadily through a main at a rate of 45 lb_m/sec. It then branches into two lines. The sum of the minor loss coefficients in each line is 2.0. Each line is constructed of galvanized iron. One line with a diameter of 1 in. is to carry one-third of the flow while the other line of 2-in. diameter is to carry the remainder over a length of 200 ft before the two streams are rejoined. What should be the length of the 1-in. line?

8.16 A straight and horizontal pipe of constant cross-sectional area of 1 ft^2 has water at 80° F flowing steadily through a length of 10 ft. The pipe may be assumed smooth.

a. If the flow is considered to be frictionless, what is the pressure drop in the liquid over the given length?

b. If a pressure drop of 0.2 lb_f/in^2 is measured, evaluate the force exerted by the fluid on the pipe.

c. Estimate the velocity of the fluid in (*b*).

8.17 Liquid Freon-12 at 70° F is pumped through 400 ft of 1-in.-diameter drawn tubing that includes a wide-open globe valve, two 90° elbows, and a return bend.

a. What is the head loss for a flow rate of 20 lb_m/min?

b. Could the minor losses be reasonably disregarded?

8.18 Water flows from an open reservoir through a horizontal 6-in.-diameter pipe for a distance of 500 ft. The minor loss coefficient at the entrance to the pipe is 0.6. The water temperature is 40°F.

a. What is the height H of the water in the open tube for a flow velocity of 10 ft/sec as shown in Fig. 8.27?

b. What is the pressure in the fluid within the pipe at the entrance?

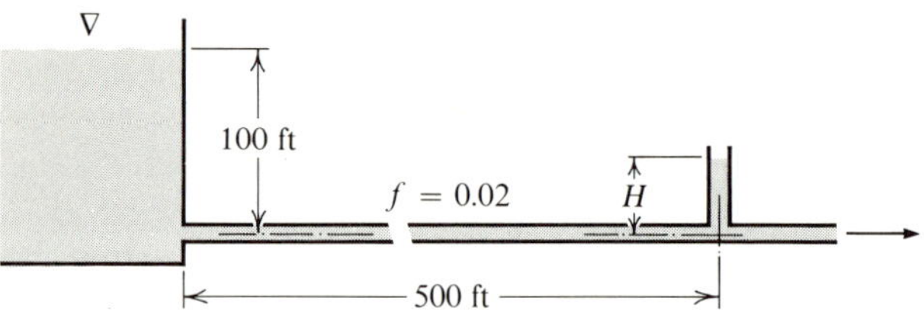

Fig. 8.27

8.19 Four ft³/sec of water at 100°F flows through the device shown in Fig. 8.28. There is 400 ft of 8-in.-diameter pipe between the pressure gages.

a. Calculate the horsepower delivered to the fluid by the pump.

b. Calculate the fluid pressure at the outlet of the pump if the suction gage is located at the entrance to the pump.

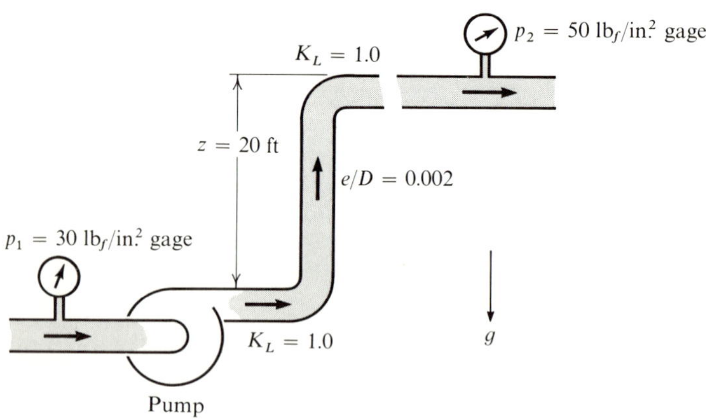

Fig. 8.28

8.20 Air flows through a galvanized duct with a square cross section. Estimate the gage reading in lb_f/in^2 at section 2 as shown in Fig. 8.29.

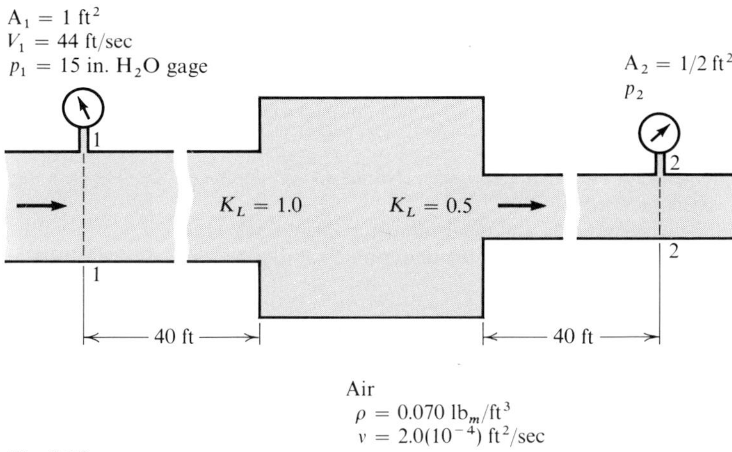

Fig. 8.29

8.21 The vertical location of the inlet to a water pump is to be determined. This is shown in Fig. 8.30. The tank is to be emptied at a rate of $\frac{1}{2}\ ft^3/sec$. The vapor pressure of the water is 0.949 lb_f/in^2

a. Estimate the maximum possible value of H in feet.

b. Estimate the maximum possible value of H in feet if there were no frictional losses.

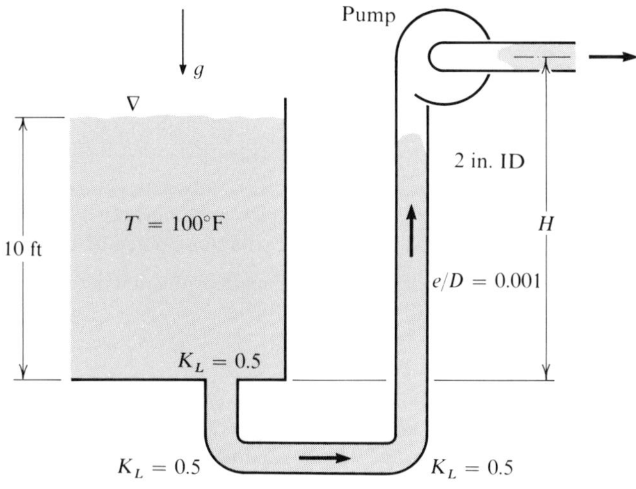

Fig. 8.30

8.22 Consider steady, turbulent, and uniform flow in an open rectangular channel of width W and height H. Establish the ratio of W/H that would yield the maximum velocity for a given cross-sectional area.

8.23 Estimate the uniform depth of water in a concrete-lined open channel with a slope of 0.0009 that carries 20,000 gal/min. The cross section is V-shaped with an angle of 60°.

8.24 A hydraulic jump occurs in a rectangular open channel and the depth doubles. The entering average velocity is 12 ft/sec.
a. Estimate the Froude number and the depth before the disturbance.
b. Estimate the upstream slope for steady uniform flow if the channel is 8 ft wide with a cement-mortar surface.

8.25 Calculate the sonic celerity in argon, nitrogen, and methane at 100° F.

8.26 Rework Example 8.15 without the use of Table F.1.

8.27 Air is to expand at a rate of 3 lb_m/sec through a convergent passage from a large tank at 100 $lb_f/in.^2$ and 350° F to a second large tank at 70 $lb_f/in.^2$ What is the smallest cross-sectional area of the nozzle?

8.28 Nitrogen passes steadily and isentropically through a variable-area passage. The flow is sonic at a 1-in.-diameter-minimum cross-sectional area. The stagnation pressure and temperature are 20 $lb_f/in.^2$ and 300° F, respectively. Determine
a. The mass flow rate.
b. The Mach number and pressure in a supersonic region of 2-in. diameter.

8.29 A perfect gas expands isentropically through a convergent-divergent nozzle. The gas has values of $c_p = 1.13$ Btu/lb_m-°R and $R = 351.3$ ft-lb_f/lb_m-°R. It enters at a low speed with pressure and temperature of 60 $lb_f/in.^2$ and 600° F and exits at 20 $lb_f/in.^2$ Calculate the exit velocity and the Mach number.

8.30 Derive an expression for the maximum mass rate of flow through an adiabatic nozzle in terms of variables A^*, p_0, and T_0. Assume that the fluid is a perfect gas with constant specific heats.

8.31 Design conditions have a gas entering a supersonic nozzle steadily at a pressure of 100 $lb_f/in.^2$ and temperature of 300°F. The monatomic gas has a specific-heat ratio of 5/3 and a gas constant of 386.0 ft-lb_f/lb_m°R. For an isentropic expansion, what is the ratio of minimum area to exit area for an exit $N_M = 2.0$?

8.32 By integration of Eq. (8.50), (8.51), or (8.52), obtain Eq. (8.60) for isentropic flow of a perfect gas with no shaft work and negligible change in gravitational potential energy.

8.33 Consider steady isentropic flow of air through a converging nozzle with a minimum area of ½ in.² The stagnation pressure and temperature are 40 $lb_f/in.^2$ and 200° F. Calculate the minimum temperature in the nozzle for
a. A receiver pressure of 34.6 $lb_f/in.^2$
b. A receiver pressure of 20 $lb_f/in.^2$

8.34 A fluid passes steadily through a converging nozzle with an area ratio of 7.26 and into a receiver maintained at a pressure of 15.0 $lb_f/in.^2$ The stagnation pressure is 50 $lb_f/in.^2$ The minimum area is 0.10 ft.² Calculate the force exerted by the fluid on the inner walls of the nozzle for isentropic flow of
a. Nitrogen with a stagnation density of 0.21 $lb_m/ft.^3$
b. An incompressible fluid with density of 1.93 slug/ft.³

8.35 A necked-down portion in a smooth 1-in. ID pipe has a diameter of 0.454 in. as shown in Fig. 8.31. If isentropic flow of a perfect gas with $\gamma = 1.4$ is sonic at the minimum cross section,

a. What is the upstream Mach number?

b. What values of the Mach number are possible downstream?

c. What are the corresponding pressures for (*b*) if p_0 is 20 lb_f/in^2?

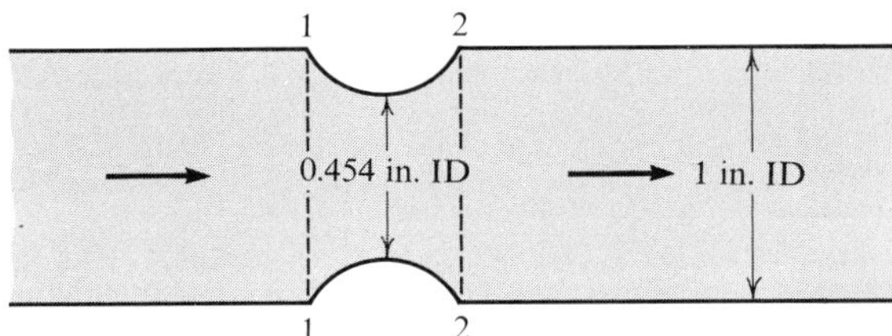

Fig. 8.31

8.36 Air flows steadily, reversibly, and adiabatically through a nozzle. A shock forms in the nozzle where $N_M = 2.18$, $p = 15$ lb_f/in^2, and $T = 140°F$. Calculate the change in stagnation pressure and specific entropy across the shock.

8.37 If a normal shock occurs in a perfect gas of $\gamma = 1.4$ at a Mach number of 2.0, what is the percent change of

a. Temperature?

b. Pressure?

c. Stagnation density?

8.38 Air enters a nozzle steadily at 100 lb_f/in^2, 300°F, and low velocity. The exit diameter is 2.2 times larger than the minimum diameter. If a shock exists at the exit section,

a. Determine the maximum mass flow rate per unit area.

b. Determine the corresponding N_M before and after the shock.

c. Sketch the variation of ρ/ρ_{01} with axial location and include the discontinuity at the shock. Label values at significant points.

8.39 The stagnation temperature and pressure of air entering a converging-diverging nozzle are 200°F and 100 lb_f/in^2, respectively. The nozzle exhausts into a region with a pressure maintained at 14.5 lb_f/in^2 The steady mass flow is 10 lb_m/sec. Assuming isentropic flow,

a. Calculate the exit area.

b. What exhaust-region pressure would produce a normal shock at the nozzle exit?

8.40 Consider the isentropic flow with a shock at the exit section in Prob. 8.39, and determine the net force exerted by the fluid on the nozzle walls between the entrance and exit. The entrance area is 3.68 times larger than the exit area.

8.41 A gas with $\gamma = 1.4$ flows steadily in an insulated and constant-area duct with $D_H = 3$ in. At one location, $N_M = 0.1$ and $p = 15.0$ lb_f/in^2 abs. Assuming $f = 0.01$, what is the location and the pressure downstream where $N_M = 0.4$?

8.42 A circular 8-in.-diameter duct receives air steadily. A normal shock wave stands in front of the duct with a corresponding Mach number of 2.92. For choked flow, *a.* Determine the velocity at the duct exit if the temperature ahead of the shock wave is $-60°$F.

b. Determine the length of the duct assuming $f = 0.015$.

8.43 Calculate the steady mass flow rate of air in choked flow through the 1-in.-diameter insulated passage shown in Fig. 8.32.

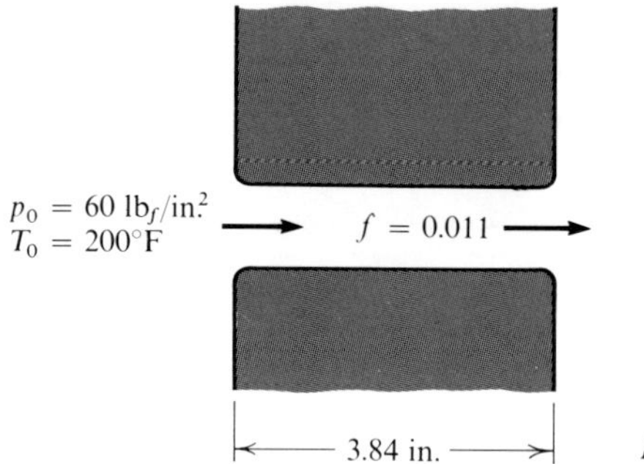

Fig. 8.32

8.44 An insulated and straight duct of triangular cross section, 1 ft on a side, has a surface with $e/D_H = 0.0002$. The air in steady flow at the entrance has $N_M = 0.30$, $p = 20$ lb$_f$/in.^{2}, and $T = 200°$F.

a. What duct length is required to yield $N_M = 0.50$?

b. What is the pressure at this location?

c. Check the value of f at the two Mach numbers mentioned.

d. Calculate the force exerted by the fluid on the duct walls between these two locations.

9

INTRODUCTION TO TURBULENCE AND BOUNDARY-LAYER FLOWS

9.1 Introduction

In this chapter a few basic features of turbulent flow will be presented, with the Navier-Stokes equation expressed in a time-averaged form applicable to these flows. With the assistance of some experimental information, techniques will be developed to predict velocity profiles for turbulent flows in pipes. The latter portion of the chapter will discuss boundary-layer flows as well as the drag on objects totally immersed in a fluid. The boundary-layer equations will be developed and applied in both differential and integral forms.

9.2 Comparison of Laminar and Turbulent Flows

In Chapter 4 a coefficient of viscosity was introduced in connection with Newton's hypothesis of flow resistance, and it was noted in Section 4.6 that a transfer of momentum at the molecular level was restricted to a well-ordered flow called laminar flow. One of the earliest experiments that demonstrated a significant difference between the laminar and turbulent flow regimes was performed by Osborne Reynolds in the latter part (1883) of the last century. He observed the flow of water through a glass tube after dye had been injected into the stream. The experiment is shown schematically in Fig. 9.1. For small flow rates, the streakline or dye filament remained relatively intact, and the transfer of momentum between the dye and the water was entirely at a molecular level. This indicated that the fluid was flowing in layers or laminas, each layer with a different velocity and sliding relative to contiguous layers; hence this regime is referred to as *laminar flow*.

At higher flow rates, the filament dispersed and mixed very thoroughly with the surrounding water before the fluid traversed the glass tube. This indicated that the mass and momentum transfer was at a macroscopic level. Migration of large chunks of fluid occurred from one region of the flow to another as well as diffusion of the much smaller fluid molecules. The regime characterized by this macroscopic transfer of momentum is referred to as *turbulent flow*. In laminar flow there is no migration of macroscopic chunks of fluid across the layers, while in turbulent flow the opposite is true.

Even though the higher flow rates were associated with turbulence in the Reynolds experiment, the characterizing parameter is the Reynolds number, which is the ratio of the inertia force to the viscosity force. There is a lower critical value of N_R at which the laminar flow begins transition to a turbulent flow regime. The experiments concerning pipe flow friction discussed in Chapter 8 indicated that the lower critical Reynolds number was approximately 2100 if the space-average velocity and inside pipe diameter are used as a characteristic velocity and a characteristic length for defining N_R. There is a higher N_R at which turbulent flow is established. In the case of pipe flow, Fig. 8.4 indicated this to be about 4000. However, laminar flow has been observed at a Reynolds number of 40,000 for the flow of an incompressible fluid in a circular conduit. In fact, it appears that the stability

Fig. 9.1 *a. Filament of dye remains "intact": laminar flow. (b) Filament of dye mixes with water: turbulent flow.*

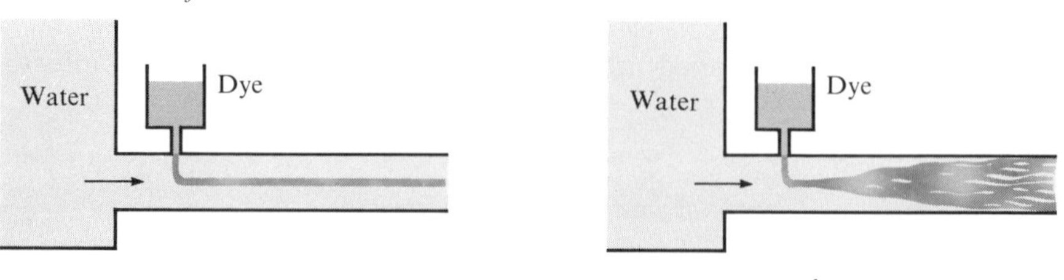

of a given laminar flow depends on the nature of disturbances imposed on the flow as well as the Reynolds number. The origin and nature of these disturbances are not completely understood. Some mathematical models are possible, and subsequent analyses have yielded informative results. These studies are classified under the general description of *hydrodynamic stability*, an area more suitable for advanced study.

With such difficulties facing the engineer, the question arises as to what can be accomplished in the way of estimating velocity profiles and shear stresses in turbulent flow. As usual, some model must be developed with the customary help of experiment, and a portion of this chapter considers several elementary possibilities. However, the definition of a few technical terms and averaging processes are necessary before analysis.

9.3 Fluctuating Variables and Time Averaging

A given function (or variable) in turbulent flow is considered to be composed of a *mean* part and a *fluctuating* part. For example, the x component of velocity V_x may be written as

$$V_x = \overline{V}_x + \tilde{V}_x \tag{9.1}$$

in which $\overline{V}_x$ is the *time-mean* part and $\tilde{V}_x$ is the *time-fluctuating* part. Figure 9.2 shows the time-mean velocity with the fluctuating part superposed. Since the fluctuation may yield an instantaneous value of V_x above (positive) or below (negative) the time-mean value of V_x, the quantity $\tilde{V}_x$ in Eq. (9.1) is added algebraically to $\overline{V}_x$, which is apparent from Fig. 9.2. It remains to

Fig. 9.2 *The variation of V_x with time at a given point in the fluid for steady turbulent flow.*

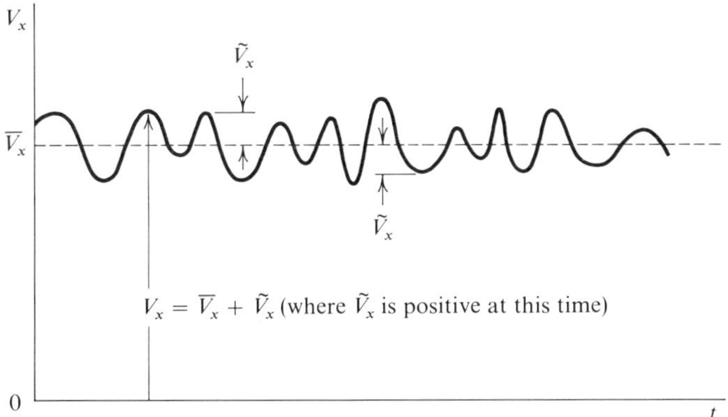

define $\overline{V}_x$, the time-mean value of V_x. This is effected by the following equation:

$$\overline{V}_x = \frac{1}{\Delta t}\int_t^{t+\Delta t} V_x\,dt \tag{9.2}$$

in which Δt is a large time interval compared to the time of a fluctuation. If $\overline{V}_x$ is independent of time at each point in the fluid, then the flow is referred to as being *steady turbulent* or *steady in the mean* with respect to the variable V_x. An unsteady turbulent flow would be one in which the mean value of the variable would vary with time at each point in the field. This text will model and examine *only steady turbulent flows of an incompressible fluid.*

9.4 Some Rules for Time Averaging in Steady Turbulent Flow

For two continuous fluid properties or flow variables M and N, each composed of time-mean and fluctuating parts,

$$M = \overline{M} + \tilde{M} \qquad \text{and} \qquad N = \overline{N} + \tilde{N}$$

the sum (or difference) of the time averages of M and N is equal to the time average of the sum (or difference) of these. This rule follows by noting that

$$\overline{M \pm N} = \frac{1}{\Delta t}\int_t^{t+\Delta t} (M \pm N)\,dt$$

from the defining equation (9.2). Because the limit of a sum (or difference) is equal to the sum (or difference) of the limits of two quantities, the right-hand side of the preceding equation may be altered to give

$$\overline{M \pm N} = \frac{1}{\Delta t}\int_t^{t+\Delta t} M\,dt \pm \frac{1}{\Delta t}\int_t^{t+\Delta t} N\,dt$$

which in turn may be written as

$$\overline{M \pm N} = \overline{M} \pm \overline{N} \tag{9.3}$$

Another useful rule in averaging is that the time average of a fluctuating component is zero, or

$$\overline{\tilde{M}} = 0 \tag{9.4}$$

a result readily obtained by noting that $\tilde{M} = M - \overline{M}$ and time averaging this equation as follows:

$$\overline{\tilde{M}} = \overline{M - \overline{M}} = \overline{M} - \overline{\overline{M}}$$

However,

$$\overline{\overline{M}} = \frac{1}{\Delta t}\int_t^{t+\Delta t} \overline{M}\,dt = \frac{\overline{M}}{\Delta t}\int_t^{t+\Delta t} dt = \overline{M}$$

so that $\overline{\tilde{M}} = \overline{M} - \overline{M} = 0$, which is what Eq. (9.4) alleges. Caution should be exercised at this point not to extrapolate this result to the time average of the product of two fluctuating components. The quantity $\overline{\tilde{M}\tilde{N}}$ does not have to vanish; indeed, it seldom is equal to zero. The verification of this last statement is left as an exercise in Probs. 9.2 or 9.3 at the end of the chapter. Another exercise left for the reader is

$$\overline{AM} = A\overline{M} \tag{9.5}$$

where A is a constant.

The last rule of averaging, which is necessary for carrying out the analyses in this chapter, is that partial differentiation with respect to a space coordinate and taking the time average of a variable are commutative operations. The validity of this may be demonstrated by time averaging the partial derivative, such as,

$$\overline{\frac{\partial M}{\partial x}} = \frac{1}{\Delta t}\int_t^{t+\Delta t} \frac{\partial M}{\partial x}\,dt = \frac{\partial}{\partial x}\left(\frac{1}{\Delta t}\int_t^{t+\Delta t} M\,dt\right)$$

since Δt and the limits of the integral are independent of space coordinates. The term in parentheses is the time average of M, and

$$\overline{\frac{\partial M}{\partial x}} = \frac{\partial \overline{M}}{\partial x} \tag{9.6}$$

which demonstrates the stated commutative property.

With the few preceding simple rules concerning the algebra and calculus of time-averaging flow variables, it is now possible to time average some basic equations in which these variables appear.

9.5 Time Averaging the Continuity Equation for an Incompressible Fluid

The continuity equation at any given instant of time for an incompressible fluid is div $\mathbf{V} = 0$. If each velocity component is replaced by a time-mean and fluctuating quantity, the following results:

$$\frac{\partial(\overline{V}_x + \tilde{V}_x)}{\partial x} + \frac{\partial(\overline{V}_y + \tilde{V}_y)}{\partial y} + \frac{\partial(\overline{V}_z + \tilde{V}_z)}{\partial z} = 0$$

carrying out the differentiation and rearranging gives

$$\frac{\partial \overline{V}_x}{\partial x} + \frac{\partial \overline{V}_y}{\partial y} + \frac{\partial \overline{V}_z}{\partial z} + \frac{\partial \tilde{V}_x}{\partial x} + \frac{\partial \tilde{V}_y}{\partial y} + \frac{\partial \tilde{V}_z}{\partial z} = 0 \tag{9.7}$$

The time average of this equation is

$$\overline{\frac{\partial \overline{V}_x}{\partial x}} + \overline{\frac{\partial \overline{V}_y}{\partial y}} + \overline{\frac{\partial \overline{V}_z}{\partial z}} + \overline{\frac{\partial \tilde{V}_x}{\partial x}} + \overline{\frac{\partial \tilde{V}_y}{\partial y}} + \overline{\frac{\partial \tilde{V}_z}{\partial z}} = 0$$

applying Eq. (9.3). Since the commutative property, Eq. (9.6), prevails and $\overline{\tilde{V}_x} = \overline{\tilde{V}_y} = \overline{\tilde{V}_z} = 0$, it is apparent that

$$\frac{\partial \overline{\overline{V}}_x}{\partial x} + \frac{\partial \overline{\overline{V}}_y}{\partial y} + \frac{\partial \overline{\overline{V}}_z}{\partial z} = 0$$

and

$$\frac{\partial \overline{V}_x}{\partial x} + \frac{\partial \overline{V}_y}{\partial y} + \frac{\partial \overline{V}_z}{\partial z} = 0 \tag{9.8}$$

which is the same equation satisfied by the velocity components V_x, V_y, and V_z. Subtracting Eq. (9.8) from (9.7) gives

$$\frac{\partial \tilde{V}_x}{\partial x} + \frac{\partial \tilde{V}_y}{\partial y} + \frac{\partial \tilde{V}_z}{\partial z} = 0 \tag{9.9}$$

which is also the same form of equation satisfied by **V**.

Before averaging the Navier-Stokes equations, a further example of the time averaging of physical equations is given.

Example 9.1

Time average the continuity equation for a compressible fluid with

$$V_y = V_z = 0 \qquad \text{and} \qquad V_x = V_x(x,t)$$

Assume that the velocity and the density are steady in the mean.

Solution. The equation of continuity for this flow is

$$\frac{\partial \rho}{\partial t} + \frac{\partial (\rho V_x)}{\partial x} = 0$$

and

$$\frac{\partial(\bar{\rho} + \tilde{\rho})}{\partial t} + (\bar{\rho} + \tilde{\rho})\frac{\partial(\overline{V}_x + \tilde{V}_x)}{\partial x} + (\overline{V}_x + \tilde{V}_x)\frac{\partial(\bar{\rho} + \tilde{\rho})}{\partial x} = 0$$

Carrying out the differentiation and performing the indicated multiplication gives

Term (1) (2) (3) (4) (5) (6)

$$\frac{\partial\bar{\rho}}{\partial t} + \frac{\partial\tilde{\rho}}{\partial t} + \bar{\rho}\frac{\partial\bar{V}_x}{\partial x} + \bar{\rho}\frac{\partial\tilde{V}_x}{\partial x} + \tilde{\rho}\frac{\partial\bar{V}_x}{\partial x} + \tilde{\rho}\frac{\partial\tilde{V}_x}{\partial x}$$

(7) (8) (9) (10)

$$+ \bar{V}_x\frac{\partial\bar{\rho}}{\partial x} + \bar{V}_x\frac{\partial\tilde{\rho}}{\partial x} + \tilde{V}_x\frac{\partial\bar{\rho}}{\partial x} + \tilde{V}_x\frac{\partial\tilde{\rho}}{\partial x} = 0$$

Term 1 is zero since the density field is steady in the mean. Upon time averaging the preceding equation, the following facts accrue:

Term 2 becomes zero since the limits on the integral defining the time average may be taken as any time t and $t + \Delta t$, and the partial differentiation may be performed after the time average is taken.
Terms 3 and 7 remain unchanged since ρ and V_x are steady in the mean.
Terms 4, 5, 8, and 9 average out to zero since their reduction terminates in taking the time average of a fluctuating quantity.
Terms 6 and 10 cannot be reduced.

The final averaged equation is

$$\overline{\tilde{\rho}\frac{\partial\tilde{V}_x}{\partial x}} + \overline{\tilde{V}_x\frac{\partial\tilde{\rho}}{\partial x}} + \bar{\rho}\frac{\partial\bar{V}_x}{\partial x} + \bar{V}_x\frac{\partial\bar{\rho}}{\partial x} = 0 \quad ◀$$

which is equivalent to

$$\frac{\partial}{\partial x}\overline{(\tilde{\rho}\tilde{V}_x)} + \frac{\partial}{\partial x}(\bar{\rho}\bar{V}_x) = 0 \quad ◀$$

Either of the last two equations could be considered a form of the answer.

Attention is drawn to the extra term $\partial/\partial x\overline{(\tilde{\rho}\tilde{V}_x)}$, in the last equation of Example 9.1. The quantity $\overline{\tilde{\rho}\tilde{V}_x}$ is referred to as an *apparent* flow. In the next section apparent stresses or apparent momentum fluxes occur when the Navier-Stokes equations are time averaged, even for the case of the incompressible fluid.

9.6 Time Averaging the Navier-Stokes Equations for an Incompressible Fluid

The x component of the constituted equation of motion is given by Eq. (4.20a), which may be altered to the following form for a steady velocity field and an incompressible fluid:

$$V_x\frac{\partial V_x}{\partial x} + V_y\frac{\partial V_x}{\partial y} + V_z\frac{\partial V_x}{\partial z} = -\frac{1}{\rho}\frac{\partial p}{\partial x} + \nu\left(\frac{\partial^2 V_x}{\partial x^2} + \frac{\partial^2 V_x}{\partial y^2} + \frac{\partial^2 V_x}{\partial z^2}\right) + (\mathbf{f}_B)_x \quad (9.10)$$

A typical term of the convective acceleration when time averaged would consist of

$$\overline{V_y\frac{\partial V_x}{\partial y}} = \overline{(\bar{V}_y + \tilde{V}_y)\frac{\partial}{\partial y}(\bar{V}_x + \tilde{V}_x)}$$

$$= \overline{\bar{V}_y\frac{\partial \bar{V}_x}{\partial y}} + \overline{\bar{V}_y\frac{\partial \tilde{V}_x}{\partial y}} + \overline{\tilde{V}_y\frac{\partial \bar{V}_x}{\partial y}} + \overline{\tilde{V}_y\frac{\partial \tilde{V}_x}{\partial y}}$$

or on further reduction,

$$\overline{V_y\frac{\partial V_x}{\partial y}} = \bar{V}_y\frac{\partial \bar{V}_x}{\partial y} + \overline{\tilde{V}_y\frac{\partial \tilde{V}_x}{\partial y}}$$

Similar terms would result upon averaging $V_x(\partial V_x/\partial x)$ and $V_z(\partial V_x/\partial z)$. The three terms on the right-hand side of Eq. (9.10) time average to

$$-\frac{1}{\rho}\frac{\partial \bar{p}}{\partial x} \qquad \nu\left(\frac{\partial^2 \bar{V}_x}{\partial x^2} + \frac{\partial^2 \bar{V}_x}{\partial y^2} + \frac{\partial^2 \bar{V}_x}{\partial z^2}\right) \qquad \text{and} \qquad (\bar{\mathbf{f}}_B)_x$$

respectively; and Eq. (9.10) when averaged becomes

$$\bar{V}_x\frac{\partial \bar{V}_x}{\partial x} + \bar{V}_y\frac{\partial \bar{V}_x}{\partial y} + \bar{V}_z\frac{\partial \bar{V}_x}{\partial z} = -\frac{1}{\rho}\frac{\partial \bar{p}}{\partial x} + \nu\left(\frac{\partial^2 \bar{V}_x}{\partial x^2} + \frac{\partial^2 \bar{V}_x}{\partial y^2} + \frac{\partial^2 \bar{V}_x}{\partial z^2}\right) + (\bar{\mathbf{f}}_B)_x - \overline{\tilde{V}_x\frac{\partial \tilde{V}_x}{\partial x}} - \overline{\tilde{V}_y\frac{\partial \tilde{V}_x}{\partial y}} - \overline{\tilde{V}_z\frac{\partial \tilde{V}_x}{\partial z}} \quad (9.11)$$

which is the same as Eq. (9.10) written for the mean flow, with three additional terms subtracted from the right-hand side of the equation. These three terms may be placed in an alternate form that more readily lends to their physical interpretation. To do this, one need only use the following identity:

$$\tilde{V}_x\frac{\partial \tilde{V}_x}{\partial x} + \tilde{V}_y\frac{\partial \tilde{V}_x}{\partial y} + \tilde{V}_z\frac{\partial \tilde{V}_x}{\partial z} = \frac{\partial(\tilde{V}_x^{\,2})}{\partial x} + \frac{\partial(\tilde{V}_x\tilde{V}_y)}{\partial y} + \frac{\partial(\tilde{V}_x\tilde{V}_z)}{\partial z} - \left[\tilde{V}_x\left(\frac{\partial \tilde{V}_x}{\partial x} + \frac{\partial \tilde{V}_y}{\partial y} + \frac{\partial \tilde{V}_z}{\partial z}\right)\right]$$

The terms in brackets of the last equation vanish since Eq. (9.9) requires that

$$\frac{\partial \tilde{V}_x}{\partial x} + \frac{\partial \tilde{V}_y}{\partial y} + \frac{\partial \tilde{V}_z}{\partial z} = 0$$

Equation (9.11) can now be written as

$$\overline{V}_x \frac{\partial \overline{V}_x}{\partial x} + \overline{V}_y \frac{\partial \overline{V}_x}{\partial y} + \overline{V}_z \frac{\partial \overline{V}_x}{\partial z} = -\frac{1}{\rho}\frac{\partial \bar{p}}{\partial x} + \nu\left(\frac{\partial^2 \overline{V}_x}{\partial x^2} + \frac{\partial^2 \overline{V}_x}{\partial y^2} + \frac{\partial^2 \overline{V}_x}{\partial z^2}\right)$$

$$+ (\overline{\mathbf{f}_B})_x + \frac{1}{\rho}\left[\frac{\partial}{\partial x}\overline{(-\rho \tilde{V}_x^2)} + \frac{\partial}{\partial y}\overline{(-\rho \tilde{V}_x \tilde{V}_y)} + \frac{\partial}{\partial z}\overline{(-\rho \tilde{V}_x \tilde{V}_z)}\right] \qquad (9.12)$$

Similarly the y and z components of the Navier-Stokes equations could be time averaged, each one yielding three additional terms similar to those in the last set of brackets of Eq. (9.12), and there would be a total of nine such terms appearing mathematically in the equations as though each one were a derivative of a stress. The quantities $\overline{-\rho \tilde{V}_i \tilde{V}_j}$ are referred to as *apparent stresses*, *Reynolds stresses*, or *turbulent stresses*.

That each of these nine quantities has the character of stress may be seen from a momentum-exchange model. Figure 9.3 shows a *mean-velocity* profile in the x, y coordinate plane. Because the flow is turbulent, a fluctuating velocity component $\tilde{V}_y$ in the y direction could cause a macroscopic chunk of fluid to migrate from some antecedent position B to a subsequent position C, which is situated at a higher velocity (and momentum) level in the fluid. If this chunk of fluid passes through the element of area A, then the mass flow is $\rho \tilde{V}_y A$, since the mean flow is in the x direction only. Upon arriving at position C there is a momentum exchange that produces a fluctuation $\tilde{V}_x$ in the x component of velocity at C.‡ The rate of momentum exchange in the x direction at C is $\rho \tilde{V}_x \tilde{V}_y A$, which is equal to the force in the

‡The momentum exchange at C could also produce fluctuations $\tilde{V}_y$ and $\tilde{V}_z$ in the other two coordinate directions.

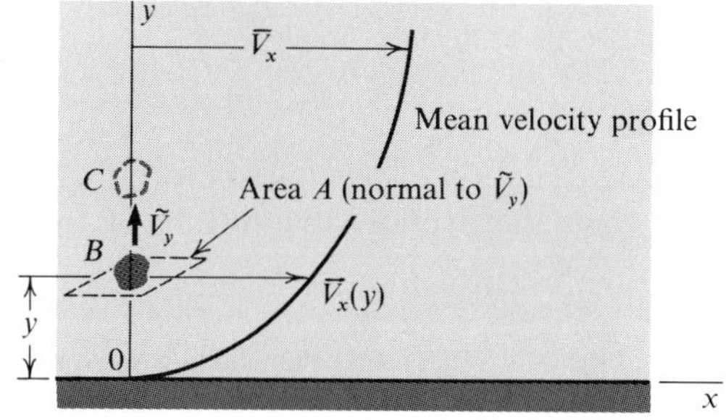

Fig. 9.3 *Mean-velocity profile for turbulent flow.*

x direction that changed this momentum. The force per unit area, which has the physical interpretation of stress, is given by $\rho \tilde{V}_x \tilde{V}_y$, and it is readily seen that each of the nine additional terms generated by the time averaging of the equations of motion is a derivative of stress or momentum flux density.

The minus sign that occurs in front of each of these nine terms is a consequence of the fact that a positive $\tilde{V}_y$ moves the chunk of fluid to a higher velocity level, and the momentum exchange produces a negative $\tilde{V}_x$. The product $\tilde{V}_x \tilde{V}_y$ is then negative, which would be true if $\tilde{V}_y$ were negative and $\tilde{V}_x$ were positive. This argument is predicated on the shape of the profile in Fig. 9.3 (note that $\overline{V}_x$ increases as y increases). However, this same condition was imposed on the derivation of the equations of motion in Chapter 3, which were constituted in Chapter 4, and time averaged here.

It now remains to find some model whereby these turbulent shear stresses may be estimated in terms of variables that are at the disposal of the designer or analyst. Unfortunately there is no general method of finding these stresses for complicated flows; however, several methods have been devised that can utilize a generalized form of experimental data and predict the velocity distribution for turbulent flows in which the velocity is a function of a single space coordinate. One of the earliest methods, and perhaps the simplest, is the Prandtl mixing-length theory presented in the next section.

9.7 *Mixing-Length Theory of Prandtl*

Prandtl[1] assumed that the *average magnitude of the velocity fluctuation* at a given location in the fluid is proportional to the *mean distance traveled* by a macroscopic particle of fluid randomly fluctuating from some antecedent location to the given location. He further assumed that the two fluctuating components of velocities at right angles to each other were of the same order of magnitude and proportional to each other. A model may be formed by using these assumptions and considering a small aggregation of macroscopic particles of fluid with a mean velocity $\overline{V}_x(y)$ fluctuating through a distance λ to a position where the mean velocity is $\overline{V}_x(y + \lambda)$ as shown in Fig. 9.4. The velocity difference in the flow field from y to $y + \lambda$ is

$$\left[\overline{V}_x(y) + \frac{d\overline{V}_x}{dy}\lambda\right] - \overline{V}_x(y) = \lambda \frac{d\overline{V}_x}{dy}$$

If λ is the mean distance traveled by the aggregation of particles before there is a transfer of momentum at $y + \lambda$, then the average magnitude of the velocity fluctuation may be taken as

[1]Ludwig Prandtl, "Essentials of Fluid Dynamics," pp. 117–118, Hafner Publishing Company, Inc., New York, 1952.

$$\overline{|\tilde{V}_x|} = \lambda \left| \frac{d\overline{V}_x}{dy} \right|$$

Since $\tilde{V}_x$ is assumed proportional to $\tilde{V}_y$,

$$\overline{|\tilde{V}_y|} = C_1 \lambda \left| \frac{d\overline{V}_x}{dy} \right|$$

in which C_1 is a constant, and

$$\overline{|\tilde{V}_x|}\ \overline{|\tilde{V}_y|} = C_1 \lambda^2 \left| \frac{d\overline{V}_x}{dy} \right|^2 = C_1 \lambda^2 \left(\frac{d\overline{V}_x}{dy} \right)^2$$

Recall that the turbulent shear stress $\tau_{xy,\,\text{turb}}$ is equal to $-\rho \overline{\tilde{V}_x \tilde{V}_y}$, so it remains to relate $\overline{|\tilde{V}_x|}\ \overline{|\tilde{V}_y|}$ to $\overline{\tilde{V}_x \tilde{V}_y}$. Recalling from Section 9.6 that $\tilde{V}_x$ and $\tilde{V}_y$ are of opposite sign, it is possible to write

$$\overline{\tilde{V}_x \tilde{V}_y} = -C_2 \overline{|\tilde{V}_x|}\ \overline{|\tilde{V}_y|}$$

with C_2 as a positive constant less than one. Then

$$\tau_{xy,\,\text{turb}} = -\rho \overline{\tilde{V}_x \tilde{V}_y} = \rho C_2 \overline{|\tilde{V}_x|}\ \overline{|\tilde{V}_y|} = \rho C_1 C_2 \lambda^2 \left(\frac{d\overline{V}_x}{dy} \right)^2$$

The constants C_1 and C_2 may be absorbed by λ to give a *mixing length* l, and

$$\tau_{xy,\,\text{turb}} = \rho l^2 \left(\frac{d\overline{V}_x}{dy} \right)^2 \qquad (9.13)$$

If this equation is to be applied, one must know how l varies with y. It appears that the problem of finding the mean-velocity profile has been traded for another problem, which consists of finding a functional relationship between the mixing length and the coordinate at right angles to the mean flow. Prandtl observed that if the mixing length may be considered unaffected by the viscosity, then the only length available is the distance from the wall; and he postulated the dimensionally correct formula

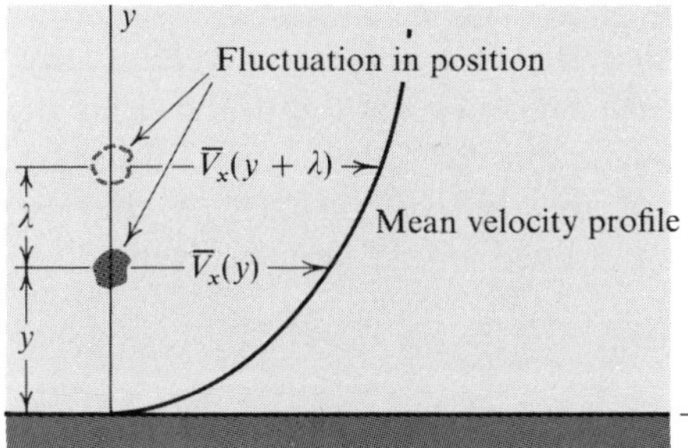

Fig. 9.4 *Profile for mean velocity $\overline{V}_x$.*

$$l = ky \tag{9.14}$$

in which k is a dimensionless constant called an *essential* or *universal, constant* of this turbulent flow problem. Whether this procedure can be employed to estimate velocity profiles will depend on how these assumptions and experimental information yield results that agree with measured velocity distributions. Some turbulent velocity profiles are generated in the next section by using Eq. (9.14).

9.8 Turbulent Velocity Profiles from Mixing Length

In Eq. (9.14) the coordinate y will be considered the distance from a wall over which fluid is flowing or the radial distance from the inside wall of a pipe through which fluid is flowing. In either case the fluid very close to the solid boundary is moving at a very low mean velocity relative to the wall, and a *laminar sublayer* is considered to exist between the wall and the turbulent main stream. In the laminar sublayer the shear stress is given by

$$\tau_{xy} = \mu \frac{dV_x}{dy} \qquad \text{for} \quad V_x = V_x(y) \tag{9.15}$$

since the time mean of V_x is V_x for laminar flow.

For the purposes of obtaining a general idea of the flow throughout its turbulent portion, Prandtl assumed that τ_{turb} in Eq. (9.13) remained constant in the turbulent region close to wall. Substituting Eq. (9.14) into (9.13) and integrating gives

$$\overline{V}_x = \frac{1}{k}\left(\frac{\tau_0}{\rho}\right)^{1/2} \ln y + \text{constant}$$

in which τ_0 is the assumed constant value of the shear stress close to the wall. The last equation is frequently written as

$$\overline{V}_x = V_*\left(\frac{1}{k}\ln y + C\right) \tag{9.16}$$

in which V_*, equal to $(\tau_0/\rho)^{1/2}$, was called the shearing-stress velocity by Prandtl, and C is simply another constant. Since V_* has the dimensions of velocity, the term in the parentheses must be dimensionless, and C must be the logarithm of a length to the extent of an additive constant. A characteristic length suggested by the variables involved would be v/V_*. Replacing C by

$$C' - \frac{1}{k}\ln\frac{v}{V_*}$$

allows Eq. (9.16) to be written as

$$\overline{V}_x = V_* \left(\frac{1}{k} \ln \frac{yV_*}{\nu} + C' \right) \tag{9.17}$$

The new dimensionless constant C' was referred to as a second universal constant by Prandtl. The experiments of J. Nikuradse[1] indicated that k is 0.40 and C' is 5.5 for flow in smooth-walled straight pipes, and so

$$\overline{V}_x = V_* \left(2.5 \ln \frac{yV_*}{\nu} + 5.5 \right) \qquad \text{for smooth-wall flow} \tag{9.18}$$

Whether or not a given wall is considered smooth may be ascertained from Fig. 9.5 (from Nikuradse). Qualitatively the diagram indicates that the curves begin to depart from the laminar flow curve at the lower critical Reynolds number of 2100. The curve with the highest relative roughness departs from the smooth-pipe curve first, while those corresponding to lower values of e/D somewhat follow the smooth-pipe locus. For very low values of

[1]J. Nikuradse, Gesetzmässigkeiten der turbulenten Strömung in glatten Rohren, *VDI—Forschungsh.* 356, 1932.

Fig. 9.5 *Nikuradse data for friction-factor variation with Reynolds number in artificially roughened pipes.*

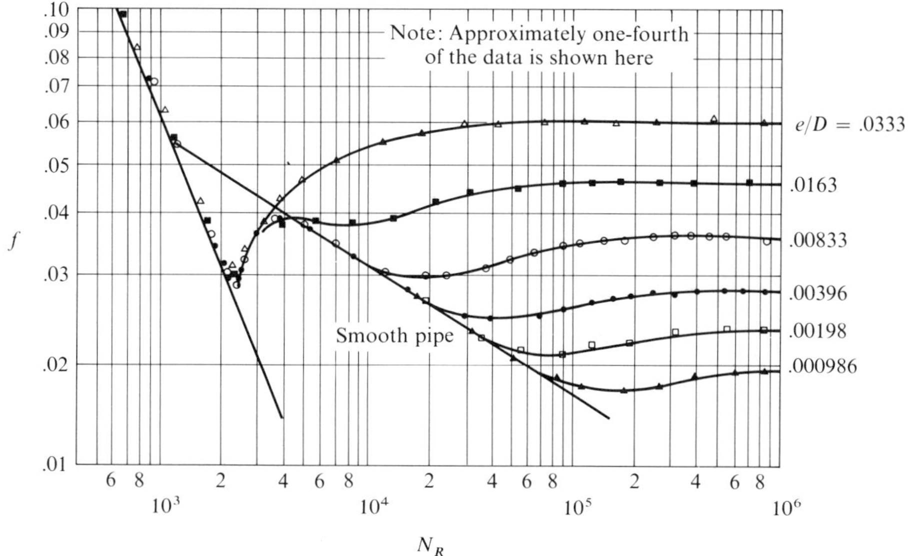

e/D, the curve is coincident with the smooth-pipe data over a large range of Reynolds numbers. Any point corresponding to a given N_R for a given e/D and which is coincident with smooth-pipe data is considered a *smooth-walled flow* and Eq. (9.18) applies for purposes of estimating the mean-velocity profile in the turbulent portion of the stream.

Figure 9.5 also indicates that there is a certain region for each relative roughness where the friction factor is independent of the Reynolds number and each curve becomes parallel to the N_R axis. A point located on this portion of any of the curves corresponds to *rough-pipe flow*, since the roughness is the controlling variable with respect to the friction factor. Equation (9.18) does not apply to rough-pipe flow; however, an argument may be extended from a previous equation:

$$\overline{V}_x = V_*\left(\frac{1}{k}\ln y + C\right) \tag{9.16}$$

Again, C must be the logarithm of a length to the extent of an additive constant. Since roughness e is the controlling factor in rough-pipe flow, it requires consideration as a characteristic length. Replacing C by

$$C'' - \frac{1}{k}\left(\ln\frac{\nu}{V_*} + \ln\frac{eV_*}{\nu}\right)$$

yields

$$\overline{V}_x = V_*\left[\frac{1}{k}\left(\ln\frac{yV_*}{\nu} - \ln\frac{eV_*}{\nu}\right) + C''\right]$$

Note that the quantity eV_*/ν is a Reynolds number corresponding to a particular pipe roughness. In the running of tests with artificially roughened pipes, Nikuradse[1] found that the pipes behaved as though they were practically smooth for $eV_*/\nu < 4$, and rough for $eV_*/\nu > 80$; C'' was found to be approximately equal to 8.5, with k still equal to 0.4. The preceding equation then reduces to

$$\overline{V}_x = V_*\left[2.5\left(\ln\frac{yV_*}{\nu} - \ln\frac{eV_*}{\nu}\right) + 8.5\right] \quad \text{for rough-pipe flow} \tag{9.19}$$

For points that do not lie either in the smooth- or rough-pipe flow regions of Fig. 9.5, several empirical formulations have been developed. These so-called transition-region equations will not be treated here. The inquisitive reader is referred to the specialized literature at this point.

[1] J. Nikuradse, Strömungsgesetze in rauhen Rohren, *VDI—Forschungsh.* 361, 1933. English translation in *Natl. Advisory Comm. Aeron. Tech. Memo.* 1292, 1950.

9.9 *Discussion of the Velocity Profiles*

The reader will observe that neither Eqs. (9.18) or (9.19) give sensible results for $y = 0$, and validity of these equations should not be expected at the wall. To find the velocity profile in the laminar sublayer requires integration of Eq. (9.15) with τ_{xy} equated to τ_0. This yields

$$V_x = \frac{\tau_0}{\mu} y + \text{a constant}$$

The constant of integration is zero because of the no-slip condition; $V_x = 0$ for $y = 0$; and in terms of V_*,

$$V_x = V_*^2 \frac{y}{v} \tag{9.20}$$

Inside of the laminar sublayer, the turbulent velocity profile gives $\overline{V}_x = 0$ and it is assumed that the laminar and turbulent velocity profiles blend smoothly.

To use either Eq. (9.18), (9.19), or (9.20) requires a knowledge of V_* or τ_0. A reasonable estimate of the shear stress at the wall can be obtained from the expression developed previously,

$$\tau_0 = \frac{1}{8} f\rho \, V_{SA}{}^2 \tag{8.24}$$

in which f is found from the Reynolds number and relative roughness by using Fig. 8.4, with the Reynolds number formed from the space-average velocity and the pipe diameter for the case of flow in pipes. The flow of a turbulent stream over a flat plate will be discussed later in the chapter where boundary-layer concepts are treated. An example will serve to illustrate the procedures employed in making calculations from the turbulent velocity profile equations.

Example 9.2

Water with a kinematic viscosity of 1.1 (10^{-5}) ft^2/sec and density of 1.94 slug/ft^3 flows at the rate of 15 ft^3/sec in an 8-in.-diameter cast-iron pipe. (*a*) Calculate the shear stress at the pipe wall; the shearing stress velocity V_*; and the velocity of the fluid at the center-line of the pipe. (*b*) At what value of yV_*/v does the turbulent velocity profile give $\overline{V}_x = 0$? (*c*) For smooth-pipe flow, what is the value of yV_*/v that corresponds to $\overline{V}_x = 0$?

Solution. (*a*) The first step is to check the Reynolds number to see if the flow is laminar or turbulent and, if turbulent, whether the flow is in the smooth-pipe, rough-pipe, or transition region.

$$V_{SA} = \frac{15 \text{ ft}^3/\text{sec}}{(\pi/4)(8 \text{ ft}/12)^2} = 43.0 \text{ ft/sec}$$

$$N_R = \frac{DV_{SA}}{\nu} = \frac{(8 \text{ ft}/12)(43.0 \text{ ft/sec})}{1.1\,(10^{-5})\text{ ft}^2/\text{sec}} = 2.60\,(10^6)$$

From Fig. 8.4, $e/D = 0.0013$ for an 8-in.-diameter cast-iron pipe. Using these values of N_R and e/D in Fig. 9.5 indicates that the point is in the rough-pipe flow region, so that Eq. (9.19) applies. The use of this equation requires calculation of the shear stress at the wall, and obtaining τ_0 from Eq. (8.24) will necessitate knowing the friction factor. Using $N_R = 2.60\,(10^6)$ and $e/D = 0.0013$ in Fig. 8.4, f is found to be 0.021, and

$$\tau_0 = \frac{1}{8} f\rho V_{SA}{}^2 = \left(\frac{1}{8}\right)(0.021)(1.94 \text{ slug/ft}^3)(43.0 \text{ ft/sec})^2 \quad ◀$$

$$= 9.5 \text{ lb}_f/\text{ft}^2$$

$$V_* = \left(\frac{\tau_0}{\rho}\right)^{1/2} = \left(\frac{9.5 \text{ lb}_f/\text{ft}^2}{1.94 \text{ slug/ft}^3}\right)^{1/2} = 2.2 \text{ ft/sec} \quad ◀$$

At this point, the Reynolds number based on the absolute roughness e may be checked against the criteria given at the end of Section 9.8, namely, the transition region determined by Nikuradse as $4 < eV_*/\nu < 80$. The value of e obtained from Fig. 8.4 was 0.00085 ft, and

$$\frac{eV_*}{\nu} = \frac{(0.00085)(2.2)}{1.1\,(10^{-5})} = 170 > 80$$

corroborating that the point is in the rough-pipe flow region. Proceeding with an estimation of the centerline velocity,

$$\overline{V}_x = V_*\left[2.5\left(\ln\frac{yV_*}{\nu} - \ln\frac{eV_*}{\nu}\right) + 8.5\right] \tag{9.19}$$

At the centerline of the pipe, $y = 4/12$ ft, and

$$\overline{V}_x = 2.2\left\{2.5\left[\ln\left(\frac{(4/12)(2.2)}{1.1(10^{-5})}\right) - \ln 170\right] + 8.5\right\} = 51.6 \text{ ft/sec} \quad ◀$$

b. For the region of rough-pipe flow, $\overline{V}_x$ will be zero when

$$2.5\left(\ln\frac{yV_*}{\nu} - \ln\frac{eV_*}{\nu}\right) = -8.5$$

and

$$\frac{yV_*}{\nu} = 5.67 \quad ◀$$

c. For smooth-pipe flow, setting

$$2.5\ln\frac{yV_*}{\nu} + 5.5 = 0$$

yields $\overline{V}_x = 0$, and

$$\frac{yV_*}{\nu} = 0.111$$ ◀

It is instructive to refer to a plot of the experimental work given in Nikuradse's report on turbulent flow in smooth pipes and the work of Reichardt[1] and Schuh in the laminar region. Figure 9.6 indicates that experimental points depart from a plot of Eq. (9.20) around $\ln(yV_*/\nu) = 1.6$, which gives the following estimate for the thickness of the laminar sublayer:

$$y_{ls} \simeq \frac{5\nu}{V_*}$$

The line, from which the constants were obtained for Eq. (9.18), has a slope of 2.5 and a $\overline{V}_x/V_*$ intercept of 5.5 when extrapolated to $\ln(yV_*/\nu) = 0$. Note that three regions may be identified from the plot: the laminar sublayer defined by $yV_*/\nu < 5$, the buffer layer given by $5 < yV_*/\nu < 30$, and the turbulent region given by $30 < yV_*/\nu$. Further insight may be gained by reference to Fig. 9.7 which shows the layers referred to above. The profile given by Eq. (9.18) and the laminar flow profile would not intersect smoothly if extended into the buffer layer where these profiles are not applicable.

[1]H. Reichardt, Die Wärmeübertragung in Turbulenten Reibungsschichten, *ZAMM* 20, Heft 6, pp. 297–328, 1940.

Fig. 9.6 Nondimensional velocity profile for turbulent flow in smooth pipes with laminar region included.

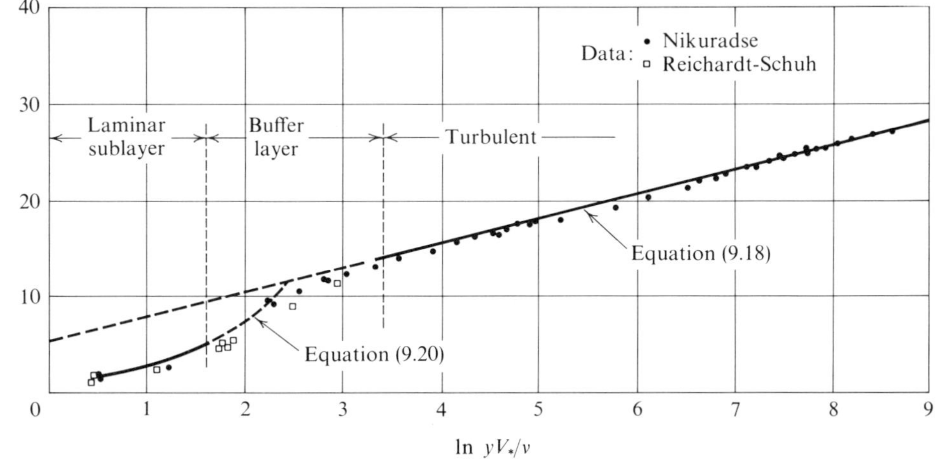

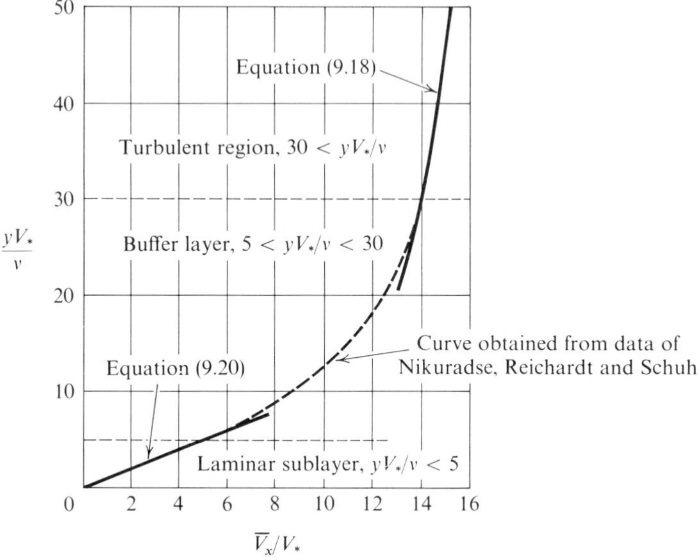

Fig. 9.7 Nondimensional velocity profile for smooth-pipe flow.

The dashed portion in the figure (smooth-pipe data from Nikuradse, and Reichardt and Schuh) connects the laminar and turbulent profiles by blending somewhat smoothly into each of these.

Example 9.3

If laminar flow is maintained in a pipe of 8-in. diameter, with a space-average velocity of 43.0 ft/sec, what is the velocity at the centerline of the pipe?

Solution. From the results of Example 4.6, the centerline or maximum velocity is twice the space-average velocity for fully developed laminar flow in a pipe.

$$V_{\text{max}} = (2)(43.0) = 86.0 \text{ ft/sec}$$ ◀

For the same space-average velocity, the turbulent profile is flatter or more obtuse than that of the laminar flow (see Example 9.2). This feature is important in the consideration of separation, which is discussed in the boundary-layer portion of this chapter.

In addition to Prandtl's mixing length, $l = ky$, von Karman[1] proposed that

$$l = K_1 \frac{d\overline{V}_x/dy}{d^2\overline{V}_x/dy^2}$$

[1]Th. von Karman, "Turbulence and Skin Friction," *J. Aero. Sci.*, vol. 1, no. 1, pp. 7–9, 1934.

for use in Eq. (9.13), which yields

$$\tau_{xy,\,\text{turb}} = \rho K_1{}^2 \frac{(d\overline{V}_x/dy)^4}{(d^2\overline{V}_x/dy^2)^2}$$

He reached these conclusions by assuming that the turbulent interchange in the neighborhood of two different points is similar and differs only by length and time scales. By further assuming that the shear stress close to the wall is constant, a logarithmic velocity profile may be derived. The constant K_1 was called a "universal constant" of turbulent exchange by von Karman; however, K_1 was found to vary from 0.38 to 0.41, which is considered suitable for a wide range of engineering applications in pipe flow problems.

Further aspects of turbulence will be considered subsequently.

9.10 Incompressible, Steady, Two-Dimensional, Laminar Boundary Layer

Consider a uniform flow at a large distance from a solid body. As the flow passes over this body, the no-slip condition must prevail at the body's surface, which means that there exists a portion of the fluid near the solid surface where the fluid velocity is reduced to zero by the viscous action of the fluid. Since the fluid changes velocity as it approaches the surface, one might expect that inertial forces come into play near the surface. Because the flow is influenced by both viscous and inertial forces in the vicinity of the solid surface, the flow there is characterized by a more complicated form of the Navier-Stokes equations than those governing the flow further away from the wall.

In order to gain a better insight into this problem, consider Eq. (8.1*b*) specialized to the case of a two-dimensional steady velocity field and with the body forces neglected:

$$V_x' \frac{\partial V_x'}{\partial x'} + V_y' \frac{\partial V_x'}{\partial y'} = -\frac{p_\infty}{\rho V_\infty{}^2} \frac{\partial p'}{\partial x'} + \frac{\nu}{L V_\infty}\left(\frac{\partial^2 V_x'}{\partial x'^2} + \frac{\partial^2 V_x'}{\partial y'^2}\right)$$

By taking the reference pressure $p_\infty = \rho V_\infty{}^2$ and basing N_R on V_∞ and L, the equation can be written as

$$V_x' \frac{\partial V_x'}{\partial x'} + V_y' \frac{\partial V_x'}{\partial y'} = -\frac{\partial p'}{\partial x'} + \frac{1}{N_R}\left(\frac{\partial^2 V_x'}{\partial x'^2} + \frac{\partial^2 V_x'}{\partial y'^2}\right) \tag{9.21}$$

and the y component of the Navier-Stokes equations would be similar in form to Eq. (9.21). These forms permit the following observation: for a large Reynolds number (large V_∞ or small viscosity, or both), the last term in Eq. (9.21) may be neglected. In this case the equations of motion take on the form of those discussed in Chapter 7 which were valid for inviscid flow which was irrotational. Regardless of how large V_∞ may be (with $1/N_R$ exceedingly small), the velocity of the fluid at the solid boundary must be zero relative to the boundary; and there is a region in the fluid close to the wall where the inertial and viscous forces are of the same order of magnitude so that the Reynolds number is of the order 1. The layer of fluid close to the solid boundary, in which these conditions prevail, is referred to as the *boundary layer*. The *thickness* δ of the layer is defined *as that distance from the boundary where the velocity is 99 percent of* V_{edge}, *the velocity in the flow outside the edge of the boundary layer.*[1] In the case of the flow over a curved surface, the value used for V_{edge} is that just outside the boundary layer at a particular station along the surface, since the velocity outside the layer will differ from V_∞ if the solid boundary is curved (see Fig. 9.8). For the flat plate, the value of V_{edge} remains constant at V_∞ since $\partial p/\partial x = 0$ in the flow outside the boundary layer. For the curved surface, V_{edge} is a function of x since $\partial p/\partial x \neq 0$. (The velocity profile within the boundary layer of Fig. 9.8*a* is shown in Fig. 9.9.)

There is another thickness, δ^*, called the *displacement thickness*, associated with boundary-layer flows. It is the distance the solid surface should be displaced normal to itself to yield the mass flow that would prevail if the flow were frictionless. For any value of x,

$$\rho\delta^* V_{edge} = \int_0^\infty \rho V_{edge}\, dy - \int_0^\infty \rho V_x\, dy$$

If the density is constant, then

$$\delta^* = \int_0^\infty \left(1 - \frac{V_x}{V_{edge}}\right) dy \tag{9.22}$$

[1]Some sources define δ on the basis of 99.2 percent of V_{edge}; however, 99 percent of V_{edge} will be used in this text.

Fig. 9.8 Boundary layer on a plane and a curved surface. (a) The flat plate is at zero incidence with respect to V_∞. The x coordinate is along the surface, and y is the co-ordinate normal to the surface. (b) Curved surface.

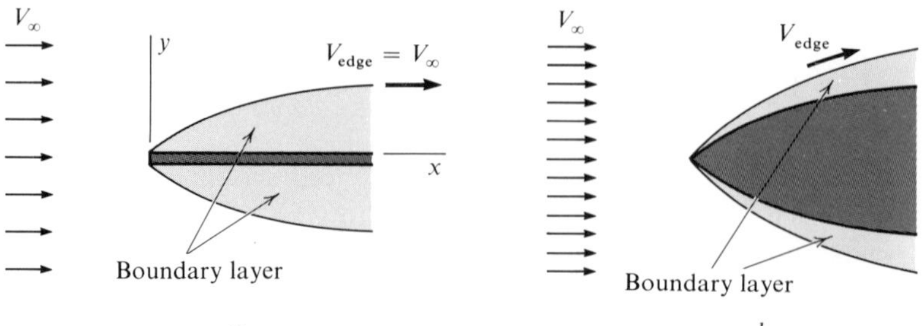

Note that δ^*, as well as δ, is a function of x; their functional dependence on x can be ascertained once the velocity field is known within the boundary layer.

9.11 Laminar Boundary Layer on Surfaces with Small Curvature

The term *small curvature* means that the radius of curvature is very large compared to the boundary-layer thickness. Since a flat surface has zero curvature, it certainly lies in the class of problems to which this section applies.

The Navier-Stokes equations (4.20*a* and *b*) reduce to

$$V_x \frac{\partial V_x}{\partial x} + V_y \frac{\partial V_x}{\partial y} = -\frac{1}{\rho}\frac{\partial p}{\partial x} + \nu\left(\frac{\partial^2 V_x}{\partial x^2} + \frac{\partial^2 V_x}{\partial y^2}\right) + (\mathbf{f}_B)_x \tag{9.23}$$

$$V_x \frac{\partial V_y}{\partial x} + V_y \frac{\partial V_y}{\partial y} = -\frac{1}{\rho}\frac{\partial p}{\partial y} + \nu\left(\frac{\partial^2 V_y}{\partial x^2} + \frac{\partial^2 V_y}{\partial y^2}\right) + (\mathbf{f}_B)_y \tag{9.24}$$

for steady and two-dimensional incompressible flow. Within the boundary layer the order of magnitude of x is larger than the order of magnitude of y except near the origin where both x and y are zero. Also $V_y < V_x$ for slightly curved surfaces. Let the order of magnitude of V_x and x be 1, and note this by $V_x = 0(1)$ and $x = 0(1)$. Then $\partial V_x/\partial x = 0(1)$, and the continuity equation indicates that $\partial V_y/\partial y = 0(1)$. Since $y = 0(\delta)$ within the boundary layer, $V_y = 0(\delta)$. The preceding orders of magnitude may be used to determine the following:

$$\frac{\partial V_x}{\partial y} = 0\left(\frac{1}{\delta}\right) \qquad \frac{\partial^2 V_x}{\partial x^2} = 0(1) \qquad \frac{\partial^2 V_x}{\partial y^2} = 0\left(\frac{1}{\delta^2}\right)$$

$$\frac{\partial V_y}{\partial x} = 0(\delta) \qquad \frac{\partial^2 V_y}{\partial x^2} = 0(\delta) \qquad \frac{\partial^2 V_y}{\partial y^2} = 0\left(\frac{1}{\delta}\right)$$

Noting the orders of magnitude of the kinematic terms in Eq. (9.23)

$$\begin{array}{ccccccccc} V_x \dfrac{\partial V_x}{\partial x} & + & V_y \dfrac{\partial V_x}{\partial y} & = & -\dfrac{1}{\rho}\dfrac{\partial p}{\partial x} + \nu\Bigg(& \dfrac{\partial^2 V_x}{\partial x^2} & + & \dfrac{\partial^2 V_x}{\partial y^2} & \Bigg) + (\mathbf{f}_B)_x \\ 0(1)\,0(1) & & 0(\delta)\,0\left(\dfrac{1}{\delta}\right) & & & 0(1) & & 0\left(\dfrac{1}{\delta^2}\right) & \end{array}$$

reveals that each term on the left-hand side of the equation is of order 1, while the term $\partial^2 V_x/\partial y^2$ is of the much larger order $1/\delta^2$. This could lead one into neglecting the terms on the left-hand side in favor of the term, $\partial^2 V_x/\partial y^2$. However, recall from Section 9.10 that the inertial forces are of the same order of magnitude as the viscous forces—the terms on the left-hand side of Eq. (9.23) represent the inertial forces per unit mass. If these terms are not neglected, then ν which multiplies $\partial^2 V_x/\partial y^2$ must be at least of order δ^2. By using this conclusion, and retaining only terms of order 1, Eqs. (9.23) and (9.24) are reduced to

$$V_x \frac{\partial V_x}{\partial x} + V_y \frac{\partial V_x}{\partial y} = -\frac{1}{\rho}\frac{\partial p}{\partial x} + \nu \frac{\partial^2 V_x}{\partial y^2} + (\mathbf{f}_B)_x \tag{9.25}$$

$$0 = -\frac{1}{\rho}\frac{\partial p}{\partial y} + (\mathbf{f}_B)_y \tag{9.26}$$

These two equations and the continuity equation

$$\frac{\partial V_x}{\partial x} + \frac{\partial V_y}{\partial y} = 0$$

which contains only terms of order 1, along with the appropriate boundary conditions, constitute the boundary-value problem that determines the unknown functions: $p(x,y)$, $V_x(x,y)$, and $V_y(x,y)$. Prandtl first discerned the necessary assumptions that led to the simplification of the equations of motion applicable to boundary-layer flows.

In the next section, the solution of the simplest possible boundary-layer flow is obtained, and the results are employed to obtain the several thicknesses δ and δ^*, as well as the drag force on the surface.

9.12 Laminar Boundary Layer on a Flat Plate

The problem assumes two-dimensional and steady laminar flow of an incompressible fluid. The plate has a zero angle of incidence with respect to the oncoming flow, and body forces are neglected. Figure 9.9 depicts the plate in the given flow. Equation (9.26) gives $\partial p/\partial y = 0$ for $(\mathbf{f}_B)_y = 0$, and $p = p(x)$ only. Then $\partial p/\partial x = dp/dx$. Recall that the pressure gradient in the x direction is zero outside of the boundary layer since the surface is flat, so Eq. (9.25) reduces to

$$V_x \frac{\partial V_x}{\partial x} + V_y \frac{\partial V_x}{\partial y} = \nu \frac{\partial^2 V_x}{\partial y^2} \tag{9.27}$$

Continuity is given by

$$\frac{\partial V_x}{\partial x} + \frac{\partial V_y}{\partial y} = 0 \tag{9.28}$$

The boundary conditions are

$$V_x = V_y = 0 \qquad \text{for } y = 0 \tag{9.29}$$

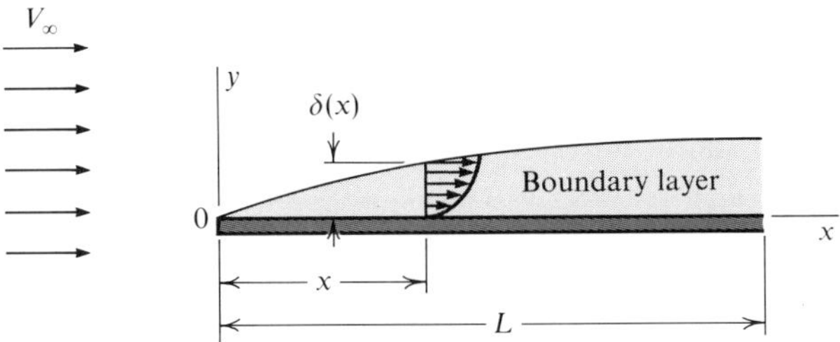

Fig. 9.9 Boundary layer on one side of a flat plate of length L. Note that δ is a function of x.

and

$$\lim_{y \to \infty} V_x = V_\infty \tag{9.30}$$

The latter condition requires the fluid velocity to be equal to the free-stream velocity at large distances from the plate. The solutions of a large class of boundary-layer problems are determined by the transformation of the partial differential equations governing the flow into ordinary differential equations. The systematic procedures whereby this simplification can be effected constitute the subject matter of specialized courses in boundary-layer flows and similarity transformations, and the interested reader should refer to the catalog of a technical library for many good references.

One method of transforming the problem in this section is offered without recourse to a systematic method; however, it is felt that the steps taken to transform Eq. (9.27) are not exceedingly arbitrary. If $V_x(\partial V_x/\partial x)$ represents the inertial effects in the boundary layer, then

$$\frac{V_x(\partial V_x/\partial x)}{\nu(\partial^2 V_x/\partial y^2)} = 0(1)$$

Since

$$\frac{\partial V_x}{\partial x} = 0\left(\frac{V_\infty}{x}\right) \qquad \text{and} \qquad \frac{\partial^2 V_x}{\partial y^2} = 0\left(\frac{V_\infty}{\delta^2}\right)$$

$$\frac{V_\infty(V_\infty/x)}{\nu(V_\infty/\delta^2)} = 0(1)$$

and

$$\frac{\delta}{x} \simeq 0(N_{R_x}{}^{-1/2}) \tag{9.31}$$

in which $N_{R_x} = V_\infty x/\nu$.

If a solution of Eq. (9.27) is to be obtained by reducing this partial differential equation to an ordinary differential equation, then both velocity components must be expressed as functions of a single function of some combination of x and y. The single function from which V_x and V_y may be obtained is the stream function previously introduced by Eqs. (7.24*a* and *b*); the combination of the x and y variables is suggested by Eq. (9.31). The single variable will be noted and defined by

$$\eta = \frac{y}{x N_{R_x}{}^{-1/2}} = y\left(\frac{V_\infty}{\nu x}\right)^{1/2} \tag{9.32}$$

One possible transformation is

$$\frac{V_x}{V_\infty} = F(\eta) \tag{9.33}$$

However, this requires relating $F(\eta)$ to the stream function ψ. Recalling that $V_x = -\partial\psi/\partial y$, Eq. (9.33) becomes

$$\frac{\partial\psi}{\partial y} = -V_\infty F(\eta)$$

and

$$\psi = -V_\infty \int F(\eta)\, dy + G(x)$$

The integral sign represents integration with respect to y; $G(x)$ is an arbitrary function of x. The indicated integration may be carried out by letting $F(\eta) = df(\eta)/d\eta$ and expressing dy in terms of $d\eta$. The latter operation is accomplished by using the definition of η in Eq. (9.32). Then $y = \eta x N_{R_x}{}^{-1/2}$, so that $dy = x N_{R_x}{}^{-1/2}\, d\eta$. Substituting into the integral for the stream function,

$$\psi = -V_\infty \int \frac{df(\eta)}{d\eta} N_{R_x}{}^{-1/2} x\, d\eta + G(x)$$

and integrating (holding x constant) yields

$$\psi = -\frac{V_\infty x}{N_{R_x}{}^{1/2}} f(\eta) + G(x)$$

$G(x)$ may be obtained by observing the no-slip condition at the plate and employing Eq. (9.33) since $V_x = V_\infty F(\eta) = 0$ for $y = 0$. Then $F(0) = 0$ since Eq. (9.32) requires η to vanish with y for $x > 0$; and

$$\psi|_{y=0} = -V_\infty \int F(0)\, dy + G(x) = G(x)$$

Now $\psi|_{y=0}$ must be a streamline (since it is coincident with the plate), and hence $\psi|_{y=0}$ is a constant. If zero is chosen as the value for this constant,[1] then $G(x) = 0$ and

$$\psi = -\frac{V_\infty x}{N_{R_x}{}^{1/2}} f(\eta) = -(V_\infty \nu x)^{1/2} f(\eta) \tag{9.34}$$

This equation and the definition $\eta = y(V_\infty/\nu x)^{1/2}$ may now be used to transform the boundary-value problem given by Eqs. (9.27) through (9.30). The velocity components obtained from the stream function will insure satisfaction of the continuity equation.

To transform Eq. (9.27), the velocity components must be expressed in terms of $f(\eta)$, η, and x. This is accomplished as follows:

$$V_x = -\frac{\partial \psi}{\partial y} = -\frac{\partial \psi}{\partial \eta}\frac{\partial \eta}{\partial y} = -\frac{\partial}{\partial \eta}[-(V_\infty \nu x)^{1/2} f(\eta)]\frac{\partial \eta}{\partial y}$$

$$V_x = (V_\infty \nu x)^{1/2}\frac{df}{d\eta}\left(\frac{V_\infty}{\nu x}\right)^{1/2} = V_\infty \frac{df}{d\eta} \tag{9.35}$$

$$V_y = \frac{\partial \psi}{\partial x} = \frac{\partial}{\partial x}[-(V_\infty \nu x)^{1/2} f(\eta)] = -(V_\infty \nu x)^{1/2}\frac{\partial f}{\partial x} - f\frac{\partial (V_\infty \nu x)^{1/2}}{\partial x}$$

$$= -(V_\infty \nu x)^{1/2}\frac{df}{d\eta}\frac{\partial \eta}{\partial x} - \frac{1}{2}\left(\frac{\nu V_\infty}{x}\right)^{1/2} f$$

$$= -(V_\infty \nu x)^{1/2}\frac{df}{d\eta}\left(-\frac{\eta}{2x}\right) - \frac{1}{2}\left(\frac{\nu V_\infty}{x}\right)^{1/2} f$$

$$= \frac{1}{2}\left(\frac{V_\infty \nu}{x}\right)^{1/2}\frac{df}{d\eta}\eta - \frac{1}{2}\left(\frac{\nu V_\infty}{x}\right)^{1/2} f$$

$$= \frac{1}{2}\left(\frac{V_\infty \nu}{x}\right)^{1/2}\left(\eta\frac{df}{d\eta} - f\right) \tag{9.36}$$

$$\frac{\partial V_x}{\partial x} = V_\infty \frac{\partial}{\partial x}\left(\frac{df}{d\eta}\right) = V_\infty \frac{d(df/d\eta)}{d\eta}\left(\frac{\partial \eta}{\partial x}\right)$$

$$= V_\infty\left(\frac{d^2 f}{d\eta^2}\right)\left(-\frac{\eta}{2x}\right) = -\frac{V_\infty \eta}{2x}\frac{d^2 f}{d\eta^2} \tag{9.37}$$

$$\frac{\partial V_x}{\partial y} = V_\infty \frac{\partial}{\partial y}\left(\frac{df}{d\eta}\right) = V_\infty \frac{d^2 f}{d\eta^2}\frac{\partial \eta}{\partial y}$$

$$= V_\infty \frac{d^2 f}{d\eta^2}\left(\frac{V_\infty}{\nu x}\right)^{1/2} \tag{9.38}$$

[1]The choice of this constant is arbitrary since the stream function determined from the velocity components is unique only to the extent of an additive constant.

and finally

$$\frac{\partial^2 V_x}{\partial y^2} = V_\infty \left(\frac{V_\infty}{\nu x}\right)^{1/2} \frac{d^3 f}{d\eta^3} \frac{\partial \eta}{\partial y} = V_\infty \left(\frac{V_\infty}{\nu x}\right)^{1/2} \frac{d^3 f}{d\eta^3} \left(\frac{V_\infty}{\nu x}\right)^{1/2}$$

$$= \frac{V_\infty{}^2}{\nu x} \frac{d^3 f}{d\eta^3} \tag{9.39}$$

Substitution for the preceding velocity components and their partial derivatives in Eq. (9.27) and simplification yields

$$\frac{d^3 f}{d\eta^3} + \frac{1}{2} f \frac{d^2 f}{d\eta^2} = 0 \tag{9.40}$$

The no-slip condition ($V_x = V_y = 0$ for $y = 0$) transforms to

$$\left.\begin{aligned} \frac{df}{d\eta} &= 0 \\ f &= 0 \end{aligned}\right\} \qquad \text{for } \eta = 0 \tag{9.41a}$$

by quick reference to Eqs. (9.35) and (9.36), respectively, and noting that $\eta = 0$ for $y = 0$ with $x > 0$. The condition $\lim_{y \to \infty} V_x = V_\infty$ transforms to

$$\lim_{\eta \to \infty} \frac{df}{d\eta} = 1 \tag{9.41b}$$

upon reference to Eq. (9.35) and noting that $\eta \to \infty$ as $y \to \infty$. The problem is now reduced to one of solving the nonlinear ordinary differential equation (9.40). However, recall that the untransformed problem contained a partial differential equation that was also nonlinear in the function $V_x(x, y)$ noted from Eq. (9.27).

The solution of Eq. (9.40) is assumed to be in the form of a power series, that is,

$$f(\eta) = A_0 + A_1 \eta + \frac{A_2}{2!} \eta^2 + \cdots + \frac{A_n}{n!} \eta^n \tag{9.42}$$

The two conditions given by Eq. (9.41a) certainly require that both A_0 and A_1 must vanish, and so

$$f(\eta) = \frac{A_2}{2!} \eta^2 + \frac{A_3}{3!} \eta^3 + \cdots + \frac{A_n}{n!} \eta^n \tag{9.43}$$

The second and third derivatives may be formed from Eq. (9.43), and these derivatives as well as $f(\eta)$ may be substituted into Eq. (9.40), the ordinary differential equation. The coefficients of each power of η may be grouped,[1] and each of these coefficients must vanish, since the left-hand side of Eq. (9.40) is zero for all values of η. In this manner one obtains a relation between the A's in the series solution; each A that is not zero is related to A_2. The first seven of these are: $A_3 = 0$, $A_4 = 0$, $A_5 = -(1/2)A_2^2$, $A_6 = 0$, $A_7 = 0$, and $A_8 = -(11/2)A_2A_5 = (11/4)A_2^3$.

Substitution of these values of A into Eq. (9.43) yields a pattern that may be noted by

$$f(\eta) = \sum_{j=0}^{\infty} \left(-\frac{1}{2}\right)^j \frac{A_2^{j+1} C_j}{(3j + 2)!} \eta^{3j+2} \qquad j = 0, 1, 2, \ldots \tag{9.44}$$

in which $C_0 = 1$, $C_1 = 1$, $C_2 = 11$, $C_3 = 375$, $C_4 = 27{,}897$, and $C_5 = 3{,}817{,}137$—the C_j's being calculated from the A's.

It now remains to determine the constant A_2, which can be accomplished by meeting the condition required by Eq. (9.41*b*). This is carried out by defining a new series, $q(\eta)$, given by the series $f(\eta)$ with $A_2 = 1$, that is,

$$q(\eta) = \sum_{j=0}^{\infty} \left(-\frac{1}{2}\right)^j \frac{C_j}{(3j + 2)!} \eta^{3j+2} \qquad j = 0, 1, 2, \ldots \tag{9.45}$$

One observes that $f(\eta) = A_2^{1/3} q(A_2^{1/3} \eta)$ by checking the series for $f(\eta)$ and $q(A_2^{1/3} \eta)$. Then

$$\frac{df(\eta)}{d\eta} = A_2^{2/3} \frac{d[q(A_2^{1/3} \eta)]}{d(A_2^{1/3} \eta)}$$

and the condition given by Eq. (9.41*b*) requires

$$\lim_{\eta \to \infty} \frac{df(\eta)}{d\eta} = 1$$

or

$$\lim_{\eta \to \infty} A_2^{2/3} \frac{d[q(A_2^{1/3} \eta)]}{d(A_2^{1/3} \eta)} = 1$$

Since the argument of the function $q(A_2^{1/3} \eta)$ is the same as the argument of $q(\eta)$ in the limit (as $\eta \to \infty$), the last equation may be written as

$$A_2^{2/3} \lim_{\eta \to \infty} \frac{dq(\eta)}{d\eta} = 1$$

[1]This is a straightforward procedure but rather lengthy because of the product $f(d^2f/d\eta^2)$ in Eq. (9.40). Obviously the more terms one carries in the expressions for f, $d^2f/d\eta^2$, and $d^3f/d\eta^3$, the more A_n's one can ascertain.

and

$$A_2 = \lim_{\eta \to \infty} \left[\frac{1}{dq(\eta)/d\eta} \right]^{3/2} \tag{9.46}$$

Calculation of A_2 can now be carried out in principle. Successively larger values of η may be assumed; from each value of η, a value of A_2 is ascertained by using Eq. (9.46) and the series $q(\eta)$ given by Eq. (9.45) to obtain the derivative $dq(\eta)/d\eta$. When the value of this constant becomes relatively insensitive to a change in η, A_2 is determined. Its value to five places is

$$A_2 = 0.33206$$

The solution for $f(\eta)$ becomes

$$f(\eta) = \sum_{j=0}^{\infty} \left(-\frac{1}{2} \right)^j \frac{(0.33206)^{j+1}\, C_j}{(3j+2)!} \eta^{3j+2} \qquad j = 0, 1, 2, \ldots \tag{9.47}$$

with the values of C_j given after Eq. (9.44).

Since the velocity components depend on the derivative of $f(\eta)$ as well as $f(\eta)$ and η [see Eqs. (9.35) and (9.36)], it is convenient to have access to a tabulation of these for various values of η. Application of the solutions and information[1] in Table 9.1 are utilized in the following examples which obtain some of the boundary-layer parameters.

Example 9.4

a. Obtain an expression for the shear stress at the surface of the flat plate in Fig. 9.9 as a function of x, the distance along the plate from its leading edge. (*b*) What is the drag force on one side of the plate of length L, expressed in terms of V_∞, μ, ρ, and L?

Solution. (*a*) The constitutive equation (4.18*d*) gives

$$\tau_{yx} = \mu \left(\frac{\partial V_x}{\partial y} + \frac{\partial V_y}{\partial x} \right) = \mu \frac{\partial V_x}{\partial y}$$

since $\partial V_y/\partial x << \partial V_x/\partial y$ in the boundary layer (except near the leading edge of the plate), and

$$\tau_{yx} = \mu \frac{\partial V_x}{\partial \eta} \frac{\partial \eta}{\partial y} = \mu V_\infty \frac{d^2 f}{d\eta^2} \left(\frac{V_\infty}{\nu x} \right)^{1/2}$$

by using the results of Eq. (9.38). The shear stress at the surface of the plate is given by

$$\tau_{yx}\Big|_{y=0} = \mu V_\infty \left(\frac{V_\infty}{\nu x} \right)^{1/2} \left(\frac{d^2 f}{d\eta^2} \right)_{y=0}$$

[1] L. Howarth, On the Solution of the Laminar Boundary Layer Equations, *Proc. Roy. Soc. London*, A 164, pp. 547–579, 1938. (Abridged table with kind permission of the publisher.)

Table 9.1

η	$f(\eta)$	$\frac{df}{d\eta}$	η	$f(\eta)$	$\frac{df}{d\eta}$
0	0	0			
0.2	0.00664	0.06641	4.6	2.88826	0.98269
0.4	0.02656	0.13277	4.8	3.08534	0.98779
0.6	0.05974	0.19894	5.0	3.28329	0.99155
0.8	0.10611	0.26471	5.2	3.48189	0.99425
1.0	0.16557	0.32979	5.4	3.68094	0.99616
1.2	0.23795	0.39378	5.6	3.88031	0.99748
1.4	0.32298	0.45627	5.8	4.07990	0.99838
1.6	0.42032	0.51676	6.0	4.27964	0.99898
1.8	0.52952	0.57477	6.2	4.47948	0.99937
2.0	0.65003	0.62977	6.4	4.67938	0.99961
2.2	0.78120	0.68132	6.6	4.87931	0.99977
2.4	0.92230	0.72899	6.8	5.07928	0.99987
2.6	1.07252	0.77246	7.0	5.27926	0.99992
2.8	1.23099	0.81152	7.2	5.47925	0.99996
3.0	1.39682	0.84605	7.4	5.67924	0.99998
3.2	1.56911	0.87609	7.6	5.87924	0.99999
3.4	1.74696	0.90177	7.8	6.07923	1.00000
3.6	1.92954	0.92333	8.0	6.27923	1.00000
3.8	2.11605	0.94112	8.2	6.47923	1.00000
4.0	2.30576	0.95552	8.4	6.67923	1.00000
4.2	2.49806	0.96696	8.6	6.87923	1.00000
4.4	2.69238	0.97587	8.8	7.07923	1.00000

However, $\eta = 0$ when $y = 0$, and reference to Eq. (9.43) indicates that

$$\left.\frac{d^2 f}{d\eta^2}\right|_{\eta=0} = A_2 = 0.33206$$

and

$$\left.\tau_{yx}\right|_{y=0} = \tau_0 = 0.33206\,\mu V_\infty \left(\frac{V_\infty}{\nu x}\right)^{1/2}$$ ◀

b. To find the drag force on the plate requires integration of the shear stress over the area of the plate.

$$F_{\text{drag}} = \int_{S_{\text{plate}}} 0.33206\,\mu V_\infty \left(\frac{V_\infty}{\nu x}\right)^{1/2} dS$$

For one side of a plate with a unit dimension in the z direction,

$$F_{\text{drag}} = 0.33206\,\mu V_\infty \left(\frac{V_\infty}{\nu}\right)^{1/2} \int_0^L x^{-1/2}\,dx$$

$$= 0.664\,(V_\infty{}^3 \mu \rho L)^{1/2}$$ ◀

This last answer may be used to express the *surface-resistance coefficient* C_f as a function of the plate Reynolds number. The definition of this coefficient is given by

$$C_f = \frac{F_{\text{drag}}}{\rho V_\infty{}^2 S/2}$$

in which S is the surface associated with the drag force. For the flat plate of length L and unit leading edge (in the z direction), $S = L$, and

$$C_f = \frac{0.664(V_\infty{}^3 \mu \rho L)^{1/2}}{\rho V_\infty{}^2 L/2} = \frac{1.328}{N_{R_L}{}^{1/2}}$$

where

$$N_{R_L} = \frac{L V_\infty \rho}{\mu}$$

The result agrees very well with experimental measurements.[1]

Example 9.5

a. How do the thicknesses δ and δ^* vary with x for the laminar boundary layer of Example 9.4? (*b*) Compare the velocity components V_x and V_y evaluated at the edge of the boundary layer, using $\eta \simeq 5$.

Solution. (*a*) From the definition of δ, $y = \delta$ where $V_x/V_\infty = df/d\eta = 0.99$. Data in the Table 9.1 indicates that η lies between 4.8 and 5.0, and interpolation between these values yields $\eta = 4.92$. Since

$$y = \eta \left(\frac{x\nu}{V_\infty} \right)^{1/2}$$

$$\delta = 4.92 \left(\frac{x\nu}{V_\infty} \right)^{1/2} \quad ◀$$

Equation (9.22) defines the displacement thickness as

$$\delta^* = \int_0^\infty \left(1 - \frac{V_x}{V_{\text{edge}}} \right) dy$$

The velocity at the edge of the boundary layer may be taken as V_∞, and it should be observed that $V_x/V_\infty \simeq 1$ for $y > \delta$, so that Eq. (9.22) may be written as

$$\delta^* = \int_0^{y>\delta} \left(1 - \frac{V_x}{V_\infty} \right) dy$$

[1] S. Dhawan, Measurements of Skin Friction, *Natl. Advisory Comm. Aeron. Tech.* Note 2567, 1952.

Changing variables in the last equation by using Eqs. (9.32) and (9.35) yields

$$\delta^* = \left(\frac{xv}{V_\infty}\right)^{1/2} \int_0^{\eta > 5} \left(1 - \frac{df}{d\eta}\right) d\eta$$

$$= \left(\frac{xv}{V_\infty}\right)^{1/2} [\eta - f(\eta)]_0^{\eta > 5}$$

The lower limit ($\eta = 0$) is zero, and δ^* may be evaluated by calculating the term in brackets for $\eta = 5$ and successively higher values of η, until this term stabilizes at a constant value. This is accomplished by taking values of η and $f(\eta)$ from Table 9.1. For

$$\eta = 5 \qquad \delta^* = 1.7167\left(\frac{xv}{V_\infty}\right)^{1/2}$$

$$\eta = 8 \qquad \delta^* = 1.72077\left(\frac{xv}{V_\infty}\right)^{1/2}$$

$$\eta = 8.8 \qquad \delta^* = 1.72077\left(\frac{xv}{V_\infty}\right)^{1/2}$$

Then

$$\delta^* = 1.721\left(\frac{xv}{V_\infty}\right)^{1/2}$$ ◀

Note that the displacement thickness for this problem is about one-third of the boundary-layer thickness of the flat plate.

b. The component of velocity normal to the plate is given by Eq. (9.36).

$$V_y = \frac{1}{2}\left(\frac{V_\infty v}{x}\right)^{1/2}\left[\eta\frac{df}{d\eta} - f\right]$$

As in (*a*), successively higher values of η may be used to evaluate the term in the brackets; the expression for V_y stabilizes with $V_y = 0.8604(V_\infty v/x)^{1/2}$. However, at $\eta \simeq 5$, $V_y = 0.8372(V_\infty v/x)^{1/2}$ and $V_x = 0.9916\, V_\infty$ so that

$$\frac{V_y}{V_x} = \frac{0.8372(V_\infty v/x)^{1/2}}{0.9916\, V_\infty} = 0.844\left(\frac{v}{xV_\infty}\right)^{1/2} = \frac{0.844}{N_{R_x}^{1/2}}$$ ◀

Again, as x becomes very small (very close to the leading edge of the plate), the solution does not yield a defined value for V_y/V_x. Recall from Example 9.4 the comment concerning $\partial V_y/\partial x \ll \partial V_x/\partial y$ except near the leading edge of the plate.

Example 9.5 indicates that there is a nonzero V_y at the edge of the boundary layer, even though this component is small for values of N_{R_x} encountered in these flows. This observation might lead one to conclude that the edge of the boundary layer is not a streamline, which is correct. This is readily seen by recalling that

$$\psi = -(V_\infty vx)^{1/2} f(\eta) \tag{9.34}$$

As η is constant at the edge of the boundary layer, so is $f(\eta)$; and ψ is a function of x, which indicates that the stream function is not constant along $y = \delta$.

A final comment about the solution presented in this section is prompted by the statement concerning similarity transformations made at the beginning of this section. From the form of the transformation,

$$\frac{V_x}{V_\infty} = F(\eta) = F\left[y\left(\frac{V_\infty}{x\nu}\right)^{1/2}\right]$$

which indicates that the velocity profiles $V_x = V_x(y)$ for various values of x are similar in the sense that there exists a scale factor for V_x and y that will reduce all the profiles to an identical profile. The factor for y is $(V_\infty/x\nu)^{1/2}$, which is proportional to $1/\delta$, and V_x may be nondimensionalized by V_∞. The flat-plate solution was first carried out by Prandtl and Blasius.[1]

9.13 *Control-Volume Analysis of a Boundary Layer*

An alternate approach to the analysis of boundary-layer flows involves the formation of the governing equations in integral form, which permits application to turbulent as well as laminar flows. It is an approximate method that requires the assumption of a reasonable velocity profile before the form is integrated. Nevertheless, the choice of reasonable profiles may be governed by experimental data and the integral forms of the boundary-layer equations used to estimate shear stresses.

Consider the two-dimensional and steady incompressible flow over a flat or slightly curved surface, and neglect body forces. Figure 9.10 defines a control volume for the application of the conservation of matter. Applying

[1]The reader interested in the genesis of boundary-layer fluid mechanics and similarity solutions is referred to the extensive bibliography contained in Hermann Schlichting (J. Kestin, tr.), "Boundary-Layer Theory," 6th ed., McGraw-Hill Book Company, New York, 1968.

Fig. 9.10 *Control volume for mass flux shown within the boundary layer. Depth of control volume is unity in the z direction.*

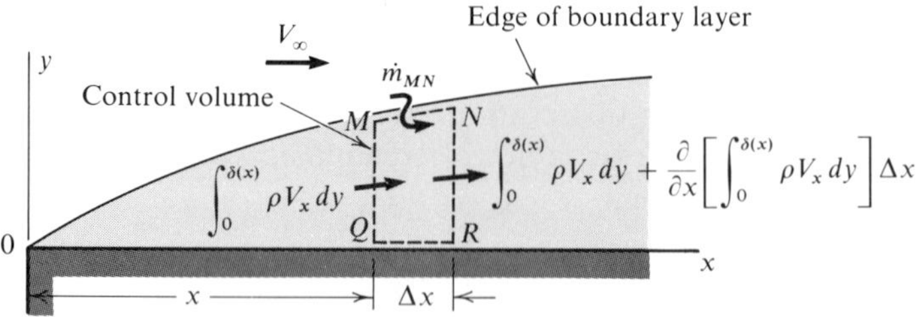

Eq. (2.1) to this control volume, the mass rate of flow across the surface MN at the edge of the boundary layer is

$$\dot{m}_{MN} = \int_0^{\delta(x)} \rho V_x \, dy + \frac{\partial}{\partial x}\left[\int_0^{\delta(x)} \rho V_x \, dy\right]\Delta x - \int_0^{\delta(x)} \rho V_x \, dy$$

and

$$\dot{m}_{MN} = \frac{\partial}{\partial x}\left[\int_0^{\delta(x)} \rho V_x \, dy\right]\Delta x \tag{9.48}$$

Recall that the edge of the boundary layer is not a streamline, and that the layer thickens as x increases because of the mass flow into the edge at $y = \delta$. Having formulated $\dot{m}_{MN}$, one may now express the linear momentum flux passing through the surface MN, since the flow outside the boundary layer has a velocity V_∞ in the x direction. Figure 9.11 shows the same control volume used in formulating Eq. (9.48), with linear momentum flux and surface forces indicated on the control surfaces. These are shown for the x direction only. Notice that the surface forces due to pressure are formulated on the surfaces MQ and NR by assuming that the pressure does not vary in the y direction. Such an assumption may be justified by considering the

Fig. 9.11 Control volumes for (a) linear momentum flux in the x direction and (b) surface forces.

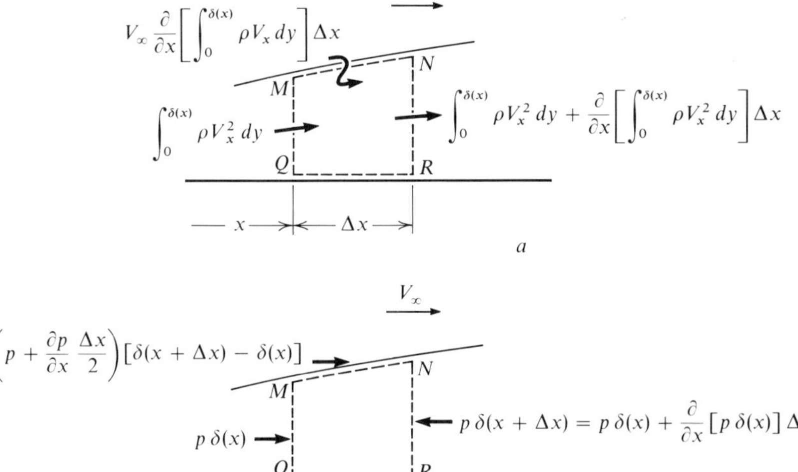

y-momentum flux and noting that $V_y < V_x$ for boundary-layer flows over flat or slightly curved surfaces. Indeed, *the negligible variation of pressure in a direction normal to these surfaces is characteristic of boundary-layer flows over this class of surfaces.* It should also be noted that the force acting on MN has been formulated by multiplication of the pressure midway between x and $x + \Delta x$ by the projection of the surface MN in the x direction. τ_0 is the shear stress at the wall. Applying Eq. (5.3) to the flow in Fig. 9.11 yields

$$p\delta(x) - p\delta(x) - \frac{\partial}{\partial x}[p\delta(x)]\,\Delta x + \left(p + \frac{\partial p}{\partial x}\frac{\Delta x}{2}\right)[\delta(x + \Delta x) - \delta(x)] - \tau_0\,\Delta x$$

$$= \int_0^{\delta(x)} \rho V_x^2\,dy + \frac{\partial}{\partial x}\left[\int_0^{\delta(x)} \rho V_x^2\,dy\right]\Delta x$$

$$- V_\infty \frac{\partial}{\partial x}\left[\int_0^{\delta(x)} \rho V_x\,dy\right]\Delta x - \int_0^{\delta(x)} \rho V_x^2\,dy$$

Substituting $\delta(x) + \partial/\partial x[\delta(x)]\,\Delta x$ for $\delta(x + \Delta x)$, carrying out the indicated multiplication, neglecting terms containing $(\Delta x)^2$, and dividing by Δx, simplifies the momentum equation to the form

$$-\delta(x)\frac{dp}{dx} - \tau_0 = \frac{\partial}{\partial x}\left[\int_0^{\delta(x)} \rho V_x^2\,dy\right] - V_\infty \frac{\partial}{\partial x}\left[\int_0^{\delta(x)} \rho V_x\,dy\right] \tag{9.49}$$

in which $\partial p/\partial x$ has been written as an ordinary derivative since $\partial p/\partial y$ has been assumed very small. Equation (9.49) is referred to as the *von Karman integral momentum equation.* For the case of zero pressure gradient and V_∞ constant, Eq. (9.49) reduces to

$$\tau_0 = \frac{\partial}{\partial x}\left[\int_0^{\delta(x)} \rho(V_\infty V_x - V_x^2)\,dy\right] \tag{9.50}$$

Equation (9.50) may be used to obtain the thickness δ as a function of x once a velocity profile is assumed since τ_0 can be found from this profile by using the appropriate constitutive equation if the flow is laminar. In the case of turbulent flow, τ_0 must be ascertained from other considerations which are discussed in the next section. The integration of Eq. (9.49) would require some knowledge of dp/dx in addition to assuming a velocity profile. The pressure gradient can be expressed as

$$-\frac{1}{\rho}\frac{dp}{dx} = V_\infty \frac{dV_\infty}{dx} \tag{9.51}$$

at the edge of the boundary layer, as the flow outside of the layer is practically irrotational. The right-hand side of Eq. (9.51) can be ascertained by solving the potential flow problem for the flow over the surface on which the boundary layer has formed. The handling of the integral forms of these flows is complicated by the existence of a pressure gradient, and this is left for more specialized or advanced work.

The details involved in the use of Eq. (9.50) are illustrated in the following example, which points out the general steps involved.

Example 9.6

The experimental data of Nikuradse given by Schlichting[1] suggest that a reasonable velocity profile for laminar flow over a flat plate of zero incidence might be of the form

$$\frac{V_x}{V_\infty} = C_1\left(\frac{y}{\delta}\right) + C_2\left(\frac{y}{\delta}\right)^3$$

in which C_1 and C_2 are constants. Assume that the fluid is incompressible. Using this profile, (*a*) obtain an expression for δ, δ^*, and τ_0 as functions of x, and compare these to the results of the exact solution of Blasius given in Examples 9.4 and 9.5. (*b*) How does the surface-resistance coefficient C_f vary with N_{R_L}?

Solution. (*a*) The first step is to determine the constants C_1 and C_2. These may be found from the following conditions:

$$V_x = 0 \qquad \text{for } y = 0$$

$$V_x = V_\infty \qquad \text{and} \qquad \frac{\partial V_x}{\partial y} = 0 \qquad \text{for } y = \delta$$

The first condition is already satisfied by the cubic form of V_x/V_∞. The last two conditions yield $C_1 = 3/2$ and $C_2 = -1/2$, and

$$\frac{V_x}{V_\infty} = \frac{3}{2}\left(\frac{y}{\delta}\right) - \frac{1}{2}\left(\frac{y}{\delta}\right)^3$$

from which

$$\tau_0 = \mu\frac{\partial V_x}{\partial y}\bigg|_{y=0} = \mu V_\infty\left(\frac{3}{2\delta} - \frac{3}{2\delta^3}y^2\right)_{y=0} = \frac{3\mu V_\infty}{2\delta}$$

Substituting for V_x and τ_0 in Eq. (9.50) gives

$$\frac{3\mu V_\infty}{2\delta} = \rho\frac{\partial}{\partial x}\int_0^\delta\left\{V_\infty^{\,2}\left[\frac{3}{2}\left(\frac{y}{\delta}\right) - \frac{1}{2}\left(\frac{y}{\delta}\right)^3\right] - V_\infty^{\,2}\left[\frac{3}{2}\left(\frac{y}{\delta}\right) - \frac{1}{2}\left(\frac{y}{\delta}\right)^3\right]^2\right\}dy$$

$$= \rho\frac{\partial}{\partial x}\left(\frac{39}{280}V_\infty^{\,2}\,\delta\right)$$

For zero pressure gradient, the velocity V_∞ at the edge of the boundary layer is not a function of x, and

[1] *Ibid.*, p. 132.

$$\frac{140}{13}\frac{\mu}{\rho V_\infty} = \delta\frac{d\delta}{dx}$$

which can be integrated to give

$$\frac{\delta^2}{2} = \frac{140}{13}\frac{\nu x}{V_\infty} + \text{constant}$$

The constant is zero, since $\delta = 0$ for $x = 0$, and

$$\delta = \left(\frac{280}{13}\frac{\nu x}{V_\infty}\right)^{1/2} = 4.64\left(\frac{\nu x}{V_\infty}\right)^{1/2} \blacktriangleleft$$

For the Blasius solution, $\delta = 4.92\,(\nu x/V_\infty)^{1/2}$; and δ for the assumed cubic profile is about 6 percent low.

The displacement thickness is found from Eq. (9.22).

$$\delta^* = \int_0^\infty\left(1 - \frac{V_x}{V_\infty}\right)dy = \int_0^\delta\left\{1 - \left[\frac{3}{2}\left(\frac{y}{\delta}\right) - \frac{1}{2}\left(\frac{y}{\delta}\right)^3\right]\right\}dy$$

$$= \frac{3}{8}\delta$$

so that

$$\delta^* = \frac{3}{8}(4.64)\left(\frac{\nu x}{V_\infty}\right)^{1/2} = 1.74\left(\frac{\nu x}{V_\infty}\right)^{1/2} \blacktriangleleft$$

which is about 1 percent high compared to the Blasius solution.

From the first part of this example,

$$\tau_0 = \mu\frac{\partial V_x}{\partial y}\bigg|_{y=0} = \frac{3\mu V_\infty}{2\delta} = \frac{3\mu V_\infty}{2}\left[\frac{1}{4.64}\left(\frac{V_\infty}{\nu x}\right)^{1/2}\right]$$

and

$$\tau_0 = 0.323\,\mu V_\infty\left(\frac{V_\infty}{\nu x}\right)^{1/2} \blacktriangleleft$$

This result is 2.5 percent below $0.33206\,\mu V_\infty(V_\infty/\nu x)^{1/2}$.

b. The surface-resistance coefficient is found from

$$C_f = \frac{\int_0^L 0.323\,\mu V_\infty(V_\infty/\nu x)^{1/2}\,dx}{\rho V_\infty{}^2 L/2}$$

and

$$C_f = \frac{1.292}{N_{R_L}{}^{1/2}} \blacktriangleleft$$

The results of this example are in fair agreement with the exact solution of Blasius. The assumption of other profiles would lead to estimates of the various boundary-layer parameters with varying degrees of disparity compared to the Blasius solution.

For laminar flow, the shear stress τ_0 at the solid boundary was found from the constitutive equation for τ_{yx} which was simplified to meet the assumptions of the problem. In general τ_0 is found from the appropriate constitutive equation when the integral forms of the boundary-layer equations are applied to laminar flow. For the case of turbulent flow, τ_0 must be estimated by some other means. A recollection of the first part of this chapter reveals that the *roughness* of the pipe wall had a direct bearing on the turbulent velocity profiles. A corresponding situation obviously exists for turbulent boundary layers. The next section introduces a method that can be applied to turbulent flows over *smooth* walls, and a few comments are offered concerning the influence of relative roughness on the boundary-layer flows.

9.14 Turbulent Boundary-Layer Flow over a Smooth Flat Plate

The work of Nikuradse has indicated that

$$\frac{\overline{V}_x}{V_{\max}} = \left(\frac{y}{R}\right)^{1/n} \tag{9.52}$$

can be used to predict turbulent velocity profiles in smooth-walled pipes. The exponent n is dependent on the Reynolds number based on a space-average velocity and the pipe diameter; n ranges from a value of 6 to 10 for a corresponding Reynolds number range of 4 (10^3) to 3.24 (10^6). The distance away from and normal to the wall is designated by y. Equation (9.52) may be used to form a space-average velocity, and an expression for $\overline{V}_x/V_{SA}$ may be generated that depends only on the exponent n since

$$V_{SA} = \frac{1}{A}\int_A \overline{V}_x\,dS = \frac{1}{\pi R^2}\int_0^R V_{\max}\left(\frac{y}{R}\right)^{1/n} 2\pi r\,dr$$

in which r is the perpendicular distance from the centerline of the pipe. Noting that $y = R - r$,

$$V_{SA} = \frac{2V_{\max}}{R^2}\int_0^R \left(\frac{R-r}{R}\right)^{1/n} r\,dr$$

and

$$\frac{V_{SA}}{V_{\max}} = \frac{2n^2}{(n+1)(2n+1)} \tag{9.53}$$

It has been found that the boundary-layer counterpart of Eq. (9.52) may be written as

$$\frac{\overline{V}_x}{V_\infty} = \left(\frac{y}{\delta}\right)^{1/n} \tag{9.54}$$

which is a reasonable velocity profile for use in the integral form (9.50). An estimate for τ_0 may be obtained by using

$$\tau_0 = \frac{1}{8} f \rho V_{SA}^2 \tag{8.24}$$

if one knows how f varies with the Reynolds number. Once the N_R range of the flow is known, the variation of f with N_R may be ascertained, as well as the value of n in Eq. (9.54). For purposes of illustrating the procedure, the following values will be considered for flow in a pipe:

$$N_R = \frac{V_{SA} D}{\nu} = 10^5 \qquad \text{and} \qquad n = 7$$

From Fig. 8.4, the variation of f with N_R on the smooth-pipe curve is

$$f = \frac{0.316}{N_R^{1/4}} \qquad 3(10^3) < N_R < 10^5 \tag{9.55}$$

Substituting this expression for f in Eq. (8.24) for the wall shear stress yields

$$\tau_0 = \frac{1}{8}\left[0.316\left(\frac{D V_{SA} \rho}{\mu}\right)^{-1/4}\right] \rho V_{SA}^2$$

For boundary-layer flow, the pipe diameter D is replaced by 2δ, and V_{SA} is eliminated through the use of Eq. (9.53) with V_{max} taken as V_∞. For $n = 7$, $V_{SA} = 0.817 V_{max} = 0.817 V_\infty$.

$$\tau_0 = \frac{0.316}{8}\left[\frac{(2\delta)(0.817 V_\infty)\rho}{\mu}\right]^{-1/4} \rho (0.817 V_\infty)^2$$

and carrying out the arithmetic,

$$\tau_0 = 0.023 \rho V_\infty^2 \left(\frac{\delta V_\infty}{\nu}\right)^{-1/4} \tag{9.56}$$

Other expressions for the wall shear stress may be generated for various values of n by using Eq. (9.53) and the proper $f - N_R$ relation from Fig. 8.4. Schlichting[1] suggests values of 6, 7, and 10 for n, which corresponds to Reynolds numbers of 4 (10^3), 10^5, and 3.24 (10^6), respectively.

Example 9.7

For $n = 7$, obtain an integrated form for δ as a function of x, the distance along a flat smooth plate.

[1] *Ibid.*, p. 563.

Solution. Applying Eq. (9.50), with

$$\frac{\bar{V}_x}{V_\infty} = \left(\frac{y}{\delta}\right)^{1/7}$$

the shear stress at the plate is

$$\tau_0 = \frac{\partial}{\partial x}\int_0^\delta \rho\left[V_\infty^{\ 2}\left(\frac{y}{\delta}\right)^{1/7} - V_\infty^{\ 2}\left(\frac{y}{\delta}\right)^{2/7}\right]dy$$

$$= \rho V_\infty^{\ 2}\frac{d}{dx}\left(\frac{7}{72}\delta\right)$$

and

$$\frac{\tau_0}{\rho V_\infty^{\ 2}} = \frac{7}{72}\frac{d\delta}{dx}$$

Using Eq. (9.56) to substitute for τ_0 yields

$$\frac{7}{72}\delta^{1/4}\,d\delta = 0.023\left(\frac{V_\infty}{\nu}\right)^{-1/4}dx$$

after separating the variables. Integrating and simplifying gives

$$\frac{7}{90}\delta^{5/4} = 0.023\left(\frac{V_\infty}{\nu}\right)^{-1/4}x + \text{a constant}$$

which is a functional relationship between δ and x, and hence an answer to the extent of the unknown constant. If it were assumed that $\delta = 0$ for $x = 0$, which was valid in the case of the laminar boundary layer (see Example 9.6), then the constant is zero, and

$$\delta = 0.38x\left(\frac{\nu}{V_\infty x}\right)^{1/5}$$ ◀

Actually, δ cannot be taken as zero at $x = 0$, for the flow in the boundary layer is laminar over a portion of the smooth plate before the flow transits to the turbulent regime (this leads into the next section which discusses transition from laminar to the turbulent flow regime). However, before leaving this example, it is worth noting that the last expression for δ indicates that the turbulent boundary layer thickens with the four-fifth power of x, while δ varies with the one-half power of x for the laminar boundary layer.

After a brief discussion of transition to turbulent flow, the problem of handling the constant of integration in the relation for δ will be reconsidered.

9.15 Transition of Boundary-Layer Flow

Over the portion of the plate near the leading edge, the boundary layer is laminar. As the flow proceeds, the layer thickens until a transition region is reached. After transition is complete, the boundary layer is turbulent with a laminar sublayer between it and the wall. An estimate of the length over which laminar flow prevails may be made by reference to experimental data[1] that indicates that transition occurs for $5\,(10^5) < V_\infty x/\nu < 10^7$.

One method of using the integral forms of the boundary-layer equations applied to the flow in Fig. 9.12 is to evaluate the drag force on the plate by assuming the flow to be wholly turbulent over its entire length L. From this, the turbulent drag of the length up to transition $L^\dagger$ is subtracted; and to the difference is added the laminar drag of $L^\dagger$. This procedure has the net effect of "subtracting out" the value of the constant[2] in the expression for δ, and as found in Example 9.7 for an assumed turbulent profile of $n = 7$. For a unit dimension of the leading edge, the drag force on one side of the plate for turbulent flow over the entire length L is

$$\underset{\text{(turb)}}{F_{\text{drag}}} = \int_0^L \tau_0\,dx = \int_0^L 0.023\rho V_\infty{}^2\left(\frac{\nu}{V_\infty \delta}\right)^{1/4} dx$$

in which τ_0 has been replaced by its equivalent from Eq. (9.56). Next, a substitution is made for δ since it is a function of x, and

$$\underset{\text{(turb)}}{F_{\text{drag}}} = \int_0^L 0.023\rho V_\infty{}^2\left[\frac{\nu}{V_\infty(0.38x)(\nu/V_\infty x)^{1/5}}\right]^{1/4} dx$$

$$= 0.037\rho V_\infty{}^2 L\left(\frac{\nu}{V_\infty L}\right)^{1/5}$$

[1]See H. Schlichting (J. Kestin, tr.), "Boundary-Layer Theory," 6th ed., p. 600, McGraw-Hill Book Company, New York, 1968, for a plot of the data of five investigators.

[2]This is approximate since the flow is not turbulent over the portion of the plate near the leading edge.

Fig. 9.12 Transition from laminar to turbulent flow in the boundary layer on a smooth flat plate.

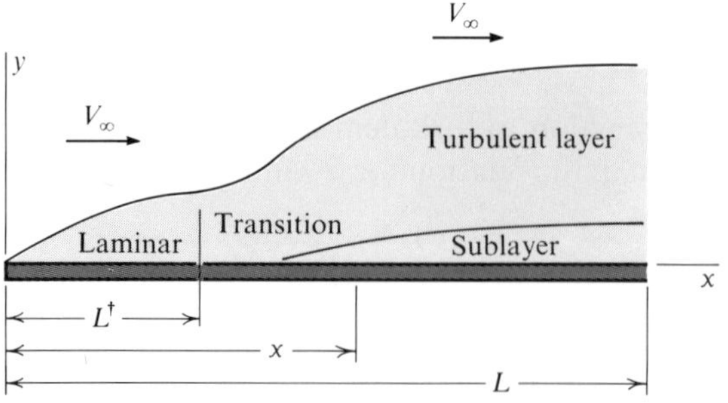

For turbulent flow up to transition, the drag is

$$F_{\substack{\text{drag}\\ \text{(turb)}}} = \int_0^{L^\dagger} \tau_0 \, dx = 0.037 \rho V_\infty^{\,2} L^\dagger \left(\frac{\nu}{V_\infty L^\dagger} \right)^{1/5}$$

For laminar flow up to transition, the drag is given in Example 9.4 as

$$F_{\substack{\text{drag}\\ \text{(lam)}}} = 0.664 (V_\infty^{\,3} \mu \rho L^\dagger)^{1/2} = 0.664 \rho V_\infty^{\,2} L^\dagger \left(\frac{\nu}{V_\infty L^\dagger} \right)^{1/2}$$

The total drag force on the plate is then given by

$$F_{\text{drag}} = 0.037 \, \rho V_\infty^{\,2} L \left(\frac{\nu}{V_\infty L} \right)^{1/5} - 0.037 \, \rho V_\infty^{\,2} L^\dagger \left(\frac{\nu}{V_\infty L^\dagger} \right)^{1/5} + 0.664 \, \rho V_\infty^{\,2} L^\dagger \left(\frac{\nu}{V_\infty L^\dagger} \right)^{1/2}$$

The surface-resistance coefficient for the plate is

$$C_{f_{\text{tot}}} = \frac{F_{\text{drag}}}{(1/2) \rho V_\infty^{\,2} L} = 0.074 \left(\frac{\nu}{V_\infty L} \right)^{1/5} - 0.074 \frac{L^\dagger}{L} \left(\frac{\nu}{V_\infty L^\dagger} \right)^{1/5} + 1.328 \frac{L^\dagger}{L} \left(\frac{\nu}{V_\infty L^\dagger} \right)^{1/2} \tag{9.57}$$

The preceding equation is in good agreement[1] with experimental data for $5\,(10^5) < V_\infty L/\nu < 10^7$, and it can be rearranged by noting that

$$\frac{L^\dagger}{L} = \frac{L^\dagger V_\infty/\nu}{L V_\infty/\nu} = \frac{N_{R_{L^\dagger}}}{N_{R_L}}$$

Then

$$C_{f_{\text{tot}}} = \frac{0.074}{N_{R_L}^{\;1/5}} - \frac{1}{N_{R_L}} (0.074 \, N_{R_{L^\dagger}}^{\;4/5} - 1.328 \, N_{R_{L^\dagger}}^{\;1/2}) \tag{9.58}$$

For a critical transition Reynolds number of $5\,(10^5)$, the total resistance coefficient becomes

$$C_{f_{\text{tot}}} = \frac{0.074}{N_{R_L}^{\;1/5}} - \frac{1750}{N_{R_L}} \tag{9.59}$$

which is valid for a smooth surface and the specified value of $N_{R_{L^\dagger}}$. Equation (9.58) applies only to a smooth wall with N_{R_L} restricted to values between $5\,(10^5)$ and 10^7.

[1] Recall that Eq. (9.55) and hence (9.56) are valid for $3\,(10^3) < DV_{SA}/\nu < 10^5$.

For rough plates (and for rough pipes) the shear stress and resistance coefficient depend on the surface roughness. A discussion of roughness effects on boundary-layer flows is presented in Schlichting.[1] Experimental data for various roughnesses are utilized in the formulation of resistance laws for flow over rough surfaces. In this case, the size of the absolute roughness e, compared to the boundary-layer thickness δ, becomes an important parameter. The appropriate formulations for rough surfaces will not be treated in this book, and the reader requiring this information is again referred to Schlichting.

An example will indicate the relative magnitudes of the drag associated with laminar and turbulent flow for flow over a smooth plate.

Example 9.8

Water at a temperature of 75°F flows steadily over a flat plate which is one foot square. The velocity outside of the boundary layer is 50 ft/sec. Consider $n = 7$, and calculate: (*a*) the surface-resistance coefficient for the entire plate; (*b*) the total drag force on one side of the plate; (*c*) the critical length $L^\dagger$; and (*d*) the drag force on the portion (one side) of the plate over which the flow is laminar.

Solution. (*a*) The plate Reynolds number is

$$N_{R_L} = \frac{V_\infty L}{\nu} = \frac{(50 \text{ ft/sec})(1 \text{ ft})}{10^{-5} \text{ ft}^2/\text{sec}} = 5\,(10^6)$$

Since $5\,(10^5) < N_{R_L} < 10^7$, Eq. (9.59) may be used to calculate the total surface-resistance coefficient for $N_{R_L\dagger} = 5\,(10^5)$.

$$C_{f_\text{tot}} = \frac{0.074}{[5\,(10^6)]^{1/5}} - \frac{1750}{5(10^6)} = 0.00304 \qquad \blacktriangleleft$$

b. The total drag force on one side of the plate is

$$F_\text{drag} = C_{f_\text{tot}}\left(\frac{1}{2}\rho V_\infty{}^2 S_\text{plate}\right)$$

$$= (0.00304)\left(\frac{1}{2}\right)\left(1.94\,\frac{\text{slug}}{\text{ft}^3}\right)\left(50\,\frac{\text{ft}}{\text{sec}}\right)^2 (1 \text{ ft})^2$$

$$= 7.4 \text{ lb}_f \qquad \blacktriangleleft$$

c. For the assumed critical Reynolds number of $5\,(10^5)$,

$$L^\dagger = \frac{N_{R_L\dagger}\nu}{V_\infty} = \frac{[5\,(10^5)](10^{-5} \text{ ft}^2/\text{sec})}{50 \text{ ft/sec}}$$

$$= 0.1 \text{ ft} \qquad \blacktriangleleft$$

[1] *Ibid.*, pp. 610–623.

d. The drag on one side of the plate over which the flow is laminar may be calculated from the results of Example 9.4.

$$\begin{aligned} F_{\substack{\text{drag} \\ (\text{lam})}} &= 0.664\,\rho V_\infty{}^2 \left(\frac{v}{V_\infty L^\dagger}\right)^{1/2} L^\dagger \\ &= (0.664)\left(1.94\frac{\text{slug}}{\text{ft}^3}\right)\left(50\frac{\text{ft}}{\text{sec}}\right)^2 \left(\frac{1}{5\,(10^5)}\right)^{1/2} (0.1\text{ ft})(1\text{ ft}) \\ &= 0.456\text{ lb}_f \end{aligned}$$

◀

Comparison of the answers in (*b*) and (*d*) indicates that a very large portion of the resistance is associated with the turbulent boundary layer. This is generally true for drag that is a consequence of the shear stress, and such resistance is frequently referred to as the *skin-friction drag*. There is another form of drag that accrues as a consequence of the pressure distribution over a surface enclosing a solid body. The reduction of this latter form of drag is sometimes favored by the presence of a turbulent boundary layer on the surface. A discussion of this leads into the consideration of the influence of pressure gradients on boundary-layer flows.

9.16 Influence of Pressure Gradients on Boundary-Layer Flow

The boundary layers discussed so far have been associated with flows over a flat or slightly curved surface. The Blasius solution for the flat plate (Section 9.12) involved a zero pressure gradient. For curved surfaces (even slightly curved), the pressure outside of the boundary layer will change with distance measured along the wall, and the pressure gradient for $y \geq \delta$ may be determined from potential flow. For the surface or wall shown in Fig. 9.13*a*, $dp/dx < 0$, which is a *favorable pressure gradient*. Figure 9.13*b* illustrates an *adverse pressure gradient* of $dp/dx > 0$. For both cases,

$$V_x \frac{\partial V_x}{\partial x} + V_y \frac{\partial V_x}{\partial y} = -\frac{1}{\rho}\frac{dp}{dx} + v\frac{\partial^2 V_x}{\partial y^2} \tag{9.25}$$

Fig. 9.13 *Boundary-layer flows with pressure gradients.*

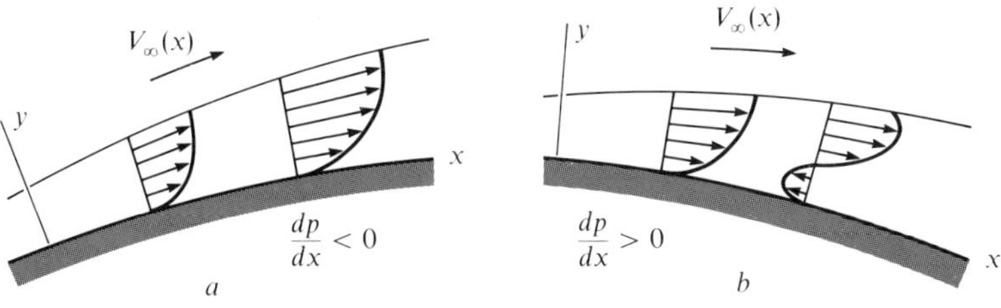

with body forces neglected and $p = p(x)$ only. Very proximate to the wall, the preceding equation indicates that

$$\frac{\partial^2 V_x}{\partial y^2} = \frac{1}{\mu}\frac{dp}{dx} \tag{9.60}$$

At the edge (or outside) of the boundary layer,

$$\left.\frac{\partial V_x}{\partial y}\right|_{y \geq \delta} > 0 \qquad (V_x \text{ almost constant for } y \geq \delta) \tag{9.61}$$

Considering first the case of the favorable pressure gradient, Eqs. (9.60) and (9.61) give, respectively,

$$\left.\frac{\partial^2 V_x}{\partial y^2}\right|_{y=0} < 0 \qquad \text{and} \qquad \left.\frac{\partial V_x}{\partial y}\right|_{y \geq \delta} > 0$$

These inequalities indicate that V_x increases throughout the entire interval $0 < y \leq \delta$. For the adverse pressure gradient, reasoning similar to that above would show that it is possible for the velocity profile to have a point of inflection.

The velocity profiles are indicated qualitatively in Fig. 9.13. In the presence of an adverse pressure gradient, the flow in the vicinity of the wall may be in a direction opposite to that of the flow outside of the boundary layer. The location along the surface where the flow begins to reverse is called the *separation point*. Downstream from this point the flow near the wall is also characterized by back flow for a continuing adverse pressure gradient as shown in Fig. 9.14. It should be noted that an adverse pressure gradient is necessary for separation; however, it may not prevail far enough along the wall to produce separation of the flow.

When separation occurs on the surface of an airfoil or any other object placed in a fluid stream, the fluid downstream of the separation point is characterized by eddies that generally result in a low-pressure region behind

Fig. 9.14 *Separation produced by an adverse pressure gradient.*

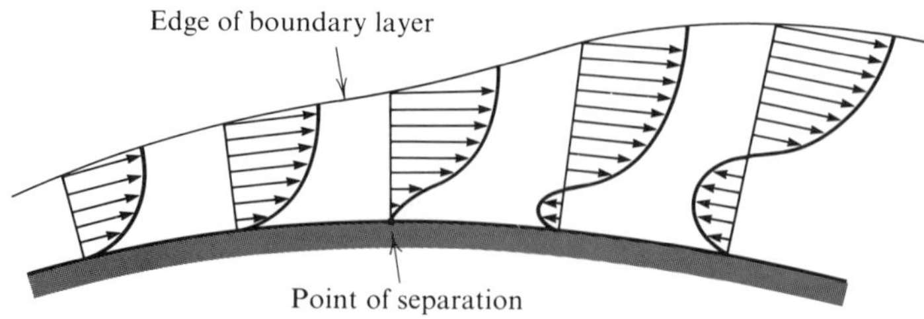

the foil or object. Figure 9.15 shows the wake behind a foil and a cylinder. The points of separation are noted by S on the sketches. The wake is a low-pressure region by comparison with the surrounding fluid; hence wakes result in a reduced pressure distribution on the downstream portion of the object and therefore increase the drag on the object. Turbulent boundary layers will stay attached to a surface over a longer distance on the body than will laminar boundary layers because the velocity close to the surface is much larger for the turbulent profile, and the force on the fluid element due to the pressure gradient cannot reverse the flow in the vicinity of the wall as easily as it could in the laminar boundary layer. It should be recalled that the turbulent velocity profile in a pipe was much flatter than the laminar profile for the same space-average velocity. The counterpart is true for boundary-layer profiles with the same velocity at the edge of the boundary layer.

Because separation increases drag, turbulent boundary layers are sometimes preferable to a laminar boundary layer, even though the latter has less skin-friction drag associated with it.

9.17 Closing Remarks

A few simple models for handling turbulent flows in a pipe and over surfaces were presented in this chapter. An introduction to boundary-layer flows was presented with some examples and comments demonstrating the relevance of these flows to the engineer.

Certainly the physical and mathematical complexities of these flows would require further study on the part of one who wishes to specialize in the field of fluid mechanics; and in this respect, the subject is not unique.

Self-Study Questions

9.1 *a.* Is the continuum concept applicable to turbulent flow?
b. What is meant by steady turbulent flow?

9.2 Consider a velocity component

$$V_x = A + B \cos Ct$$

Fig. 9.15 *Separation points and wakes.*

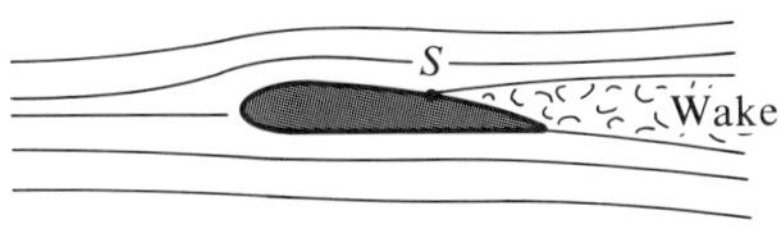

a

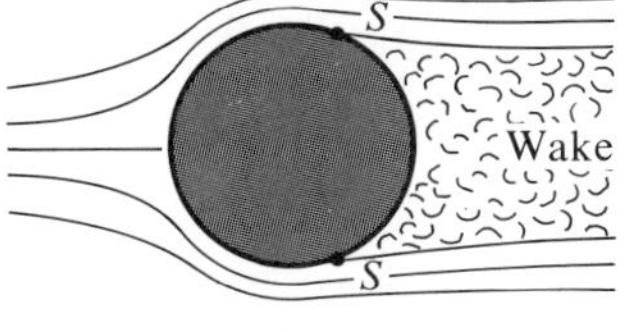

b

where A, B, and C are constants.

a. Is this velocity consistent with steady turbulent flow?

b. Show that $\overline{\tilde{V}_x} = 0$.

c. Is turbulent flow really periodic or random?

9.3 What are four rules of time averaging?

9.4 If density and velocity are steady in the mean for a flow field, determine whether the following are "equal to" or "may not be equal to" zero:

$$\overline{\frac{\partial \tilde{\rho}}{\partial t}} \qquad \overline{\bar{\rho}\frac{\partial \tilde{V}_x}{\partial x}} \qquad \overline{\tilde{\rho}\frac{\partial \bar{V}_x}{\partial x}} \qquad \overline{\tilde{\rho}\frac{\partial \tilde{V}_x}{\partial x}}$$

9.5 What physical interpretation may be given to each of the following for steady turbulent flow:

a. $\overline{\tilde{\rho}\tilde{V}_x}\, dy\, dz$ *b.* $\left[\overline{\tilde{\rho}\tilde{V}_x} + \dfrac{\partial(\overline{\tilde{\rho}\tilde{V}_x})}{\partial x}\, dx\right] dy\, dz$ *c.* $\dfrac{\partial(\overline{\tilde{\rho}\tilde{V}_x})}{\partial x}$

9.6 What physical interpretation may be given to each of the following for steady turbulent flow of an incompressible fluid:

a. $-\rho\overline{\tilde{V}_x\tilde{V}_x}\, dy\, dz$ *c.* $-\rho\overline{\tilde{V}_x\tilde{V}_y}\, dx\, dz$

b. $\dfrac{\partial(-\rho\overline{\tilde{V}_x{}^2})}{\partial x}\, dx\, dy\, dz$ *d.* $\dfrac{\partial(-\rho\overline{\tilde{V}_x\tilde{V}_y})}{\partial x}\, dx\, dy\, dz$

9.7 Briefly explain why C_2 is a positive constant less than 1 in

$$\overline{\tilde{V}_x\tilde{V}_y} = -C_2\,\overline{|\tilde{V}_x|}\,\overline{|\tilde{V}_y|}$$

an expression developed in Section 9.7.

9.8 What postulates were made concerning the turbulent flow region in order to obtain Eq. (9.16)?

9.9 Is it reasonable to expect that Eq. (9.19) does not reduce to Eq. (9.18) when e is reduced to smooth-pipe values?

9.10 Sketch extensions of the velocity profile from the turbulent and laminar sublayer into the buffer-layer region as shown in Fig. 9.7.

9.11 Show that the mixing length proposed by von Karman reduces to that of $l = ky$ if a logarithmic turbulent velocity profile is assumed.

9.12 *a.* Distinguish between boundary-layer thickness and boundary-layer displacement thickness.

b. Why may the upper limit of ∞ of the integral in Eq. (9.22) defining displacement thickness be replaced by δ?

9.13 With reference to Fig. 9.9, what is the order of magnitude for each of the following:

a. $x \qquad y \qquad V_x \qquad V_y \qquad \dfrac{\partial V_x}{\partial x} \qquad V_x\dfrac{\partial V_x}{\partial x}$

b. $\dfrac{\partial^2 V_y}{\partial x^2} \qquad \dfrac{\partial^2 V_x}{\partial y^2} \qquad \dfrac{1}{\rho}\dfrac{\partial p}{\partial x} \qquad \nu\dfrac{\partial^2 V_x}{\partial y^2} \qquad \nu\dfrac{\partial^2 V_y}{\partial x^2}$

9.14 Does a velocity potential ϕ exist for the stream function ψ given by Eq. (9.34)? Why or why not?

9.15 Briefly outline how the boundary-layer equations of motion were solved by Blasius for laminar flow over a flat plate.

9.16 Sketch qualitatively the variation of shear stress at a wall, the wall area, and the total drag on the wall as a function of x.

9.17 Show that a local surface-resistance coefficient (at a given x) defined as

$$\frac{\tau_{xy}|_{y=0}}{(1/2)\,\rho V_\infty^{\,2}}$$

is equal to $C_f/2$ where C_f is defined in Example 9.4.

9.18 Can the stream function defined by Eq. (9.34) be used within and outside the laminar boundary layer? Why or why not?

9.19 Does the slope of the boundary-layer edge shown in Fig. 9.9 compare with the δ variation obtained in Example 9.5?

9.20 Express a physical meaning for each term in Eq. (9.49).

9.21 Carry out the integration of Eq. (9.50) in Example 9.6.

9.22 Use an assumed linear velocity profile, $V_x = V_\infty y/\delta$, with Eq. (9.50) to obtain an expression for $\delta(x)$.

9.23 Verify Eq. (9.55) by using Fig. 8.4.

9.24 Why isn't $\tau_0 = 0$ for turbulent flow past a flat smooth plate, since

$$\left.\frac{\partial \overline{V}_x}{\partial y}\right|_{y=0} = 0$$

using Eq. (9.52), as opposed to the result of Eq. (9.56)?

9.25 Considering the existence of laminar and turbulent boundary layers, is the corresponding drag exerted on a flat smooth wall
a. Doubled if the length is doubled?
b. Increased if transition occurs earlier?

Problems

9.1 Verify that Eq. (9.5) is correct.

9.2 Show that the time average of $\tilde{M}\tilde{N}$ is not generally zero. Hint: Let M and N be sinusoidal functions of time.

9.3 Consider the velocity components at a given point in a fluid to be

$$V_x = 3 + 2\cos t \qquad \text{and} \qquad V_y = 2 + \cos t$$

a. Show a plot of V_x and V_y versus t.
b. What are the values of

$$\overline{V_x} \qquad \overline{\tilde{V}_y} \qquad \overline{\tilde{V}_x\tilde{V}_y} \qquad \overline{V_xV_y}$$

c. What is the minimum time increment for time averaging?

9.4 If $V_x = 8 + 10 \sin \pi t$, and $V_y = 2 \sin \pi t$ for a given point in a fluid, evaluate the following:

$$\overline{V}_x \quad \overline{\tilde{V}}_y \quad \overline{\tilde{V}_x \tilde{V}_y} \quad \overline{\tilde{V}}_x \overline{\tilde{V}}_y \quad \overline{V_x V_y}$$

9.5 Show that Navier-Stokes equations reduce to the following for steady turbulent and incompressible flow in the x direction between parallel plates, assuming that $\overline{V}_x = f(y)$ is the only velocity component of the time-mean flow:

$$\frac{1}{\rho}\frac{\partial \bar{p}}{\partial x} = \nu \frac{\partial^2 \overline{V}_x}{\partial y^2} + (\bar{\mathbf{f}}_B)_x - \frac{\partial}{\partial y}(\overline{\tilde{V}_x \tilde{V}_y}) - \frac{\partial}{\partial x}(\overline{\tilde{V}_x^2})$$

$$\frac{1}{\rho}\frac{\partial \bar{p}}{\partial y} = (\bar{\mathbf{f}}_B)_y - \frac{\partial(\overline{\tilde{V}_y^2})}{\partial y} - \frac{\partial}{\partial x}(\overline{\tilde{V}_x \tilde{V}_y})$$

9.6 Use Eq. (9.18) to obtain an expression for $\overline{V}_x - \overline{V}_{x,\max}$ for turbulent flow in a smooth pipe. Show that Eq. (9.19) yields an identical expression for a rough pipe. $\overline{V}_{x,\max}$ is the velocity at the centerline of the pipe.

9.7 Estimate the thickness in inches of the laminar sublayer and the location from the wall where a turbulent region begins for the flow in Example 9.2.

9.8 Consider incompressible turbulent flow in a smooth and straight circular pipe of radius R and

a. Show that

$$V_{SA} = V_*(C_1 \ln \frac{RV_*}{\nu} + C_2)$$

with $C_1 = 2.5$ and $C_2 = 1.750$.

b. Obtain the following functional relation:

$$\frac{1.13}{f^{1/2}} - \frac{1}{2}\ln f + 1.03 = \ln N_R$$

c. Is the result in (*b*) consistent in the range of experimental evidence given in Fig. 9.5? Also, compare with smooth-pipe values in Fig. 8.4.

9.9 Water at 100°F flows steadily in straight drawn tubing of 8-in. diameter at a Reynolds number of 10^4.

a. Estimate the velocity 1 in. from the wall and compare with the space-average velocity.

b. Estimate the thickness of the laminar sublayer for this flow.

9.10 The shear stress in turbulent flow is sometimes considered as the sum of time steady and fluctuating effects, so that

$$\tau_{xy} = (\nu + \varepsilon)\frac{d(\rho \overline{V}_x)}{dy}$$

where ε is called the kinematic eddy viscosity. Using Prandtl mixing-length theory and experimental results for smooth-pipe flow of an incompressible fluid,

a. Show that ε, unlike ν, is a function of the flow.

b. Show that

$$\frac{\varepsilon}{\nu} = \frac{0.4 y V_*}{\nu}$$

c. Use Fig. 9.6 to estimate the minimum value of ε/ν for the turbulent flow in Example 9.2.

9.11 Air flows steadily in a straight section of drawn tubing of 8-in. diameter at a Reynolds number of $5.5\,(10^4)$. Estimate the mean velocity 1 in. from the wall and the maximum mean velocity at a section where the time-mean pressure and temperature is 20 $\text{lb}_f/\text{in.}^2$ and 100°F, respectively.

9.12 Obtain the ratio δ^*/δ for an assumed linear velocity profile in a laminar boundary layer of an incompressible fluid moving steadily past a flat plate.

9.13 Calculate the drag force per square foot exerted by air at 1 atm and 200°F flowing steadily past a flat plate with $V_\infty = 10$ ft/sec,
a. If the area is located $0 < x < 1$ ft.
b. If the area is located $3 < x < 4$ ft.

9.14 Water flows steadily over a flat plate as illustrated in Fig. 9.9. V_∞ is 2 ft/sec. Evaluate the surface-resistance coefficient for a section of unit depth between $x = 2$ ft and $x = 3$ ft, if
a. The liquid temperature is 50°F.
b. The liquid temperature is 150°F.

9.15 Evaluate ψ at δ for $x = 10$ ft in laminar flow of air past a flat plate. The pressure, temperature, and free-stream velocity are 1 atm, 100°F, and 10 ft/sec, respectively. What is the volume flow rate of air into the boundary layer between $x = 0$ and $x = 10$ ft?

9.16 Use the von Karman integral momentum equation to obtain an expression for δ, δ^*, τ_0, and C_f (each expressed in terms of N_{R_x}) for flow with zero pressure gradient over a flat plate assuming an incompressible Newtonian fluid and a velocity profile of $V_x = K_1 y + K_2 y^2$ where K_1 and K_2 are constants for the flow. Compare the results with Example 9.6.

9.17 Repeat Prob. 9.16 for an assumed velocity profile of

$$V_x = K_1 y e^{-K_2 y}$$

9.18 Obtain an expression for boundary layer thickness δ if the velocity profile is

$$\frac{V_x}{V_\infty} = \frac{3}{2}\left(\frac{y}{\delta}\right) - \frac{1}{2}\left(\frac{y}{\delta}\right)^3$$

and the free-stream velocity increases linearly for $x > 0$, as

$$V_\infty(x) = C_1 x + C_2$$

with C_1 and C_2 positive constants for the given incompressible flow.

9.19 Derive the following integral expression of the first law of thermodynamics by using a control volume similar to that shown in Fig. 9.10. The upper limit of y is δ_T, the thermal boundary-layer thickness where $T = 0.99\,T_\infty$.

$$\frac{d}{dx}\left[\int_0^{\delta_T} (T_\infty - T)\,V_x\,dy\right] = \frac{\kappa}{\rho c_p}\left.\frac{\partial T}{\partial y}\right|_{y=0}$$

9.20 Obtain an expression for wall shear stress exerted by an incompressible fluid in steady flow past a flat and smooth plate assuming a turbulent profile with $n = 6$ and a pipe $N_R = 4(10^3)$. See development of Eq. (9.56).

9.21 Obtain an expression for the boundary-layer thickness for zero pressure gradient in Prob. 9.20.

9.22 Repeat Prob. 9.20 using $n = 10$ and $N_R = 3(10^6)$; and with the information in Fig. 8.4 verify that

$$\tau_0 = 0.01\,\rho V_\infty{}^2\left(\frac{\delta V_\infty}{\nu}\right)^{-1/6.35}$$

9.23 Use the von Karman integral equation to obtain an expression for $\delta(x)$ for flow with zero pressure gradient using the result of Prob. 9.22.

9.24 Obtain the drag of 50°F water on a flat surface using $N_{R_{L\dagger}} = 5(10^6)$, $V_\infty = 15$ ft/sec, and $L = 200$ ft. The width of the surface is 40 ft.

A

FORMULATION OF STOKES' CONSTITUTIVE RELATIONS

Stokes' assumptions in relating stress components to rates of strain are:

1. Each of the stress components is linearly proportional to the strain rates.
2. The fluid is isotropic, and hence there are no preferred directions.
3. In the absence of strain rates, the normal stresses must reduce to the pressure, and the shear stresses must vanish.

On the basis of assumption 1, the most general linear relations are

$$
\begin{aligned}
\sigma_{xx} &= C_{11}e_{xx} + C_{12}e_{yy} + C_{13}e_{zz} + C_{14}\gamma_{xy} + C_{15}\gamma_{xz} + C_{16}\gamma_{yz} + C_{17} \\
\sigma_{yy} &= C_{21}e_{xx} + C_{22}e_{yy} + C_{23}e_{zz} + C_{24}\gamma_{xy} + C_{25}\gamma_{xz} + C_{26}\gamma_{yz} + C_{27} \\
\sigma_{zz} &= C_{31}e_{xx} + C_{32}e_{yy} + C_{33}e_{zz} + C_{34}\gamma_{xy} + C_{35}\gamma_{xz} + C_{36}\gamma_{yz} + C_{37} \\
\tau_{xy} &= C_{41}e_{xx} + C_{42}e_{yy} + C_{43}e_{zz} + C_{44}\gamma_{xy} + C_{45}\gamma_{xz} + C_{46}\gamma_{yz} + C_{47} \\
\tau_{xz} &= C_{51}e_{xx} + C_{52}e_{yy} + C_{53}e_{zz} + C_{54}\gamma_{xy} + C_{55}\gamma_{xz} + C_{56}\gamma_{yz} + C_{57} \\
\tau_{yz} &= C_{61}e_{xx} + C_{62}e_{yy} + C_{63}e_{zz} + C_{64}\gamma_{xy} + C_{65}\gamma_{xz} + C_{66}\gamma_{yz} + C_{67}
\end{aligned}
\tag{A.1}
$$

in which the 42 C's are constants that must be determined from the assumptions and experimental evidence. According to assumption 2, there are no preferred directions, so that the constants in Eqs. (A.1) must be the same for rotation of the coordinate axes. This may be formalized as

$$
\begin{aligned}
\sigma_{x'x'} &= C_{11}e_{x'x'} + C_{12}e_{y'y'} + C_{13}e_{z'z'} + C_{14}\gamma_{x'y'} + C_{15}\gamma_{x'z'} + C_{16}\gamma_{y'z'} + C_{17} \\
\sigma_{y'y'} &= \quad | \qquad | \qquad | \qquad | \qquad | \qquad | \qquad | \\
\sigma_{z'z'} &= \quad | \qquad | \qquad | \qquad | \qquad | \qquad | \qquad | \\
\tau_{x'y'} &= \quad | \qquad | \qquad | \qquad | \qquad | \qquad | \qquad | \\
\tau_{x'z'} &= \quad | \qquad | \qquad | \qquad | \qquad | \qquad | \qquad | \\
\tau_{y'z'} &= C_{61}e_{x'x'} + C_{62}e_{y'y'} + C_{63}e_{z'z'} + C_{64}\gamma_{x'y'} + C_{65}\gamma_{x'z'} + C_{66}\gamma_{y'z'} + C_{67}
\end{aligned}
\tag{A.2}
$$

Substitution of Eqs. (A.1) in Eq. (4.3*a*) gives

$$\begin{aligned}\sigma_{x'x'} = {} & l_1{}^2(C_{11}e_{xx} + C_{12}e_{yy} + C_{13}e_{zz} + C_{14}\gamma_{xy} + C_{15}\gamma_{xz} + C_{16}\gamma_{yz} + C_{17}) \\ & + m_1{}^2(C_{21}e_{xx} + C_{22}e_{yy} + C_{23}e_{zz} + C_{24}\gamma_{xy} + C_{25}\gamma_{xz} + C_{26}\gamma_{yz} + C_{27}) \\ & + n_1{}^2(C_{31}e_{xx} + C_{32}e_{yy} + C_{33}e_{zz} + C_{34}\gamma_{xy} + C_{35}\gamma_{xz} + C_{36}\gamma_{yz} + C_{37}) \\ & + 2l_1m_1(C_{41}e_{xx} + C_{42}e_{yy} + C_{43}e_{zz} + C_{44}\gamma_{xy} + C_{45}\gamma_{xz} + C_{46}\gamma_{yz} + C_{47}) \\ & + 2l_1n_1(C_{51}e_{xx} + C_{52}e_{yy} + C_{53}e_{zz} + C_{54}\gamma_{xy} + C_{55}\gamma_{xz} + C_{56}\gamma_{yz} + C_{57}) \\ & + 2m_1n_1(C_{61}e_{xx} + C_{62}e_{yy} + C_{63}e_{zz} + C_{64}\gamma_{xy} + C_{65}\gamma_{xz} + C_{66}\gamma_{yz} + C_{67})\end{aligned}$$

The coefficients of the rates of strain in the previous equation may be grouped as follows:

$$\begin{aligned}\sigma_{x'x'} = {} & (l_1{}^2C_{11} + m_1{}^2C_{21} + n_1{}^2C_{31} + 2m_1l_1C_{41} + 2l_1n_1C_{51} + 2m_1n_1C_{61})\,e_{xx} \\ & + (l_1{}^2C_{12} + m_1{}^2C_{22} + n_1{}^2C_{32} + 2m_1l_1C_{42} + 2l_1n_1C_{52} + 2m_1n_1C_{62})\,e_{yy} \\ & + (l_1{}^2C_{13} + m_1{}^2C_{23} + n_1{}^2C_{33} + 2m_1l_1C_{43} + 2l_1n_1C_{53} + 2m_1n_1C_{63})\,e_{zz} \\ & + (l_1{}^2C_{14} + m_1{}^2C_{24} + n_1{}^2C_{34} + 2m_1l_1C_{44} + 2l_1n_1C_{54} + 2m_1n_1C_{64})\,\gamma_{xy} \\ & + (l_1{}^2C_{15} + m_1{}^2C_{25} + n_1{}^2C_{35} + 2m_1l_1C_{45} + 2l_1n_1C_{55} + 2m_1n_1C_{65})\,\gamma_{xz} \\ & + (l_1{}^2C_{16} + m_1{}^2C_{26} + n_1{}^2C_{36} + 2m_1l_1C_{46} + 2l_1n_1C_{56} + 2m_1n_1C_{66})\,\gamma_{yz} \\ & \quad + l_1{}^2C_{17} + m_1{}^2C_{27} + n_1{}^2C_{37} + 2m_1l_1C_{47} + 2l_1n_1C_{57} + 2m_1n_1C_{67} \qquad \text{(A.3)}\end{aligned}$$

A second expression for $\sigma_{x'x'}$ in terms of the rates of strain referred to the x, y, z coordinate system may be obtained by substitution of $e_{x'x'}$, $e_{y'y'}$, $e_{z'z'}$, $\gamma_{x'y'}$, $\gamma_{x'z'}$, and $\gamma_{y'z'}$ given by Eqs. (4.12) into the first equation of the set (A.2). Making this substitution and grouping the coefficients of the strain rates gives

$$\begin{aligned}\sigma_{x'x'} = {} & (C_{11}l_1{}^2 + C_{12}l_2{}^2 + C_{13}l_3{}^2 + C_{14}2l_1l_2 + C_{15}2l_1l_3 + C_{16}2l_2l_3)\,e_{xx} \\ & + (C_{11}m_1{}^2 + C_{12}m_2{}^2 + C_{13}m_3{}^2 + C_{14}2m_1m_2 + C_{15}2m_1m_3 \\ & \qquad + C_{16}2m_2m_3)\,e_{yy} \\ & + (C_{11}n_1{}^2 + C_{12}n_2{}^2 + C_{13}n_3{}^2 + C_{14}2n_1n_2 + C_{15}2n_1n_3 + C_{16}2n_2n_3)\,e_{zz} \\ & + [C_{11}l_1m_1 + C_{12}l_2m_2 + C_{13}l_3m_3 + C_{14}(l_1m_2 + l_2m_1) \\ & \qquad + C_{15}(l_1m_3 + l_3m_1) + C_{16}(l_2m_3 + l_3m_2)]\,\gamma_{xy} \\ & + [C_{11}l_1n_1 + C_{12}l_2n_2 + C_{13}l_3n_3 + C_{14}(l_1n_2 + l_2n_1) \\ & \qquad + C_{15}(l_1n_3 + l_3n_1) + C_{16}(l_2n_3 + l_3n_2)]\,\gamma_{xz} \\ & + [C_{11}m_1n_1 + C_{12}m_2n_2 + C_{13}m_3n_3 + C_{14}(m_1n_2 + m_2n_1) \\ & \qquad + C_{15}(m_1n_3 + m_3n_1) + C_{16}(m_2n_3 + m_3n_2)]\,\gamma_{yz} + C_{17} \qquad \text{(A.4)}\end{aligned}$$

Equations (A.3) and (A.4) must be identical for all geometrically admissible values of the nine direction cosines; therefore the coefficients of corresponding rates of strain in each equation may be equated to one another. The corresponding terms containing only the C's may be equated also. Doing this provides the following relations among the 42 constants:

$$l_1^2C_{11} + m_1^2C_{21} + n_1^2C_{31} + 2l_1m_1C_{41} + 2l_1n_1C_{51} + 2m_1n_1C_{61}$$
$$= C_{11}l_1^2 + C_{12}l_2^2 + C_{13}l_3^2 + C_{14}2l_1l_2 + C_{15}2l_1l_3 + C_{16}2l_2l_3$$

$$l_1^2C_{12} + m_1^2C_{22} + n_1^2C_{32} + 2l_1m_1C_{42} + 2l_1n_1C_{52} + 2m_1n_1C_{62}$$
$$= C_{11}m_1^2 + C_{12}m_2^2 + C_{13}m_3^2 + C_{14}2m_1m_2 + C_{15}2m_1m_3 + C_{16}2m_2m_3$$

$$l_1^2C_{13} + m_1^2C_{23} + n_1^2C_{33} + 2l_1m_1C_{43} + 2l_1n_1C_{53} + 2m_1n_1C_{63}$$
$$= C_{11}n_1^2 + C_{12}n_2^2 + C_{13}n_3^2 + C_{14}2n_1n_2 + C_{15}2n_1n_3 + C_{16}2n_2n_3$$

$$l_1^2C_{14} + m_1^2C_{24} + n_1^2C_{34} + 2l_1m_1C_{44} + 2l_1n_1C_{54} + 2m_1n_1C_{64}$$
$$= C_{11}l_1m_1 + C_{12}l_2m_2 + C_{13}l_3m_3 + C_{14}(l_1m_2 + l_2m_1)$$
$$+ C_{15}(l_1m_3 + l_3m_1) + C_{16}(l_2m_3 + l_3m_2)$$

$$l_1^2C_{15} + m_1^2C_{25} + n_1^2C_{35} + 2l_1m_1C_{45} + 2l_1n_1C_{55} + 2m_1n_1C_{65}$$
$$= C_{11}l_1n_1 + C_{12}l_2n_2 + C_{13}l_3n_3 + C_{14}(l_1n_2 + l_2n_1)$$
$$+ C_{15}(l_1n_3 + l_3n_1) + C_{16}(l_2n_3 + l_3n_2)$$

$$l_1^2C_{16} + m_1^2C_{26} + n_1^2C_{36} + 2l_1m_1C_{46} + 2l_1n_1C_{56} + 2m_1n_1C_{66}$$
$$= C_{11}m_1n_1 + C_{12}m_2n_2 + C_{13}m_3n_3 + C_{14}(m_1n_2 + m_2n_1)$$
$$+ C_{15}(m_1n_3 + m_3n_1) + C_{16}(m_2n_3 + m_3n_2)$$

$$l_1^2C_{17} + m_1^2C_{27} + n_1^2C_{37} + 2l_1m_1C_{47} + 2l_1n_1C_{57} + 2m_1n_1C_{67} = C_{17}$$

One may solve for the constants in terms of C_{11}, C_{12}, and C_{17} by assigning values to the direction cosines compatible with the following constraints:

$$l_1^2 + m_1^2 + n_1^2 = l_2^2 + m_2^2 + n_2^2 = l_3^2 + m_3^2 + n_3^2 = 1$$
$$l_1^2 + l_2^2 + l_3^2 = m_1^2 + m_2^2 + m_3^2 = n_1^2 + n_2^2 + n_3^2 = 1$$
$$l_1l_2 + m_1m_2 + n_1n_2 = l_2l_3 + m_2m_3 + n_2n_3 = l_1l_3 + m_1m_3 + n_1n_3 = 0$$
$$l_1m_1 + l_2m_2 + l_3m_3 = m_1n_1 + m_2n_2 + m_3n_3 = n_1l_1 + n_2l_2 + n_3l_3 = 0$$

The results of this procedure are

$$C_{11} = C_{22} = C_{33} \qquad C_{44} = C_{55} = C_{66} = \frac{1}{2}(C_{11} - C_{12})$$

$$C_{17} = C_{27} = C_{37} \qquad C_{12} = C_{21} = C_{13} = C_{31} = C_{23} = C_{32}$$

and all other constants are zero. By substituting these results in Eqs. (A.1), the constitutive equations may be written as

$$\sigma_{xx} = C_{11}e_{xx} + C_{12}e_{yy} + C_{12}e_{zz} + C_{17}$$
$$\sigma_{yy} = C_{12}e_{xx} + C_{11}e_{yy} + C_{12}e_{zz} + C_{27}$$

$$\sigma_{zz} = C_{12}e_{xx} + C_{12}e_{yy} + C_{11}e_{zz} + C_{37}$$

$$\tau_{xy} = \frac{1}{2}(C_{11} - C_{12})\,\gamma_{xy}$$

$$\tau_{xz} = \frac{1}{2}(C_{11} - C_{12})\,\gamma_{xz} \qquad \text{(A.5)}$$

$$\tau_{yz} = \frac{1}{2}(C_{11} - C_{12})\,\gamma_{yz}$$

Assumption 3 requires the normal stress to reduce to the pressure in absence of deformation, and $C_{17} = C_{27} = C_{37} = -p$, so that only two constants in Eqs. (A.5) are yet to be determined. A relation between these two may be obtained by reference to the simple one-dimensional flow between two parallel plates. Equation (4.14) can be written as $\tau_{xy} = \mu(\partial V_x/\partial y)$. The constitutive equations require

$$\tau_{xy} = \frac{1}{2}(C_{11} - C_{12})\,\gamma_{xy} = \frac{1}{2}(C_{11} - C_{12})\frac{\partial V_x}{\partial y}$$

since $V_y = 0$ for the flow in Fig. 4.5. This indicates that $(1/2)(C_{11} - C_{12}) = \mu$ or $C_{11} = 2\mu + C_{12}$. Substituting these values of C_{11}, C_{17}, C_{27}, and C_{37} into Eqs. (A.5) gives

$$\sigma_{xx} = 2\mu e_{xx} + C_{12}(e_{xx} + e_{yy} + e_{zz}) - p$$

$$\sigma_{yy} = 2\mu e_{yy} + C_{12}(e_{xx} + e_{yy} + e_{zz}) - p$$

$$\sigma_{zz} = 2\mu e_{zz} + C_{12}(e_{xx} + e_{yy} + e_{zz}) - p \qquad \text{(A.6)}$$

$$\tau_{xy} = \mu\gamma_{xy}$$

$$\tau_{xz} = \mu\gamma_{xz}$$

$$\tau_{yz} = \mu\gamma_{yz}$$

These relations are given in the text as Eqs. (4.16). It should be pointed out that application of the transformation of rates of strain to any of the remaining five stress components (the development here used only the component $\sigma_{x'x'}$) would yield no additional information concerning the 42 constants.

B

SUMMARY OF SOME VECTOR OPERATORS AND BASIC EXPRESSIONS IN SEVERAL COORDINATE SYSTEMS

Cartesian.

Material derivative.

$$\frac{D}{Dt} = V_x \frac{\partial}{\partial x} + V_y \frac{\partial}{\partial y} + V_z \frac{\partial}{\partial z} + \frac{\partial}{\partial t}$$

Gradient operator.

$$\mathbf{grad} = \frac{\partial}{\partial x}\mathbf{i} + \frac{\partial}{\partial y}\mathbf{j} + \frac{\partial}{\partial z}\mathbf{k}$$

Divergence operator (on **A***).*

$$\text{div } \mathbf{A} = \frac{\partial A_x}{\partial x} + \frac{\partial A_y}{\partial y} + \frac{\partial A_z}{\partial z}$$

Curl operator (on **A***).*

$$\mathbf{curl\ A} = \left(\frac{\partial V_z}{\partial y} - \frac{\partial V_y}{\partial z}\right)\mathbf{i} + \left(\frac{\partial V_x}{\partial z} - \frac{\partial V_z}{\partial x}\right)\mathbf{j} + \left(\frac{\partial V_y}{\partial x} - \frac{\partial V_x}{\partial y}\right)\mathbf{k}$$

Laplacian operator.

$$\text{div } \mathbf{grad} = \frac{\partial^2}{\partial x^2} + \frac{\partial^2}{\partial y^2} + \frac{\partial^2}{\partial z^2}$$

Constitutive relations for a Newtonian fluid.

$$\sigma_{xx} = -p - \frac{2}{3}\mu \text{ div } \mathbf{V} + 2\mu \frac{\partial V_x}{\partial x}$$

$$\sigma_{yy} = -p - \frac{2}{3}\mu \operatorname{div} \mathbf{V} + 2\mu \frac{\partial V_y}{\partial y}$$

$$\sigma_{zz} = -p - \frac{2}{3}\mu \operatorname{div} \mathbf{V} + 2\mu \frac{\partial V_z}{\partial z}$$

$$\tau_{xy} = \mu\left(\frac{\partial V_y}{\partial x} + \frac{\partial V_x}{\partial y}\right)$$

$$\tau_{xz} = \mu\left(\frac{\partial V_z}{\partial x} + \frac{\partial V_x}{\partial z}\right)$$

$$\tau_{yz} = \mu\left(\frac{\partial V_z}{\partial y} + \frac{\partial V_y}{\partial z}\right)$$

Navier-Stokes equation of motion for a compressible fluid with variable viscosity.

$$\rho\frac{DV_x}{Dt} = -\frac{\partial p}{\partial x} - \frac{\partial}{\partial x}\left(\frac{2}{3}\mu \operatorname{div} \mathbf{V} - 2\mu\frac{\partial V_x}{\partial x}\right) + \frac{\partial}{\partial y}\left[\mu\left(\frac{\partial V_y}{\partial x} + \frac{\partial V_x}{\partial y}\right)\right] + \frac{\partial}{\partial z}\left[\mu\left(\frac{\partial V_x}{\partial z} + \frac{\partial V_z}{\partial x}\right)\right] + \rho(\mathbf{f}_B)_x$$

$$\rho\frac{DV_y}{Dt} = -\frac{\partial p}{\partial y} - \frac{\partial}{\partial y}\left(\frac{2}{3}\mu \operatorname{div} \mathbf{V} - 2\mu\frac{\partial V_y}{\partial y}\right) + \frac{\partial}{\partial x}\left[\mu\left(\frac{\partial V_x}{\partial y} + \frac{\partial V_y}{\partial x}\right)\right] + \frac{\partial}{\partial z}\left[\mu\left(\frac{\partial V_y}{\partial z} + \frac{\partial V_z}{\partial y}\right)\right] + \rho(\mathbf{f}_B)_y$$

$$\rho\frac{DV_z}{Dt} = -\frac{\partial p}{\partial z} - \frac{\partial}{\partial z}\left(\frac{2}{3}\mu \operatorname{div} \mathbf{V} - 2\mu\frac{\partial V_z}{\partial z}\right) + \frac{\partial}{\partial y}\left[\mu\left(\frac{\partial V_y}{\partial z} + \frac{\partial V_z}{\partial y}\right)\right] + \frac{\partial}{\partial x}\left[\mu\left(\frac{\partial V_x}{\partial z} + \frac{\partial V_z}{\partial x}\right)\right] + \rho(\mathbf{f}_B)_z$$

Cylindrical.

Material derivative.

$$\frac{D}{Dt} = V_r\frac{\partial}{\partial r} + \frac{V_\theta}{r}\frac{\partial}{\partial \theta} + V_z\frac{\partial}{\partial z} + \frac{\partial}{\partial t}$$

Gradient operator.

$$\mathbf{grad} = \frac{\partial}{\partial r}\boldsymbol{\varepsilon}_r + \frac{1}{r}\frac{\partial}{\partial \theta}\boldsymbol{\varepsilon}_\theta + \frac{\partial}{\partial z}\boldsymbol{\varepsilon}_z$$

Divergence operator (on **A***).*

$$\operatorname{div} \mathbf{A} = \frac{1}{r}\frac{\partial}{\partial r}(rA_r) + \frac{1}{r}\frac{\partial A_\theta}{\partial \theta} + \frac{\partial A_z}{\partial z}$$

Curl operator (on **A***).*

$$\textbf{curl A} = \left(\frac{1}{r}\frac{\partial A_z}{\partial \theta} - \frac{\partial A_\theta}{\partial z}\right)\boldsymbol{\varepsilon}_r + \left(\frac{\partial A_r}{\partial z} - \frac{\partial A_z}{\partial r}\right)\boldsymbol{\varepsilon}_\theta + \frac{1}{r}\left(\frac{\partial (rA_\theta)}{\partial r} - \frac{\partial A_r}{\partial \theta}\right)\boldsymbol{\varepsilon}_z$$

Laplacian operator.

$$\text{div }\textbf{grad} = \frac{1}{r}\frac{\partial}{\partial r}\left(r\frac{\partial}{\partial r}\right) + \frac{1}{r^2}\frac{\partial^2}{\partial \theta^2} + \frac{\partial^2}{\partial z^2}$$

Constitutive relations for a Newtonian fluid.

$$\sigma_{rr} = -p - \frac{2}{3}\mu \text{ div }\mathbf{V} + 2\mu\frac{\partial V_r}{\partial r}$$

$$\sigma_{\theta\theta} = -p - \frac{2}{3}\mu \text{ div }\mathbf{V} + 2\mu\left(\frac{1}{r}\frac{\partial V_\theta}{\partial \theta} + \frac{V_r}{r}\right)$$

$$\sigma_{zz} = -p - \frac{2}{3}\mu \text{ div }\mathbf{V} + 2\mu\frac{\partial V_z}{\partial z}$$

$$\tau_{r\theta} = \mu\left(\frac{1}{r}\frac{\partial V_r}{\partial \theta} + \frac{\partial V_\theta}{\partial r} - \frac{V_\theta}{r}\right)$$

$$\tau_{rz} = \mu\left(\frac{\partial V_r}{\partial z} + \frac{\partial V_z}{\partial r}\right)$$

$$\tau_{\theta z} = \mu\left(\frac{1}{r}\frac{\partial V_z}{\partial \theta} + \frac{\partial V_\theta}{\partial z}\right)$$

Navier-Stokes equation of motion for a compressible fluid with variable viscosity.

$$\begin{aligned}\rho\left(\frac{DV_r}{Dt} - \frac{V_\theta{}^2}{r}\right) = {} & -\frac{\partial p}{\partial r} + \frac{\partial}{\partial r}\left(2\mu\frac{\partial V_r}{\partial r} - \frac{2}{3}\mu \text{ div }\mathbf{V}\right) \\ & + \frac{1}{r}\frac{\partial}{\partial \theta}\left[\mu\left(\frac{1}{r}\frac{\partial V_r}{\partial \theta} + \frac{\partial V_\theta}{\partial r} - \frac{V_\theta}{r}\right)\right] + \frac{\partial}{\partial z}\left[\mu\left(\frac{\partial V_r}{\partial z} + \frac{\partial V_z}{\partial r}\right)\right] \\ & + \frac{2\mu}{r}\left(\frac{\partial V_r}{\partial r} - \frac{1}{r}\frac{\partial V_\theta}{\partial \theta} - \frac{V_r}{r}\right) + \rho(\mathbf{f}_B)_r\end{aligned}$$

$$\begin{aligned}\rho\left(\frac{DV_\theta}{Dt} + \frac{V_r V_\theta}{r}\right) = {} & -\frac{1}{r}\frac{\partial p}{\partial \theta} + \frac{1}{r}\frac{\partial}{\partial \theta}\left(\frac{2\mu}{r}\frac{\partial V_\theta}{\partial \theta} + \frac{2\mu V_r}{r} - \frac{2}{3}\mu \text{ div }\mathbf{V}\right) \\ & + \frac{\partial}{\partial z}\left[\mu\left(\frac{1}{r}\frac{\partial V_z}{\partial \theta} + \frac{\partial V_\theta}{\partial z}\right)\right] + \frac{\partial}{\partial r}\left[\mu\left(\frac{1}{r}\frac{\partial V_r}{\partial \theta} + \frac{\partial V_\theta}{\partial r} - \frac{V_\theta}{r}\right)\right] \\ & + \frac{2\mu}{r}\left[\frac{1}{r}\frac{\partial V_r}{\partial \theta} + \frac{\partial V_\theta}{\partial r} - \frac{V_\theta}{r}\right] + \rho(\mathbf{f}_B)_\theta\end{aligned}$$

$$\rho \frac{DV_z}{Dt} = -\frac{\partial p}{\partial z} + \frac{\partial}{\partial z}\left(2\mu \frac{\partial V_z}{\partial z} - \frac{2}{3}\mu \operatorname{div} \mathbf{V}\right) + \frac{1}{r}\frac{\partial}{\partial r}\left[\mu r\left(\frac{\partial V_r}{\partial z} + \frac{\partial V_z}{\partial r}\right)\right] + \frac{1}{r}\frac{\partial}{\partial \theta}\left[\mu\left(\frac{1}{r}\frac{\partial V_z}{\partial \theta} + \frac{\partial V_\theta}{\partial z}\right)\right] + \rho(\mathbf{f}_B)_z$$

C

UNIT CONVERSION AND PROPORTIONALITY FACTORS

Mass.

1 slug = 32.17 lb_m

1 lb_m = 453.59 grams

Length.

1 in. = 2.54 cm

Volume.

1 ft^3 = 7.48 gal

Force.

1 lb_f = 32.17 lb_m-ft/sec^2

1 lb_f = 1 slug-ft/sec^2

1 dyne = 1 gram-cm/sec^2

Pressure.

1 atm = 14.70 $lb_f/in.^2$

Energy.

1 Btu = 778.16 ft-lb_f = 252.16 cal

1 erg = 1 dyne-cm

Power.

1 hp = 550 ft-lb_f/sec = 2545 Btu/hr = 746 watts

D

SHORT TABLE OF ERROR FUNCTION‡

η	0	2	4	6	8
0.0	0.0000	0.0226	0.0451	0.0676	0.0901
0.1	0.1125	0.1348	0.1569	0.1790	0.2009
0.2	0.2227	0.2443	0.2657	0.2869	0.3079
0.3	0.3286	0.3491	0.3694	0.3893	0.4090
0.4	0.4284	0.4475	0.4662	0.4847	0.5027
0.5	0.5205	0.5379	0.5549	0.5716	0.5879
0.6	0.6039	0.6194	0.6346	0.6494	0.6638
0.7	0.6778	0.6914	0.7047	0.7175	0.7300
0.8	0.7421	0.7538	0.7651	0.7761	0.7867
0.9	0.7969	0.8068	0.8163	0.8254	0.8342

‡ $\operatorname{erf} \eta = \dfrac{2}{\pi^{1/2}} \displaystyle\int_0^{\eta} e^{-\eta^2}\, d\eta$

η	0	2	4	6	8
1.0	0.8427	0.8508	0.8587	0.8661	0.8733
1.1	0.8802	0.8868	0.8931	0.8991	0.9048
1.2	0.9103	0.9155	0.9205	0.9252	0.9297
1.3	0.9340	0.9381	0.9419	0.9456	0.9490
1.4	0.9523	0.9554	0.9583	0.9611	0.9637
1.5	0.9661	0.9684	0.9706	0.9726	0.9745
1.6	0.9763	0.9780	0.9796	0.9811	0.9825
1.7	0.9838	0.9850	0.9861	0.9872	0.9882
1.8	0.9891	0.9899	0.9907	0.9915	0.9922
1.9	0.9928	0.9934	0.9939	0.9944	0.9949
2.0	0.9953	0.9957	0.9961	0.9964	0.9967
2.1	0.9970	0.9973	0.9975	0.9977	0.9980
2.2	0.9981	0.9983	0.9985	0.9986	0.9987
2.3	0.9989	0.9990	0.9991	0.9992	0.9992
2.4	0.9993	0.9994	0.9994	0.9995	0.9995
2.5	0.9996	0.9996	0.9997	0.9997	0.9997
2.6	0.9998	0.9998	0.9998	0.9998	0.9998
2.7	0.9999	0.9999	0.9999	0.9999	0.9999
2.8	0.9999	0.9999	0.9999	0.9999	1.0000

Courtesy of Donald Shoop.

E

PHYSICAL PROPERTIES OF FLUIDS

Table E.1 Properties of gases at low pressure and room temperature

	c_p Btu-lb$_m^{-1}$-°R^{-1}	κ Btu-ft^{-1}-hr^{-1}-°R^{-1}	μ lb$_f$-sec-ft^{-2}	R ft-lb$_f$-lb$_m^{-1}$-°R^{-1}
Air	0.240	0.0151	$3.75(10^{-7})$	53.3
Argon	0.125	0.0102	$4.66(10^{-7})$	38.7
Carbon dioxide	0.201	0.0096	$3.04(10^{-7})$	35.1
Freon-12	0.142	0.0056	$2.57(10^{-7})$	12.8
Helium	1.24	0.0867	$4.10(10^{-7})$	386.0
Methane	0.532	0.0198	$2.25(10^{-7})$	96.3
Nitrogen	0.248	0.0150	$3.61(10^{-7})$	55.1
Oxygen	0.219	0.0155	$4.19(10^{-7})$	48.3
Water vapor	0.445	0.0105	$2.2\ (10^{-7})$	85.8

Where c_p = specific heat at constant pressure
κ = thermal conductivity
μ = absolute viscosity
R = specific gas constant

Table E.2 Properties of saturated liquids at room temperature

	ρ	μ	c_p	κ	E_T	E_s	β
	lb_m-ft^{-3}	lb_f-sec-ft^{-2}	Btu-lb_m^{-1}-°R^{-1}	Btu-ft^{-1}-hr^{-1}-°R^{-1}	in^2-lb_f^{-1}	in^2-lb_f^{-1}	°F^{-1}
Carbon tetrachloride	99.5	2.02 (10^{-5})	0.204	0.0599	73.1 (10^{-7})	50.1 (10^{-7})	0.69 (10^{-3})
Freon-12	82.0	5.28 (10^{-6})	0.235	0.0399	—	—	2.3 (10^{-4})
Glycerine	78.7	3.13 (10^{-2})	0.580	0.166	14.5 (10^{-7})	—	0.28 (10^{-3})
Kerosene	51.2	4.17 (10^{-5})	—	0.087	54.3 (10^{-7})	49.4 (10^{-7})	0.81 (10^{-3})
Mercury	845	3.25 (10^{-5})	0.033	4.85	2.58 (10^{-7})	2.42 (10^{-7})	0.10 (10^{-3})
Oil, light	56.8	8.63 (10^{-4})	0.44	0.077	—	—	0.38 (10^{-3})
Water, pure	62.2	1.67 (10^{-5})	0.998	0.352	31.7 (10^{-7})	30.6 (10^{-7})	0.115 (10^{-3})

Where ρ = density
μ = absolute viscosity
c_p = specific heat at constant pressure
κ = thermal conductivity
E_T = isothermal compressibility
E_s = isentropic compressibility
β = thermal-expansion coefficient

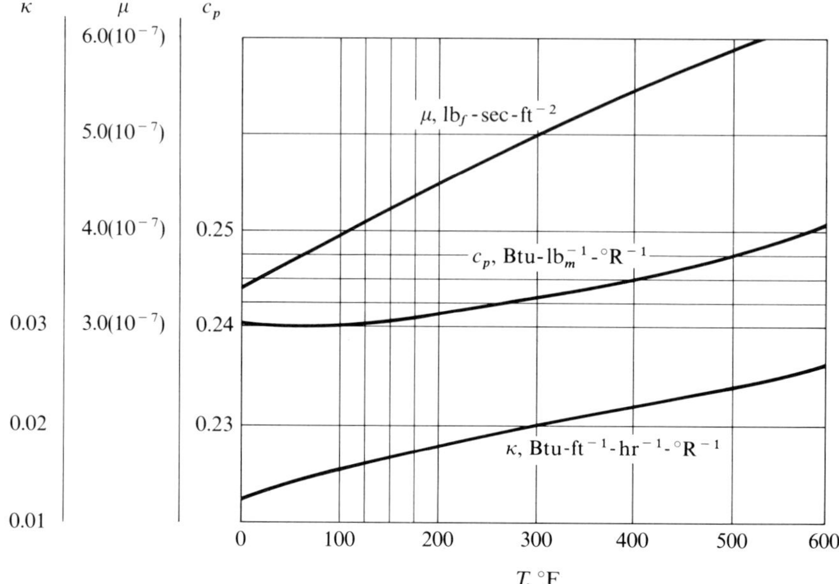

Fig. E.1 *Physical properties of air at low pressure.*

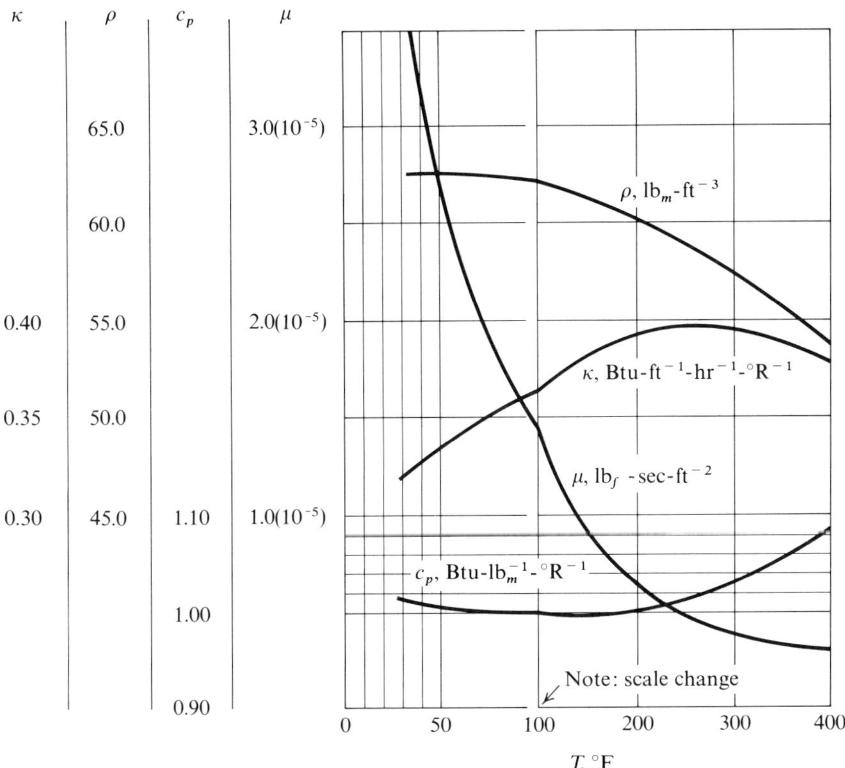

Fig. E.2 Physical properties of saturated liquid water.

Bibliography for Physical Properties of Fluids

Gray, Dwight E. (ed.), "American Institute of Physics Handbook," McGraw-Hill Book Company, New York, 1963.

Hilsenrath, J., et al., Tables of the Thermal Properties of Gases, *Natl. Bur. Std. Circ.* 564, 1955.

"International Critical Tables," McGraw-Hill Book Company, New York, 1929.

Keenan, Joseph, and Joseph Kaye, "Gas Tables," John Wiley & Sons, Inc., New York, 1948.

Reid, Robert C., and Thomas K. Sherwood, "The Properties of Gases and Liquids," McGraw-Hill Book Company, New York, 1958.

Weast, Robert C., (ed.), "Handbook of Chemistry and Physics," 49th ed., Chemical Rubber Company, Cleveland, Ohio, 1968–1969.

F

ONE-DIMENSIONAL COMPRESSIBLE FLOW DATA FOR A PERFECT GAS WITH CONSTANT SPECIFIC HEAT

Table F.1 Isentropic flow **(γ = 1.4)**

N_M	A/A^*	p/p_0	ρ/ρ_0	T/T_0
0.06	9.67	0.9975	0.9982	0.9993
0.08	7.26	0.9955	0.9968	0.9987
0.10	5.82	0.9930	0.9950	0.9980
0.12	4.86	0.9900	0.9928	0.9971
0.14	4.18	0.9864	0.9903	0.9961
0.16	3.67	0.9823	0.9873	0.9949
0.18	3.28	0.9776	0.9840	0.9936
0.20	2.96	0.9725	0.9803	0.9921
0.22	2.71	0.9668	0.9762	0.9904
0.24	2.50	0.9607	0.9718	0.9886
0.26	2.32	0.9541	0.9670	0.9867

Table F.1 **Isentropic flow (γ = 1.4)** (*Continued*)

N_M	A/A^*	p/p_0	ρ/ρ_0	T/T_0
0.28	2.17	0.9470	0.9619	0.9846
0.30	2.04	0.9395	0.9564	0.9823
0.32	1.92	0.9315	0.9506	0.9799
0.34	1.82	0.9231	0.9445	0.9774
0.36	1.74	0.9143	0.9380	0.9747
0.38	1.66	0.9052	0.9313	0.9719
0.40	1.59	0.8956	0.9243	0.9690
0.42	1.53	0.8857	0.9170	0.9659
0.44	1.47	0.8755	0.9094	0.9627
0.46	1.42	0.8650	0.9016	0.9594
0.48	1.38	0.8541	0.8935	0.9559
0.50	1.34	0.8430	0.8852	0.9524
0.52	1.30	0.8317	0.8766	0.9487
0.54	1.27	0.8201	0.8679	0.9449
0.56	1.24	0.8082	0.8589	0.9410
0.58	1.21	0.7962	0.8498	0.9370
0.60	1.19	0.7840	0.8405	0.9328
0.62	1.17	0.7716	0.8310	0.9286
0.64	1.15	0.7591	0.8213	0.9243
0.66	1.13	0.7465	0.8115	0.9199
0.68	1.11	0.7338	0.8016	0.9153
0.70	1.09	0.7209	0.7916	0.9107
0.72	1.08	0.7080	0.7814	0.9061
0.74	1.07	0.6951	0.7712	0.9013
0.76	1.06	0.6821	0.7609	0.8964
0.78	1.05	0.6691	0.7505	0.8915
0.80	1.04	0.6560	0.7400	0.8865
0.82	1.03	0.6430	0.7295	0.8815
0.84	1.02	0.6300	0.7189	0.8763
0.86	1.02	0.6170	0.7083	0.8711
0.88	1.01	0.6041	0.6977	0.8659
0.90	1.01	0.5913	0.6870	0.8606
0.92	1.01	0.5785	0.6764	0.8552
0.94	1.00	0.5658	0.6658	0.8498
0.96	1.00	0.5532	0.6551	0.8444
0.98	1.00	0.5407	0.6445	0.8389
1.00	1.00	0.5283	0.6339	0.8333
1.02	1.00	0.5160	0.6234	0.8278
1.04	1.00	0.5039	0.6129	0.8222
1.06	1.00	0.4919	0.6024	0.8165
1.08	1.01	0.4800	0.5920	0.8108
1.10	1.01	0.4684	0.5817	0.8052
1.12	1.01	0.4568	0.5714	0.7994
1.14	1.02	0.4455	0.5612	0.7937
1.16	1.02	0.4343	0.5511	0.7879
1.18	1.02	0.4232	0.5411	0.7822

Table F.1 **Isentropic flow (γ = 1.4)** (*Continued*)

N_M	A/A^*	p/p_0	ρ/ρ_0	T/T_0
1.20	1.03	0.4124	0.5311	0.7764
1.22	1.04	0.4017	0.5213	0.7706
1.24	1.04	0.3912	0.5115	0.7648
1.26	1.05	0.3809	0.5019	0.7590
1.28	1.06	0.3708	0.4923	0.7532
1.30	1.07	0.3609	0.4829	0.7474
1.32	1.08	0.3512	0.4736	0.7416
1.34	1.08	0.3417	0.4644	0.7358
1.36	1.09	0.3323	0.4553	0.7300
1.38	1.10	0.3232	0.4463	0.7242
1.40	1.11	0.3142	0.4374	0.7184
1.42	1.13	0.3055	0.4287	0.7126
1.44	1.14	0.2969	0.4201	0.7069
1.46	1.15	0.2886	0.4116	0.7011
1.48	1.16	0.2804	0.4032	0.6954
1.50	1.18	0.2724	0.3950	0.6897
1.52	1.19	0.2646	0.3869	0.6840
1.54	1.20	0.2570	0.3789	0.6783
1.56	1.22	0.2496	0.3710	0.6726
1.58	1.23	0.2423	0.3633	0.6670
1.60	1.25	0.2353	0.3557	0.6614
1.62	1.27	0.2284	0.3483	0.6558
1.64	1.28	0.2217	0.3409	0.6502
1.66	1.30	0.2151	0.3337	0.6447
1.68	1.32	0.2088	0.3266	0.6392
1.70	1.34	0.2026	0.3197	0.6337
1.72	1.36	0.1966	0.3129	0.6283
1.74	1.38	0.1907	0.3062	0.6229
1.76	1.40	0.1850	0.2996	0.6175
1.78	1.42	0.1794	0.2931	0.6121
1.80	1.44	0.1740	0.2868	0.6068
1.82	1.46	0.1688	0.2806	0.6015
1.84	1.48	0.1637	0.2745	0.5963
1.86	1.51	0.1587	0.2686	0.5910
1.88	1.53	0.1539	0.2627	0.5859
1.90	1.56	0.1492	0.2570	0.5807
1.92	1.58	0.1447	0.2514	0.5756
1.94	1.61	0.1403	0.2459	0.5705
1.96	1.63	0.1360	0.2405	0.5655
1.98	1.66	0.1318	0.2352	0.5605
2.00	1.69	0.1278	0.2300	0.5556
2.02	1.72	0.1239	0.2250	0.5506
2.04	1.75	0.1201	0.2200	0.5458
2.06	1.78	0.1164	0.2152	0.5409
2.08	1.81	0.1128	0.2104	0.5361
2.10	1.84	0.1094	0.2058	0.5313

Table F.1 Isentropic flow **(γ = 1.4)** *(Continued)*

N_M	A/A^*	p/p_0	ρ/ρ_0	T/T_0
2.12	1.87	0.1060	0.2013	0.5266
2.14	1.90	0.1027	0.1968	0.5219
2.16	1.94	0.0996	0.1925	0.5173
2.18	1.97	0.0965	0.1882	0.5127
2.20	2.00	0.0935	0.1841	0.5081
2.22	2.04	0.0906	0.1800	0.5036
2.24	2.08	0.0878	0.1760	0.4991
2.26	2.12	0.0851	0.1721	0.4947
2.28	2.15	0.0825	0.1683	0.4903
2.30	2.19	0.0800	0.1646	0.4859
2.32	2.23	0.0775	0.1609	0.4816
2.34	2.27	0.0751	0.1574	0.4773
2.36	2.32	0.0728	0.1539	0.4731
2.38	2.36	0.0706	0.1505	0.4688
2.40	2.40	0.0684	0.1472	0.4647
2.42	2.45	0.0663	0.1439	0.4606
2.44	2.49	0.0643	0.1408	0.4565
2.46	2.54	0.0623	0.1377	0.4524
2.48	2.59	0.0604	0.1346	0.4484
2.50	2.64	0.0585	0.1317	0.4444
2.52	2.69	0.0567	0.1288	0.4405
2.54	2.74	0.0550	0.1260	0.4366
2.56	2.79	0.0533	0.1232	0.4328
2.58	2.84	0.0517	0.1205	0.4289
2.60	2.90	0.0501	0.1179	0.4252
2.62	2.95	0.0486	0.1153	0.4214
2.64	3.01	0.0471	0.1128	0.4177
2.66	3.06	0.0457	0.1103	0.4141
2.68	3.12	0.0443	0.1079	0.4104
2.70	3.18	0.0430	0.1056	0.4068
2.72	3.24	0.0417	0.1033	0.4033
2.74	3.31	0.0404	0.1010	0.3998
2.76	3.37	0.0392	0.0989	0.3963
2.78	3.43	0.0380	0.0967	0.3928
2.80	3.50	0.0368	0.0946	0.3894
2.82	3.57	0.0357	0.0926	0.3860
2.84	3.64	0.0347	0.0906	0.3827
2.86	3.71	0.0336	0.0886	0.3794
2.88	3.78	0.0326	0.0867	0.3761
2.90	3.85	0.0317	0.0849	0.3729
2.92	3.92	0.0307	0.0831	0.3696
2.94	4.00	0.0298	0.0813	0.3665
2.96	4.08	0.0289	0.0796	0.3633
2.98	4.15	0.0281	0.0779	0.3602
3.00	4.23	0.0272	0.0762	0.3571
3.02	4.32	0.0264	0.0746	0.3541

Table F.1 ***Isentropic flow*** **(γ = 1.4)** (*Continued*)

N_M	A/A^*	p/p_0	ρ/ρ_0	T/T_0
3.04	4.40	0.0256	0.0730	0.3511
3.06	4.48	0.0249	0.0715	0.3481
3.08	4.57	0.0242	0.0700	0.3452
3.10	4.66	0.0234	0.0685	0.3422
3.12	4.75	0.0228	0.0671	0.3393
3.14	4.84	0.0221	0.0657	0.3365
3.16	4.93	0.0215	0.0643	0.3337
3.18	5.02	0.0208	0.0630	0.3309
3.20	5.12	0.0202	0.0617	0.3281
3.22	5.22	0.0196	0.0604	0.3253
3.24	5.32	0.0191	0.0591	0.3226
3.26	5.42	0.0185	0.0579	0.3199
3.28	5.52	0.0180	0.0567	0.3173
3.30	5.63	0.0175	0.0555	0.3147
3.32	5.74	0.0170	0.0544	0.3121
3.34	5.84	0.0165	0.0533	0.3095
3.36	5.96	0.0160	0.0522	0.3069
3.38	6.07	0.0156	0.0511	0.3044
3.40	6.18	0.0151	0.0501	0.3019
3.42	6.30	0.0147	0.0491	0.2995
3.44	6.42	0.0143	0.0481	0.2970
3.46	6.54	0.0139	0.0471	0.2946
3.48	6.66	0.0135	0.0462	0.2922
3.50	6.79	0.0131	0.0452	0.2899
3.52	6.92	0.0127	0.0443	0.2875
3.54	7.05	0.0124	0.0434	0.2852
3.56	7.18	0.0120	0.0426	0.2829
3.58	7.31	0.0117	0.0417	0.2806
3.60	7.45	0.0114	0.0409	0.2784
3.62	7.59	0.0111	0.0401	0.2762
3.64	7.73	0.0108	0.0393	0.2740
3.66	7.87	0.0105	0.0385	0.2718
3.68	8.02	0.0102	0.0378	0.2697
3.70	8.17	0.0099	0.0370	0.2675
3.72	8.32	0.0096	0.0363	0.2654
3.74	8.47	0.0094	0.0356	0.2633
3.76	8.63	0.0091	0.0349	0.2613
3.78	8.79	0.0089	0.0342	0.2592
3.80	8.95	0.0086	0.0335	0.2572
3.82	9.11	0.0084	0.0329	0.2552
3.84	9.28	0.0082	0.0323	0.2532
3.86	9.45	0.0080	0.0316	0.2513
3.88	9.62	0.0077	0.0310	0.2493
3.90	9.80	0.0075	0.0304	0.2474
3.92	9.98	0.0073	0.0299	0.2455
3.94	10.16	0.0071	0.0293	0.2436

Table F.1 ***Isentropic flow*** **(γ = 1.4)** *(Continued)*

N_M	A/A^*	p/p_0	ρ/ρ_0	T/T_0
3.96	10.34	0.0069	0.0287	0.2418
3.98	10.53	0.0068	0.0282	0.2399
4.00	10.72	0.0066	0.0277	0.2381
4.10	11.71	0.0058	0.0252	0.2293
4.20	12.79	0.0051	0.0229	0.2208
4.30	13.95	0.0044	0.0209	0.2129
4.40	15.21	0.0039	0.0191	0.2053
4.50	16.56	0.0035	0.0174	0.1980
4.60	18.02	0.0031	0.0160	0.1911
4.70	19.58	0.0027	0.0146	0.1846
4.80	21.26	0.0024	0.0134	0.1783
4.90	23.07	0.0021	0.0123	0.1724
5.00	25.00	0.0019	0.0113	0.1667

Courtesy of Donald Shoop.

Table F.2 ***Normal shock*** **(γ = 1.4)**

N_{M_1}	N_{M_2}	p_2/p_1	p_{02}/p_{01}	ρ_2/ρ_1	T_2/T_1
1.00	1.0000	1.0000	1.0000	1.0000	1.0000
1.02	0.9805	1.0471	1.0000	1.0334	1.0132
1.04	0.9620	1.0952	0.9999	1.0671	1.0263
1.06	0.9444	1.1442	0.9998	1.1009	1.0393
1.08	0.9277	1.1941	0.9994	1.1349	1.0522
1.10	0.9118	1.2450	0.9989	1.1691	1.0649
1.12	0.8966	1.2968	0.9982	1.2034	1.0776
1.14	0.8820	1.3495	0.9973	1.2378	1.0903
1.16	0.8682	1.4032	0.9961	1.2723	1.1029
1.18	0.8549	1.4578	0.9946	1.3069	1.1154
1.20	0.8422	1.5133	0.9928	1.3416	1.1280
1.22	0.8300	1.5698	0.9907	1.3764	1.1405
1.24	0.8183	1.6272	0.9884	1.4112	1.1531
1.26	0.8071	1.6855	0.9857	1.4460	1.1657
1.28	0.7963	1.7448	0.9827	1.4808	1.1783
1.30	0.7860	1.8050	0.9794	1.5157	1.1909
1.32	0.7760	1.8661	0.9758	1.5505	1.2035
1.34	0.7664	1.9282	0.9718	1.5854	1.2162
1.36	0.7572	1.9912	0.9676	1.6202	1.2290
1.38	0.7483	2.0551	0.9630	1.6549	1.2418
1.40	0.7397	2.1200	0.9582	1.6897	1.2547
1.42	0.7314	2.1858	0.9531	1.7243	1.2676
1.44	0.7235	2.2525	0.9476	1.7589	1.2807
1.46	0.7157	2.3202	0.9420	1.7934	1.2938

Table F.2 ***Normal shock*** **(γ = 1.4)** *(Continued)*

N_{M_1}	N_{M_2}	p_2/p_1	p_{02}/p_{01}	ρ_2/ρ_1	T_2/T_1
1.48	0.7083	2.3888	0.9360	1.8278	1.3069
1.50	0.7011	2.4583	0.9298	1.8621	1.3202
1.52	0.6941	2.5288	0.9233	1.8963	1.3336
1.54	0.6874	2.6002	0.9166	1.9303	1.3470
1.56	0.6809	2.6725	0.9097	1.9643	1.3606
1.58	0.6746	2.7458	0.9026	1.9981	1.3742
1.60	0.6684	2.8200	0.8952	2.0317	1.3880
1.62	0.6625	2.8951	0.8877	2.0653	1.4018
1.64	0.6568	2.9712	0.8799	2.0986	1.4158
1.66	0.6512	3.0482	0.8720	2.1318	1.4299
1.68	0.6458	3.1261	0.8639	2.1649	1.4440
1.70	0.6405	3.2050	0.8557	2.1977	1.4583
1.72	0.6355	3.2848	0.8474	2.2304	1.4727
1.74	0.6305	3.3655	0.8389	2.2629	1.4873
1.76	0.6257	3.4472	0.8302	2.2952	1.5019
1.78	0.6210	3.5298	0.8215	2.3273	1.5167
1.80	0.6165	3.6133	0.8127	2.3592	1.5316
1.82	0.6121	3.6978	0.8038	2.3909	1.5466
1.84	0.6078	3.7832	0.7948	2.4224	1.5617
1.86	0.6036	3.8695	0.7857	2.4537	1.5770
1.88	0.5996	3.9568	0.7765	2.4848	1.5924
1.90	0.5956	4.0450	0.7674	2.5157	1.6079
1.92	0.5918	4.1341	0.7581	2.5463	1.6236
1.94	0.5880	4.2242	0.7488	2.5767	1.6394
1.96	0.5844	4.3152	0.7395	2.6069	1.6553
1.98	0.5808	4.4071	0.7302	2.6369	1.6713
2.00	0.5774	4.5000	0.7209	2.6667	1.6875
2.02	0.5740	4.5938	0.7115	2.6962	1.7038
2.04	0.5707	4.6885	0.7022	2.7255	1.7203
2.06	0.5675	4.7842	0.6928	2.7545	1.7369
2.08	0.5643	4.8808	0.6835	2.7833	1.7536
2.10	0.5613	4.9783	0.6742	2.8119	1.7705
2.12	0.5583	5.0768	0.6649	2.8402	1.7875
2.14	0.5554	5.1762	0.6557	2.8683	1.8046
2.16	0.5525	5.2765	0.6464	2.8962	1.8219
2.18	0.5498	5.3778	0.6373	2.9238	1.8393
2.20	0.5471	5.4800	0.6281	2.9512	1.8569
2.22	0.5444	5.5831	0.6191	2.9784	1.8746
2.24	0.5418	5.6872	0.6100	3.0053	1.8924
2.26	0.5393	5.7922	0.6011	3.0319	1.9104
2.28	0.5368	5.8981	0.5921	3.0584	1.9285
2.30	0.5344	6.0050	0.5833	3.0845	1.9468
2.32	0.5321	6.1128	0.5745	3.1105	1.9652
2.34	0.5297	6.2215	0.5658	3.1362	1.9838
2.36	0.5275	6.3312	0.5572	3.1617	2.0025
2.38	0.5253	6.4418	0.5486	3.1869	2.0213

Table F.2 *Normal shock* (γ = 1.4) (*Continued*)

N_{M_1}	N_{M_2}	p_2/p_1	p_{02}/p_{01}	ρ_2/ρ_1	T_2/T_1
2.40	0.5231	6.5533	0.5401	3.2119	2.0403
2.42	0.5210	6.6658	0.5317	3.2367	2.0595
2.44	0.5189	6.7792	0.5234	3.2612	2.0788
2.46	0.5169	6.8935	0.5152	3.2855	2.0982
2.48	0.5149	7.0088	0.5071	3.3095	2.1178
2.50	0.5130	7.1250	0.4990	3.3333	2.1375
2.52	0.5111	7.2421	0.4911	3.3569	2.1574
2.54	0.5092	7.3602	0.4832	3.3803	2.1774
2.56	0.5074	7.4792	0.4754	3.4034	2.1976
2.58	0.5056	7.5991	0.4677	3.4263	2.2179
2.60	0.5039	7.7200	0.4601	3.4490	2.2383
2.62	0.5022	7.8418	0.4526	3.4714	2.2590
2.64	0.5005	7.9645	0.4452	3.4937	2.2797
2.66	0.4988	8.0882	0.4379	3.5157	2.3006
2.68	0.4972	8.2128	0.4307	3.5347	2.3217
2.70	0.4956	8.3383	0.4236	3.5590	2.3429
2.72	0.4941	8.4648	0.4166	3.5803	2.3642
2.74	0.4926	8.5922	0.4097	3.6015	2.3858
2.76	0.4911	8.7205	0.4028	3.6224	2.4074
2.78	0.4896	8.8498	0.3961	3.6431	2.4292
2.80	0.4882	8.9800	0.3895	3.6636	2.4512
2.82	0.4868	9.1111	0.3829	3.6838	2.4733
2.84	0.4854	9.2432	0.3765	3.7039	2.4955
2.86	0.4840	9.3762	0.3701	3.7238	2.5179
2.88	0.4827	9.5101	0.3639	3.7434	2.5405
2.90	0.4814	9.6450	0.3577	3.7629	2.5632
2.92	0.4801	9.7808	0.3517	3.7821	2.5861
2.94	0.4788	9.9175	0.3457	3.8012	2.6091
2.96	0.4776	10.0552	0.3398	3.8200	2.6322
2.98	0.4764	10.1938	0.3340	3.8387	2.6555
3.00	0.4752	10.3333	0.3283	3.8571	2.6790
3.02	0.4740	10.4738	0.3227	3.8754	2.7026
3.04	0.4729	10.6152	0.3172	3.8935	2.7264
3.06	0.4717	10.7575	0.3118	3.9114	2.7503
3.08	0.4706	10.9008	0.3065	3.9291	2.7744
3.10	0.4695	11.0450	0.3012	3.9466	2.7986
3.12	0.4685	11.1901	0.2960	3.9639	2.8230
3.14	0.4674	11.3362	0.2910	3.9811	2.8475
3.16	0.4664	11.4832	0.2860	3.9981	2.8722
3.18	0.4654	11.6311	0.2811	4.0149	2.8970
3.20	0.4643	11.7800	0.2762	4.0315	2.9220
3.22	0.4634	11.9298	0.2715	4.0479	2.9471
3.24	0.4624	12.0805	0.2668	4.0642	2.9724
3.26	0.4614	12.2322	0.2622	4.0803	2.9979
3.28	0.4605	12.3848	0.2577	4.0963	3.0234
3.30	0.4596	12.5383	0.2533	4.1120	3.0492

Table F.2 *Normal shock* (γ = **1.4**) (*Continued*)

N_{M_1}	N_{M_2}	p_2/p_1	p_{02}/p_{01}	ρ_2/ρ_1	T_2/T_1
3.32	0.4587	12.6928	0.2489	4.1276	3.0751
3.34	0.4578	12.8482	0.2446	4.1431	3.1011
3.36	0.4569	13.0045	0.2404	4.1583	3.1273
3.38	0.4560	13.1618	0.2363	4.1734	3.1537
3.40	0.4552	13.3200	0.2322	4.1884	3.1802
3.42	0.4544	13.4791	0.2282	4.2032	3.2069
3.44	0.4535	13.6392	0.2243	4.2178	3.2337
3.46	0.4527	13.8002	0.2205	4.2323	3.2607
3.48	0.4519	13.9621	0.2167	4.2467	3.2878
3.50	0.4512	14.1250	0.2129	4.2609	3.3151
3.52	0.4504	14.2888	0.2093	4.2749	3.3425
3.54	0.4496	14.4535	0.2057	4.2888	3.3701
3.56	0.4489	14.6192	0.2022	4.3026	3.3978
3.58	0.4481	14.7858	0.1987	4.3162	3.4257
3.60	0.4474	14.9533	0.1953	4.3296	3.4537
3.62	0.4467	15.1218	0.1920	4.3429	3.4819
3.64	0.4460	15.2912	0.1887	4.3561	3.5103
3.66	0.4453	15.4615	0.1855	4.3692	3.5388
3.68	0.4446	15.6328	0.1823	4.3821	3.5674
3.70	0.4439	15.8050	0.1792	4.3949	3.5962
3.72	0.4433	15.9781	0.1761	4.4075	3.6252
3.74	0.4426	16.1522	0.1731	4.4200	3.6543
3.76	0.4420	16.3272	0.1702	4.4324	3.6836
3.78	0.4414	16.5031	0.1673	4.4447	3.7130
3.80	0.4407	16.6800	0.1645	4.4568	3.7426
3.82	0.4401	16.8578	0.1617	4.4688	3.7723
3.84	0.4395	17.0365	0.1589	4.4807	3.8022
3.86	0.4389	17.2162	0.1563	4.4924	3.8323
3.88	0.4383	17.3968	0.1536	4.5041	3.8625
3.90	0.4377	17.5783	0.1510	4.5156	3.8928
3.92	0.4372	17.7608	0.1485	4.5270	3.9233
3.94	0.4366	17.9442	0.1460	4.5383	3.9540
3.96	0.4360	18.1285	0.1435	4.5494	3.9848
3.98	0.4355	18.3138	0.1411	4.5605	4.0158
4.00	0.4350	18.5000	0.1388	4.5714	4.0469
4.10	0.4324	19.4450	0.1276	4.6245	4.2048
4.20	0.4299	20.4133	0.1173	4.6749	4.3666
4.30	0.4277	21.4050	0.1080	4.7229	4.5322
4.40	0.4255	22.4200	0.0995	4.7685	4.7017
4.50	0.4236	23.4583	0.0917	4.8119	4.8751
4.60	0.4217	24.5200	0.0846	4.8532	5.0523
4.70	0.4199	25.6050	0.0781	4.8926	5.2334
4.80	0.4183	26.7133	0.0721	4.9301	5.4184
4.90	0.4167	27.8450	0.0667	4.9659	5.6073
5.00	0.4152	29.0000	0.0617	5.0000	5.8000
5.10	0.4138	30.1783	0.0572	5.0326	5.9966

Table F.2 **Normal shock (γ = 1.4)** (*Continued*)

N_{M_1}	N_{M_2}	p_2/p_1	p_{02}/p_{01}	ρ_2/ρ_1	T_2/T_1
5.20	0.4125	31.3800	0.0530	5.0637	6.1971
5.30	0.4113	32.6050	0.0491	5.0934	6.4014
5.40	0.4101	33.8533	0.0456	5.1218	6.6097
5.50	0.4090	35.1250	0.0424	5.1489	6.8218
5.60	0.4079	36.4200	0.0394	5.1749	7.0378
5.70	0.4069	37.7383	0.0366	5.1998	7.2577
5.80	0.4059	39.0800	0.0341	5.2236	7.4814
5.90	0.4050	40.4450	0.0318	5.2464	7.7091
6.00	0.4042	41.8333	0.0297	5.2683	7.9406

Courtesy of Donald Shoop.

Table F.3 **Fanno flow (γ = 1.4)**

N_M	T/T^*	p/p^*	p_0/p_0^*	V/V^*	$f(L_{crit}/D_H)$
0.02	1.19990	54.7701	28.9421	0.02191	1178.45
0.04	1.19962	27.3817	14.4815	0.04381	440.35
0.06	1.19914	18.2508	9.6659	0.06570	193.03
0.08	1.19847	13.6843	7.2616	0.08758	106.72
0.10	1.19760	10.9435	5.8218	0.10943	66.922
0.12	1.19655	9.1156	4.8643	0.13126	45.408
0.14	1.19531	7.8093	4.1824	0.15306	32.511
0.16	1.19389	6.8291	3.6727	0.17482	24.198
0.18	1.19227	6.0662	3.2779	0.19654	18.543
0.20	1.19048	5.4555	2.9635	0.21822	14.533
0.22	1.18849	4.9554	2.7076	0.23984	11.596
0.24	1.18633	4.5383	2.4956	0.26141	9.3865
0.26	1.18399	4.1850	2.3173	0.28291	7.6876
0.28	1.18147	3.8820	2.1656	0.30435	6.3572
0.30	1.17878	3.6190	2.0351	0.32572	5.2992
0.32	1.17591	3.3888	1.9219	0.34700	4.4468
0.34	1.17288	3.1853	1.8229	0.36822	3.7520
0.36	1.16968	3.0042	1.7358	0.38935	3.1801
0.38	1.16632	2.8420	1.6587	0.41039	2.7055
0.40	1.16279	2.6958	1.5901	0.43133	2.3085
0.42	1.15911	2.5634	1.5289	0.45218	1.9744
0.44	1.15527	2.4428	1.4739	0.47293	1.6915
0.46	1.15128	2.3326	1.4246	0.49357	1.4509
0.48	1.14714	2.2314	1.3801	0.51410	1.2453
0.50	1.14286	2.1381	1.3399	0.53453	1.0691
0.52	1.13843	2.0519	1.3034	0.55482	0.91741
0.54	1.13387	1.9719	1.2702	0.57501	0.78662
0.56	1.12918	1.8976	1.2403	0.59507	0.67357
0.58	1.12435	1.8282	1.2130	0.61500	0.57568

Table F.3 Fanno flow **(γ = 1.4)** *(Continued)*

N_M	T/T^*	p/p^*	p_0/p_0^*	V/V^*	$f(L_{crit}/D_H)$
0.60	1.11940	1.7634	1.1882	0.63481	0.49081
0.62	1.11433	1.7026	1.1656	0.65449	0.41720
0.64	1.10914	1.6456	1.1451	0.67402	0.35330
0.66	1.10383	1.5919	1.1265	0.69342	0.29785
0.68	1.09842	1.5413	1.1097	0.71267	0.24978
0.70	1.09290	1.4934	1.09436	0.73179	0.20814
0.72	1.08727	1.4482	1.08057	0.75076	0.17215
0.74	1.08155	1.4054	1.06815	0.76958	0.14113
0.76	1.07573	1.3647	1.05700	0.78825	0.11446
0.78	1.06982	1.3260	1.04705	0.80677	0.09167
0.80	1.06383	1.2892	1.03823	0.82514	0.07229
0.82	1.05775	1.2542	1.03047	0.84334	0.05593
0.84	1.05160	1.2208	1.02370	0.86140	0.04226
0.86	1.04537	1.1889	1.01787	0.87929	0.03097
0.88	1.03907	1.1584	1.01294	0.89703	0.02180
0.90	1.03270	1.12913	1.00887	0.91459	0.014513
0.92	1.02627	1.10114	1.00560	0.93201	0.008916
0.94	1.01978	1.07430	1.00311	0.94925	0.004815
0.96	1.01324	1.04854	1.00137	0.96634	0.002056
0.98	1.00664	1.02379	1.00033	0.98324	0.000493
1.00	1.00000	1.00000	1.00000	1.00000	0

BIBLIOGRAPHY FOR GENERAL READING

Aris, Rutherford: "Vectors, Tensors, and the Basic Equations of Fluid Mechanics," pp. 1–133, Prentice-Hall, Inc., Englewood Cliffs, N.J., 1962.

Chapman, Alan J.: "Heat Transfer," The Macmillan Company, New York, 1960.

Chow, Ven Te: "Open-Channel Hydraulics," McGraw-Hill Book Company, New York, 1959.

Coleman, B. D., H. Markovitz, and W. Noll: "Viscometric Flows of Non-Newtonian Fluids," Springer-Verlag New York, Inc., New York, 1966.

Fredrickson, Arnold G.: "Principles and Applications of Rheology," Prentice-Hall, Inc., Englewood Cliffs, N.J., 1964.

Hinze, J. O.: "Turbulence," McGraw-Hill Book Company, New York, 1959.

Holman, J. P.: "Heat Transfer," 2d ed., McGraw-Hill Book Company, New York, 1968.

Jeans, Sir James: "An Introduction to the Kinetic Theory of Gases," Cambridge University Press, London, 1960.

Kinsman, Blair: "Wind Waves," Prentice-Hall, Inc., Englewood Cliffs, N.J., 1965.

Knudsen, James G., and Donald G. Katz: "Fluid Dynamics and Heat Transfer," McGraw-Hill Book Company, New York, 1958.

Landau, L. D., and E. M. Lifschitz: "Fluid Mechanics," Addison-Wesley Publishing Company, Inc., Reading, Mass., 1966.

Pai, Shih-I: "Viscous Flow Theory I—Laminar Flow," D. Van Nostrand Company, Inc., Princeton, N.J., 1957.

———: "Viscous Flow Theory II—Turbulent Flow," D. Van Nostrand Company, Inc., Princeton, N.J., 1957.

Reynolds, William C.: "Thermodynamics," 2d ed., McGraw-Hill Book Company, New York, 1968.

Robertson, James M.: "Hydrodynamics in Theory and Application," Prentice-Hall, Inc., Englewood Cliffs, N.J., 1965.

Rotty, Ralph M.: "Introduction to Gas Dynamics," John Wiley & Sons, Inc., New York, 1962.

Rouse, Hunter, (ed.): "Advanced Mechanics of Fluids," John Wiley & Sons, Inc., New York, 1959.

——— and Simon Ince: "History of Hydraulics," Dover Publications, Inc., New York, 1963.

Schlichting, Herman: "Boundary-Layer Theory," 6th ed., McGraw-Hill Book Company, New York, 1968.

Sears, William R., (ed.), and Milton Van Dyke (assoc. ed.): "Annual Review of Fluid Mechanics," vol. 1, Annual Reviews, Inc., Palo Alto, Calif., 1969.

Shames, Irving H.: "Mechanics of Fluids," McGraw-Hill Book Company, New York, 1962.

Streeter, Victor L., (ed.-in-chief): "Handbook of Fluid Dynamics," McGraw-Hill Book Company, New York, 1961.

INDEX